AF518433

DEMOCRATISATION OF WATER

DEMOCRATISATION OF WATER

Editors

B.K. Thapliyal

S.S.P. Sharma

P. Shiv Ram

U. Hemant Kumar

SERIALS PUBLICATIONS

NEW DELHI

for

राष्ट्रीय ग्रामीण विकास संस्थान

(ग्रामीण विकास मंत्रालय, भारत सरकार का संगठन)

राजेन्द्रनगर, हैदराबाद-500 030. (भारत)

NATIONAL INSTITUTE OF RURAL DEVELOPMENT

(Ministry of Rural Development, Govt. of India)

Rajendranagar, Hyderabad–500 030. (India)

First Published - 2008

ISBN: 978-81-8387-162-4

Published by
SERIALS PUBLICATIONS
4800/24, Ist Floor, Ansari Road
Darya Ganj, New Delhi-110002 (India)
Phone : 23245225. Fax : 91-11-23272135
E-mail: serials@satyam.net.in

CONTENTS

PART II
MANAGEMENT OF IRRIGATION

PART III
DEVELOPMENT AND MANAGEMENT OF DRINKING WATER

PREFACE

Water, till recently, was regarded as a free gift of the nature, but has now become a priced commodity. The growing need for water has also led to over exploitation of both surface as well as ground water resources. It has also created the situation where the natural gift of water is not easily accessible to all sections of population at affordable price and in desirable quantity. All these factors together have resulted in water scarcity and creating competition for water, which is a major threat to future socio-economic development in the country.

In rural areas, while supply and access to safe drinking water by all sections of the population is a major concern of the rural development process, equally important is access to irrigation water by the small and marginal farmers. Nearly three fourth of the total irrigation is by the minor irrigation sources and more than 80 percent of it is from ground water sources.

Multiplicity of problems in water management sector has forced the policy-makers to think about sustainable ways of managing the precious natural resource like water. One of the most talked about solution to these problems has been the democratisation of water management as against the centralised systems of water management.

Democratisation means many things in different contexts. However in the context of water, democratisation of water can be perceived in terms of control, supply and distribution, and access to it with equity by all sections of the population.

Though there are many stake holders for water, a major portion of the total harnessed water, which accounts for nearly 90 percent is used for agriculture which is the primary occupation of about three fourth of the rural population. Majority of the small and marginal farmers are dependent on minor irrigation. The various Rural Development Programmes implemented earlier have helped millions of Small and Marginal Farmers to have access to minor irrigation. This process is continuing through the programmes like SGSY.

The rate at which agricultural land is being diverted to non – agricultural uses, particularly by the urbanization process and infrastructure projects, the experts claim that by 2020 there will be only 100 million hectares of good quality land left for agriculture and the food grain demand will shoot over 250 million tons. This equation can be solved only through increased productivity and this will be possible with increase in irrigation.

Drinking water supply in rural areas is another major Rural Development Programme. In the past four decades about 55,000 Crore Rupees were invested to provide safe drinking water to the people living in 1.37 million rural habitats and the process is continuing to cover remaining habitats numbering around 60,000.

Meeting the water needs of the increasing population, food demand and industry is going to create water crisis unless we prepare ourselves to meet these challenges and start working on the strategies and programmes for development and conservation of water resources, management of supply and distribution with equity on sustainable manner to ensure water security for all.

This edited volume contains papers presented by the experts, academics, researchers and activists in the foundation day seminar of the Institute. The purpose of foundation day celebrations, besides reviewing the Institutes activities and contribution to the rural development policies and programmes in general and design its future course of activities in consonance with the emerging new challenges relating to rural development with specific focus on poverty alleviation, is also an occasion to address the topical issues which have direct bearing on the social and economic well being of the rural people and the rural areas. Water, a natural resource, which was freely available, is becoming a critical resource and access to it is also getting difficult day by day. In fact, slowly, water is becoming a market commodity and thus the problem of equity and access, specially for the poor is going to be the major concern. It is in this context, "Democratisation of Water" as a concept and in practice is currently gaining considerable importance.

I hope that the present volume will be of immense value to all those interested in the subject in general and to those who are concerned with policy formulation, planning and implementation of projects related to drinking water supply, irrigation and industrial uses. This volume will also provide valuable reference material to the scholars and researchers concerned with water.

V. S. Sampath
Director General, NIRD
Hyderabad

ACKNOWLEDGEMENTS

This volume on "Democratisation of Water" is a collection of 35 papers contributed by experts, academics, researchers and activists actively engaged in various aspects of water management all over the country. These papers were presented and discussed in a National Seminar on "Democratisation of Water" organised by the National Institute of Rural Development on the occasion of its Foundation Day Celebrations. The papers are grouped in to three themes, namely, Water Policy, Development and Management of Irrigation Water and, Development and Management of Drinking Water.

Water till recently considered as a free gift of the nature, has now become a critical resource due to increasing demand caused by increasing population for drinking purposes and food production and expanding industrial activities. While the availability of usable water for human needs is limited, pollution and environmental degradation due to human negligence are further aggravating the water crisis.

The editors of this volume express their gratitude and thanks Shri V S Sampath, Director General of the institute, who has provided us this opportunity to organise the seminar on such a topical issue, which is not only going to be the critical issue globally in general, and for India in particular because of increasing demand for water due growing population and expanding industrial activities.

The Hon'ble Union Minister for Rural Development, Dr. Raghuvans Prasad Singh graced the occasion by inaugurating the seminar and also raised many critical issues relating to drinking water, minor irrigation and conservation of water to augment water resources in the country. We are beholden to him. Ms. Sunita Basant, Secretary to the Government of India, Department of Drinking Water and Sanitation, Ministry of Rural development was kind enough to deliver valedictory address to the seminar. The editors are thankful to her.

At the institute, the authors are grateful to all faculty colleagues who helped us in screening a large number of papers for the seminar as well as for publication. We are thankful to Ms Shanti Sri and Ms Padma Sri, stenos with the center for Agrarian Studies and Disaster Mitigation who have ably bore the burden of typing and correcting the papers and providing all necessary ministerial support.

Last, but not the least, the authors are thankful to the non-academic staff at the institute who extended their valuable support in various ways in conducting the seminar and also bringing out this volume.

We hope that this volume will provide valuable literature on the subject as well as reference material to the academics and the researchers for further studies of the subject and help policy makers and project planners and managers to overcome the threat of water crisis in the country.

Editors

PART I

WATER POLICY: PROBLEMS AND PROSPECTS

1

DEMOCRATISATION OF WATER CONCEPT, STATUS AND ISSUES

B.K. Thapliyal

INTRODUCTION

Access to adequate water is one of the leading factors limiting development in India. Agricultural, industrial and domestic uses are competing more and more for a limited supply. The agricultural sector continues to dominate water use owing to its continued importance to the Indian economy, while industrial demands are increasing as the sector continues to grow as shown can be seen in the table given below. Domestic needs claim only a small portion of the annual water withdrawals as access to adequate water and sanitation supplies remains low throughout the country. Greater access and an improvement in the socio-economic situation in rural India will likely result in higher demand for water from this sector in the coming years.

Agriculture remains central to the Indian economy and it therefore receives the greater share of the annual water allocation. According to the World Resources Institute (2000), 91.2% of India's utilisable

* B.K. Thapliyal, Prof. & Head (CAS & DM), SSP Sharma, Prof. & Head (CWLM), P Shivram, Associate Prof. and U Hemantha Kumar, Asst. Prof., NIRD Hyderabad.

Annual Withdrawals (cubic kilometer per year)

Total (km3/ per year)	*Per capita (m3 per person)*	*Withdrawals by sector (per cent)*		
		Agriculture	*Industry*	*Domestic*
645.8	635	91.2	4.0	4.8

Source: World Resources, 2005 (FAO, 2000).

water is devoted to this sector, mostly in the form of irrigation. The necessity of irrigation for agricultural production is great due to the unpredictable nature of the monsoon. In regions completely dependent on rainfed agriculture, a weak monsoon season can result in drought conditions that can lead to reduced yields or even total crop failure. Even during normal monsoon seasons, farmers are usually able to produce only one crop per year and yields are generally much lower than in irrigated areas. Indeed, the productivity of irrigated agriculture per unit of land has been estimated at seven times that of rainfed agriculture (World Bank, 1999). Massive investment in irrigation in the past fifty years has shown in an expansion of the gross irrigated area from 23 million hectares in 1951 to over 90 million hectares in 1997, and plans exist to continue developing irrigation infrastructure over the coming years (World Bank, 1999). This growth in irrigated area, along with improvements in farming technologies and plant genetics, has been responsible for the incredible growth in crop production over this period. The increase in production has also contributed greatly to the national economy and to India's food security.

However, irrigation expansion has also placed greater demands on surface and groundwater resources. Groundwater alone accounts for 39% of the water used in agriculture and surface water use which often comes at the expense of other sectors such as the industrial and domestic supply. At the other extreme, flood conditions can be equally devastating to the agricultural sector and requires thoughtful planning in terms of drainage and the construction of flood control structures. Development projects such as dam and canal construction were devised to help mitigate the emphasis on industrial development and international trade. Water consumption for this sector has consequently risen and will continue growing at a rate of 4.0% per year (World Bank, 1999). According to the World Bank Report, demand for industrial, energy production and other uses will rise from 67 billion m^3 to 228 billion m^3 by 2025. Hydro-electric generation already accounts for a large percentage of the water demand from this sector. The potential in effect of the monsoon on rivers and seasonal streams. For the most part, they have been successful at reducing the impact of floods in some areas, although their effectiveness is limited in the face of exceptional rainfall events.

In the past several decades, industrial production has increased in India owing to liberalisation of the economy. As the industrial sector is poised to make rapid growth, the demand for water from this sector is also going to be many folds.

Though demand from the domestic sector has remained low and accounts for only 4.8% of the annual freshwater withdrawals in India (World Resources Institute, 2000), increasing population and rapid urbanization is going to increase considerably. Recent data from the World Bank indicates that demand over the next twenty years will double from 25 billion m^3 to 52 billion m^3.

Besides meeting the demand for water by various sectors, quality, quantity, sustainability, equity, etc., are several other issues to be considered. The challenge before the nation is to evolve 'Democratic Frame' in the access, use and governance of water resources and 'Citizenship of Water' as a means for water security.

Multiplicity of problems in water management sector has forced the policy-makers to rethink over the more sustainable ways of managing the precious natural resource like water. One of the most talked about solution to these problems has been the democratisation of water management as against the centralised systems of water management.

DEMOCRATISATION OF WATER

Democratisation involves multiple issues in this contexts. However in the context of water, democratisation of water can be perceived in terms of control, distribution, access and equity. While the constitutional provisions for the control of surface water resources, indicates that it is both central and state subject, the increasing inter–state conflict in sharing river water shows that there is ambiguity in it and therefore there is need to clearly define the jurisdiction of Center and States in the control of water resources.

In the case of ground water resources, though some states have evolved regulations, but without any enforcement and therefore ground water is almost under private control. As a consequence, there are areas where ground water table have declined to alarming levels and thus creating water shortage all around. Due to these factors, the water belongs to those who have land and resources and the landless poor and resources are at the mercy of the government for even drinking water. In other words there is also problem of distribution of and access to water by the resource poor people.

To resolve these problems there is need to look for appropriate means to manage water and democratisation of water could be one of the important initiative in this direction. Thus in the final analysis, the democratisation of water can be defined in terms of Government and people's participation in the development and management of water resources and distribution of water for different uses.

NATIONAL WATER POLICY ON WATER SHARING: EMERGING CONFLICTS

As water becomes increasingly scarce, national conflicts are expected to rise. Agriculture, industry and domestic needs (drinking water) are in competition for the resource that is so desperately needed for further development and management. These conflicts arise mostly when the resource becoms scarce. There is peace in times of good rains. However, when the monsoon fails, the conflict erupts. Hither to government initiations are pertained to reach a crisis which requires a quick fix, rather than pursuing a long term solution. For example, the Cauveri dispute has been a serious issue since 1974 between Tamilnadu and Karnataka. Similarly, Andhra Pradesh and Karnataka are also in tension over Paragodu water sharing. The ultimate sufferers are small and medium sized farmers due to lack of water. Crops are withering as tempers flare. As a result of this desperation, threats have turned into acts of violence. Numerous rivers in India are shared by two or more states.

The National Policy on Water (NPW–2002) has failed to tackle the problem. The NPW outlines that water sharing/distribution amongst the states should be guided by a national perspective with due regard to water resources availability and needs within the river basin. Necessary guidelines, including for water deficit states, and even outside the basin, need to be evolved for facilitating future agreements amongst the basin states. The Inter-State Water Disputes Act of 1956 may be suitably reviewed and sutiably amended for timely adjudication of water disputes referred to the Tribunal. This has so far not happened in spite of stringent efforts made by the Central Government.

Water and the Constitutional Provisions

The relevant provisions of the Indian Constitution are

- Entry 17 in the State List,
- Entry 56 in the Union List, and
- Article 262.

The first provision makes water a state subject, but qualified by Entry 56 in the Union List, which states: "Regulation and development of inter-state rivers and river valleys to the extent to which such regulation and development under the control of the Union is declared by parliament by law to be expedient in the public interest." Article 262 explicitly grants parliament the right to legislate over the matters in Entry 56, and also gives it primacy over the Supreme Court. As documented by Ramaswamy Iyer (1994), parliament has not made much use of Entry 56. Various River Authorities have been proposed, but not legislated or established as bodies vested with powers of management. Instead, river boards have been created with only advisory powers.

Hence, the state governments dominate the allocation of river waters. Since rivers cross state boundaries, disputes are inevitable. The Inter-State Water Disputes Act of 1956 was legislated to deal with conflicts, and included provisions for the establishment of tribunals to adjudicate where direct negotiations have failed. However, states have declined to accept the decisions of the tribunals. Therefore, arbitration is not binding. Significantly, the court verdicts have also been ignored on some occasions. The Central Government has sometimes intervened directly as well, but in the most intractable cases, such as the sharing of the Ravi-Beas water sharing among Haryana, Jammu and Kashmir, Rajasthan, and Punjab, the Central intervention has been unsuccessful.

The National Water Resources Council (1983), National Water Policy (1987) and National Water Development Agency (1992) emphasizes an integrated and environmentally sound basis for developing national water resources, but provides no specific recommendations for institutions to achieve this and does not meet any of the required criteria, as well as it does not provide specific mechanisms for dispute resolution, it does not delegate sideways to achieve commitment possibilities, and it does not have any statutory force. The main issues related to these aspects can be listed as under:

- The essential problem faced by different states or groups within a state is that the initial allocation of water is sub optimal due to changing circumstances.
- The conflict or disagreement over Inter-state river waters in India as an attempt to influence or determine the initial allocation of property rights over water, by methods such as lobbying.
- Political objectives dominate: political considerations for the center, beyond maximizing the joint welfare of the two parties to the dispute.
- Incomplete Information: estimates of costs and benefits in general, as they enter the utility functions privately and subjectively, may not be objectively verifiable. This complicates matters further.
- The issue of inter-state water allocation, while it involves special legal and technical features, has been croweded by some of the general problems of Indian federalism.
- From a federal perspective, with respect to water, it has been extensively pointed out that water is in the State List of the Constitution (Entry 17), but that the entry there is qualified, "subject to the provisions of Entry 56 of List I" (the Union List). This has meant a high level of explicit or implicit "horse-trading" among the center and states that are potentially key elements of a central coalition.

At this stage, the present Seminar will be addressed the issues and seek possible solutions regarding water-dispute settlement mechanisms, including a national water commission independent of daily political pressures, a federated structure incorporating river basin authorities and water user associations, and fixed time periods for negotiation and adjudication.

Pricing of Water: Issues and Options

Irrigation water rates supplied from canals are fixed at the State level on a crop-area basis in nominal rupee terms, and remain unrevised for years, sometimes for decades in almost all the states. Poor revenues from public irrigation systems created at a high cost are among the many factors responsible for the severe resource constraint faced by most State Governments. The result is poor maintenance of systems, which tend to fall into disrepair, and an unreliable and unsatisfactory service. The Successive Commissions and Committees (Irrigation Commissions, Finance Commissions and Vaidyanathan Committee) have stressed the importance of recovering at least the O&M costs if not the capital-related charges, but these have not been acted upon. The National Water Policy 2002 stated that irrigation water rates should be such as to cover the O&M costs fully and make a partial contribution towards capital-related charges (the extent of this was not specified) but this too, like many other principles laid down in the NWP, remained non-operationalised. Some of the issues related to this aspect are:

- Whether the charge is to recover the annual O&M cost or the objective to move towards full cost recovery. This was not cleared by different commissions and committees.
- The pricing and their revisions are subject to review and approval by an independent regulatory authority rather on the lines of other public utilities.
- Whether poor farmers would be able to pay increased water charges.
- Private sector participation in irrigation financing is an idea that has been favoured by both the Planning Commission and the Ministry of Water Resources.
- There is also the basic conflict between regarding water as an 'economic good' in which tradable rights can be created (which is a pre-requisite for private corporate investments) and regarding it as a life-supporting element and therefore a basic right (now recognized as a human right in the UN context). Many would argue that water is more appropriately treated as a common pool resource to be managed by the community, than as a commercial commodity. It is difficult to reconcile those two conflicting approaches in regard to water.
- There is a more fundamental issue of whether it is possible to define enforceable property rights in water.
- ***There is an equity consequence of low pricing***: the profligate use of water by the head-reach farmers results in the non-availability of water for the tail-end farmers.

The tail ender issue is more complex. The specific causes for tail-enders problems are : (a) unauthorized drawing of water (13 projects); (b) absence of outlet gates/ structures (in 10 projects); (c) poor maintenance, seepage and reduced carrying capacity due to weeds (8 projects); and (d) design/construction faults and unlined canals (7 projects).

The present debate is on the paucity of funds for new investment is forcing State and Central Governments to explore alternative sources of financing for developing and rehabilitating irrigation projects, such as the tapping of the capital of private investors, private borrowers through revenue-backed instruments, and the mobilization of local capital through investments by farmers in the systems that serve them.

The revised National Water Policy (2002) also advocates private sector participation in the planning, development and management of water resources projects wherever feasible. According to one version it has been mention that the private sector participation is expected to help in introducing innovative ideas, generating financial resources, speeding the construction process, introducing corporate management models, and improving service efficiency. But many of the social activists and academicians across the country criticizes the entry of private sector participation and state that the water prices must go up because private investors must get a return on their investments and will not solve the water problem for the poor.

In practice, private sector participation is likely to encounter several problems: large investments, long gestation periods, modest or low returns, high financial risks, likely political interference in project management and water-rate assessment, the political influence of farmers' lobbies, the difficulties in raising water rates and collecting them effectively, and so on. The much-advocated idea of private sector participation has made hardly any headway in power projects, and there is no basis for thinking that it would be more successful in irrigation projects. At this juncture the present seminar may address these issues and find possible solutions to correcting the pricing structure could facilitate both the generation of resources for expansion and maintenance of the sector as well as facilitate optimization of use of water in the sector.

GROUNDWATER

Since mid sixties, the encouragement through institutional finance and rural electrification the number of groundwater resources and its exploitation has increased tremendously. In India, while 76 percent of the operational holdings are small and marginal farms of less than 2 ha. They operate only 29 percent of the area. They constitute 38 percent of net irrigated area by well and account for 35 percent of the tube wells fitted with electric pump sets. Thus, in relation to operational area, small and marginal farmers are well represented in groundwater irrigation in greater proportion than their representation in landholdings. Another reason for acceleration of groundwater resources is that the productivity of canal irrigated areas is not the same as productivity of groundwater irrigated areas. Land irrigated from groundwater has notably higher productivity than from canals. The situation of over extraction and the resultant environment degradation is a consequence of lack of appropriate and adequate policies (Policy failure) for managing the subsurface water resources

The difficulties arise in groundwater sector essentially in two inter related contexts: the prevailing law governing the ownership of groundwater, and the emergence of water markets -

- Under Indian law, the ownership of land carries with it the ownership of the groundwater under it, subject to regulation and control by the state. And "there is no limitation on how much groundwater a particular landowner may draw". It follows that only those owning land can have rights over groundwater; the landless (including communities, tribal and other, who may have been using certain natural resources for centuries) can have no such rights.
- Further, this legal position leads to inequities of various kinds: a rich farmer can install power-driven tube wells or bore wells in his land and their operation can make dug wells in the neighbourhood run dry; he can sell water so extracted to his poorer neighbours even though the water may come from a common aquifer running under their lands; and he can deplete the aquifer through excessive exploitation. The easement right makes regulation difficult. Easement right is not only problem. In principle a well owner who is adversely affected by deepening or other

action by a neighbour has the legal right to seek restraint or compensation from the latter. That this right is not easy to enforce makes its empty.

- Even if you dissociate rights to extract underground resources from ownership of land, the task of monitoring compliance and enforcing regulation to ensure equitable sharing and sustainable use is far too difficult in terms of effort and cost. That is the reason for emphasizing economic incentives to check over extraction and introducing some form of community regulation of groundwater extraction and use at the local level. That too is fraught with many difficulties.

Many studies have observed that - *in the process of negotiation for water rights, power relations are very important:* the elites are more likely to negotiate better water rights for themselves than the less powerful. This is because the elites control the decision-making processes that legitimize the rules for allocation and distribution of water. It is against that legal background that water markets have emerged in the context of groundwater extraction through tube wells and bore wells. National Water Policy on Groundwater (1987&2002) was not clear about these issues. The NWP 2002, states that–

- There should be a periodical re-assessment on a scientific basis of the ground water potential, taking into consideration the quality of the **water** available and economic viability.
- Exploitation of ground water resources should be so regulated as not to exceed the recharging possibilities, as also to ensure social equity. Ground water recharge projects should be developed and implemented for augmenting the available supplies.
- Integrated and coordinated development of surface water and ground water and their conjunctive use, should be envisaged right from the project planning stage and should form an essential part of the project.
- Over exploitation of ground water should be avoided near the coast prevent ingress of sea water into sweet water aquifers.

These concerns are not directly addressing the issues at present the groundwater sector is facing and as a consequence–

- The first victims in this process are small and marginal farmers. The impact of resource degradation on these farmers is in two ways. Firstly, while small and marginal farmers dominate the ownership of wells in general and open wells in particular, medium and large farmers dominate the ownership of tube wells/bore wells. Secondly, one of the interesting observations in recent years is that of late bore well technology is becoming cheaper making it size (owned land) neutral, though the process may be slow. As a result these farmers are also investing substantial amounts of money on bore wells. Such investments become unviable in the event of well failure. Even the recent policies in water management fail to take these problems into account. For instance, in Andhra Pradesh and many other states, groundwater is totally left out of the purview of the water user association legislation. There are no efforts to integrate well and tank irrigation at the policy level.
- As far as institutional approaches to water resource management are concerned, groundwater is a neglected area. In fact, groundwater does not find a place even in the constitution and the consequent acts or tribunals fail to take comprehensive view of water resources. Groundwater rights are customarily attached to land ownership; hence groundwater management is totally left to the private initiatives.

Therefore there is uregent need that groundwater should be brought under pricing mechanism. While, cost based power tariff is useful in checking over exploitation, adding scarcity rent of water to the tariff would be more appropriate. A pre-condition for this is to minimise the risk and uncertainties in groundwater and power availability. More importantly, institutional arrangements such as making groundwater a real common pool resource and exploiting it on a community basis are critical for equitable distribution and sustenance of the resource.

There are also serious criticisms about groundwater estimation and development methodologies. *The estimation done by Groundwater Estimation Committee* is based on water balance methodology. And also assumes that 30 percent of the gross draft is presumed to go as return seepage to the groundwater regime in all the districts of the country. One important conceptual weakness of the methodology is that it treats development units as independent units (constant state) and ignores dynamic interactions. *In the light of the foregoing, discussion in relation to groundwater have to be two fold:* (i) the legal position on the ownership of groundwater needs to change, with aquifers being regarded as common pool resources in which only limited use rights (and not property rights) are conferred on individuals or institutions; and (ii) water markets need to be carefully regulated with reference to equity, resource conservation and environmental sustainability. We hope the present seminar may address these issues and find possible solutions.

NATIONAL POLICY ON DRINKING WATER

National Policy on Water (1987) has stated that adequate drinking water facilities should be provided to the entire population both in urban and in rural areas by 1991. Irrigation and multipurpose projects should invariably include a drinking water component, wherever there is no alternative source of drinking water. Drinking water needs of human beings and animals should be the first charge on any available water. In the planning and operation of systems, water allocation priorities should be broadly as follows:-Drinking water, Irrigation, Hydro-Power, Navigation and Industrial and other uses. The policy stated that the top priority is for drinking water. However we have not been able to evolve an integrated water use system which provides priority for drinking water.

Issues

Based on above discussions, the following issues concerning National Water Policy are identified :

- Ownership and Control of both surface and ground water resources.
- Legal frame work (deficiencies in the existing legislation with specific reference to equity, conservation and sustainability)
- Equity and Efficiency in the use of water for Agriculture, Industry and domestic purposes.
- Water Resources Planning through non-conventional methods such as inter-basin transfers, artificial recharge of ground water and desalination of brackish water and traditional water conservation practices.

- Institutional Mechanism for development and management of water resources.
- "Water Citizenship" as an agenda towards achieving water security, and
- Pollution control for water resources

DEVELOPMENT AND MANAGEMENT OF IRRIGATION

From the beginning of the First Five Year Plan, the net area irrigated increased from 20.9 mha to 57.2 mha by 1999-2000, indicates an increase of nearly 275 percent, while the agriculture production increased four times during the same period. Though this appears to be satisfactory achievement, but when we analyse the growth pattern through different sources of irrigation, the picture emerges as unsatisfactory. The net area under medium and major irrigation was increased from 8.61 mha in 1952-53 to 17.25 mha by the year 1999-2000. The increase has not been commensurate with the volume of investment incurred on medium and major irrigation projects. While the area irrigated under tanks and other sources have rather decreased, area irrigated by ground water sources increased by 5.6 times. While increasing urbanization and infrastructure projects, coupled with degradation of land, the area under agriculture is shrinking, while the demand for food is on the increase due to high rate of population growth. To meet ever increasing food demand, the farm productivity should continue to increase and this is possible through intensive farming for which irrigation is must.

Problems of Medium and Major Irrigation Projects

The potential for irrigation growth is also limited under the existing situation. Besides very high costs, the vast difference in irrigation potential created under medium and major irrigation projects and actual area irrigated is one of the major concern. The efficiency of irrigation through canals is also another area of concern. Maintenance and operation of medium and major irrigation systems vis–vis pricing and revenue generation through irrigation cess continues to be matter of debate. Due to all these factors, now there are many who think that medium and major irrigation projects are not only financially unviable but also cause adverse impact on the eco-system and environment. Though there are many issues relating to medium and major irrigation projects, a few important issues which need to be addressed immediately to make surface irrigation more efficient and productive, are:

- Modernisation of major and medium irrigation systems to improve transportation and distribution of water and evolve proper irrigation technologies for improving efficiency of canal irrigation.
- Introduce realistic price mechanism so that the revenue generated should not only meet the maintenance and operation costs but also a part of the capital cost.
- Decentralisation of management of irrigation systems and involvement of farmers in the management of maintenance of the system and distribution of water for irrigation.
- The proposed interlinking of rivers is still passing through the controversy of financial and environmental feasibility. While all the proposed projects may not be feasible, there are definitely some feasible projects and which can be further evaluated and put to action.

Minor Irrigation and Ground Water

Nearly 59 % of irrigation and 85 % of drinking water is drawn from ground water sources. This clearly indicates over dependence on the ground water resources, and as a result, the water table in many areas decreased alarmingly. There are 7063 development blocks and it is estimated that two to three percent of them have over exploitation of ground water. The problem is severe in Maharashtra, Rajasthan, Gujarat, Punjab, Haryana, Uttar Pradesh, Karnataka, Andhra Pradesh and Tamilnadu (Singh, 2002). In a study of six states (Gujarat, Haryana, Karnataka, Punjab, Rajasthan and Tamilnadu), it was observed that between 1989 and 1995 the proportion of dark blocks had increased from 16% to 26% (Selvarajan, 2001). In Punjab and Haryana, about 50% of the blocks fall under dark category.

Ground water is the most important source of irrigation for small and marginal farmers as they are well represented in the use of this resource through open and bore wells. The medium and large farmers, who generally go for tube–wells, the number of which is rising day by day, and has become a threat for the small and marginal farmers in general and the ground water resources in particular.

Tank irrigation is the most ancient source of irrigation. During 1950-51, its share in the total net irrigated area was 3.6 mha, but by 1999-200 it has reduced to 2.7 mha, indicates sorry story of this system.

Keeping above factors in view there are some important issues which need special attention:

- Tank Irrigation has declined due to various reasons, most important being–lack of maintenance causing silting, illegal encroachment and pollution.
- Minor Irrigation–Regulation of ground water use
- Pricing of water for irrigation (with specific reference to high value and high water consumption crops)
- Institutional Mechanism both traditional and modern (Decentralisation and delegation of powers and functions to local bodies and farmers associations) in the maintenance and management of irrigation water.
- Watershed Management–Ground water development

DEVELOPMENT AND MANAGEMENT OF DRINKING WATER

Drinking water supply and sanitation are the subjects of State Governments. These are included in the Eleventh Schedule of the Constitution of India among the subjects that are entrusted to Panchayats by the States. Since, the First Five Year Plan onwards, both the Government of India and the State Governments have substantially invested about Rs. 55,000/- Crores in rural water supply sector for providing potable water to the rural people. As a result, more than 37 lakh hand pumps and 1,73,000/- piped water supply systems have been installed in providing access to potable drinking water, with full coverage of 1.37 million habitats, partial coverage of about 44,500 habitats, crediting the country with one of the largest rural drinking water supply networks in the world. However the problem of safe drinking water doesn't end here as there are many problems relating to sustainability, management of maintenance and operation, distribution and equity, etc. Some of the issues are identified as under.

Accessibility to Safe Drinking Water

According to the National Family Health Survey (1998-99), 74.5 per cent of urban and 25 per cent of rural households have access to piped water supply. Hand pump as a source of drinking water is the second important source as 18.1 per cent of urban and 47.3 per cent of rural households depend on it. The remaining household still depends on unsafe water sources. A large number of habitations (2,16,794 habitations) are affected with water quality problems like fluoride, arsenic, iron, brackishness, etc. Thereby, the problem of rural drinking water supply has not been fully solved. Though Rajiv Gandhi Technology Mission for Drinking Water is on the job to solve this problem, still much remains to be done. There is a feeling that the State Governments are not very serious about this problem.

There is no permanent mechanism in place for periodical checking the quality of the water and in most cases user communities are unaware about the actions required to keep water sources and supply system clean and free from contamination.

Fully Covered Villages Slipping back to 'Not Covered/Partially Covered' Category

During 2005-2006, the status of coverage of rural habitations with drinking water facilities is that more than 95 per cent of habitations have been 'fully covered' with one or two water systems. Are these fully covered villages supplying safe drinking water? The 'partially covered' habitations are 3.13 per cent followed by 'not-covered' (0.98 per cent) habitations, which are negligible. However, a large number of habitations, which are fully covered earlier, have slipped back as 'not covered'/'Partially covered' due to a number of factors which are as follows - Sources are becoming dry because of over exploitation of ground water; increasing population resulting in lower per capita availability of drinking water. In addition, factors like further deepening of sources leading to quality problems, demand from other competing sectors like agriculture, industry, etc., pose threat to sustainability of drinking water supply schemes as well as increasing the water quality problem villages. The quality problem villages are increasing every year.

Operation and Maintenance (O&M)

Not only hundred percent capital investment on drinking water is being met through Government funding, but to a considerable extent the state has to bear the burden of Operation & Maintenance (O&M) as the water charges collected from the user community is a fraction of the investment required for the O&M of the water supply system. Due to inadequate funding, poor O & M of the water facility it is becoming defunct in many cases. Under Swajaldhara water supply schemes are expected to be planned, designed, implemented and managed by the Village Panchayats and therefore it is expected that the user community contributes for O&M of the water supply facility. However, the research studies revealed that the extent of community participation and user management is below standard. It is therefore necessary to evolve a mechanism to generate adequate funds through the user community without putting undue burden on the village poor to sustain the facility efficiently. The Government cannot go on funding O & M of the water supply facility.

As per the 73rd Amendment to the Constitution of India, the subject of rural water supply vests with the Panchayti Raj Institutions (PRIs). The Panchayats are to play a major role in providing safe drinking water and managing the water systems and sources. Only a few states have delegated powers to PRIs. In order to reduce the Government burden it is necessary to devolve powers to the local government to share the responsibility of managing water supply systems.

Privatization vs Right to Water

At the national and international level the issue of privatization of water services are being debated. Water for basic needs has been identified as a primary public good and a human right-and not as a commodity to be traded in the open market. As a public good, it is felt that water must be managed as a social responsibility of the government. However increasing cost of providing and maintaining water supply, it would be difficult for the government to shoulder the responsibility indefinitely. It is in this context there is need to evolve a system either Public–Public Partnership or Public–Private Partnership that reduces the financial burden of the government on the one hand and preserves the right to water by common masses on the other.

Gender Concerns

Since the collection of drinking water in the villages is primarily done by women, they need to be involved in planning, implementation, location and distribution system, and O&M of water supply schemes.

Capacity Building for Water Engineers in Social Development

Most of the Engineers responsible to plan and execute drinking water systems look at it from engineering point of view and are not aware of the social aspects of the same. It is therefore necessary to undertake capacity building of the Water Engineers to make them Social Engineers as well.

Need for Water legislation to Protect and Preserve Traditional Water Systems

The traditional structures have been providing drinking water to the people even during the scarcity period. However, the traditional water systems like temple tanks, village tanks, rivulets are neglected and at present they are in the verge of collapse. Therefore, these structures need to be preserved and protected through legislation.

Issues

Based on above factors, the main issues pertaining to drinking water can be identified as under :

- The majority of the household still depends on unsafe water sources. The problem of rural drinking water supply and sanitation has not been fully resolved.
- Resource constraint and magnitude of the problem increased day by day, hence, the changes took place in policy during Five Year Plans periods is inevitable. There is no systematic strategy for safe drinking water.
- Involvement of Private–public partnership in safe drinking water is not attractive is not practical.
- Traditional arrangements through which rural communities secured their water supplies—community and private wells, tanks, rivers—have gradually declined partly with the spread of modernity and rising expectations

and partly because of the government efforts to deliver piped water supply.

POLLUTION CONTROL, PROTECTION OF WATER AND ENVIRONMENT

India has significant ground and surface water pollution problems in both its urban and rural areas. Industrial and agricultural wastes contribute to the problem and only 10 percent of sewage is treated leaving the rest to contaminate groundwater or surface water flows. Pollution is suspected as a cause in the deaths of 400,000 to 500,000 children under five caused by diarrhea each year. During the past few decades' Indian industries have registered a quantum jump, which has contributed to high economic growth but simultaneously it has also given rise to severe environmental pollution. Consequently, ambient air and water quality is seriously affected which is far lower in comparison to the international standards. The problem is worse in the case of water pollution. It is found that one-third of the total water pollution comes in the form of effluent discharge, solid wastes and other hazardous wastes. Untreated or allegedly treated effluents have increased the level of toxins like cyanide and chromium up to 20 times the safe level in 22 critically polluted areas of the country. The surface water is the main source of industries for waste disposal. It is found that almost all rivers are polluted in most of the stretches by some industry or the other. Although all industries function under the strict guidelines of the Central Pollution Control Board (CPCB) but still the environmental situation is far from satisfactory. Different norms and guidelines are given for all the industries depending upon their pollution potentials. In India there are sufficient evidences available related with the mismanagement of industrial wastes? Consequently, at the end of each time period the pollution problem takes menacing concern.

The analysis shows that most of the direct pollution is generated by the sectors of the highly polluting category. Other moderately or low-polluting sectors are responsible for only small proportion of pollution generation. Sectors such as, dairy, beverages, textile, paper, leather, rubber, heavy chemicals, fertilizers, drugs, synthetic fiber, etc, are responsible for most of the pollution in the country.

Community Action and Informal Regulation

In addition to the action of the different administrative agencies, some kind of pollution regulation can be enforced by direct action of affected communities. According to Goldar and Banerjee (2002) the two channels of informal regulation are (1) to report violation of legal standards to the regulatory institutions (where such standards and institutions exist), and (2) to put pressure on regulators (politicians and administrators) to tighten their monitoring and enforcement.

Groundwater Quality Problems in India

The main quality problem encountered with ground water in India is due to excess fluoride, arsenic, iron, nitrate and salinity. Nitrate contamination is mainly anthropogenic, due to the use of fertilizers and discharge of fecal material. Salinity may have different origin, but the most common is the infiltration of brackish water in a fresh aquifer due to the over exploitation of this aquifer.

Pollution in Urban Environment

Almost all of our cities produce volumes of solid and liquid waste and unfortunately only a fraction of it is treated before leaving it to drains which ultimately mix with river water as also seep into soil and pollute sub-soil water. Though each state has Pollution Control Board (SPCB) but remain ineffective in controlling pollution. In the tenth plan document from the Indian Planning Commission sewage alone was reported to be responsible for 80% of the total water pollution in the country.

Pollution through Domestic Waste

No figures are available about the non-point source pollution due to domestic wastewater discharge in rural areas. According to the Central Statistical Organisation (CSO), 3.15 % of the rural population had access to sanitation services in 1993. This left around 563.6 million people living in rural areas having no access to toilets. Global numbers about this kind of pollution would not be very useful anyway, since the effect of such pollution is essentially local, and intimately linked with the local practice of water fetching and hygiene.

Pollution by Agricultural Run-offs

Pollution by agricultural run-offs has two main effects on the environment. Pesticides may be responsible for poisoning. They are especially difficult to remove from freshwater, and thus, can be found in municipal or bottled water, even after conventional treatment. A study from the CSE recently drew the alarm about the concentration in pesticides such as organ chlorines and organophosphates that were exceeding the WHO standards in almost all the Indian brands of bottled water (Down to Earth, February 2003). In the case of fertilizers, they have an indirect adverse impact on the water resources. Indeed, by increasing the nutritional content of the watercourses, fertilizers allow organisms to proliferate. These organisms may be disease vectors, or algae. The proliferation of algae may slower the flow in the watercourses, thus increasing again the proliferation of organisms and sedimentation. In spite of these well-known adverse effects, and the worrying growth of fertilizer and pesticide use in the agricultural sector, is another matter of concern from pollution point of view. Chart given below shows the picture of major environmental costs in the country.

Summary of the Major Annual Environmental Costs for India
[Total = US$ 9.7 billion]

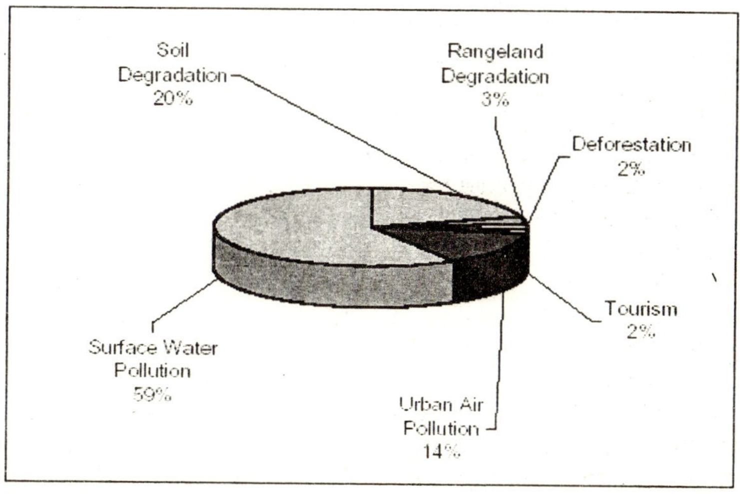

Source: Brandon & Homman, 1995

CONCLUSION AND ISSUES

The measurement of pollution in river flows and underground reservoirs developed during the last decades and it is now possible to have an overview of the ambient water quality in the country thanks to a network of around 500 monitoring stations. The information available is expressed using the ambient quality standards defined by the CPCB and does not allow a real analysis of the water pollution problem, nevertheless it shows clearly that pollution is exceeding the standards in many places. The only attempt at giving a nation-wide cost of water pollution is the study carried out by Brandon & Homman [1995]. We observe that the figures provided are only rough estimates. Moreover, the cost provided is only deduced from the overall figures on diarrheal diseases occurrence in the country, and lets aside many other significant costs. Nevertheless, even if underestimated, the figure of 4.53% of GDP for the total cost of pollution, with surface water pollution accounting for 59% of it has the merit of being clear. India cannot afford not to deal with water pollution. The rough results provided by Brandon & Homman may be enriched with the comparison with similar although more local studies. Three other studies were reviewed in this report, each of them applying a different method in order to put a price on environmental change. The study carried out by Misra (1999) showed results from a survey using the contingent valuation method (CVM) in order to assess the willingness to pay of local population from in the surrounding of an industrial area for an improvement in ambient water quality. CVM was also applied in an attempt at valuing the benefits from the Ganga Action Plan (Markyanda & Murty, 2000). A method more similar to the one applied by Brandon & Homman was used in Appasamy et. al (2002) were several type of costs (health, agriculture and fisheries productivity) were calculated in the case of the Noyyal river basin in Maharashtra. The second type of studies reviewed dealt with costs of pollution reduction. Most of the academic work done in this field deals with industrial pollution. Several methods have been applied in order to estimate pollution abatement function for different sectors of the Indian industry. Nevertheless, very few information is available on the cost of controlling other types of pollution such as domestic pollution and especially agricultural pollution.

REFRENCES

Bandopadhyay J., (1987), *Need For Ecological Water Resource Policy,* EPW, No. 50, Dec. 1987.

Desarda H. M., *Water –its Economics and Politics* in Survey of Environment, The Hindu Annual Publication, 1994, Madras.

GOI, Ministry of Water Resources, *National Water Policy* (1987), 2003.

GOI, Ministry of Water Resources, *Integrated Water Resources Development–A Plan of Action;* the report of the National Commission on Integrated Water Resource development Plan, 1999.

GOI, Ministry of Water Resources, task Force on Interlinking of Rivers, 2003.

GOI, Central Water Commission (1999), Report.

International Water Management Institute (2001), Basin Level Use and Productivity of Water, Research Report.

Iyer Ramaswamy R., *Water Resource Planning, Changing Perspectives,* EPW, Vol. 34, No. 9, 1999.

Iyer Ramaswamy R., *World Commission on Dams and India,* EPW, 23 June, 2001.

Rudrappan D., *Ground Water Management, Environment Security and Agriculture Developmen,* Southern Economist, Vol. 42, No 5, p.5; 2003.

United Nations (1997), Report of the UN Water Conference on Waterand the Environment development Issues for the 21st Century, Dublin, Ireland.

Vincent J. R., *et.al. Valuing Environment Benefits in Developing Countries,* Special Report, East Lansing, USA, Michigan State University.

World Bank, *India : Water Resource Management – The Irrigation Sector,* (1999), New Delhi, Allied Publishers.

2

THE RIGHT TO WATER TOWARDS DEMOCRATIZATION AND PARTICIPATORY GOVERNANCE

C.R. Bijoy

INTRODUCTION

The 'right to water' is quietly being reinterpreted to mean 'right of access to water'– a process that necessarily requires the 'citizen' to be reduced to a mere 'consumer' for the emerging high-growth 'water market'. The traditional and customary notions of natural rights to water subject to realities of social relationships and social distancing had been superimposed with state control guided by the Constitution to promote natural justice and equity. Instead, this in effect relegated ownership and control of water to the state, the propertied (landed) class and the market/business (capital).

The new regime that has emerged with liberalization and globalization attempts to further disenfranchise vast sections of the citizenry who are the marginalized and powerless. With intensification of the use of water–a result of the water intensive agriculture, urbanization, and industrialization–the modern water management systems and its structures and technologies, the runaway decline in both quantity and quality and the rapidly escalating demand for water by the priority high-growth centres, the focus is on development of water resources and improvement of efficiency in supply to meet these rapidly growing but differential demands. This has in turn increased and intensified conflict around water. Rather than focusing on ways to review the mechanism for resolving disputes as well as in addressing the diverse needs through a democratic process, i.e. an overhaul of the water governance system, the approach continues to seek techno-managerial solutions in developing water resource and their distribution. However, this does not take into consideration that the citizen is not merely a consumer or a potential consumer but in fact the right holder. A review and scope for another governance system for water is attempted here.

The National Water Policy 2002 advocates a participatory approach to management of water resources. The scope of participation is determined by (a) the constitutional and statutory frame (b) the legislative process (c) the governance approach (d) conflicts and (e) scope of resolution.

CONSTITUTIONAL AND STATUTORY FRAMEWORK

- *Article 15(2)* of the Constitution explicitly states that no citizen shall 'on grounds only of religion, race, caste, sex, place of birth or any of them' be subject to any disability, liability, restriction or condition with regard to 'the use of wells, tanks, bathing ghats'. This refers to equal access to water.
- *Article 21* (Protection of life and personal liberty) speaks of the right to life. This has been liberally interpreted by the Supreme Court to include all facets of life. The directive principles of state policy (DPSP), which the Constitution in Article 37 declares to be non-justiciable, recognize the principle of equal access to the material resources of the community. Article 37 further states that the provisions in Part IV of the Constitution (which set out the DPSP) 'shall not be enforceable by any court, but the principles therein laid down are nevertheless fundamental in the governance of the country and it shall be the duty of the State to apply these principles in making laws'.
- *Article 39 (b)* mandates that 'the State shall, in particular, direct its policy towards securing that the ownership and control of the material resources of the community are so distributed as best to subserve the common good.'

- *Article 51-A (g)* casts a fundamental duty on every citizen of India 'to protect and improve the natural environment including forests, lakes, rivers, wild life and to have compassion for living creatures.'

DISTRIBUTION OF LEGISLATIVE POWERS

The constitutional and statutory intent is to be translated through legislation with legislative powers are distributed between the Parliament of India and the Legislative Assemblies of the different States.

- *Article 245*: Parliament makes laws for the whole or any part of the country and a legislature of the State for the whole or any part of the territory of the State. The topics of legislation on which the Parliament has exclusive powers to make laws are set out in List I (Union List) of the Seventh Schedule and the topics over which the State legislatures have exclusive powers are set out in List II (State List) of the Seventh Schedule. List III is the 'Concurrent List' enumerating the topics over which both the Parliament and State legislatures can make laws.

    Entry 56 in the Union List pertains to 'regulation and development of inter-State rivers and river valleys to the extent to which such regulation and development under the control of the Union is declared by Parliament by law to be expedient in the public interest.' In exercise of this power, the Parliament enacted the River Boards Act, 1956, although this law has remained largely inoperative.
- *Article 262:* The aspect of resolving dispute between States in relation to sharing of river waters finds place in this Article which envisages the creation of an exclusive tribunal for the purpose. Parliament may by law provide for adjudication of any dispute or complaint with respect to the use, distribution or control of the waters of, or in, any inter-State river or river valley. This empowers the Parliament to set up suitable machinery for adjudication of any dispute or complaint with respect to the use, distribution, or control of the waters of inter-state rivers and river valleys. Accordingly, the Parliament has enacted the Inter-State Water Disputes Act, 1956.
- *Article 252:* The Parliament can, under this Article, make laws even on topics in respect of which it has no powers, provided

the legislatures of two or more States resolve that Parliament should make such law. Thus, the Water (Prevention and Control of Pollution) Act, 1974, a law on a topic relatable to Entries 6 and 17 of the State List was passed. Entry 6 in the State List reads: 'public health and sanitation; hospitals and dispensaries.' Entry 17 reads: 'water, that is to say, water supplies, irrigation and canals, drainage and embankments, water storage and water power subject to the provisions of Entry 56 of List I. The justification for a central law to tackle the growing problem of pollution of rivers and streams was the inadequate and unsatisfactory nature of local laws. In the context of water pollution, another significant central legislation is the Environment (Protection) Act, 1996, under which the notifications issued by the Ministry of Environment and Forests in exercise of the powers under Section 3 of the said Act, lays down the regulatory framework for activities along the vast coastline of India [the Coastal Regulation Zone (CRZ) Notification]. The Government of India has also set up the Central Ground Water Authority by a notification under this very statute, to regulate the existing indiscriminate use of ground water in various parts of the country and in particular the major urban metropolises.

- *State Legislations*: The legislatures of the different States have enacted a large number of statutes that touch upon the various aspects of the control, regulation and distribution of water. Examples are laws that relate to canals and irrigation, [such as The Northern India Canal and Draining Act, 1873, The Bombay Irrigation Act, 1879, The Rajasthan Irrigation and Drainage Act, 1954, Karnataka Irrigation Act, 1965]; use of water sources [such as The Kumaon and Garhwal Water (Collection, Retention and Distribution) Act, 1975]; water sewerage and drainage [such as The Utter Pradesh Water Supply and Sewerage Act, 1975], and ground water [such as Madras Metropolitan Area Ground Water (Regulation) Act, 1987, Kerala Ground Water (Control and Regulation) Act, 2002, Karnataka Ground Water (Protection and Regulation for Drinking Water) Act, 2003].
- The aspects of collection of taxes and cesses on the use of water are also covered by legislations enacted by both Parliament as

well as State legislatures [The Water (Prevention and Control of Pollution) Cess Act, 1977].

THE APPROACH IN LEGISLATIONS

Command and Control Model Governance

The command and control model of governance is essentially colonial in character, a model for subjugation of peoples, their territories and resources. This model, a colonial legacy, continues to be retained and fostered in independent India in spite of the Constitution. The model, the predominant frame of Indian governance, therefore increases the resource related conflicts between the state and people.

Under the command and control model, a centralized authority to enforce a regulatory system is setup which grants permissions for use or extraction of water from the identified source. The power to grant licenses spawns other bureaucratic mechanisms which facilitate arbitrary decisions susceptible and vulnerable to influence besides red-tapism and corruption in the chain of command. This model is largely impervious or unresponsive to the wishes of democratic aspirations. This approach and model of regulation and control in these legislations runs contrary to the objective of decentralized and community based control of water and the right to water recognised in the Constitution in the 73rd Amendment Act, 1992, by which the Panchayat as an institution of self-government, comprised of elected members, was formally recognized.

Self-Governance

- *Article 40 under Part IV of the Directive Principles of State Policy:* Organisation of village panchayats - The State shall take steps to organise village panchayats and endow them with such powers and authority as may be necessary to enable them to function as units of self-government.
- *Part IX and Article 243B*: Part IX of the Constitution deals with panchayats and Article 243B mandates that 'there shall be constituted in every State, panchayats at the village, intermediate and district levels in accordance with the provisions of this Part'. The provisions introduced by this amendment provide for administrative and legislative powers of the Panchayat.

- *Article 243G*: Under this Article, the legislature of a State can make a law to empower the Panchayat to prepare plans 'for economic development and social justice' apart from other matters specified in the Eleventh Schedule to the Constitution. The topics in the Eleventh Schedule include minor irrigation, water management and watershed development (Entry 3), fisheries (Entry 5), drinking water (Entry 11), waterways (Entry 13), health and sanitation (Entry 23), public distribution system (Entry 28) and maintenance of community assets (Entry 29). But this scheme of decentralization of responsibilities is yet to be effectively operationalised and the undemocratic bureaucratic command and control model dominates the Panchayat. However, Part IX Article 243 on Panchayats gives pre-eminence to the State rather than village panchayats as envisaged under Article 40.
- *Article 244*: The extension of the provisions of Part IX of the Constitution relating to the Panchayats to the Scheduled Areas as referred to in Clause (1) of Article 244 (Provisions of the Panchayats (Extension to the Scheduled Areas) Act, 1996) goes even further directing that the state laws shall be in consonance with traditional management practices of community resources (Section 4.a); every Gram Sabha shall be competent to safeguard and preserve the community resources (Section 4.d), approve the plans, programmes and projects (Section 4.e.i), identify and select beneficiaries (Section 4e.ii), and certify utilisation of funds (Section 4.f); plan and manage minor water bodies (Section 4.j); control local plans and resources (Section 4.m.vii). Further the Panchayats at the higher level are not to assume the powers and authority of any Panchayat at the lower level or of the Gram Sabha (Section 4.n). These reflect the pre-eminence of the village panchayat, especially the gram sabha, as envisaged in Article 40.

Judicial Pronouncements on the Right to Water

- *Article 13, Article 32 and Article 226*: The Constitution of India gives powers to the High Courts and the Supreme Court of India to issue writs to strike down unconstitutional legislative and executive action. Under Article 13 of the Constitution, any enacted law that is inconsistent or contrary to the fundamental rights enumerated in Part III of the Constitution would be liable to be declared as void. The Supreme Court of India, under Article 32 of the Constitution, and each of the High Courts, under Article 226 of the Constitution, can issue writs to declare the unconstitutionality of statutes that are challenged by way of Writ Petitions.

- '*The right to life* includes the right to live with human dignity and all that goes with it, namely, the bare necessaries of life such as adequate nutrition, clothing and shelter and facilities for reading, writing and expressing oneself in diverse forms, freely moving about and mixing and commingling with fellow human beings. The magnitude and components of this right would depend upon the extent of economic development of the country, but it must, in any view of the matter, include the bare necessities of life and also the right to carry on such functions and activities as constitute the bare minimum expression of the human self.[1]'
- '*Water is the basic need* for the survival of human beings and is part of the right to life and human rights as enshrined in Article 21 of the Constitution of India and can be served only by providing source of water where there is none.[2]'
- The court has, in the context of water pollution, mandated the cleaning up of water sources including rivers[3], the coastline[4] and even tanks and wells. The court said: 'It is important to notice that the material resources of the community like forests, tanks, ponds, hillock, mountain etc. are nature's bounty. They need to be protected for a proper and healthy environment which enables people to enjoy a quality of life which is the essence of the guaranteed right under Article 21 of the Constitution of India.[5]'
- The concern over pollution of ground water by unregulated discharge of effluents has led the court to issue mandatory directions for clean up by the polluter and restitution of the soil and ground water[6].
- The court has also applied the 'precautionary principle' to prevent the potential pollution of drinking water sources consequent upon the setting up industries in their vicinity[7].
- The court has recognized that water is a community source which is to be held by the State in public trust in recognition of its duty to respect the principle of inter-generational equity. The court declared that 'our legal system – based on English

common law – includes the public trust doctrine as part of its jurisprudence. The State is the trustee of all natural resources which are by nature meant for public use and enjoyment. Public at large is the beneficiary of the seashore, running waters, air, forests and ecologically fragile lands. The State as a trustee is under a legal duty to protect the natural resources. These resources meant for public use cannot be converted into private ownership.'[8]

SOME CRITICAL ISSUES

There are a number of obstacles that require to be overcome before the effective and equal access to water to every citizen is made a reality.

The Sharing of River Waters

- Among the early disputes regarding the sharing of waters was the one between the States of Gujarat, Madhya Pradesh, Maharashtra and Rajasthan over the sharing of the waters of the river Narmada. The dispute was referred to the Inter-State Water Disputes Tribunal (IWDT) constituted under Article 262 of the Constitution. An award was made at the end of a deçade long determination. But this led to further disputes. A final award was made in the year 1977[9].
- The disputes between the States of Tamil Nadu and Karnataka over the sharing of the waters of river Cauvery has also been referred to a similar tribunal in the year 1992. After continuous hearings spread over 14-15 years, and many riots in between this period, the award was finally announced in 2007, though the Karnataka state has declined to accept it.
- Disputes between Punjab, Haryana and Himachal Pradesh[10] over the sharing of the waters of the Ravi and Beas rivers have proved problematic despite the award of the tribunal. The state of Punjab refused to honour the award of the tribunal. This led Haryana to approach the Supreme Court for its enforcement.[10] When the States are not prepared to respect the rule of law, the rights of farmers to river water for irrigation are violated.

The above disputes point to the States prioritizing political expediency over their obligations to honour their commitments under the Constitution.

Subversion of the Constitutional Scheme

- At the community level, there is a conflict with (a) the State and (b) the village level administrative bodies, over the control of water sources. This indicates the state's attempt to render the constitutional scheme of decentralization of power over water sources unworkable.
- The State can also legislatively avoid any contractual obligation it might have to assure availability of fixed quantities of water for irrigation purposes to cultivators[11] for purposes of providing 'benefit to a larger area and more people'.[12]
- On providing of equal access to water at the community level, the caste based graded inequality can prevent the socially disadvantaged groups from accessing common water resources, or render water unusable. S.3 (xiii) of the Scheduled Castes and Scheduled Tribes (Prevention of Atrocities) Act, 1989 which makes a punishable offence the act of a person, not being a member of a Scheduled Caste or Scheduled Tribe, who 'corrupts or fouls the water of any spring, reservoir or any other source ordinarily used by the members of the Scheduled Castes or Scheduled Tribes so as to render it less fit for the purpose for which it is ordinarily used'.[13]
- The traditional water managers at the village level find themselves marginalized with the recent spate of legislation by the States failing to provide any role for them.[14]
- There are other users of water like fisher folk, whose livelihoods face a crisis on account of the changing priorities of the State as for instance applied to communities living around a protected forest area over the issue of traditional access of such communities to forest resources including fish in rivers and reservoirs.[15]
- The claims by industrial units that their right to carry on trade and business be accorded priority over the right to a clean environment resulting in conflicts has invariably been resolved by the court leaning in favour of the latter. The court drawing extensively on international environmental law employed the 'polluter pays' principle and the 'precautionary principle'. The court has stated that the industry 'has no right to destroy the ecology, degrade the environment and pose a health hazard. It cannot be permitted to expand or even continue with the present production unless it tackles by itself the problem of

pollution created by it.'[16] The polluter pays principle has been applied by the Court.[17] The court has pointed out that the inadequacies of science in being able to assimilate the impact of environmental harm had led to the replacement of the 'assimilative capacity' rule with the 'precautionary principle'. This led to the 'special principle of burden of proof in environmental cases where burden as to the absence of injurious effects of the actions proposed is placed on those who want to change the status quo' and 'an order of exemption carelessly passed, ignoring the `precautionary principle', could be catastrophic.'[18]

However, the Supreme Court, in cases[19], did not take recourse to the precautionary principle by stating that 'once the State Government has taken all precautions to ensure that the impact on the environment is transient and minimal, a court will not substitute its own assessment in place of the opinion of persons who are specialists and who may have decided the question with objectivity and ability'.

There is a major thrust to privatize the distribution of water for drinking purposes in both the urban and rural areas.[20] There was also an instance of sale of even a river.[21] In instances of industries which use water as a major raw material, the extent to which community level decision making bodies can control the activities of corporations has come up with the state questioning the legality of the domain of community bodies.[22] The bottled water industry is a high-growth sector. This has added another problematic dimension to the already stressed availability of water to communities for their basic survival needs.

There is a distinct trend with the replacement of the state's obligation to ensure availability to the community of basic minimum quantities of affordable quality water, with water as commodity of commerce. Strangely, water is also seen as an instrument to curb population growth.[23] This is violative of equal access to common resources. The Courts have often attributed disproportionate blame on poor households, often forcibly evicting them in favour of industrial units in pollution of water bodies.[24] The utilitarian logic pervades the states decisions dominating or brushing aside issues of equity, justice, ecology and sustainability. The larger good is often cited by the courts to explain away large scale displacements with rehabilitation of the oustees itself being interpreted thus: 'At the rehabilitation sites they will have more, and better, amenities than those enjoyed in their tribal hamlets. The gradual assimilation in the mainstream of society will lead to betterment and progress.'[25] The Court defends the state in such instance with 'when a decision is taken by the government after due consideration and full application of mind, the court is not to sit in appeal over such decision' and explains such displacement as necessary for the 'larger good'.[26]

Towards A People's Governance of Water

The law is yet to clarify on matters related to rights over water and governance of water. The sources of law are statutory provisions, case laws or precedential court decisions, doctrines and principles deriving from the British common law system, international agreements, personal law, and customary law and practices. The right to ground water is linked to right to land supported by the legal construction under Indian Easements Act of 1882 by which there is almost absolute ownership of ground water by the property holders. At the same time, the state as eminent domain has power over natural resources.

The British Colonizers introduced the concept of eminent domain of the state that all resources that were 'unclaimed' or not 'used' belonged to the empire. These laws still remain as statutory laws and constitute the framework for governance, creating both a barrier for democratization of the society and a wedge between the state and the people leading to intense increasing conflicts.

In the federal set up, constitutionally, the subject-matter like 'water' and 'land' are within the legislative power of the states under Art 246 and List II of the Seventh Schedule. 'Water' is expressed as "water supplies, irrigation and canals, drainage and embankments, water storage and water power" under Entry 17 of List II. Apart for a reference to an exception for inter-state rivers, it is the state that is the exclusive authority to make laws with respect to 'water'. The Ministry of Water Resources, Government of India, had issued model bills for the state to enact centralizing the authority with the state governments. Under this, State Ground Water Authorities are to manage and develop ground water resources.

On the other hand, the Panchayat Raj Act 1993 in pursuance to Article 40 provides for 'drinking water' and 'minor irrigation, water management and watershed development' in the Eleventh Schedule as subject-matters over which responsibility can be devolved from state to village level. Subsequently, it is up to each and every state to pass regulations on the authority of the Panchayats on the 29 listed subjects. Accordingly, duties of the village Panchayats are 'maintenance of traditional drinking water sources, preservation of ponds and other water tanks, setting up and management of water supply schemes'[27].

Matters such as natural resource management including water are very much a community domain exercised through customary laws and practices. The Constitution recognizes customs and customary practices. The term 'law' in Article 13 includes 'customs' and 'usages' having the force of law but not infringing upon any of the fundamental rights conferred by part III of the Constitution. In reality, customary laws are subjected to being in consonance with the statute made laws which are still largely a colonial legacy. Article 21 on fundamental rights confers Right to Life, not merely of animal existence but life with human dignity, and right to livelihood except according to a due process of law. The state under Article 39(b) is duty bound to sub-serve the common good. Part IV-A of the Constitution imposes a duty (and not authority) on the citizens to protect and improve the natural environment including forests, lakes and rivers which are explicit directives to be promoted through legislations.

It is in the V Schedule Area that the Panchayat Raj (Extension to Scheduled Areas) Act 1996 made a decisive departure introducing a frame to decolonize the colonial legislative frame with comprehensive powers for 'traditional management practices of community resources', 'safeguard and preserve community resources' which include 'planning and management of minor water bodies'. 'Community resources' implies water. The assertion of the constitutional intent of community or people-centric State rather than the State versus people paradigm resulted from Adivasis struggles. Consequently, the Gram Sabhas (the people in the village/hamlet) rather than the Panchayat is to be in primary command with the restriction placed on the structures at the higher level to not assume the powers and authority of any Panchayat at the lower level or of the Gram Sabha paving way legally for the opening up to a people's participatory democratic governance system. Yet, the dichotomy between the sovereign right of peoples and the sovereign right of the state to natural resources persists alongside, a contradiction, which in many places has led the state to wage war on her own peoples. The sovereign right of nation that is her peoples is clearly on a collision course with the State. The statutory laws emanating from the Constitutional frame and those that are the colonial legacy are on a collision course. There is a strong reluctance to dismantle the colonial frame with the nexus between the bureaucratic structure and the dominant political class.

Solutions for problems of democracy lie in further democratization. The challenge is to move away from decentralization to that of evolution of non-centralized people's participatory governance system.

The Gram Sabha, the fundamental and primary natural unit of governance in the panchayat structure, shall be the competent authority to determine the use, control and management of ground water and minor water bodies within its geographical jurisdiction. Gram Sabha shall have the command over ground water for 'traditional management practices of ground water resources', 'safeguard and preserve ground water resources', 'planning and management of minor water bodies' and 'management of water supply schemes'. The Gram Sabha, while exercising its powers to 'safeguard and preserve ground water resources and minor water bodies', shall have the powers to ensure water to its members, both in sufficient quality and quantity i.e. safe and adequate water. In exercise of this authority, the Gram Sabha shall reasonably restrict the use of ground water and minor water bodies to ensure safety and adequacy of water. The Gram Sabha shall have the authority to take necessary steps and penalize the violators in case of violations that threaten the quality and quantity of water resources.

All development activities falling within the geographical jurisdiction of the Gram Sabha shall be carried out with the express consent and approval of the Gram Sabha with a view to protect, conserve, sustainably use and develop the said water resources. Resolution by Gram Sabha to cancel such consent shall automatically lead to cancellation of license/approval issued by the appropriate bodies/authorities. The Gram Sabha shall reasonably restrict or ban the use of ground water and/or minor water bodies as the case may be, if the Gram Sabha so considers that such use constitutes a threat to quantity and/or quality of water.

To enable the Gram Sabha to carry out these responsibilities and functions to the fullest extent, the Gram Sabha shall call forth the services and assistance of the state Ground Water Authority, the Ground Water Department and the State Pollution Control Board, the Village Panchayat, the Block Panchayat and/or the District Panchayat as the case may be and other relevant bodies, which shall be made available.

To enable the Gram Sabha to exercise this natural right, it is therefore necessary that relevant State laws such as (a) the state Panchayat Raj Acts; (b) the state ground water Acts (c) the water pollution prevention and control Acts and (d) any other relevant laws as are applicable are to be amended. The 73rd Amendment, the Panchayat Raj Act 1993, should be reviewed and amended conferring the command over natural resources to the gram sabhas.

NOTES

1. Francis Coralie Mullin V. The Administrator, Union Territory of Delhi 1981 (2) SCR 516.
2. Narmada Bachao Andolan V. Union of India (2000), 10 SCC 664 at 767.
3. For orders relating to the pollution on the river Ganga, see M.C. Mehta v. Union of India AIR 1988 SC 1037, 1115 and (1997) 2 SCC 411. For an important decision regarding closure of a hotel resort which was polluting the Beas river in Himachal Pradesh, see M.C. Mehta v. Kamal Nath (1997) 1 SCC 388.
4. S. Jagannath V. Union of India (1997), 2 SCC 87.
5. In Hinch Lal Tiwari V. Kamala Devi (2001), 6 SCC 496.
6. Bhavani River-Shakti Sugars Ltd. (1998), 6 SCC 335. In Indian Council for Enviro-Legal Action V. Union of India (1995) 3 SCC 77, a compensation package was worked out for farmers affected by their only source of irrigation, a river in Andhra Pradesh, was polluted by discharge of untreated effluents by industries alongside its banks.
7. A.P.Pollution Control Board v.Prof. M.V.Nayudu (1999) 2 SCC 718 and A.P.Pollution Control Board (II) v. Prof. M.V.Nayudu (2001) 2 SCC 62.
8. M.C. Mehta V. Kamal Nath (1997), 1 SCC 388.

9. Narmada Bachao Andolan v. Union of India (2000), 10 SCC 664 at 767.

10. State of Haryana V. State of Punjab (2004), 12 SCC 673.

11. This was demonstrated in the context of the Parambikulam Aliyar Project undertaken by the State of Tamil Nadu to ensure supply of water for agricultural operations in some of the areas of Coimbatore district. Although the State government entered into an agreement with the cultivators to assure supply of a minimum quantity of water for irrigation purposes, it made a legislation ten years later seeking to expand the area over which the said water for irrigation purposes would be made available.

12. Parambikulam A.P.O.A. Association V. State of Tamil Nadu (1997), 7 SCC 626 at 632.

13. This offence is A similar provision is s. 4 (iv) of the Protection of Civil Rights Act, (1955), Which Renders Punishable the act of a Person who, on the ground of 'Untouchability', Enforces against any other Person any Disability with Regard to 'the use of, or access to, any River, Stream, Spring, Well, Tank, Cistern, Water-tap or other Watering Place, or any bearing ghat'.

14. These traditional managers are called by different names: in Maharashtra they are the Havaldar, Jagliya or Patkari, in Ladakh the Churmun, in Tamil Nadu, Andhra Pradesh and Karnataka, they are known as Neerkattis. Their essential function was to apportion water among cultivators accessing a common water resources, be it a tank or a river. It is unfortunate that both the Andhra Pradesh Farmers Management of Irrigation Systems Act, 1997 and the Tamil Nadu Farmers Management and Irrigation Systems Act, 2000 do not provide for any role for the Neerkattis even while they envisage the creation of associations of water users at the village level.

15. Imposed under the Wildlife (Protection) Act, 1972.

16. Vellore Citizens Welfare Forum V. Union of India, (1996), 5 SCC 647.

17. Indian Council for Enviro-Legal Action V. Union of India (1996), 3 SCC 212 in which residents of village Bichhri in Rajasthan who were denied access to Drinking Water on Account of Pollution of the Wells by Seepage of Toxic Untreated Wastes Produced by Chemical Factories, were held Entitled to Restitution and Compensation. Also S. Jagannath V. Union of India (1997), 2 SCC 87, In Re: Bhavani River-Shakti Sugars Ltd. (1998), 6 SCC 335 and in Indian Council for Enviro-Legal Action V. Union of India (1995), 3 SCC 77 a Compensation Package was Worked out for Farmers Affected by their only Source of Irrigation, a River in Andhra Pradesh, was Polluted by Discharge of Untreated Effluents

by Industries Alongside its Banks. This Precautionary Principle was Applied by the Supreme Court to Strike Down the Notification Issued by the Government of Andhra Pradesh Exempting an oil Industry Located in the Vicinity of two Major Water Reservoirs.

18. A. P. Pollution Control Board (II) V. Prof. M.V. Nayudu (2001), 2 SCC 62 at 79. More recently, the Supreme Court declared as Illegal the Action of the Government of Karnataka Excluding the land Brought Under Mining by the Kudremukh Iron Ore Company Ltd. Excluded from the Purview of Notification Issued under s.35 (1) of the Wildlife (Protection) Act 1972 Declaring the Kudremukh National Park to be a National Park: T.N. Godavarman Tirumalpad V. Union of India (2002), 8 SCALE 204.

19. Essar Oil Ltd. V. Halar Utkarshi Samiti (2004), 2 SCC 392.

20. While Maharashtra has Enacted a law which paves the way for privatisation, the Draft Legislation in Delhi has met with stiff resistance from Water Users.

21. The exclusive right to 23.5km of the Sheonath River was handed over to a Private Company in the State of Chhattisgarh under the Build-own-operate-transfer Agreement by a state owned industrial development corporation illustrates the negative fallout of such a move. The waters were to be sold by the company to the state owned corporation which in turn would sell it to the ultimate users in the nearby industrial township. The immediate impact was adverse to the Fisher folk who were barred from fishing in the River and to small farmers who depended on the river. With protest, the Government announced the cancellation of the agreement.

22. Judgment related to the Hindustan Coca-Cola Beverages (P) Ltd. V. Perumatty Grama Panchayat dated December 16, 2003 of K. Baakrishnan Nair, J. of the Kerala High Court in Writ Petition (C) No. 34292 of 2003 and Judgment Dated April 7, 2005 of the Division Bench of the Kerala High Court in Writ Appeal No. 215 of 2004. The Case is in the Supreme Court.

23. Maharashtra Water Resources Regulatory Act (2005), in S.12 (11) States: 'Notwithstanding anything contained in this Act, a person having more than two children shall be required to pay one and half times the normal rates of water charges fixed…' which places an unconscionable restriction on the availability of affordable water by linking the payment for water to the size of the family.

24. Order dated March 3, (2003), of the Delhi High Court in Writ Petition 2112 of 2002 (Wazirpur Bartan Nirmata Sangh). The Court ordered the Removal of Unauthorised Encroachments along the Banks of the River Yamuna Flowing Through Delhi, Which the Court

Considered as the Main Cause of the Pollution While the City's Industrial Units, Many of them Unauthorized on the River Banks, Which the Real Cause of Pollution by Discharging of Large Quantities of Untreated Effluents into it were Left Untouched.

25. Narmada Bachao Andolan V. Union of India (2000), 10 SCC 664.
26. Ibid.
27. As for Instance the Third Schedule Under Sec. 166 of the Kerala Panchayat Raj Act 1994.

REFERENCE

Muralidhar. S. (2006), The Right to Water. An Overview of The Indian Legal Frame, in Eibe Reidel & Peter Rothen eds., The Human Right to Water, (Berlin: Berliner Wissenschafts-Verlag, 2006), p. 65-81.

Bijoy, C.R. (2006), Kerala's Plachimada Struggle. A Narrative on Water and Governance Rights, *Economic & Political Weekly,* October 14-20, Vol. XLI No. 41, p. 4332-39.

3

PATHWAY TO THE WATER DEMOCRACY IN INDIA

H.M. Desarda[*]

INTRODUCTION

The economics and politics of the world during the past few centuries, particularly after the World War II, is overwhelmingly dominated by the growth mania– a craze for faster and faster growth. Despite the strong warning of the consequences of undifferentiated and undirected growth the limits to growth are not acceptable to the powers– that– be!. On the contrary geo-political power struggle is guided by the ideology of growth without any limits. The political economy of the globalisation has put it on a all pervasive global political agenda.

As a consequence, the control of resources is the major concern, almost an obsession, of the global power-game. Alas, this has pushed the world to the brink of disaster. Everyday the ecological footprints are becoming more and more gruesome and are threatening the very survival of human kind and security of the

* Prof. Desarda, Retired Professor of Economics, is a former Member of Maharashtra State Planning Board. Recently the Rural Development Ministry of GOI has nominated him as a Member of Monitoring and Vigilance Committee.

planet. There is a pervasive plunder, exploitation and destruction of the natural resources. Hopefully the two citations in the beginning of the paper should convey the critique of the present path and pattern of water resource planning and policy vehemently pursued in India.

Unless and until we unequivocally recognise this grim reality, we can not seriously grapple with the water challenges before India and the world today. This reality demands that we discuss and debate the issues that has bearing on our livelihood and well-being. The essence of democracy is participation. As such, the water democracy means participatory water resource development and management from micro-watershed to river basin.

For this we have to clear the conceptual cobwebs. First and foremost, we have to redefine development. Without that the prefix 'Sustainable' will not be meaningful. Indeed, we are afflicted by the disease of development. The task before us is to diagnose the disease. Precisely, this is the role of democratisation - be it water or any other life and death issue.

The paper aims at to present some ideas to provoke (informed) debate. With this in view the issues in the realm of water resource development which are critical to Jal Swaraj, the water democracy are discussed in the context of the present policies, programmes and projects of water resources development in India.

Given the present national context and centrality of the water problem in ensuring the equitable and sustainable development, a plan, of the holistic development and management of the country's water resources is outlined. Infact, it is an alternative approach : The ecological and ethical perspective of growth and development. As of now, the techno-engineering model of growth, including the water resource development is resource-squandering and destructive. Therefore, there is overarching need for the alternative pathway.

THE TASK

The foremost challenge before India today is *"the equitable and sustainable development of our water resources"*. Recurrent drought and floods coupled with mass poverty, chronic unemployment and pervasive under nutrition is the most pressing and paramount national problem. However, instead of *indulging* in the usual rhetoric of truant rains and doling out funds in adhoc and arbitrary manner for drought and/or flood mitigation and/or the related relief measures, the Union Government should urgently launch a nation-wide Watershed Development Programme through the National Rural Employment Guarantee, with the involvement of all the state governments and active participation of the panchayats and people. The main objective of this should be the eradication of the scourge of man-made famine, floods, hunger and poverty permanently. Simultaneously, providing good quality drinking water and sanitation to each and every person must be the basic goal of the *water democracy.*

A PERSPECTIVE

In our opinion, *water* and *employment* is the *key* to build India in accordance with the vision and values of founding fathers and mothers, which are succinctly enunciated in the Constitution of India, as the Fundamental Rights and the Directive Principles of the State Policy. "The objective of our freedom movement is to wipe every tear from every eye', said Gandhiji. And he gave us the *talisman,* to put the last and lowliest at the center of all our endeavors. Very aptly Jawaharlal Nehru observed, 'so long as there are tears, we shall not relent.' Indeed, the philosophy and pathway of the watershed development is the significant substantial step-the talisman-to solve the major national problem we face in the 60th year of the Independence and after completion of the Ten Five Year Plans. Candidly speaking, it is the tragic consequence of the resource-destructive growth and governance model we are pursuing in the name of the national development.

The author of this paper has formulated the following proposals and action plan after visiting many states and interacting with social scientists, land and water experts, social activists and the policy planners across the length and breadths of this continental country from the Thar desert, Indo-Gangetic plains, Brahmaputra Valley and peninsular India; from the Kutch to Kamrup and Kashmir to Kanyakumari. Agreeably, the scientific and societal approach to achieve this objective is : *micro-watershed development.* Monsoon management and community-based rainwater harvesting with the focus on *holistic* development of water resources is the major thrust of this strategy.

Quintessence, we should follow a socio-ecological-ethical pathway to equitable and sustainable development. Indeed, that alone can ensure democratisation of water and usher in the vibrant water democracy.

We should always remember that India's water resources are mighty. The quantum of water India receives through the precipitation is 4000 billion cubic meter (BCM), of this annual average run-off is 1953 BCM. Of course, there is a marked *spatial* and *temporal* variation in precipitation and water availability. We are familiar with paradox of the floods and droughts. The flood ravages in Brahmaputra valley and Gangetic plains; and droughts in the arid and semi arid region of the western and peninsular India pose a big challenge. Paradoxically enough, this year we had devastating flood in the arid region of Rajasthan and severe drought in the Brahmaputra valley. The local, regional and global level climate changes are showing not only their symptoms but horrendously devastating manifestations.

After the independence we have built more than 4000 large dams entailing an investment of Rs. 5000 billion at the current price level and cost norms. Still day by day, the water-scarcity is becoming more and more acute. Surely, the present *supply-side* approach is not a solution to the 'scarcity' problem we are facing.

The need of the hour is to come out of the *trap* of the cult of *gigantism* and techno-engineering growth mania. Indisputably, this is the root cause of India, a water rich country, becoming a water-short nation. As a consequence, we are fast heading towards *water-conflicts* between regions/states, sectors, groups and persons. Despite the Supreme Court's direction, the tug of war and tensions we have witnessed between the two riparian states, Tamilnadu and Karnataka, over *sharing* of the Cauvery waters and Punjab's attempt to wriggle-out of the Inter-State Water Award should awaken the nation. The prevalence of whole-gamut of the highly ingenious indigenous rainwater harvesting systems - which are succinctly documented in the CSE publication *Dying Wisdom* - are fast falling into *disuse* needs to be revived and rebuilt.

If areas of the low and scanty rains like the Thar desert were able to *manage* their meager water resources wisely there should not be any difficulty in providing water to each and every Indian, any where in our sub-continent. The global experience tells us that there are local solutions to ward off the water-scarcity. Infact, water is not only what we see as rains and storage on surface but the vast *link* from the ocean to rains, snowmelts, dew and all allied water-entities. Agreeably, the soil moisture is most strategic part of the water regime. Instructively, over 40 percent of our total rainwater is *stored* in soil-profile. Hence, we have to carefully grasp the complex character of the hydrological cycle. Above all, the role of the vital eco-entities like the crop lands, grazing lands, forests and entire range of the biotic and biomass system which helps to augment, renew and recycle the vital life source : *water,* the synonym for the life ! Well, water is used to extinguish fire but when water itself catches the fire the situation becomes volatile, violent and threatens national peace and progress. Perhaps, that is what we are witnessing in India today.

The chief cause of this is the avarice and greed, which is making us crazy about the grid of the water and electricity. As Gandhiji's famous thought reminds us, "There is enough for everyone's need but not for anybody's greed". As long as we fail to make this as our conscious national value and vision, we will not find a lasting solution to any problem including the most pivotal one like water. With the aforesaid diagnosis, perception and perspective, albeit very short and succinct, we proceed to present the broad contours and components of the action-agenda, including the planning and policy proposals that needs to be pursued with the active involvement of the people, to build and strengthen the water democracy in India.

THE PLAN: CONCEPTUAL FRAMEWORK AND OPERATIONAL STRATEGY

A *master plan* for initiating and streamlining the watershed works throughout the country is outlined below. The conceptual framework and the operational strategy of the plan is indicated in brief. In this regard, it is necessary and useful to begin with the critique of the prevailing policy and planning.

Bid Good-bye to Gigantism

As a first step we need to move beyond the present large dam-centric planning and policy of water resource development, which has distorted the water resource planning and is draining the exchequer excessively. Secondly, the drought-prone and flood prone area programmes and scattered projects of watershed development should become a part of broad-based national movement for rainwater harvesting. The major focus of this should be integrated development of the land, water and biomass resources. For this we have to adopt a *holistic* approach.

Now therefore *in-situ* rainwater harvesting is the cost effective pathway. As such, the prevailing perception about the *storage* needs to be thoroughly reviewed and radically restructured. We should always remember that the *protective* irrigation for drought - proofing is different from the perennial irrigation. The large dams and mega storages with a vast network of canals are required for the Green-Revolution oriented agricultural development. And also for the water supplies to the metropolises and water-squandering industries and the elitist Westernised *lifestyles.* Hence, we have to *overhaul* the present water policy and planning paradigm. This is the primary pre-requisite to initiate the national resource-conserving and pro-people water resource planning. We have to clearly grasp the ecological and ethical challenge we are facing and the genesis and consequences of this.

Resource Literacy and Employment

In the framework of long-term perspective the watershed works should be dovetailed with the employment Programme which is statutorily guaranteed and linked to 'food and watershed works scheme' The edifice of this should be community empowerment through employment and entitlement. Systematic compaigns for public awareness should be launched. The major plank of which will be resource-literacy, care and conservation of prime natural resources-land, water, flora, fauna and biodiversity. Simultaneously, a training of the panchayat members and those who wish to participate as barefoot planners should be initiated at every village and micro-watershed level.

Clearly and carefully the spirit of voluntarism and professionalism has to be blended beyond the paternalistic and parasitic NGO-cult. Exceptions apart, the present government / non-government alliance generally excludes the community and creates an army of middle-men in the delivery system, who are often outfits of the vested interests. Therefore, a plea for power to the people through the Panchayat, Gramsabha and watershed community. The phenomenon of the treadmill 'NGOisation' is antithetical to the organic development of peoples' power.

Given the fact that the programme needs to be launched in thousands of villages simultaneously, there is a pressing need for forging a community-based institutional arrangement, which the panchayat-raj system *alone* can provide across the length and breadth of the country. Notwithstanding the present politicking prevalent at the panchayats, there is no other viable and desirable alternative but to accept it as a part of democratic-schooling. Now there is an urgent need to *redefine* and recast the growth and development as concepts and categories. Boldly, we have to *de-mystify* and *decentralise* the meaning and process of development and make it pro-poor pro-women, and pro-nature.

Utilising the Labour-Force

The vital favourable factor in the wake of drought and other calamities leading to the crop failure is labour availability, which must be seen as an opportunity for creating durable social assets. Moreover, we should remember that the currency notes are meaningless *sans* labour resource and social capital in terms of community participation. This is the basic core around which the programme should be designed and implemented. Hitherto this has been missing element and that's why nearly 5000 billion rupees spent on irrigation projects, DPAP and such other programmes have failed to rescue the agriculture from 'the vagaries of the monsoon'. With this clear perspective, an alternative approach should be properly worked out. This means a paradigm shift in the prevalent drought-mitigation and also macro-economic policies of the growth and globalisation.

The Dimension

This scheme should aim at an employment of 20 million workers for 200 days (or 40 million workers for one hundred days as estimated/envisaged under the NREGS) at a wage of rupees 50 in cash and two and half kilogram of grains per day. This would entail an expenditure of Rs. 20,000 Crore in cash plus 10 million tones of food worth Rs. 5000 Crore, at the minimum procurement price. Thus, a sum total of Rs. 40000 Crore investment (wage plus material cost at the rate of Rs. 100 per-person day) would generate a work of 4000 million person-days in a year. Meticulously and meaningfully it should be used for creation of durable social assets. Broadly, this includes the works of soil conservation, moisture retention, afforestation, groundwater recharge, water harvesting structures etc. Retarding the *runoff* and enhancing the *recharge* is the only option to augment the water availability. The noteworthy aspect of this is that the returns to the investment are the most positive and quick. As such, it is most viable economic programme on the benefit-cost criteria. After passing of the landmark legislation, the National Rural Employment Guarantee Act (NREGA) and launching of the scheme in the 200 (out of 607) district the outlay of eleven thousand Crore is made in the budget of 2006-07. This is the huge step forward. Now the only question is: to turn these financial outlays into the physical *outcome*. Precisely, this is the thrust of the plan outlined in this paper.

Certainly, it is not a dole distribution or charity programme but a very sound employment and production programme, which fulfill the consideration of economic efficiency, social justice and ecological sustainability.

The Ridge to Valley Methodology

Going by the current state of (degradation) watersheds it would require on an average around 100 person days of physical work per hectare to carry-out the watershed works. This means around 20 million hectares (mh) can be *treated* and tackled in the year. True, as per the local or watershed wise availability of the number of labourers it would have to be phased out over 2 years. In any case, once initiated the entire watershed must be completed in a period of 3 years. The work should be carried out strictly *sequentially* from the 'ridge to valley' in each and every micro-watershed within the area of sub-watershed in that river basin. This alone can ensure that the *size, scale & sequence* of *impounding* water is appropriate. Regrettably, at present it is highly topsy-turvy and this is the chief cause of the water-scarcity and imbalance in water availability in the country. Therefore an *alternative approach*, has to be adopted which is integrated and holistic.

Need to stop the Degradation

According to the National Commission of Agriculture (1976) 175 mh of land in the country is degraded and is in need of urgent *treatment* and improvement. Even at the scale suggested here, it would take 9 years to cover the watershed works to arrest wind, water and biotic - pressure - related erosion and degradation. The erosion, degradation and damage has been going on unabated for years causing an annual soil-loss of 12 billion tonnes. Indeed, it is a colossal national loss of precious natural resource. As such, the task of conserving the resource is most challenging.

As indicated in the (draft) Approach Paper of the 11th Five Year Plan at Rs.10000 per hectare estimated 80 million hectares needing urgent treatment would require Rs.80,000 Crore. The funding through the NREGS is the major source for this.

The well-being of the present generation as well as posterity depends on addressing the *number one* national problem. Surely this calls for a National Movement with the clarion call, "Mitti Bachao, Desh Bachao' - save soil to save the country. Undoubtedly, the *life* and *livelihood* of people directly depend on health of the soil and natural resources. In short, the overriding national objective should be to minimise the soil-loss. That would maximize the moisture and mitigate the water-scarcity and the crop failure. Ideally we should plan for a zero soil loss. Well, this should be the goal to highlight the gravity of the consequences of the colossal soil losses.

Curiously, we face, at the national level, a paradox of floods and drought simultaneously. Not only the traditional flood prone areas in Brahmaputra and Gangetic valleys are more intensely ravaged by the flash floods but even arid and semi-arid regions like the Thar desert are experiencing floods, the chief cause of this is: indiscriminate and ruthless *deforestation* in the catchment and mountain regions. As per the estimates some years ago the silt carried by the Brahmaputra is 735 million metric ton every year. As of now, the situation is much worse and mind boggling. The amount of soil loss on account of the destruction of the vegetative cover-trees and grasses - in each and every river basin, rivulets and watercourse is causing tremendous soil loss and as a consequence it reduces the recharge. The co-relation between *rainfall, run-off* and *recharge,* the three fundamentals of water, is diverged and disturbed. It directly influences the water availability in each micro, milli and mega watershed in any river basin. Unless and until we take care of this, the water-scarcity can never be tackled. We have to see it in totality and deal with the basic issues. Unequivocally, this is the paramount task before the nation today.

Ecological Fundamentals

It should be observed and examined carefully that the formation of an *inch of soil* takes several hundred years. Further, we have to reckon with the fundamental fact: margin of the life on earth squarely depends on the nine inches of the top soil, just that, which we are so callously squandering through the mind-boggling *erosion* that is going on unabated for several centuries. Therefore, we have to put our house in order to protect the soil resources without any laxity or excuse. After all, the protection of the *biophysical foundations* of growth is the most pressing challenge before the country today. The threat posed to the *life-support-system* is most alarming and hence, we have to put our acts together to protect the same from the indiscriminate deterioration and degradation of the resources on which hinges well-being of the present and future generations.

Diabolically, our planners, policy-makers, the industrial tycoons and the army of the economists and experts, who are the protagonist of the *market* and are in the 'service' of the state and noveau-riche, *see* the solution to every problem in globalisation. Forgetting altogether the pressing problems of the mass poverty, illiteracy, ill-health, hunger and ubiquitous unemployment, day in and day out they keep on chanting the mantra : our *macro-economic fundamentals are sound!* The degradation, denudation and decline of natural resources is not a matter of serious concern and consideration to them. Needless to say that it is not only a *myopic* view but tantamount to criminal attitude to the eco-system. What a pity that our *ruling elite* and the moneyed classes care so little about the state of our planet. It is the base of the world-view of the modernity and growth mania, which poses a big threat to our survival and well-being. As a perceptive critique of the water resources management in India observes,

"Water resource management strategy in India has so far been guided by the philosophy that water is strictly a fixed resource and we cannot really destroy it on any significant scale. We have successfully destroyed much of our water resources in our streams and rivers, fields and tanks, wells and ponds. Water is not available automatically as a renewable resource because the ecological professes that maintain and give stability to the water cycle govern its renewability. Disrupted water cycles can turn water from an abundant renewable resource into a vanishing non-renewable resource."

Precisely, this is what has happened during the past 50 years of the water resource planning in India. We have forgotten altogether the *casualty* of the renewal of the resources and the result is : pervasive water scarcity. Therefore, we have to seriously rethink and restructure the present approach of the water resource planning and overhaul the project model.

Alternative Approach

Whether it is the large irrigation and power projects or watershed works or any other programme of economic and social development, it cannot succeed just by pumping in more and more money. Obviously "More of the same" is no answer at all. To avoid the past mistakes the need of the hour is social motivation, transparency and accountability. That alone can break the crooked chain of corruption, which has turned development projects and programmes into *kick-back* industry! That is the real name of the game ! This chronic disease cannot be cured without an effective decentralisation of political power and devolution of the economic resources. Therefore a central-stage should be accorded to the Gramsabha and the Panchayat institutions.

At the national level the Prime Minister and in the states the Chief Minister should personally monitor the watershed development programme. That will accord it the due priority and put it on high national action agenda. The main focus should be the community mobilisation, social participation and empowerment and the role of the government has to be facilitative in terms of providing institutional framework and finance and allow the community to execute works, keeping in view the location-specificity and drawing upon collective wisdom of the community. Hence, the plea for adopting *socio-ecological pathway.*

Bhoomi, Jal Aur Jangal Vikas Sena

In addition to the availability of *funds,* the most crucial factor is effective organisation of the labour force to do the actual work. Obviously an institutional framework on the pattern of the National Literacy Mission is essential. Therefore, there should be a National Watershed Development Mission. Infact, it can be linked to the proposed National Rainfed Area Development Authority, which the Prime Minister announced in his Independence Day address in 2005. It should look-after the production and maintenance activities. Further, we suggest a creation of special task-force named as "Bhoomi, Jal Aur Jangal Vikas Sena' - *land army* - to carry out the watershed works in a planned and effective way. Infact, it would give the labour-force, particularly the landless labour and poor peasants a sense of security and also provide a *steady supply* of labour to complete the work in orderly and scientific manner, with due regard to the location-specificity of the drought-prone areas in different agro-climatic zones, topography and terrain.

Power to the People

The control and the command for development and management of natural resources should be expressly at the hands of the village and watershed community. The composition of this should clearly provide for the representation of the major interest groups like land-less labour, women, small farmers, artisans and some elders, who can exercise the balancing influence. In short, the finance and functions should be developed and decentralised without any ambiguity. That should take care of the community being at the center-stage and bureaucracy and technocracy will work in close co-operation with the cross-sections of the stake-holders. The spirit of functioning as a true *public servant* should be the prime criteria in choosing the personnel from the cadre.

Towards Meaningful Education

Given the present trend of 'jobless growth' large majority of youths do not have any interest in so called 'higher education'. Sadly, the universities and colleges have become *dens* of ignorance, nuisance and parasitism of the worst kind. Hence, it would be worthwhile to declare 'examination holiday' for a year or two and *involve* the students and the teachers along with the drop-out youths (who have a tendency to shun the physical work and are ashamed to *soil* their hands) in the national watershed development movement. Certainly, it would be an unique opportunity to *engage* them physically and intellectually in the task of nation-building on the land. Indeed, it would make them really resource literate and employable and not just educated (?) unemployables (diploma degree holders) crowding the employment exchanges, craving for the clerical white collar jobs!

Pervasive parasitism, which has become the hallmark of the elite, urbane, and so-called educated member of (the civil?) society needs to be consciously changed. Therefore, they should welcome this opportunity to work with community and learn the value of *shrama pratistha,* dignity of labour, which will help them to overcome their alienation from nature and society. By actively participating in this unique venture they would learn the true meaning of life, nature and productive work-a much needed 'watershed' in education-for-life! Indeed, this should be the meaning of the *education for National development'* which has been the pious wish of the Kothari Commission on Education.

We Have the Resources

Notwithstanding many disturbing and diversionary trends in our development journey, we have the financial, technical and administrative skills as well as physical resources to *fund* such a strategic *nation-building* programme. Notably, the outlay of Rs.40,000 Crore it requires is *slightly more than one (yes only 1%) percent of India's GDP.* Thoughtfully, we should be able to *garner* this amount by levying a surcharge on the income-tax and/or through a special professional tax for the purpose on the pattern of Maharashtra State.. Well, if we cannot muster and mobilise bare one percent of nation's GDP for such a vital task then all tall-talk of peoples' planning and policy of the *inclusive* growth is utterly *futile.* Notably, the most critical missing factor in our economic policy is the *moral capital.* Obviously, through the social mobilisation and community action we have to consciously build this vital moral capital. If we fail to act on this forthwith, the acute suffering of masses will lead to anarchy, violence and disorder. We have to see the writings on the wall and *act* with a sense of national urgency, here and now!

Movement for the Right to Resources

In order to create wider social support and involve civil society in this national programme, a district level advisory committee of social visionaries, artists, literary figures and social analysts/ activists should be created to guide, supervise and evaluate the watershed development programme. Social commitment coupled with political will, administrative skill and social mobilisation in the spirit of the Freedom Movement is necessary to ensure the success of this pivotal programme. In short, we have to launch a vigorous *national movement* for watershed development through the employment guarantee. A tie has come to take it from a stage of *project to movement*, so that it replicates fast and ensure quality and efficiency. Ultimately, we should strive to create *Jal Sanskriti*, the water culture, which promotes health and happiness of everyone. To educate the public opinion and create societal awareness, we have suggested the wider resource-literacy campaign, focusing in particular on *Jal Saksharta*, the water literacy, which would emphasize the care, conservation and availability augmentation simultaneously. And it should be based on the clear concept on the right to all livelihood resources-land, water, flora and fauna.

TOWARDS THE TRUE SWARAJ

All in all, the objective of creating employment for the burgeoning labour force during the Eleventh plan in non-formal and non-industrial sector can be made viable, productive and meaningful (not just the salaried jobs!) only if it is properly linked to the watershed development programme and watershed plus activities of the value addition through the diversification of agriculture and allied activities. Thus it can be very well interwoven with the planned programmes of agricultural and rural development in the 11th plan. Hence, this plea for *adopting* the plan of the watershed development outlined above as the *nodal* programme for national regeneration with an elan of *people's movement* for community-based rainwater harvesting. This *alone* can ensure *sustainable* and *equitable* development for every Indian and make India a truly developed and civilized society, free from the greed and violence.

This is the true meaning and message of the water democracy which in turn will deepen and strengthen the foundation of social and political democracy, in India. Indeed, this can enable us to make the growth more *inclusive,* which is the vital concern of the 11th Five Year Plan. As such, the right to resources and in particular the access and entitlement to the water resources will effectively empower the people. More crop per drop can give impetus to the broad based agricultural growth. Thus, the water democracy holds the key to solve the basic problems that besiege, the toiling people of India and put it on pathway of the equitable and sustainable development and democratic social order and right to the basic livelihood resources including clear air and water envisioned in the Constitution of India. It can help us attain the objective of power to the people through the right to livelihood resources *(Jal-Jangal-Jamin Adhikar)* - a real *Swaraj* through the *Jal Swaraj.*

REFERENCES

Agrawal A. and Narain S., (eds), (1997), *Dying Wisdom,* Centre for Science Environment, New Delhi (hereafter referred as CSE).

Agrawal A., etal, (2001), *Making Water Everybody's Business,* CSE, New Delhi.

Ahmed Imtiaiz, etal, (1997), *Power and People– A South Asian Manifesto on the Politics and Knowledge of Water,* Fellowship in South Asian Alternatives, Colombo.

Alvares C., (1992), *Science, Development and Violence,* Oxford University Press (hereafter refererd as OUP) Delhi.

Asian Development Bank, (1998), *The Bank's Policy on Water: Working Paper,* Manila.

Bandopadhyay J, (1987), *Need for Ecological Water Resources Policy,* Economics and Political Weekly, Mumbai (hereafter referred as EPW) No.50, Dec.12, 1987.

Baviskar A., (1995), *In the Belly of the River,* OUP, Delhi.

Beard Daniel, (1994), 'Address to the ICOLD Conference', (mimeo), (in capacity of Commissioner US Bureau of Reclamation) Quoted in Partick McCully, *Silenced Rivers.*

Boulding K., (1964), *The Meaning of the 20th Century* Harper & Row, New York.

Brown Lester etal, (1991), *Saving the Planet,* World Watch Institute, Alert Series, WW Norton and Co., New York.

Carson Rachel, (1962), *The Silent Spring,* Houghton Miffon, Boston, (1987 Reprint).

Chambers Robert, (1988), *Managing Canal Irrigation,* Oxford - IBH, New Delhi.

Chambers Robert, etal, (1991), *To the Hands of Poor-Water and Tress,* Oxford-IBH, New Delhi.

Chopra K. and Gulati S. (2001), *'Migration, CPRs and Environmental Degradation : Inter linkages in India's Arid and Semi - Arix regions',* Sage Publications, New Delhi.

Chopra K., etal, (1990), *Participatory Development and Common Property Resoures,* Sage, New Delhi.

Constanza R. (ed), (1991), *Ecological Economics: The Science and Management of Sustainability,* Colombia Press, New York.

Daly Harman, (1987), *Economics, Ecology and Ethics,* WH Freeman & Co. New York.

Daly H. and Cobb J., (1989), *For the Common Good Redirecting the Economy towards Community, the Environment and a Sustainable Future,* Beacon Press, Boston.

Darryl D.'Monte, (1985), *Temples or Tombs?* CSE, New Delhi.

Desarda HM, (1994), Water-its Economics and Politics' in the *Survey of Environment, The Hindu Annual Publication,* Madras.

Desarda HM, (1997), Towards an Alternative Agricultural Vision, in *Bhupat M. Desai (ed), Agricultural Development Paradigm for the Ninth Plan Under New Economic Environment,* Oxford-IBH Publishing Co. New Delhi.

Desarda HM (1998), National Water Policy–Some Societal Issues, in *National Water-Policy Agricultural Scientist's Pereptions,* National Academy of Agril. Science, Delhi.

Dharmadhikary S., (1995), *The World Bank and Sardar Sarovar Project* in PIRG, Delhi.

Dhawan BD (ed), (1990), *Big Dams, Claims, Counter Claims,* Commonwealth, New Delhi.

Dreze J, etal, (1997), *The Dam and the Nation: Displacement and Resettlement in the Narmada Valley,* OUP, Delhi.

Dobson A. (ed), (1991), *The Green Reader,* Mercury House INC Sanfrancisco.

Dogra B. (1992), *Development, that Displaces the Poor,* Third World Network Features.

Fadaka J. (1978), *Superdams, The Dream That Failed.*

Gadgil Madhav and Guha Ramchandra, (1998), *Ecology and Equity,* Penguin Books.

Gandhi M.K., (1909), *Hind Swaraj,* Navjivan, Ahmedabad, (Reprint 1982)

George S., (1977), *How the Other Half Dies,* Penguin, London.

Georgescu Roegen, (1971), *The Entropy Law and the Economic Process,* Cambridge, Harvard University Press.

Gleick P. (ed) (1999), "The Biennuial Report of Freshj Water Resourecs', *The World Water,* 1998-99, Island Press, Okland California.

Goldsmith Edward, (1998), *Learning To Live With Nature: The Lessons of Traditional Irrigation,* The Ecologist, Vol. 28, No.3, May/June 1998.

Goldsmith E. and Hildyard N., (1984), *The Social and Environmental*

Impacts of Large Dams, Volume 1 and 2, Wadegridge Ecological Centre, Cornwall.

Goodland Robert, (2000), *Social and Environmental Assessment to Promote Sustainability, Environment Management Series Paper No. 74,* World Bank, Washington D.C.

Goodland Robert, etal, (1991), *Environmentally Sustainable Economic Development: Building on Brundtland,* UNESCO, Paris.

Government of India (for Short GOI) *Various Ministers and Planning Commission*-Referred as PC-New Delhi.

GOI, Ministry of Agriculture, 200 *Warsa, Jan Sahbhagita, Guidelines for Watershed Development Project for Rainfed Areas.*

GOI, (1987), *National Water Policy,* Minsitry of Water Resources (hereafter MOWR).

GOI, MOWR, (1994), *Report of the Five Member Group on Sardar Sarovar Project.*

GOI, Ministry of Power, (1997), *Report of the Expert Committee on the Environmental and Rehabilitation Aspects of the Tehri Hydro-Electric Project,* Ministry of power.

GOI, MOWR, (1997), *Report of the Working Group on Participatory Irrigation Management for the Ninth Plan.*

GOI, MOWR, (1999), *Integrated Water Resources Development– A Plan for Action, The Report of the National Commission on Integrated Water Resources Development Plan.*

GOI, Planning Commission, (1989), *Agro-Climatic Zonal Planning,* New Delhi.

GOI, PC, (1998), *Ninth Five Year Plan 1997-2002 Volue I & II,* Planning Commission.

GOI, PC, (2001), *Draft Approach Paper of the Tenth Five Year Plan* (approved by NDC).

GOI, PC, (2006), *Towards Faster and More Inclusive Growth* (Drafth Approach Paper, 11th plan), PC, New Delhi.

Government of Maharashtra, Planning Department, (1990), *Report of the Eighth Plan Study Group on Agricultural, Horticulture and Co-operation.* (Chairman: H.M. Desarda) Mumbai.

Guha Ramchandra, (2000), *Environmentalism: A Global History* OUP, New Delhi.

Gyawali D. and Dixit A., (1999), *Mahakali Impasse and Indo-Nepal Water Conflict,* EPW Vol. 34 No. 9.

Indian Society of Ecological Economics, (2001), *Ecological Economics for Sustainable Development,* Academic Foundation, New Delhi.

Iyer Ramaswamy R., (1998), *Water Resource Planning, Changing Perspectives,* EPW, 12 December, 1998.

Iyer Ramaswamy Rd, (2001), *World Comission on Dams and India* EPW, June, 23, 2001.

Jain LC, (1998), *Poverty, Environment and Development,* EPW, February, 13, 1998.

Jain LC, (2000), *Dams Vs Drinking Water, Exploring the Narmada Judgement,* Parisar, Pune.

Kala Pabola, (2001), *In the Spaces of Erasure, Globalisation, Resistance and Narmada River,* EPW, June, 2, 2001.

Kamarappa JC, (1945), *The Economy of Permanence,* All India Village Industries Association, Sarva Seva Sangh (Reprint 1980).

Leopold A., (1989), *A Sand Country Almanac,* Ballantine Books, New York.

McCullyP, (1996), *Silenced Rivers: The Ecology and Politics of Large Dams,* Zed, London.

Mehta L., (1999), *The Political Ecology of Water in Kutch. Dry Land Blindness of Indian Planners,* EPW, (Based on Authors Doctoral Dissertation at IDS, Sussex).

Morse B. and Berger T., (1992), *Sardar Sarovar: The Report of the Independent Review, Resources Future International, Ottawa.*

Moore L. and Collins J., (1977), *Food First: The Myth of Scarcity, Souvenier Press,* London.

Naess Arne, (1973), *The Shallow and Deep, Long-Range Ecology Movement,* in Inquiry No. 16 Reprinted in The Green Reader, 1991, Mercury House INC Sanfranssisco.

Narmada Bachao Andolan, (1999), *Big Dams in Narmada Valley: The need for Alteranatives,* NBA, Badwani.

National Research Council, (1989), *Alternative Agriculture,* National Academy Press, Washington D.C.

Navalawala BN, (2001), *World Comission Dams: Biased?* EPW March, 24, 2001.

Nehru Jawaharlal, Speech at the Annual Meeting of Central Board of Irrigation and Power, Speech Reported in *Irrigation and Power* Vol. XVI No.1, January, 1959.

Needham J., (1971), *Science and Civilisation in China,* Vol. 1 to 4, Cambridge University Press.

Norgoaard Richard, (1999), *Beyond Growth and Globalisation* (mimeo) IEG, Delhi.

Postel Sandra, (1989), *Water for Agriculture-Facing the Limits,* World Watch Paper 93, Washington D.C.

Postel Sandra, (1997), *Last Oasis : Facing Water Scarcity,* W.W. Norton, New York.

Rao CHH, (2000), *Watershed Development in India, Recent Experience and Emerging Issues,* EPW, Vol.35, No.45.

Rao CHH, (2001), *Some Reflections on the Sustainable use of Water for Irrigation in Indian Agriculture,* (mimeo) Presidential Address to the Indian Society for Ecological Economics, Bhopal.

Rao KL, (1979), *India's Water Wealth,* Orient Longman, New Delhi.

Reddy Amulya, (1987), *New Development Oriented Approach to Energy Planning,* Indian Institute of Science, Bangalore.

Reisner M, (1986), *Cadillac Dessert The American West and Its Disappearing Water,* Penguin Books, New York.

Repetto Robert, (1986), *Skimming the Water* World Resources Institute Washington D.C.

Roy Arundhati, (1999), *The Greater Common Good,* India Book Distributors, Mumbai.

Sant, et al, (1998), *Reassessing The Role of Large Dams in Meeting Power Demands,* Prayas, Pune, (WCD Submission Eco. 0131).

Schumacher EF, (1973), *Small is Beautiful* Blond and Briggs Ltd. London.

Serageldin I., (1995), *Towards Sustainable Management of Water Resources,* World Bank.

Sengupta Nirmal, (1985), *Irrigation: Traditional Vs Modern,* EPW, Spl. No. August, 1985.

4

RISING POPULATION, WATER SCARCITY AND CONFLICTS IN RURAL INDIA

*S.S.P. Sharma**

INTRODUCTION

The importance for water for life itself, for food production, and for societal activities and production in general makes it a fair hypothesis that development would meet the greatest difficulties in region where water is scarce. The fact is that currently most of the poorest countries are located in the zone where part of the year is dry and/ or recurrent drought years are part of the climate; moreover, due to high evaporative demand of a warm and dry atmosphere, the rainfall has particularly low efficiency in these areas. Thus, there is one resource, which we cannot neglect for freshwater availability. Water is not only necessary for life, but in its biological functions it is a non-substitutable resource. It composes two-thirds of the human body and four-fifths of the human brain (Falkenmark, 1997).

The human population has increased rapidly and significantly particularly in the last 300 years from around 500 million to 6.4

* Professor & Head, Center for Water and Land Management, NIRD, Hyderabad.

billion in the year 2002. India crossed 1000 million in 2000. The latest projection of world population by 23rd century will be around 11 billion, which is twice the level existing in 1995. This almost doubles the present world population means, of course, halving the absolute per capita supply of all resources, including water. The demand for water by 2025 has been estimated at 1027 Billion Cubic Meter (BCM). It requires to be increased from 520 BCM in 1997 to more than 1000 BCM by 2025.

WATER AS A RESOURCE AND ITS SCARCITY

There is a strong emotional feeling that water is just too precious and essential for human survival to be treated as merely a commodity. Water is, however, a very practical and necessary resource and there can be little doubt that within modern economies, at whatever level of development, water is a commodity and it is almost always for economic purpose that we use water. Of course, at the same time that it is an economic commodity, water is also a splendid part of the natural environment and some of the world's most attractive landscapes, but then so are many other resources, like land, most societies do treat as economic goods.

As a natural resource, water is scarce or are by definition in economic theory, all resources used to satisfy human wants, as we use all resource, to satisfy our wants and because our wants are, more or less unlimited then we have a problem of choice, of choosing how to employ resources to satisfy which demands among the large number of individual and social wants.

There is a widespread and growing opinion that the world faces an imminent and ever more serious water crisis. This crisis will be so marked as to cause serious economic, political and social repercussions. There are many papers and articles which suggest that the threat of increasing water scarcity is becoming so grave as has been stated in a relatively balanced recent reports from the United Nations, to be 'weakening one of the resources bases on which human society built' (WMO1997). Other recent reports are less cautious and speak of increases in water scarcity leading to 'conflicts among users, regions and countries' (IPCC, 1997).

In the absolute sense, In spite of climate change, water is not becoming scarcer. It is true that the amount of water is finite, but it is not only finite, it is inherently stable. We are not using up the water resources, because we cannot use it up. Water is not exhaustible in the sense that coal, oil, any metal ore or other non-renewable natural resources are. The amount of water available on earth has not changed for a very long time. Why then there is this concern about water security? The present concern with growing water scarcity seems to be due to the interaction of a multiplicity of very different factors. Among these factors the most important seem to be the following:

1. The concept of water as finite and limited resources;
2. Tradition of a bureaucratic and centralized approach to water management;
3. Concern with the continuing growth of the human population;
4. The increasingly widespread economic change and growth in economy is reflected in the expansion of industrial, urban and irrigation demands for water.

There is a tendency among many commentators to accept a Malthusian view of the relationship between water and population For what has been important which has been strengthened by the uncertainties due to climate change caused by global warming.

Water is a renewable though finite resource and, unfortunately, not evenly distributed. Water in adequate quantity and quality is not available whereever it is needed. Some facts: about 97 per cent of all water on earth is saltwater which is unsuitable for drinking or growing crops. Of the remaining 3 per cent most is locked away in the polar ice caps. Only 0.3 per cent of the total freshwater reserves on earth are found in rivers and lakes. Only this fraction of a percent along with groundwater—form the bulk of freshwater. Roughly, on average about 10 per cent of the available quantity is used for drinking, 20 per cent for industrial uses, and 70 percent consumed in agriculture (Gleick, 1993). Therefore what is important to man is not the existence of the water in the oceans, but the hydrological cycle which begins with the evaporation from the oceans of salt-free water, much of which falls on land in the form of rain or snow before flowing back, by one means or another, to the oceans to begin the process over again.

There have been numerous alarming predictions regarding immediate or impending water scarcity. The UN has estimated that: some 80 countries comprising 40 per cent of the world's population, are already suffering from serious water shortages and that, in many case, the scarcity of water resources has become the limiting factor to economic and social development (UN, 1977).

The circulation of freshwater through the hydrological cycle is a major element in the maintenance of the natural environment, as we know it. The hydrological cycle is made up of number of components, in particular snow, runoff from precipitation, soil moisture, ground water and temporary storage in lakes or aquifer. The total amount of freshwater accessible from these sources for human use is, however, only a very small proportion of the total amount of freshwater that is found on earth. This is because the largest proportion of freshwater is trapped in the permanent ice and snow of the polar ice caps (Lee, 1999).

In the use of water, the global sectoral shares of households, industry and agriculture have been about 8 per cent, 23 per cent and 69 per cent respectively. In India, 93 per cent of freshwater used for agriculture and livestock and out of the balance 7 per cent, the domestic consumption accounts for 3 per cent and industrial use and power generation the remaining 4 per cent. The total freshwater used for human activity is drawn partly from ground water source (Table 1)

The ground water, which is rechargeable, supplies the water requirement of one-third of the world's population and is the main source of supply in rural areas in many parts of the world. In India the aggregate replenishible ground water potential has been estimated to be 431km^3 per year. Out of the total ground water potential, 71 km^3 of water is used for industrial and domestic use and most of it leaves the local ecosystem after use and flows through the streams and rivers and moves towards sea.

India has an excellent network of river systems, which have been categorized into 20 major river basins, as listed in Table 3. There are 12 major river basins, inter-state in nature, with a combined catchments area of 256 mha (million hectares), 46 medium basins of catchments varying sizes covering an area of about 25 mha, and other water bodies including tanks and ponds draining a total area of 7 mha. The total water resources (Surface and ground water) have been estimated at about 2300 billion cubic Meters (BCM), of which the surface water resources are estimated by the Central Water Commission at 1870 BCM, and the ground water potential assessed by the Central Ground Water Board to be about 431 BCM. The total utilizable flow is estimated as 1086 BCM (surface water 690.03 BCM and 395.6 BCM. of Ground water), of which 37% surface water and 38 % ground water have been utilized so far. Apart from being inter-State, the Indus, Ganga and Brahmaputra-Barak basins are also international river basins.

Table 1
Fresh Water Withdrawal

Country	Annual freshwater withdrawal 1995 (a) Total (Cubic Kilometers)	As percentage of total water resources	Per capita (cubic meters) Domestic	Others
1. Haiti	0.0	0.4	2	5
2. Burundi	0.1	2.8	7	13
3. Namibia	0.2	0.3	7	103
4. Jamaica	0.3	3.9	11	148
5. Paraguay	0.4	0.1	16	93
6. Nepal	2.7	1.6	6	144
7. Srilanka	6.3	14.6	10	493
8. Bangladesh	22.5	1.0	7	213
9. Egypt	56.4	97.1	67	889
10. Uzbekistan	82.2	76.4	165	3,956
11. Japan	90.8	16.6	125	610
12. Russian Federation	117.0	2.7	1334	656
13. Pakistan	153.4	32.8	21	2,032
14. India	380.0	18.2	18	594
15. China	460.0	16.4	28	433
16. United States	467.3	18.9	244	1,626

Source: Anon 1997, The State in a Changing World, World Development Report 1997, OUP, Washington, *pp. 232-233*

India receives nearly 3800 billion cubic meters (BCM) of fresh water annually through rainfall and snowfall. Most of the rainfall is concentrated in 100 to 120 days during monsoon (June–September). There are also wide variations in the annual rainfall from region to region, 100 mm in parts of western Rajasthan to as high as 11,000mm in Cherrapunji in eastern part of Meghalaya (GOI, 1999: 12). More importantly, nearly 60 percent of India's fresh water resources are concentrated in the Ganga-Brahmaputra-Meghna basin (Table 3), which accounts for 33 percent of the country's total geographical area. The water resources of the basin of westward flowing rivers south of Tapi accounts for 11 percent of total fresh water resources whereas the basin occupies 3 percent of the country's geographical area. The remaining of 64 percent of the country's geographical areas endowed with only 29 percent of the total fresh water resources. There are regions where the average per capita fresh water availability is far below 500 CM per annum (for instance, many areas in north Gujarat, Saurashtra, Kachchh, Western Rajasthan). Thus, the distribution and availability of water is not uniform over space. Similarly, there are wide year-to-year fluctuations in the rainfall in the country. Due to heavy investment made in this area, the water supply in rural as well as in urban area has increased. Water supply coverage in rural areas has increased from 30.9 percent in 1981 to 86.7 percent in 1998. The coverage in urban areas has been increased from 77.8 percent to 90.6 percent during the same period. There is considerable variation among the states in regard to the coverage of population by safe water supply. The achievements in terms of water supply coverage are relatively high (about 70 percent or higher) in Gujarat, Haryana, Karnataka, Maharashtra, Punjab and West Bengal, and relatively poor (below 40 percent) in Kerala and Orissa (Mathur; 1998).

Water Scarcity: Regional Perspective in India

Based on per capita renewable water availability, India—the second most populous country in the world—has enough water to meet its people's needs. Despite an estimated availability of 2,464 cubic meters per person per year, nearly 900 million people suffer severe water shortages due to uneven availability of water from region to region and the season.

India's vulnerability to regional water scarcity is well illustrated by the case of Rajasthan, a state in northwest India. Situated in one of the most inhospitable arid zones in the world, Rajasthan's northwest corner extends into the vast Thar Desert. With a wide range of temperatures and an unpredictable monsoon climate, drought and desertification are common, and water is a scarce commodity. More than five per cent of India's population, Rajasthan can claim only one percent of the country's water resources, in the form of groundwater, limited rainfall, and a restricted share of waters through canals.

Population growth in Rajasthan, as in the rest of India, has been rapid. By the year 2005, Rajasthan's population is projected to increase to almost 59 million. In 1990 per capita water use in the state was 562 cubic meters, a level nearly commensurate with absolute scarcity. With the projected increases in human numbers for the coming decade, acute shortages are imminent.

Even those who live in areas of high rainfall in India often face drought because landscapes have been denuded. Soil is compacted and most rainfall runs off before it can sink into the ground, increasing flooding. The region of Cherrapunji in Meghalaya, for example, receives among the highest levels of mean rainfall recorded in the world, yet because of intense seasonal rainfall and the fact that the area's forests have been degraded, the area suffers from water shortage during dry months.

PROBLEM OF EXPANDING POPULATION AND WATER

There is widespread fear that with the increasing world population at some 90 million annually, heavy pressure will be exerted on water quantity and quality by expanding agricultural production alongside growing urban, industrial and environmental claims. Agricultural water users will undoubtedly press to retain claims to their historical share of relatively inexpensive water (mainly through subsidies). To aggravate matters, in many countries, agriculture has also been a major source of non-point pollution, responsible for degrading surface and ground water with a broad range of substances. Global warming, as well as other human land use activities, such as deforestation and desertification may intensify water scarcity. In certain regions, there will be reduction in the water supply by about 10 per cent, resultantly doubling the amount of CO_2 concentrations, by the middle of the next century (Abrahamson, 1989). In fact, the relative scarcity of natural resources has been in continuous decline over the last two centuries. There is, however, the apparent exception of fresh water. For fresh water, the Malthusian argument appears to hold, no increase in the natural supply is possible. The balance of the hydrological cycle acts to fix quantity of fresh water available in nature. It follows, therefore, that, water alone among all the natural resources, meets the necessary conditions to support the Malthusian growth mode (Barnett, 1960):

1. In the long-term fresh water resources will become scarce once population rises beyond a certain critical level;
2. The critical population/resources level will be reached with exponential population growth, whatever, the fertility

rate, as long as the rate of population growth exceeds the maintenance level;

3. The amount of fresh water is fixed as was Malthus fixed land;
4. The advance of technology and institutional solutions is too slow to affect the decline in marginal returns arising from population growth;
5. Total social output will show diminishing marginal returns in respect to the water resource.

The human population has increased rapidly and significantly particularly in the last 30 years from around 500 million to an estimated almost 6 billion in the year 2000. Extrapolating these historical rates of increase into the future would suggest that a population-related catastrophe, is indeed in the human future, but this may not be the case. The most recent population projections prepared by the United Nations provide an estimate that the world population, will stabilize around the end of the 23rd century, at slightly less than 11billion.This is around twice the level existing in 1995, but smaller population than estimated in earlier projections (United Nations, 1998). This almost doubling the present world population means, of course, halving the absolute per capita supply of all resources, including water.

In AD 1650 human population numbered around 0.5 billion and was growing at the rate 0.3 percent per annum corresponding to a doubling time of 250 years. By AD 1900 the world population had reached 1.6 billion with a growth rate of 0.5 per cent per year. By 1970 world population became 3.6 billion and with a growth of 2.1 per cent per year with substantive acceleration of population growth due to fall in death rate. Between 1971 and 1991 world population further rose to the level of 5.4 billion (Table 2) accompanied by the fall of rate of growth of population from the level of 2.1 per cent in 1971 to 1.7 per cent in 1991. While global birthrate was falling since the middle of the 20th century, it was falling slower than the death rate (Sengupta, 2001). With a population that is 16 per cent of the world, India has 2.45 per cent of the world's land resources and 4 per cent of its water resources (Iyer, 2001).

Future growth of population, due to the regional differences in fertility rates, will be accompanied by changes in relative distribution unless there are large migratory movements. It is projected that the population of Europe will decline so that by the year 2150 it will be 18 per cent lower than it was in 1995. The largest increase in population will be in Africa, which will quadruple its population to 2.8 billion. In the Americas and most of Asia population increases will be similar to the world average, except in China where the population is only projected to increase by one-third.

Here, it appears to be significant evidence from the already apparent changes in fertility rates that the human population will not continue to increase as it has done so far so long. This conclusion may be different to accept after decades of concern over the expansion of the human population, but accept it we must if we are to make sensible decisions about the future. This element in the Malthusian model may not hold in the future. Changes in demographic growth may make a major contribution to the reconsideration of the impact of population on the future demand for water, as for other natural resources.

Table 2
Population Growth and Projections

OECD	*Average annual growth of Population (%)*			*Population (millions)*		
Countries	*1965-80*	*1980-90*	*1989-2000*	*1990*	*2000*	*2025*
1. Australia	1.8	1.5	1.4	17	20	23
2. Canada	1.3	1.0	0.8	27	29	32
3. Japan	1.2	0.6	0.3	124	128	128
4. USA	1.0	0.9	0.8	250	220	307
5. UK	0.2	0.2	0.2	57	59	61
SAARC Countries						
6. Bangladesh	2.6	2.3	1.8	107	128	176
7. Bhutan	1.6	2.1	2.4	1	2	3
8. India	2.3	2.1	1.7	850	1006	1348
9. Nepal	2.4	2.6	2.5	19	24	37
10. Pakistan	3.1	3.1	2.7	112	147	240
11. Srilanka	1.8	1.4	1.1	17	19	24

Source: Calculated from World Development Report, 1992

WATER DEMAND AND ITS MANAGEMENT

According to long-range United Nations population projections, India's population could, under the low-growth projection, stabilize and eventually shrink in size after 2025, which would actually increase per capita water availability. Under the medium and high projections, however, the country would move into conditions of water stress and water scarcity, respectively.

Figure 1
Annual Renewable Fresh Water Available Per Person Under Three Long-Range United Nations Population Projections for India

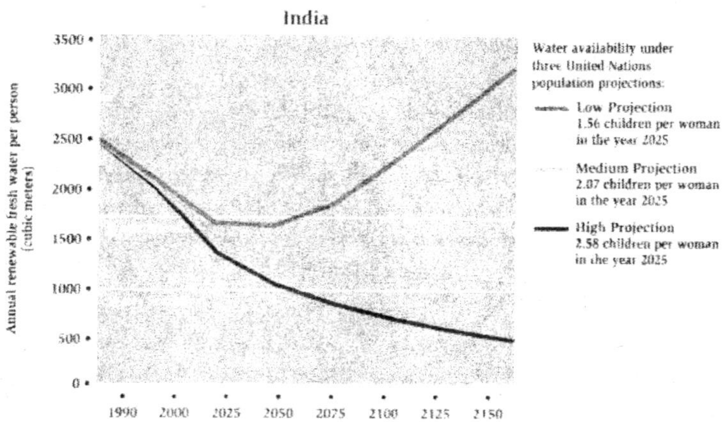

The definition of water stress is the relationship in a country between water availability and water withdrawals for human use. Countries are classified into four groups, low water stress and high water stress, medium-high water stress and high water stress. A country suffers from high water stress when more than 40 per cent of the available water is withdrawn for use. The report describes such countries as suffering from serious water scarcity with increasing reliance on desalination and using groundwater at a rate faster than it can be replenished. Water scarcity in such cases, it is claimed, can become a limiting factor for economic growth.

In 1995,according to the estimate obtained, a number of countries already could be classified to be suffering from high water stress. These countries were mainly in North Africa and Western Asia, although Belgium and South Korea also fell into this group.

The assumption that the demand for water will increase with population growth has logic, although it may not hold, but it is also commonly assumed that the demand for water will increase with economic growth. This assumption is not quite so impelling.

It has been accepted for sometime that the proposition that price of services of natural resources must rise relative to the services of reproducible capital over time as a consequence of economic growth is demonstrably false (Schultz, 1961). It is false because demand for natural resources does not necessarily increase with economic growth.

The question then - is there any evidence that would provide any indication of the relationship between the use of water resources and economic growth? Giving an answer to this question is difficult given the discussed precarious nature of the evidence available on the demand for water. Fortunately, however, there is one relatively reliable recent source available. A source, which suggests that for water, as for other resources, the demand does not necessarily increase with economic growth. Consequently, therefore, growth may not result in greater scarcity of the resource, even in the absence of market allocation and real prices.

Household Demand

Meeting the growing household water needs is fast becoming a crucial issue due to population growth in urban and metropolitan conglomerates. There has been a corresponding increase in the number of works on issues associated with residential water demand. Gottlieb (1963) estimated residential water demand in Kansas; Hanke and More (1982) for Malmo, Sweden; Reziza (1991) for Saudi Arabia; Hansen(1996) for Copenhagen and Deudy, Nguyen and Davis (1997) for Adelaide, to cite a few examples. The primary aim in estimating municipal water demand was and still is to determine the price elasticity of demand, and the impact of other variables, such as income. Water authorities are typically interested in the responsiveness of use to water prices. The answer might indicate, for instance, whether supply or demand management is an appropriate policy.

Water Management

Management of water resource is... one of the greatest challenges facing the present generation ...the dimension and implications of today's freshwater problem is the source of a potential major crisis of global society and the environment...(German Advisory Council on Global Change, 1999).

According to ancient and global tradition of water management as a public activity, it would not be surprising to find many alternatives. During most of 20th century, it was the centralized system of water management, identified as scientific management. It became the preferred form of water management and dominated approaches towards water management in most countries. The bottom-up users managed organizations did not disappear, but for most of the 20th century they have been very much marginal to the main trends in water management. Gradually the state occupied most water management space until its abrupt retreat in the middle of the 1980s. Under the influence of socialist ideas, public sector investment and the role of government in the economy as a whole grew continually until very recently in nearly all developed and developing countries. Currently, not only governments, but also international institutions have shown increasing concern to improve the efficiency of water management. For example, the World Bank has seriously paid attention to this issue. In its recent annual report on world economic development, the World Bank has discussed the issue of the efficiency of investment and the effectiveness of management in the public sector at some length (Frederiksen, 1992). Similarly, the World Resources Institute in its recent report on natural resource management has emphasized the importance of improving management capabilities to make existing investments more profitable (World Resource Institute, 1993). The result of this reconsideration of the role of the government and the public sector in the economy has been a wide spread revival in the faith of markets and the private sector. This trend towards the market has been felt within water management and water resource administration, as in other areas of the economy.

In India, utilizable water availability per capita per year is estimated at 1250m^3 (CWC, 1999). Severe water shortages occur in many regions seasonally, particularly as a result of uneven distribution of water resources over time and space. Thus the main problem is inadequate availability of water where and when it is needed. Provision of clean drinking water to all is a major policy goal, which is yet to be achieved. Upto March 1993, 78 per cent of rural and 85 per cent of urban Indians had access to safe potable water. That still leaves some 200 million persons without safe drinking water. On the other hand, sanitation services are available to only 48 per cent of the urban and about 3 per cent of the rural population (Central Statistical Organization, 1997).

In recent times, water quality has deteriorated due to rapid industrialization, population growth, and intensive agriculture as they generate considerable amount of polluted water. Critical parameters affecting water quality are high biochemical oxygen due to domestic and human waste is the cause of many water – borne diseases. Water quality degradation is increasingly becoming a source of conflict among upstream and downstream users.

An estimated 80 per cent people in the country use groundwater for their domestic needs (UNICEF1998). India's total ground water resources are estimated at 431.8km^3(CSO, 1997). Average level of groundwater development in India is 32 %. Some states (e.g., Punjab 94 %, Haryana 84 %, Tamil Nadu 60 %, Rajasthan 51 %) have exploited groundwater resources to a greater extent than some other states (Gujarat 41 %, UP 38 %, Andhra Pradesh 24 &, Bihar 19 &) (CSO, 1997). Irrigation is using 90 per cent of the ground water (table-4) whereas domestic needs account for just 6 per cent of the total volume pumped (Parikh. et al,1999).

Piped water is supposed to be the most reliable, safest and easiest source of drinking water available in the country. Access to piped water has declined during the 1990s,in general and in urban areas in particular. In urban areas it has declined from 72.1 per cent to 70.1 per cent. The second best sources, tube wells and hand pump are found to be compensating for this decline. Dependence on tube wells and hand pumps increased from 17.2 per cent in 1988 to 21.3 percent in 1998 in urban areas. As per the 1998 estimates, still 4.9 per cent of the urban households suffer from inadequacy of drinking water (Table 4 and 5). Of these 17.2 per cent of the urban households do not take any measures to compensate this scarcity, either by borrowing or buying. That is, at the aggregate level about 3 per cent of the population manage with inadequate drinking water (Reddy, 2001).

Table 3
Water Resources Potential in the River Basins of India

Name of the river basin	*Average Annual Surface Water Potential (BCM)*	*Estimated Utilizable Flow excluding Ground Water (BCM)*	*Total replenishable Ground Water Resources (BCM)*	*Pop. in 1991*	*Per Capita available Surface Water M³*	*Per Capita available Surface and Ground Water M³*
1. Indus (up to border)*	73.31	46.00	26.49	41.90	1750	2382
(a) Ganga*	525.02	250.00	170.99	356.80	1471	1951
(b) Brahmaputra* Barak & Others*	585.60	24.00	53.91	35.24	16617	18147
2. Godavari*	110.54	76.30	40.65	53.98	2048	2801
3. Krishna*	78.12	58.00	26.41	60.78	1285	1720
4. Cauvery*	21.36	19.00	12.30	29.33	728	1148
5. Pennar*	6.32	6.86	4.93	9.70	652	1160
6. East flowing Rivers between Mahanadi & Pennar	22.52	13.11	18.22	23.60	954	831
7. East flowing Rivers between Pennar and Kanyakumari	16.46	16.73		45.20	364	
8. Mahanadi*	66.88	49.99	16.46	26.60	2514	3133
9. Brahmini & Baitarani*	28.48	18.30	4.05	9.77	2915	3329
10. Subernarekha*	12.37	6.81	1.82	9.46	1308	1500
11. Sabarmati*	3.81	1.93		10.58	360	
12. Mahi*	11.02	3.10	18.42	10.48	1052	1120
13. West Flowing Rivers of Kutch, Saurashtra, including Luni	15.10	14.98		22.10	683	
14. Narmada*	45.64	34.50	10.83	14.70	3105	3842
15. Tapi*	14.88	14.50	8.27	14.80	1005	1564
16. West Flowing Rivers from Tapi to Tadri	87.41	11.94		25.80	3388	
17. West flowing Rivers from Tadri to Kanyakumari	113.53	24.27	17.69	32.60	3483	3744
18. Area of Inland drainage in Rajasthan desert	NEG			7.10		
19. Minor River Basins draining into Bangladesh & Mayanmar	31.00			2.10	14762	147.62
Total	1869.35	690.31	431.44	842.62	2218	2731

Table 4
Access to Piped Water In India
(Percentage of Households)

NSS Round/Year	*Urban*	*Rural*
44th round (1988)	72.1	15.5
49th round (1993)	70.4	18.9
54th round (1998)	70.1	18.7

Source: National Sample Survey (54th round), Drinking Water, Sanitation and Hygiene in India, July 1999

Table 5
Proportion of Households Reporting Inadequacy of Drinking Water by Source

	State	*Urban*			
		Tap	*Tube Well*	*Well*	*Others*
1.	Andhra Pradesh	47.2	49.5	25.5	43.5
2.	Assam	76.6	84.1	67.1	77.7
3.	Bihar	64.6	70.4	26.5	58.2
4.	Gujarat	81	48.8	50.5	78.7
5.	Haryana	84.4	76.7	0	82.8
6.	Karnataka	66.2	16.2	70.6	60.2
7.	Kerala	81.4	50.8	65.8	71
8.	Madhya Pradesh	63.6	49.8	31	59.1
9.	Maharashtra	81.7	45.7	27.9	78.2
10.	Orissa	68.5	17.4	52	45.7
11.	Punjab	94.5	93.1	100	94
12.	Rajasthan	92	26.5	19.3	81.4
13.	Tamilnadu	48.2	63.8	30.1	48.9
14.	Uttar Pradesh	88.3	71.7	23	77.1
15.	West Bengal	57.4	31.4	73	47.8

Source: National Sample Survey (54th round), Drinking Water, Sanitation and Hygiene in India, July 1999.

Quality Aspects of Water

Hircheifer, DeHaven and Milliman (1960) listed important water quality variables which affect demand: taste, odour, colour, softness, corrosiveness, temperature bacteria content, organic matter, dissolved salts, and acidity. In general terms, the problem of selecting the optimal quality mix can be described by the utility maximization of a representative household. The household utility is a function of the vector commodities consumed, *y*, and a vector of water quantities of different qualities, *w*

$$U = U(y_i, \ldots, y_n, w_i, \ldots, w_m) \quad (1)$$

The household budget constraint is given by

$$\sum_{i=1}^{n} y P_{yi} + \sum_{j=1}^{m} w_1 P_{wi} = M \quad (2)$$

When there is block pricing of water, the price is no longer an exogenous variable but is determined by the total quantity of water used by the household. This implies that equation (2) becomes a piece-wise linear function (Howitt and Hanneman, 1995). Maximization of (1) subject to (2) gives the demand functions for every commodity, and of water with given quality.

Economic studies dealing specially with urban water quality issues are quite scarce. This stems partly from the fact that water quality is a public good, and there are no data for the quality components of urban water. One way to overcome the lack of relevant data would be to employ hypothetical market techniques, such as contingent valuation, to elicit valuations of the quality component.

Public Policy for Irrigation Management

In the area of irrigation and water management, the state should feel more concerned itself with overall public policy to create conducive socio-economic environment for the efficient and productive use of water. The state should divest itself from the tasks of managing the systems, because its failures have been conspicuous in this area which accounts for the major opportunities available for improving water productivity (CGIAR 2001).

Governments in the states and at the center should accord highest priority to the renovation/modernization of the existing systems which account for nearly 40 percent of the irrigated area from the major and medium irrigation projects. The rate of return from such investments would be very high when compared to the projects creating new irrigation potential. Modernisation of the delivery systems and the distribution channels for the existing projects would have a high pay- off, as they would facilitate a clear definition of property rights or entitlements of farmers and their effective enforcement. They would also facilitate the adoption of measures by the water users to improve water-use-efficiency and productivity.

Price and procurement policy of the government has an important bearing on the efficiency and productivity of water use through its impact on cropping pattern. A glaring illustration of the price policy for rice and wheat in the recent period which has resulted in the accumulation of over 60 million tones of food grain–nearly thrice the actual requirement–with the Food Corporation of India (FCI). It has been estimated that for each hectare of paddy raised under irrigated conditions, as much as 16,000 tones of irrigation water is applied to the semi-arid north-western Indian states (Dhawan 2001), from where bulk of the rice is procured by the FCI. One can thus imagine how many billion tones of scarce water has been wastefully used over the years in the semi-arid plains of India, which could have been saved or diverted to higher productivity uses if the price and procurement policies were better aligned with the emerging domestic and export demand for food grain.

The experimental work backed by the ICAR(Indian Council of Agricultural Research) found that 25 – 40 per cent saving in irrigation water could be achieved through intermittent submergence and transplanting paddy seedlings at about the time of onset of monsoon rains. Similarly, water savings of the order of 50 per cent in the case of drip irrigation and 25 per cent in the case of sprinkler irrigation can be realized (Dhawan 2001). But, only about 1 per cent of irrigated area in the country is presently covered by drip and sprinkler methods of irrigation. These technologies are not adopted by the farmers because water is available at a very low cost or is even free. The incentives are in fact perverse inasmuch as the capital cost of drip and sprinkler irrigation is prohibitively high for the Indian farmer due to various taxes and high interest rate (Dhawan 2001).

The debate on water rates in the country so far has centered round the revenue accruing to the government for covering the Operation and Maintenance charges. The objective of conserving water needs to be brought into the forefront in these discussions. Given that water rates have depreciated very much in real terms, it is not certain that a significant rise in these rates within the feasible range would induce the farmers to save water under the existing system for management. On the other hand, a system of rationing water supplies in water – scarce regions even with moderate charges can induce farmers to increase water productivity, provided the supplies are reliable in terms of quantity and timings as in the case of the Warabandi system in north–western India (Berkoff 1990; IWMI 2001). In any case, raising water rates significantly would not be politically feasible unless it is linked with the performance of irrigation services and made explicit that revenues so raised would be reinvested locally to upgrade the services.

Water Users' Associations and panchayats should be fully empowered and induced to charge and collect water and electricity rates from the farmers on the basis of the volume of actual consumption where metering is possible, or, on the basis of the quantities of water received or electricity consumed as estimated by the farmers' associations where metering is uneconomical and otherwise infeasible. The revenues so collected should be allowed to be retained entirely by the Water Users' Associations under the aegis of panchayats for maintenance and development of irrigation services, with possible supplementation from the state governments.

DYNAMICS OF RURAL CONFLICTS IN INDIA

Participants at the World Water Forum in the western Japanese city of Kyoto said that water is a far more important commodity than the oil. "Water is an inalienable human right. Water is life," said Mikhail Gorbachev, former president of the Soviet Union. By 2050, water shortages brought on by booming populations, pollution and global warming, will affect between two billion and seven billion people around the world out of a projected population of about 9.3 billion.

A new study by three economists, two based in Argentina and one at the Haas School in Berkeley, California, finds that water privatization in Argentina brought increase in productivity and profitability. Overall, childhood mortality fell by 8 percent in areas where water services were privatized, and by 24 percent in the poorest areas.

Inter-State Conflicts

Within India, it should be noted that the river waters of the Ganga basin as a whole have not been allocated to the basin States of Uttarakhand, Himachal Pradesh, Haryana, Rajasthan, Delhi, Uttar Pradesh, Bihar, Jharkhand, & West Bengal, except for specific sub

- basin or project-specific agreements among or between the States. The major sub - basins of the Ganga on the right bank above the delta bifurcation are the Yamuna, including the Chambal River, the Tons, the Karamnasa, and the Son. The Right Bank of the Ganga below its confluence with the Son is drained by a number of small rivers. The important left bank tributaries of the Ganga above the delta bifurcation are the Ramganga, the Gomati, the Ghaghra, the Sarda (Mahakali in Nepal on which a treaty on water sharing has been signed) a major right bank tributary of the Ghaghra, the Gandak, the Buri Gandak and the Kosi (formed by three major tributaries of Sun Kosi, Arun and Tamur). The Mahananda, a left bank tributary, Which rises and flows in India for 65 % of its catchment, joins the Ganges/Padma tributary in Bangladesh. Bhagirathi, the right bank tributary of the Ganga in its last stretches in India, receives drainage from the Dwarka, Ajay, Damodar, Rupnarain, and Haldi rivers, before it joins the sea through the Hooghly estuary. The Ganga and its left bank tributaries - Yamuna, Ghaghra, Gandak and Kosi - have their origins in the glaciers in the Great Himalayas, and therefore have a significant snow melt contribution which sustains the river flow. Interstate disputes over river waters are becoming intractable, and the constitutional conflict-resolution mechanisms do not seem to be working well. Conflicts could also arise between uses, between users, between areas, between classes. There is possibility of conflicts between the people and the state, as has happened in the case of Narmada and Tehri.

Causes of Conflicts

The major causes of conflicts in river Water sharing can be grouped in two categories, technical and non-technical. A List for each of these is provided below.

Technical

- Disagreement on water availability, especially during the lean season;
- Disagreement on the basic hydrological data and the actual present utilization of water;
- Disagreement on the present and future water requirements for various uses and the basis thereof;
- Lack of openness and transparency in the exchange of data and information;

- Differences in the approach for planning, design, construction, and operation of joint projects on Trans-boundary rivers;
- Different interpretations of operative clauses and sub-clauses in existing agreements/treaties/tribunal awards.

Non-Technical

- Disagreement on riparian rights and basis thereof;
- Disagreement on the basis and modalities of water sharing;
- Violation of agreements by one party or the other;
- Rigid political or administrative stands.

As the disputes progress, the two categories get so intertwined that they cannot be separated and resolved on technical merits alone.

Principles Used for Settlement

Several options at dispute settlement have been employed, some with excellent results and some with continuing rancor and legal battles in the courts. The following broad picture emerges as to the guidelines adopted in settling the sharing of water.

- The Principles of equitable distribution of water availability assessed at agreed locations on the main river and or its tributaries (on 75%) dependability or average availability have been followed in the Inter-State water disputes. The Helsinki Rules on Equitable distribution and now the UN Law on Non - Navigational Uses of International Water Courses are widely referred to. Even in the case of International Treaties and Agreements and Treaties these principles are followed.
- There is no set guidelines prescribed under the National Water Policy, though efforts are currently being made to evolve a set of guidelines acceptable to all the States.
- The ground water resources in the River Basins in different States or Countries have not been taken into account in the assessment of basins' water resources for distribution or sharing among the riparian States or Countries. However, each State within an Inter-State basin has the right to use the ground water resources for beneficial purposes. The Law on ground water is still evolving, and Authorities are in the process of being established in the States on the basis of guidelines prescribed by the Central Government, which has established the Central Ground Water Authority.

- Changes of State boundaries due to the reorganization of States have brought even past agreements in dispute. This has resulted in establishing Tribunals to settle the sharing arrangements.
- Specific trans-basin diversions have been permitted in Inter-State Agreements/Tribunal Awards, but not as a general rule.
- Each basin State is free to use the Inter-State basin water allocated to it in a general sharing arrangement or on specific projects and regulate within its boundaries, to enjoy the benefits of that water within that State in a manner not inconsistent with the specific agreements or orders of the Tribunal or the Agreement.
- Invariably, a formula for sharing of excesses or shortages in any water year over the allocated shares, are specified. Percentages of return flows are also accounted in working out water assessments and its sharing.
- Non-riparian States have been allocated water for utilization for drinking water, irrigation and other beneficial uses on the basis of Agreements of the riparian States considering the established water shortages and hardship in such States or towns or cities.
- Review Period of Tribunal Awards or Agreements has been invariably specified.

The Settlement Mechanism

An appreciation of how, in the past, the settlement mechanism has worked is given below.

- Over 130 Agreements have been evolved on the sharing of Inter-State River waters or on specific projects. All these agreements have used the negotiation route, with the Central Government playing the pivotal role under the Constitutional Laws, Acts, and Statutory Rules. Most of these Agreements have worked well since they were done with the willing consent of the Party States to the Inter-State Basin.
- Inter-State statutory organizations for specific projects or basins have been set up under the Central Government with the participation of the basin States and beneficiary States with defined roles. These have been set up either by mutual agreements or under the direction of the Tribunal Awards. Interstate organizations have also been set up by mutual

agreement of only two States without Central Government participation. The Dispute settlement mechanism prescribed under Article 256 of the Constitution has been used under specific reference(s) made by the States for settlement of disputes under the provisions of the Inter-State Water Disputes (ISWD) Act. The Tribunals have taken a long period to settle the disputes referred to them, and some have been unable to give the Award for over ten years for various reasons.

Even after Awards were given, the issues of people affected by the project, and non - consultation with the Stakeholders in the planning process, have stalled implementation. This has resulted in the Supreme Court intervening in the Public Interest Litigations (PILs) filed by the NGOs and affected Parties. The States reopened even settled issues such as the height of the dam. The Tribunal has not suggested corrective steps in such cases and the Implementation. Mechanisms set up under the direction of the Tribunal have lacked adequate powers.

- The set date for the reopening for review of the water sharing awards given by the Tribunal has opened up a race among the States to implement projects in a haphazard way to establish prescriptive water rights on the particular date. This has led to overreaching project-agreed provisions of storage, resulting in serious disputes between States. This has led to litigation being filed by the aggrieved Parties in the Supreme Court for adjudication.
- Water being a 'State Subject' the role of the Central Government is limited. Modifications in the Statuary provisions have been debated, but no tangible solutions have been possible.
- It is increasingly being recognized that maintaining a certain minimum flow in the rivers during the lean season months for ecological considerations is necessary, and provisions have been made for the same in the new agreement (Upper Yamuna) and treaty (Mahakali Treaty) signed in recent years.
- Since most of the river basins of India are Inter-State in character, the Central Organizations viz., the Planning Commission and the Ministry of Water Resources with its technical attached organization, the Central Water Commission, have exercised a very well set schedule of techno-economic clearance guidelines in approving the Inter-State

projects planned by the States for implementation under the Five Year Plans. This procedure has been institutionalized, even though it is time consuming. This route of clearance ensures that projects on the Inter-State rivers are not taken up without an agreement on water sharing in general, or project specific sharing in particular, of the waters of the river basin. There is a loophole in this, since the clearance is required only if the State wants Central Plan funding for the project. Otherwise, the State can go ahead with the project if funds are not a constraint. In that case, the aggrieved States can seek judicial intervention to stop the project.

CONCLUSION

Efforts have been made to explore the relation of water scarcity with the increasing population. The distribution of water and the distribution of human population are uneven and do not always coincide. The issue of scarcity must be given due consideration in water management, but should not be exaggerated into crisis. Water is un evenly distributed among uses in many parts of the world as well as in India also and this will not be easy or quickly remedied. It is, however, exactly our failure to the market to price water which is producing these problems in the distribution of water among users. The problem must be tackled, correctly pricing the resource, not by massive intervention by governments that, in many cases, are incapable of managing the most tasks required of them. This does not mean that government interventions in water management should never be contemplated. It underlines the need to target intervention on the areas where market failures are most pronounced, to pay attention to the costs and benefits, and to design mechanisms to maximise the benefits in relation to the costs.

Encouraging private partner in the management of water for irrigation in the country can mitigate the misuse and the wastage of water in agriculture sector. Because, the public policy has proved failure in checking the wastage and misuse of water in the states. Therefore, there is a dire need of rationalisation of the water distribution system and for the various uses also. Water is the state subject in the country which is another aspect and does not tackle the problem of rising conflicts. Conflicts among the users and in between different states should be dealt with the regulatory mechanism. The need to pass some legislation does not solve the issues unless the transparent policies and the people 's participation is not encouraged in view of the demand and supply of water as an economic good and not as free good.

REFERENCES

Abrahamson, D.E. (ed) (1989), The Challenge of Global Warming, Island Press.

Barnett, H.J. (1960), Malthusianism and Conservation: their role as Origins of the Doctrine of Increasing Economic Scarcity of Natural Resources, in Demographic and Economic Change in Developing Countries, Special Conference Seriesno.11,Princeton University Press, PP423-456;

Berkoff, D. J. W. (1990), Irrigation Management on the Indo-Gangetic Plain, World Bank Technical Paper No. 129, The World Bank.

Bohm et al. (1993), Sustainability of Potable Water Supplies in the Philippines, Water Resources Research, 29, 1955-63,

Central Water Commission (1999), Report, Government of India, New Delhi;

Consultative Committee for International Agricultural Research (CGIAR), Technical Advisory Committee (2001), Water and CGIAR, Discussion Paper.

Cropper, M. Land Freeman, A.M. (1991), Environmental Health Effects, in J.B. Braden and C.D. Colstad (eds) Measuring Demand for Environmental Quality, North Holland; Elsevier Science Pub;

Daudy, G, T. Nguyan and C. Davis (1997), Estimating Residential Water Demand in the Presence of Free Allowances, Land Economics, 73, (1) 125-39.

Deevey, Edward (1960), The Human Population, Scientific American, 203(9).

Dhawan,B. D. (2001), Technological Change in Indian Agriculture: Diffusion of Water Economising Technologies and Practices (mimeo) Institute of Economic Growth, Delhi.

Flankenmark, M. (1997), A Water Perspective on Population, Environment and Development, in Parthdas Gupta and Karl-Goren Maler (eds) The Environment and Emerging Development Issues, Clarendon Press, Oxford.

German Advisory Council on Global Change (WBGU) (1999), World in Transition: Ways Towards Sustainable Management of Fresh Water Resources, Annual Report 1997, Scientific American, 203(9).

Gleick, P (1993) An introduction to global fresh water issues' in P. Gleick (ed) Water in Crisis— A Guide to the World Freshwater Resources, Oxford University Press, Oxford.

G ottlieb, M. (1963), Urban Domestic Demand for Water: a Kansas case study, Land Economics, 39(May), 204-10.

Haneman, W.M. (1994), Valuing the Environment Through Contingent Valuation, Journal of Economic Perspective, 8, 19-43;

Hanke, S. and L. de More (1982), Residential Water Demand: a Pooled Time Series, Cross Section Study of Malmo, Sweden, Water Resources Bulletin, 18 (August) 621-71.

Hansen, I.G. (1996), Water and Energy Price Impacts on Residential Water Demand in Copenhagen, Land Economics, 72, (1) 66-79.

Hirsh leifer, J.J., et al. (1960), Water Supply, Economics, Technology and policy Chicago, IL: University of Chicago Press.

International Panel on Climate Change (IPCC) (1997), The Regional Impacts of Climate Change: Summary for Policy Makers (SPM), Electronic Version.

International Water Management Institute (IWMI) (2001), Basin Level Use and Productivity of Water: Examples from South Asia, Research Report 49.

Iyer, R.R. (2001), Water: Charting a Course for the Future, Economic and Political Weekly, Vol.36, No.13, pp-1116.

Johnson, P. (1995), Evaluating Healtj Risks: An Economic Approach, Cambridge University Press, Cambridge; Lean, G. and D. Hinrichsen (1994) Atlas of the Environment, New York, Harper Perennial.

Lindberg, K. at al. (1997), Contingent Valuation of Rural Tourism Development with Tests of Scope and Mode Stability, Journal of Agriculture and Resource Economics; 22, 44-60.

Reziza, OSA (1991), Residential Water Usage: a Case Study of the Major Cities in the Western Region of Saudi Arabia, Water Resource Research, 27, (May) 667-71.

Reddy, V. Ratna (2001), Declining Social Consumption in India, EPW, July 21.

Schultz, T.W. (1961), Connections Between Natural Resources and Economic Growth, in J. J. Spengler (ed), Natural Resources and Economic Growth, Conference Paper, Washington, Resources for the Future.

Sengupta, R.P. (2001), Ecology and Economics, An Approach to Sustainable Development, Oxford University Press, Delhi.

United nations, Population Division (1998), World Population Prospects: the 1996 Revision, New York.

United Nations (1997), Report of the UN Water Conference on Water and the Environment: Development Issues for the 21st Century, Dublin, Ireland;.

Vincent J. R.et al. (1991), Valuing Environment Benefits in Developing Countries, Special Report, East Lansing. USA, Michigan State University.

Whitehead, J.C., and Houtven, G.V. (1997), Methods of Valuing the Benefits of Safe Drinking water Act: Review and Assessment, Greenville, USA; East Carolina University.

Whittington, D. et al. (1990), Estimating the Willingness to Pay for Water Services in Developing Countries: a Case Study of the use of Contingent Valuation Surveys in the Southern Haiti; Economic Development and Cultural Change, 38, 293.

Whittingten, D. et al. (1993), Household Sanitation in Kunati, Ghana; World Development, 21, 235-43; World Meteorological Organization (1997) Comprehensive Assessment of the Freshwater Resources of the World, Geneva.

5

NATIONAL WATER POLICY OPTIONS ON INTERLINKING OF RIVERS–DIMENSIONS AND ISSUES

*V. Madhava Rao, R.R. Hermon & P. Kesava Rao**

INTRODUCTION

It was Colonel Arthur Cotton, who successfully executed many river projects for irrigation purpose and control of floods in the Kaveri, Godavari-Krishna deltas in Andhra Pradesh and in Mahanadi in Orissa in 1858, and even drawn up a plan to connect the Indian subcontinent through a grid of navigation and irrigation canals, a peninsular system, which would link Karachi in the northwest to Madras, which was the first edifice of river linking proposal in the country. Subsequently, however it was evolved by Dr K.L. Rao's proposal of a Ganga-Cauvery link and Captain D.J. Dasturs Garland Canal. The proposal envisaged 30 links - 16 peninsular and 14 Himalayan which were identified by the National Water Development Agency in 1982.

The 'Ganga-Cauvery Link' proposal was not appreciated on the grounds of the very large financial and energy costs involved,

* Head, Geomatic Cell and Assistant, Professors, NIRD, Hyderabad

while the 'Garland Canal' was dismissed as impractical, because it was thought to be technically unsound. The current thinking on interlinking of rivers erupted with the setting up of National Water Development Agency establishment in 1982 in pursuance of the 'National Water Perspectives' brought out by the Ministry of Irrigation in 1980, with two main components namely the Himalayan Rivers component and the Peninsular Rivers component.

The Himalayan component envisages a number of links, including some within the Ganga system (Kosi-Ghagra, Gandak-Ganga, Ghagra-Yamuna, Sarda-Yamuna, etc), some between neighbouring rivers in the Brahmaputra system (Manas-Sankosh-Teesta), a couple between those two systems (Teesta-Ganga, and an alternative Brahmaputra-Ganga link), one long link from Sarda to Sabarmati through the Yamuna and Rajasthan, one from the Ganga to Subernarekha *via* Damodar and then on to Mahanadi, and a few others. The general idea is to transfer waters from 'surplus' eastern rivers to 'deficit' central, western and southern regions.

The Peninsular Rivers component again involves a number of links, of which the most important would be those connecting Mahanadi, Godavari, Krishna, Pennar and Cauvery. The idea is to transfer the surpluses estimated to exist in the Mahanadi and the Godavari to the deficit southern basins (Cauvery, Vaigai). Other links in the Peninsular component would include Ken-Betwa, Parbati-Kalisindh-Chambal, Par-Tapi-Narmada, Damanganga-Pinjal, etc. Another idea is the partial diversion of certain rivers flowing into the Arabian Sea eastwards to link with rivers flowing into the Bay of Bengal (Bedti - Varda, Netravati – Hemavati, Pamba – Achankovil – Vaippar).

FEASIBILITY ISSUES

The river linking proposal threw a number of issues relating to techno-economic feasibility and environmental aspects. The disputes among inter-state basins still pose a big hurdle. It is envisaged that river linking projects in Kerala would benefit more Tamil Nadu than Kerala. It was estimated that over 4.5 lakh people would be adversely affected and over 2,000 hectares of forest land on the periphery of the Periyar wildlife sanctuary and the Thoni-Achankovil range would have to be utilised if the project comes through. The ecological injustice involved in blocking the flow of the river endangering all life forms in the flowing river and of the people dependent on the river and related ecosystems are equally important issues.

One of the primary questions concerning the policy planners and economists is related to the huge investments in the massive project and the probable income flows. The lending and the time frame of benefit flows are also debatable questions to be addressed. The opinions of the feasibility of linking the rivers in India is sharply divided. Those against linking the techno-economic feasibility and also apprehend that such a project would cause enormous environmental damage apart from the problems of displacement of the population. The legal objections include implications from the point of view of federal structure of the country as well as commitments to Bangladesh on Farakka. Doubts are expressed on the proposed completion of interlinking by 2012 as against the year 2035 suggested by the Center for peninsular rivers, and 2043 suggested for the Himalayan rivers. There are also no clear cut indications from the Planning Commission on the project report, which was expected to be completed by 2006.

As water is a state subject, interlinking of rivers needs a national debate and Constitutional amendments urging the Center to go ahead with existing river linking proposals, otherwise a legal crisis may erupt regarding the implementation of the river linking programme. Tamil Nadu is the only State that has filed affidavits in support of interlinking the rivers. India has world's 40 per cent population and has 264 largest river basins, but the funds required will be about Rs 5,60,000 crores which will either have to be raised through international agencies or some other mode of private funding.

The high population growth and increasing water demands for various basic and developmental purposes have forced engineers and planners to contemplate and propose more comprehensive, complex and ambitious plans for water resources systems. The development, conservation and efficient use of water forms one of the main elements in the development planning. The water resources are limited considering the future demands.

Rainfall is mostly confined to the monsoon season and is unevenly distributed both over space and time, even during the monsoon season, resulting in frequent droughts in some parts of the country while floods in some other parts of the country. The runoff water, which falls into the sea, could be conserved in various storage reservoirs and used to maintain sustainable supply of water for various uses.

Critics question the rationale of the river linking project stating that the plan is based on the simple and deeply flawed belief that rivers have surplus water and that floods and droughts can be banished by technical solutions alone. This belief is stated to be grounded in the troubled legacy of hydraulic management in the sub-continent dictated by a supply-side approach, which ignores the complexities inherent in river ecosystems. Further, it is stated that no socio-economic criteria appear to have been laid down for evaluating the proposals and the classified data is hardly available for analysis. The water logging and other environmental impacts may cause more adverse than the normal water resources development.

LESSONS

International Water Resources Law under the Helsinki Rules, 1966, and the UN Convention on the Non-navigable Uses of International Water-Courses, 1997, spells out guiding principles in international water conflicts.

China embarked on the South-North Water Transfer Project (SNWTP) for water transfer between river basins, to combat floods and droughts. Generally, dams are being constructed for dealing with floods and for assured irrigation. For India, the Chinese Model of tackling floods and droughts seems to be a good lesson to emulate for solving it's problems. The SNWTP basically aim at alleviating floods in the south and drought in the north of China by transferring water from surplus to scarce river basins. There are three canals that link the Yangtze river in the south to the 3-H river basin in the north, drained by the Yellow, Hai and the Huai rivers.

The investment for the SNWTP has been estimated at $50 billion, and it is expected to be completed by the year 2050. The idea was coined by Chairman Mao in 1952, the project was initiated in December 2002. It is estimated to transfer up to 44 billion cubic meters of water per year through the Eastern, Central and Western canals. The 3-H river basin has acute water shortage though has a high density of population as 2 out of 5 Chinese live in this region and account for about 40 per cent of China's cultivated area, contributing 31 per cent of its gross industrial output with only 10 per cent of China's water resources.

Compared to China, the river linking project of India looks more ambitious and investment prone. The cost of the river linking project will be to the tune of $120 billion, and the completion of major projects in the stipulated time frame is doubtful, keeping experiences of handling such projects in India due to social, political and economic problems. Besides, the growing rivalry among states over water sharing and other issues. There are also international water sharing disputes among neighbouring countries like Nepal, Bangladesh and Pakistan, which challenge the implementation of such a massive river linking project.

Floods in eastern India and droughts in western and southern India almost every year, made the country to look for river linking projects, to bring in water sharing, catering to drinking water and irrigation, water resources conservation and overall economic development of rural India. However, the huge investment, ecological problems, displacement of people and rehabilitation and other problems pose many challenges. However the benefits estimated from US $ 120 billion project to link 37 Himalayan and Peninsular rivers, which will handle 178 km^3 of inter-basin water transfer per year, through 12,500 km of canals would generate 35 gigawatts of hydro-power and irrigate an additional 35 million hectares of land.

Though India envisages great benefits from the river linking project in terms of irrigation, flood control, navigation, fishery development, drinking water and the like, the livelihood of 100 million people downstream in Bangladesh, who depend on the Ganges and Brahmaputra, may be threatened. Further, there were issues relating to international water disputes with Pakistan, Nepal and Bhutan also.

The rich biodiversity resources in Sundarbans may be destroyed due to shortage of water from the Ganges and Brahmaputra after the river linking project is implemented. Professor Peter Cullen, a leading water resource expert warns that 'any scheme to redirect waterways inland to irrigate marginal farming areas would have adverse affects on salinity levels and coastal industries such as prawn fisheries.

The global population projected to touch 7.9 billion by 2020, which is about 50 per cent larger than that in 1990 (Dyson, 1996), which may exert six-fold increase in the number of living in conditions of water stress - from 470 million today to 3 billion in 2025 (Postel, 1999). The World Bank Report (1992) brings out the issue of ecological disaster as a consequence of excessive extraction of water for irrigation from the Amu Darya and Syr Darya rivers, where river runoff into the sea fell from an average 55 cubic kilometers a year during 1950s to zero in the early 1980s.

In Latin American Chile, water rights and judicious use of water has been the prime concern. Even in affluent USA, there are efforts to decommission dams that are no longer serving any purpose and greater efforts are being made for river restoration (Gleick, 2000).

China has almost half as much arable land per capita as compared to India and is consciously trying to stabilise the use of water in agriculture and increase the efficiency of water use for growth of crops. This lesson may have significance for India from the point of improvement in the efficiency of the use of irrigation water. The World Bank Irrigation Sector Report on India (World 1999:11) too suggests the efficient use of irrigation practices for food security.

The worldwide shared water resources are covered by over 2,000 bilateral agreements on various aspects of navigation, research, fishing, water quotas and flood control (Mc Neely, : 1999) and the principles of international law have to be considered while allocating water within a river basin to avoid any *international* water disputes regarding the extent of upstream and downstream use of water. While India and Nepal want to exploit *the Ganga-Brahmaputra* river huge water resources, Bangladesh wants *minimised* effect on floods during monsoon months; and water shortages during dry months.

CONCLUSION

Though very laudable idea, the project of interlinking rivers is not an easy task as it has several issues relating to social, political, economic and environmental concerns. Lessons from abroad seek to relook at the project itself. As such, instead river linking, there are new approaches which are being thought to conserve the river basins and the hydrological cycle for sustained environmental protection. There are many other alternative approaches such as conservation and improvement in the management of water sources, their exploitation and distribution, efficiency in water usage particularly in irrigation and protection of water from pollution.

REFERENCES

Bandopadhyay, Jayanta and Perveen, Shama The Interlinking of Indian Rivers Questions on the Scientific, Economic and Environmental Dimensions of the Proposal.

Linking Rivers in China: Lessons for India.

Ministry of Water Resources, Government of India.

National Water Grid, Dr. K.L. Rao (2003), India to Build Water Grid to Divert River Waters: Inviting Disaster for Bangladesh, The New Nation.

Water Diversion Plan Threatens California's Salton Sea, National Geographic Today, September 24, 2002.

World Bank (1999), India: *Water Resource Management- The Irrigation Sector, New Delhi*: Allied Publishers.

Himalayan Component

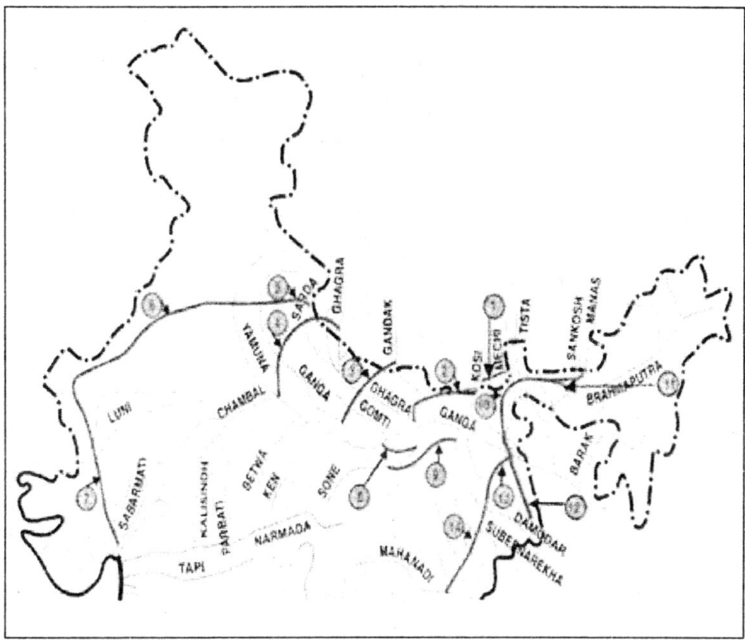

1. Kosi - Mechi link
2. Kosi - Ghaghara link
3. Gandak - Ganga link
4. Ghaghara - Yamuna link
5. Sarda - Yamuna link
6. Yamuna - Rajasthan link
7. Rajasthan - Sabarmati link
8. Chunar - Sone Barrage link
9. Sone dam - Southern tributaries of Ganga link
10. Manas - Sankosh - Tista - Ganga link
11. Jogighopa - Tista - Farakka link
12. Farakka - Sunderbans link
13. Ganga - Damodar - Subernarekha link
14. Subernarekha - Mahanadi link

Peninsular Component

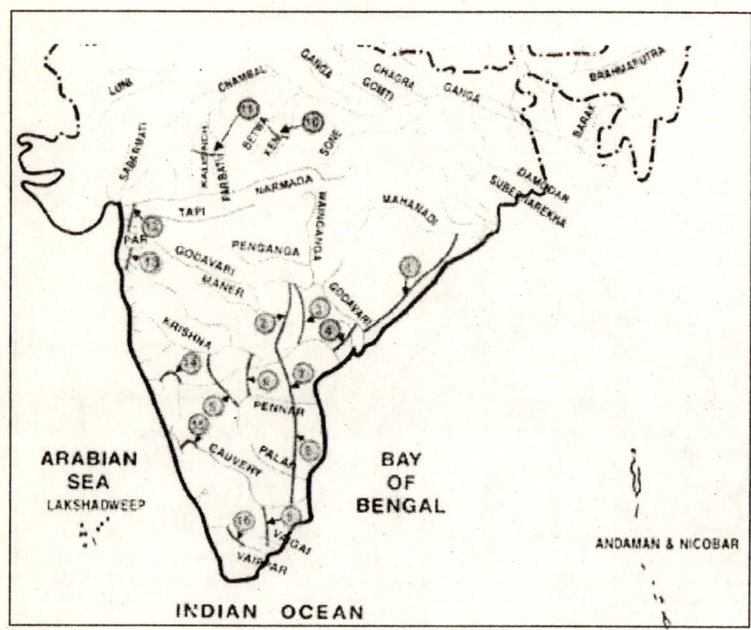

1. Mahanadi - Godavari Link
2. Inchampalli - Nagarjunasagar Link
3. Inchampalli - Pulichintala Link
4. Polavaram - Vijayawada Link
5. Almatti - Pennar Link
6. Srisailam - Pennar Link
7. Nagarjunasagar - Somasila Link
8. Somasila - Grand Anicut Link
9. Kattalai – Vaigai - Gundar Link
10. Ken - Betwa Link
11. Parbati - Kalisindh - Chambal Link
12. Par - Tapi - Narmada Link
13. Damanganga - Pinjal Link
14. Bedti - Varada Link
15. Netravati - Hemavati Link
16. Pamba – Anchankovil - Vaippar Link

6

INEQUITY IN WATER DISTRIBUTION & INTERVENTIONS TOWARDS SECURITY

*T.K. Giri & N. Upadhyay**

INTRODUCTION

Water is the very basis of life and is the foundation for human survival and development. It is the most precious gift of the earth that has ever been bestowed upon us. Human body consists of 81% water; 18% carbon, calcium and nitrogen compounds; and 1% odour molecules. Out of total water available on earth 97% is seawater, 2% is locked in ice caps and glaciers, and 1% lies too far underground to retrieve. Out of the eye-popping 336 million cubic miles of water that is on the earth only 1/3% is fresh water, which we can drink, and 3/4 of this is locked up in the Antarctic ice cap. Without this minute percentage of water a human being can only live for 10 days at the longest, for our bodies is 65% water. Sustainable and equitable use of water over millennia has been ensured by cultural adaptation to water availability through water conservation technologies, agricultural systems and cropping patterns adapted to different climatic zones, and conservation-

* North Eastern Regional Centre, Jawahar Nagar; Khanapara Guwahati

based life styles. With the passage of time water is becoming more and more scarce commodity against the ever-increasing demand for it. By the year 2025 it is estimated that over 1/3 of the worlds freshwater will be gone. In the last few decades the consequences of population growth, indiscriminate industrialization and rapid urbanization, and the associated consumerism have interfered with the natural hydrological cycle of rainfall, soil moisture, groundwater, surface water and storage of all sizes. This has led to overuse, abuse and pollution of our vital water resources and has disturbed the quality and the natural cleansing capacity of water. Already 26 countries, home to 232 million people are considered to be "water-scarce", meaning they don't have 1,000 cubic meters of water for one person for one day. Not only are countries water-scarce but also a lot of their drinking water is contaminated. With this background an attempt has been made in the present paper to look into the broader perspectives of water resources management in the country through peoples participation *(UNO 2004).*

WATER AND MANAGEMENT ISSUES IN INDIA

Water and Poverty

As ever, it is the poor and vulnerable communities who are most adversely affected human activities play a significant role in increasing risk and vulnerability towards natural disaster and there are also strictly man-made disasters, such as oil-spills and toxic run-off that render great damage to our precious water resources *(UNO 2004).* 1.3 billion people lack access to adequate supply of safe water and 25 million people die each year by diseases caused by using unsafe water and lack of water for sanitation and hygiene. Half of them are young children.

Water is central to our hopes of eradicating poverty and achieving sustainable development. Thus there is a need for sustainable and effective Water Resource Management (WRM) strategies. Sustainable WRM means rational and effective management at economically viable and ecologically sustainable development of available resources in the contemporary setting without causing deleterious effects on the basic life support system and endangering the resource base which is already under stress. It also means the rationality in the water distribution to ensure the availability of the resource in right quantity and desired quality to meet the needs of different consumptive sectors at the right time rather than yielding to the pressure of the interested groups or water-lords.

Water and Health

Water is one of the earth's most precious and threatened resources. Industrialised countries have decreased their level of water related diseases through good water management. In developing countries water related diseases are rampant and those blights the lives of the poor. Poor environmental sanitation and unsafe drinking water has been one of the major health problems in India. Diarrhea, dysentery, typhoid fever, intestinal helminthiasis, infective hepatitis are endemic in the country. Much of this suffering is needless. Health provides an effective gateway for development and poverty alleviation. Improving water management is a powerful tool that can be used by individuals, communities and households to protect their own health.

WATER AND AGRICULTURE

Over the last half-century, irrigation water management protected the world from devastating food shortages and the threat of mass starvation by significant productivity gains in agriculture through Green Revolution. But the next 30 years will throw up new challenges. As world population grows - to an estimated 8,300 million in 2030 - agriculture must respond to changing patterns of demand for food, combat food insecurity and poverty in rural areas, and compete for scarce water with other users. The water needs of humans and animals are relatively small - the average human drinks about four liters a day. But producing the same person's daily food can take up to 5,000 liters of water. That is why the production of food and fiber crops claims the biggest share of freshwater withdrawn from natural sources for human use, or some 70% of global withdrawals (*FAO 2004).*

People and Potable Water

With an average annual rainfall of 1,170 mm, India is one of the wettest countries in the world. At one extreme are areas like Cherrapunji, in the northeast, which is drenched each year with 11,000 mm of rainfall, and at the other extreme are places like Jaisalmer, in the west, which receives barely 200 mm of rain.

Though the average rainfall is adequate, nearly three-quarters of the rain pours down in less than 120 days, from June to September. The country gets about 420 million hectare-meters (mhm) of precipitation annually, of which 20 mhm is contributed by rivers flowing in from neighbouring countries. Net evapo-transpiration losses are nearly 200 mhm. About 135 mhm is available on the surface and the remaining recharges groundwater. The availability of water is highly uneven in both space and time. Precipitation is mainly confined to about three or four months (5 to 60 rainy days) in the year and varies from 100 mm in the western parts of Rajasthan to over 10000 mm at Cherrapunji in Meghalaya. Availability of water in our country is presented in Table 1.

Table 1
Water Availability in India

1.	Average annual water resource potential	1869 BCM
2.	Utilizable Surface water resources	690 BCM
3.	Replenishible Ground water resource	432 BCM
4.	Total utilizable water resource	1122 BCM

There is little consensus on the issue of exploitable precipitation. Estimates range from 85 to 105 mhm. Even if the lower value of 85 mhm is taken into account, domestic consumption is no more than 10 mhm. The remaining goes to irrigation. But, with demand outpacing the exploitable potential, the maximum usable water supply of 105 mhm will be inadequate to meet the growing demand by 2025.

Supply of Safe Drinking Water

The basic criteria for safe drinking water supply for human consumption are:

- (i) It should be free from disease causing organisms which would constitute a health hazard.
- (ii) It should be clean, colourless and palatable.
- (iii) It should not contain chemical substances in such concentrations as to cause a health hazard when consumed over a lifetime or economic loss.
- (iv) The location, construction, operation and maintenance of water supply, its sources, reservoirs, treatment and

distribution must exclude all potential sources of pollution and contamination.

Once a water supply has been created, it is necessary to ensure its sustainability physically, technologically and financially. In India, nearly 80% of rural population depend upon ground water for meeting their drinking and other domestic needs. The per capita demand of domestic water supply is increasing over the years. This coupled with the steep increase in demand for irrigation. It is difficult to estimate the proportion of population that has access to clean drinking water. At the time of the First Five-Year Plan, 6% of the rural population and some 48% of the urban population had access to safe drinking water. There has been a dramatic increase in coverage, and by 1994-95, as much as 82% of the rural population was covered. According to the Center for Science and Environment, about 81% of the country's total population has access to safe water.

Inequality in Water Supply

Although about 88% of the population, on average, has been covered with organized drinking water supply in class I cities and class II towns of the country, there is a huge disparity in quantity of water supplied. Of the 299 class I cities, only 77 cities have cent percent water supply coverage. The per capita water supply also ranges from as low as 9 liters per capita per day (lpcd) in Tuticorin to as high as 584 lpcd in *Triuvannamalai (CPCB 2000a)*. Besides an inequitable distribution of water in a given city, the supplies are erratic with water quality degrading continuously over time. Rural water schemes have remained the scourge of planners since Independence *(State of the Environment 2001)*.

Despite massive resource allocation during the last nine Five-Year Plans, there were as many as 61,747 problem villages in the country towards the end of 1997. Interestingly, the country started out with a figure of 150,000 problem villages in 1972; this rose dramatically to 231,000 in 1980. According to the latest statistics, about 15,000 habitations in the country were reported to be without any source of potable water; some 200,000 habitats were partially covered by drinking water schemes; and 217,000 villages reported problems with the quality of water.

Water Stress

In several parts of the country there is steady decline in the ground water level due to over exploitation (annual extraction exceeding the annual recharge) ground water especially for irrigation. Water stress is becoming acute in both urban and rural situations. Not only the quantity but also the quality of water supplied or available is being questioned. At one extreme, water is being wasted in urban areas and by industries; on the other, the rural poor lack access to safe water. According to experts, the usable water resources in several river basins will eventually be exhausted, most surface water will be polluted, and environmental deprivation will be universal. Water scarcity has paved the way for the emergence of the bottled water industry worth over Rs. 1,000 Crore *(Sharma 2004).*

The Tenth Five Year Plan identified several issues that need to be tackled in the water sector including irrigation, flood control and command area development as also the requirement for rural and urban drinking water supply and rural and urban sanitation. In the year 2002-03 and 2003- 04, 39,240 and 34,256 habitations have been respectively covered under portable drinking water supply in rural areas. Of the balance 75,607 habitations, 5,759 are yet to be covered and the rest are partially covered. In urban areas, out of a total of 53.69 million households, 36.86 million households had access to water supply from taps (Parikh, et al 2005).

Despite of several policy measures undertaken under the tenth plan period and adoption of revised water conservation policy – 2002 still today over 100 millions of people in the country have to spent more than billions of hours to fetch the portable water for domestic usage. About half of the villagers do not have clean and safe drinking water and sanitation facilities.

While India has 16 per cent of the world's population, its share in the world's fresh water availability is only 4 per cent. A per capita availability of 1700 m3 is required in order to be free of water stress, while availability below 1000 m3 is termed as water scarcity. Per capita availability was about 5200 m3 in 1951 but it had fallen to 2200 m3 in 1991 and further to 1820 m3 in 2001, reflecting the effect of rising population. It is expected to fall further to 1340 m3 in 2025 and 1140 m3 in 2050. Average availability is therefore likely to fall below the water-stress level in the near future. Nine river basins with 20 Crore population are already under water scarce condition *(Kathpalia et al 2002).*

Water and Disaster

In India today, floods and droughts continue to play havoc with livelihoods. Pollution levels are rising, ground water levels are falling, water-borne diseases are rampant and precious water sources are drying up. This is in spite of numerous schemes being launched every year with crores of rupees from Government and other large donors to solve these issues. Conflicts on water rights are escalating between states, between agriculture and industrial needs, between rural and urban needs. But India is also a country where modern technologies and scientific research are helping devise sustainable, low-cost alternatives. Where traditional practises are being revived successfully in drought-stricken communities. Where innovations in social and institutional arrangements are leading to new, equitable models for water usage.

Every year water-related disasters including floods, droughts, hurricanes, typhoons and tropical cyclones inflict a terrible toll on human life and property, causing innumerable waterborne diseases, affecting millions of people and provoking crippling economic losses. By 2025, up to 2/3rd of world population may experience moderate or severe water stress, mainly in developing countries.

Water Pollution

In India, faecal contamination of water is a major issue, especially where human and animal wastes are not adequately collected and treated. Although this applies to both rural and urban areas, the situation is probably more critical in fast-growing cities. Organics and Heavy Metals resulting from untreated sewerage of industrial activities entered into the river and water bodies through various point and non point sources causes water pollution and eutrophication. Increased salination of water due to evapo-transpiration losses and mining discharges are another problem of water quality control. In India, the percentage of sewered population is almost negligible in rural areas and is quite meager (0 to 50%) in most medium and small towns. As a result, the contamination of groundwater by pollution from unsewered areas is one of the most important environmental problems facing the country. A high proportion of the rural population in agricultural areas in India obtain their domestic water supplies from shallow, private bore holes, which suffer the impact of nitrate pollution due to fertilizer chemicals (e.g., nitrate) and pesticides from regular intensive cultivation of crops.

NEED FOR WATER CONSERVATION MEASURES

More than 97.5 % of the water available is saline and unfit for human consumption or industrial or agricultural use unless sea/ brackish water is purified. Natural resources of water are fast depleting. Deforestation and urbanization is adversely affecting the natural rainfall. Too much uncertainty and unpredictability of the rains. Most of the ground water is contaminated with poisonous substances such as arsenic, lead etc. in many parts of the world especially India. Increasing water productivity by water conservation measures is the need of the hour. The emphasis should shift from supply management to demand managements. Only by managing demand, rather than ceaselessly striving to meet, it there is hope for truly secure and sustainable water in future *(Mazumdar 2006)*.

Any method that saves water is a conservation method. Since ancient times, farmers and herders living in areas of limited rainfall practiced a variety of water conservation measures like:

- Rain water harvesting
- Harvesting of surface runoff
- Ground water recharge
- Desalination of salt water; etc.

DECENTRALIZATION THE WAY FORWARD

Without water security, it is impossible to ensure food and energy security, which are essential for development and poverty alleviation. After decades, we can now talk of food surplus, and hence, food security (and to a certain extent, energy security), but all will be lost if we face a water crisis. Simply put, a good harvest means a good year for all Indians. For this, water security is essential and a water crisis needs to be avoided at all costs.

One of the possible solutions to make rural water supply programme sustainable is to decentralise their operation & maintenance by making the beneficiaries and panchayats, stakeholder in the system. For this to be successful the panchayats should either be able to collect funds from the beneficiaries or they should themselves be able to collect funds from the beneficiaries or they should themselves be financially so sound that they can provide funds for O & M. A. sustainable solution is possible only if the local beneficiary population is motivated to become a stakeholder by sharing a part of the capital cost and also shoulder the full responsibility of the capital cost and also shoulder the full responsibility of operation and maintenance as well as replacement of the schemes thus installed.

Participatory irrigation management and community water harvesting structure may be given top priority for effective production of usable water. This exercise must be governed by community-based decision making systems and institutions, and enabling legal and financial measures, which promote community action. This can be achieved is by deepening systems of participatory democracy and expanding people's participation at the village-level as much as possible *(Goswami 2004).* A well-administered water rights system with active participation by farmers has the ability to gradually eliminate the institutional vacuum currently afflicting our water economy.

There is no single strategy to solve the ominous water problem in India. Site-specific strategies have to be found out and implemented in different regions depending on the water problems faced by the particular region due to bad governance or inefficient use. In some areas, two or more strategies will need to be clubbed together to ensure water security as the types of water problems vary all over India. Several innovation strategies in combination are required to be applied to achieve the desired results of sustainable development and management of available natural resources to offset the weaknesses of each other. A set of several optional strategies like: Supply Management, Demand Management, Management of impacts on environmental system including that of urbanization and Tran boundary Water Management etc may be considered

CONCLUSION

Though the resources are not distributed evenly in time and space, in many parts of the world, large gaps in coverage has nothing to do with water scarcity. In fact, for universal access to 50 liters of water per person by 2015 we would require only one per cent of global water withdrawals. The world has enough water, but so far the political will and financial commitments to provide the poor with access to it is missing. For effective natural resources management and sustainable fresh water development management procedure should be to intertwine with the water resources problems with social problems including demographic, economic, cultural, physical, ecological and environmental aspects of the society at different levels looking at both qualitative and quantitative aspects simultaneously with a view to evolve an environment-friendly and ecologically sound planning effort in the management. In other words, the approach is to be directed to understand adequately the nature and influence of various interacting forces on different components of the hydrologic cycle and select the key elements to which the system behaviour is sensitive and which are amenable to management.

REFERENCES

"UN and Water" Message of the UN Secretary-General Kofi Annan World Day for Water 22 March 2004), 'Water and Agriculture' FAO and Water Management 2004.

Management of Fresh water Resources: State of the Environment (2001), A Country Report by MOEF, GOI 2002.

Drinking Water Supply and Sanitation (Rural & Urban India); Planning Commission, Govt. of India, 2002.

Parikh Jyoti, Parikh Kirit and Vijay Laxmi, India Development Report 2004-05.

Sharma. S. (2004), India News and Features, Infochange Together.

Kathpalia, G. N. and Kapoor, R. (2002), Water Policy and Action Plan for India (2020), an alternative, Developmental Research and Communication Group, New Delhi.

Mazumdar, A. (2006), "Managing Rural Water Supply' Training Programme on Water Resources Management, AIIPHE, Kolkata.

Goswami A.B. (2005), 'Water Resources Management Dynamics" Ph.D. Thesis 2003; Jadavpur University, Kolkata.

IRRIGATION, WATER MANAGEMENT & DROUGHT RISKS IN ORISSA SOME POLICY ISSUES

*B.K. Sahu**

INTRODUCTION

Water is crucial for human life and key to production on the land. In the absence of proper water management and control supply of water, the cumulative impact of water scarcity and production risks becomes very high especially in arid and semi-arid regions leading to increasing vulnerability among poor and weaker sections. Water mismanagement and its consequences have become ever complex over period due to increase in competition of water sues. On the other hand recurrence of drought and flood has been observed both in low rainfall as well as high rainfall regions and in plains as well as in hilly regions. However, water scarcity in many parts of the country has led to increase in poverty and vulnerability.

In the dry areas, without access to assured irrigation, rainfall is a dominant production risk. The spatial spread of rainfall varies too. In fact, drought like situation becomes common across the

* Asst. Professor, Indian Institute of Foreign Trade (Deemed University), New Delhi-110016, India, *E-mail: basanta_sahu@yahoo.com*

country mainly due to lack of controlled irrigation and inadequate water management. In the water resource rich regions such as Orissa the situation of water scarcity become more serious in recent years resulting in severe drought. Impact of drought has been many fold in recent years despite the fact that there more policy intervention in water sector.

Meteorologically drought is understood as the degree of dryness and the duration of the dry period or quantity of rainfall and its spread. According to IMD (Indian Meteorological Department) drought is a situation when the rainfall is less than 25 percent of the normal rainfall. In recent years the frequency and spread of drought has increased and consequently its impact in terms of production shocks, increasing vulnerability, socio-economic deprivation and sever agrarian crisis. However, the intensity of drought risks can be reduced by appropriate interventions - mainly through water management, irrigation development and adequate drought preparedness. An integrated and comprehensive water policy may help in improving water management and reduction of adverse impact of water such as droughts and floods on human well-being. The impact of these extreme consequences of water situations varies with the time scale depending on the nature and structure of irrigation. Longer spread and more frequent occurrence of these calamities in many parts of the country in recent years need more careful attention of policy makers.

Orissa is predominantly an agrarian economy and grossly depend on the quantum and spread of rainfall. Almost each year more than 10 million people and 11,914 villages in various districts reel under drought. Recurrent of drought in the state is mainly due to by erratic rainfall, poor water management and inadequate irrigation development. Many a time crops got damaged in just a few weeks by a lack of precipitation and high temperatures at critical crop-development stages. Though such short periods are not officially considered as droughts but it affects the farm production and productivity. Lack of preparedness in case of long spell deficient precipitation affects important components of the water supply, such as streams, shallow groundwater tables, and small lakes and reservoirs that will affect other components of the water supply such as deep groundwater tables, and large lakes and reservoirs. Under this situation important measures for expansion of irrigation network and water management become center stage in policy decision.

CONSEQUENCES OF WATER SCARCITY: INCREASING VULNERABILITY AND POVERTY

In the absence of proper water management actions drought has widespread impact on the ecology, economy and human development in a monsoon economy like Orissa. In short run the major impact of droughts is loss of crop production and the consequent human misery - food shortages, loss of employment, migration, deprivation of basic needs, shortages of drinking water, fodder and fuel wood. The coping strategies of the people under stress of water scarcity consist of distress sale of assets, reduced consumption, increased borrowings, migration etc. all tend to add to the human misery. In long run, agriculture suffers in terms of loss of soil moisture, reduced fertility of soil and reduced agricultural productivity and unstable agricultural growth, fluctuating incomes and employment which discourage farmers to invest in agriculture (Rao, Ray and Subbarao 1988). Then poverty tends to get concentrated in the water scarcity regions.

Inappropriate water management leading to drought can increase production risk, poverty and vulnerability (Gaiha and Imai; 2003). The coping strategy followed by the household under this situation may lead to concentration of human poverty due to (a) long hours spent on collection of fodder, fuel and water, (b) drop out from schools, (c) distressed migration, which is a major coping strategy, adversely affects attendance of children in school and (d) forced migration. This adverse impact, resulting in low human capital, would further reduce the opportunities of the poor to upward mobility and increase their vulnerability.

Water scarcity has severe impact on environment and ecology-depletion and degradation of natural resources, degradation of common property resources and increasing fallow. Social impact of water scarcity affects women the worst and the traditional social security system tends to decline. It results in lesser dependence on kith and kin or lesser sharing risks, family dislocations and decline in social relations (Chen 1991).

Therefore, careful policy interventions for water management and supply of controlled water requires regions attention now than before if drought risks are to be minimized. In this paper we have focused on important development in water management and irrigation in the state of Orissa during last few years and have critically analyzed major policy changes in water sector to reduce production risks. It is note us or they that slow pace of irrigation, poor water management in state might have affected growth in agriculture which further affected overall growth rate.

WATER SCARCITY, IRRIGATION AND AGRICULTURE RISKS: CONSEQUENCES & COPING STRATEGY

About, three fourth of the rural poor are dependent on agriculture and about two third of total cultivable land are rain-fed. In recent years uncertainties in agricultural income have increased because of highly uncertain production environment largely because of water related problems like lack of control water supply or water logging. Though at household level different strategies are adopted to mitigate this production risk for a stable income and consumption, it is often hard on the part of the poor to cope with consequences of man made water scarcity. A distinction is usually made between risk reducing strategies that the farmer adopts ex-ante to lower risk and risk roping strategies that the farmer adopts ex-post.

There are several ways to compensate for shortfalls in farm income - sell stored produce, liquidate assets, borrow for consumption, receive transfers from relatives, change jobs and/or increase their labour market participation and migrate in search of work. In their choices, farm households will try to protect their assets so as to minimize the adverse impact on their future livelihoods. Households might adopt a diversified portfolio of agricultural activities, adopt technologies, such as inter-cropping or drought- resistant crops, and contractual arrangements such as sharecropping that reduce the variance of income, or diversify their activities through migration or local non-agricultural employment. In the absence of controlled water supply any of these ex ante actions might be costly, so that the households would be sacrificing income, on average, in order to assure a less risky stream of income.

Under this situation managing risks of water scarcity can be possible at community level. Community interactions, sharing social customs, economic exchange based on reciprocity and informal borrowing and lending on flexible terms are instances of community level mechanisms that can help farm households to cope with adversity. However, in the absence of formal insurance markets joint community effort may fail to cope with risks when all are in similar situation. At the community level, exchanges and informal credit are the principal traditional risk coping strategies but all these are covariate nature of agricultural risks. ICRISAT studies show that village level risk sharing is able to mitigate a large portion of idiosyncratic risk. Nonetheless, some idiosyncratic risk remains and poorer households are considerably more vulnerable than richer households. Furthermore, what is disturbing is that risk pooling at the level of village seems weaker than by caste and kinship groups.

WATER MANAGEMENT, PRODUCTION RISKS & TECHNOLOGICAL CHANGE

The production risks can be transformed by developments in the agricultural economy-impact of technological change on production risks and the government response to production risks in the form of water management, expansion of irrigation, drought proofing and crop insurance. Usually HYV crops are high water intensive and consume more input. This is one of the reasons why change in production practices and displacement of traditional varieties by HYV varieties has been slow even in a high water resource state like Orissa. However, the impact of improved varieties on production risks has been controversial. Improved varieties do well in assured irrigated environments. So water resources, water management, pest management and control water supply remains vital for growth areas under improved varieties. Although there is little conclusive evidence to suggest that these have increased the risks of production the adoption of new technologies does carry some long-term risks in terms of poor irrigation, soil depletion and genetic uniformity.

CONSEQUENCES OF WATER SCARCITY

Our field experience shows that water scarcity has severely affected agricultural production, employment and poverty, environment, education and health as well as on the long term growth of the state economy. Some short run impact includes mainly on agriculture in terms of increase in fallow areas, decline in productivity and production of crops, high uncertainty about the Rabi crop, consumption of seeds, mortgaging and distress selling of assets, and increased debts. These impacts are translated into reduced employment and food insecurity; which result in reduced incomes and reduced food consumption of people. The coping strategies adopted by people include (1) reduction in food consumption and in the number of meals, (2) falling back on fruits, vegetation etc. available free in surrounding areas, and (3) changing food distribution in the family with women eating the least food. This affects the nutrition and health status of people highly adversely.

Impact on Education and School Attendance: As children migrate with families or are engaged in the coping strategies of households, they do not attend the school. The critical situation of food also prevents children from attending schools. One observes drastic reduction in school attendance.

Impact on Environment: Since the drought affected people fall back on surrounding common natural resources for survival, these resources get badly depleted and degraded during droughts. Deforestation and cutting of trees for food and for fuel wood, exploitation of biomass for fodder etc. lead to sever depletion of natural resources. Also, open fields without vegetation are subjected to erosion and degradation.

Long term Impact: In long run water scarcity and peoples coping strategies have left their mark on the people and the economy. Indebtedness, permanent migration and loss of asset under distress cell are the direct impact and it takes several years to come to normalcy. In other words, the drought proneness of the state has proved to be a constraint to (1) poverty reduction, (2) agricultural growth and (3) the overall economic growth.

WATER SCARCITY, CROP PRODUCTION AND COPING STRATEGIES IN ORISSA

The experience of water problems in Orissa is more or less same in nature but differ in magnitude as compared to other states. Delay in monsoon or long dry spells during the rainy season or a sudden end of the rainy season, causing crop failures or drastic reduction in yields. The fluctuating agriculture and the consequent fluctuating incomes have discouraged farmers to invest much in agriculture, with the result that agricultural productivity is quite low in the state. Though two third of workforce is engaged in agriculture, it does not meet the requirement of food in the state

(Sarap 2003). Studies have shown that one major reason for the low yields is lack of water supply as well as water logging. Most of the factors are linked with the water management and irrigation issue.

Paddy is the main crop, which covers more than 70% of the cultivated area and about 80 percent of the area under paddy is rainfed. Non food crops are grown on less than 5 percent of the cultivated area as food crops. In most areas farmers take only one crop - Kharif crop in a year and double cropping area is less than 10 percent of total crop area. Reduction in crop production and frequent crop failures reported in the state reduces employment drastically and brings a lot of suffering to people. People are forced to adopt coping strategies like mortgaging and selling of assets, reduced food consumption, borrowings at high interest rates, migration etc. Some regions have become perennially drought prone and are included in the DPAP programme of the government. In fact, 28 of the total 30 districts are declared drought prone during 2002.

Since about 80 percent of the workers are engaged in agriculture, droughts have a widespread impact on the economy. The other factors which contribute to the severity of droughts are the poor crop diversification, inadequate measures to store rain water, low double cropping and low employment opportunities in the non farm sectors. The adverse impact of droughts is observed to be high because most farmers grow rice as one crop without any crop rotation. Of course farmers do not have much access to new crops or new varieties of crops easily. Not many farmers use strategies like mixed cropping or suitable agricultural practices to minimize crop losses. Farmers have poor facilities to store water for irrigation.

Under this situation many people depend on public relief programmes. But these are not reliable in the sense that their timings, duration and location are not always favourable.

The relief works could not provide enough work or wages to people. The works are not always located in nearby places and are not available when needed. People are therefore forced to migrate as a massive scale in search of work. Women and children are the worst sufferers of water scarcity-women's burden of work increases considerably and children are usually pushed into relief works by their parents. The usual coping strategies of the affected population include increased working hours, more time spent on collection of fodder, wood, fruits, leaves etc. and selling of livestock. Mortgage assets, distress sale and massive migration including land are common coping strategy.

WATER MANAGEMENT AND DROUGHT INTERVENTION IN INDIA: AN OVERVIEW

Water related problem like flood and droughts are not new to India so as the policies for fighting them. There was an alarming increase in the frequency of famines during the British Period (Bhatia 1967). Famines or scarcities were defined as "events of significant losses in agricultural production arising from inadequate rainfall or floods leading to severe unemployment and suffering of, particularly agricultural labourers and small farmers. In recent years the approach of the Central Government towards water management and droughts has four major components: (i) Scarcity relief works, (ii) Drought prone area development programmes, (iii) Watershed development, and (iv) Promotion of dry farming agriculture as a part of agricultural policy.

However, the experiences show that such policy has not emerged as an effective solution to the water problem. Major policy flaw lies in lack of integration of all these components into a comprehensive water policy. According to the present approach, managing water is the primary responsibility of state governments. The Center's role has been restricted to one of providing resources to States and acting in a supportive manner. Recently, some policy measures on water and irrigation management at the state level have been devised, there is a need to realize that different regions need different the treatment and central water policy needs to be integrated primarily with the state level policy. Several state governments have made some beginning in this direction. Unfortunately Orissa being one of the high rainfall receiving states has suffered from worst situation of water in recent years due to poor irrigation and water management system.

So far the Government of India has released Rs. 1136.23 Crores for DPAP programmes (between 1995-96 and 2002-2003) out of which Orissa has received only Rs. 36.37 Crores. The table indicates that there has been a significant increase in the release of the funds since 2000-2001. The government of India has sanctioned 16268 watershed development projects up to April 2003 out of which about 3728 projects are almost complete. As each project covers 500 ha area, the total area treated so far is 1864000 ha, which is a small fraction of the total drought prone area of the country. Orissa has completed 161 watershed projects.

Table 1
DPAP Coverage in Selected States

States	*No. of Districts*	*No. of Blocks*	*Area (Sq. km)*
1. Andhra Pradesh	11	94	99218
2. Bihar	6	30	9533
3. Chhattisgarh	8	29	21801
4. Jharkhand	14	100	34843
5. Madhya Pradesh	23	105	89101
6. Orissa	8	47	26178
7. Rajasthan	11	32	31969
8. West Bengal	4	36	11594
9. All India Total	183	971	745914

Source: Department of Rural Development, Government of India, New Delhi, 2002

WATER RESOURCES, IRRIGATION, AGRICULTURE AND DROUGHT IN ORISSA

Orissa is one of the poorest and one of the wettest states in India with about 1480 mm of the average annual rainfall accompanied by perennial rivers, lakes, reservoirs and good ground water conditions. The rainfall comes mainly from South-West monsoons between June and October, with the average number of rainy days being around 70. However, the rainfall has been erratic, unevenly distributed particularly at the beginning and end of the monsoon. Also, the state receives cyclonic rainfall frequently that affects the entire economy. The state is subjected to frequent disasters like droughts, cyclone and floods. Development of irrigation has been slow and inadequate over period. No notable improvement in irrigation network taken place in the state during last few decades. Orissa is fast gaining the status of being India's most disaster prone states. According to official statistics, during the last 100 years, there were floods in 49 years, drought in 30 years and cyclone in 10 years, sparing the state only in 10 years as normal (4th Sept 2003, The Economic Times). It is important to note that frequency and severity of such calamities has drastically increased over the years with increasing geographic coverage.

Table 2
Releases of Fund DPAP (Rs in Crores)

State Total		*1995-96*	*1996-97*	*1997-98*	*1998-99*	*1999-00*	*2000-01*	*2001-02*	*2002-03*
1. Andhra Pradesh	21.07	26.20	22.92	22.91	26.71	47.60	40.67	48.55	256.63
2. Bihar	7.25	0.34	1.15	2.38	2.30	1.00	2.42	2.50	19.34
3. Madhya Pradesh	19.39	21.19	8.93	8.82	14.02	24.95	43.61	47.21	188.12
4. Orissa	4.04	2.95	0.64	2.75	0.46	6.82	9.70	9.01	36.37
5. Rajasthan	6.44	1.52	4.19	1.73	3.86	9.81	11.95	14.31	53.81
6. West Bengal	2.30	0.28	2.6	0.00	2.09	1.35	3.18	1.08	12.88
Total	118.95	109.98	90.75	73.00	94.99	189.58	208.99	249.99	1136.23

Source: Department of Rural Development, Government of India, New Delhi, 2002

Not surprisingly, some of the poorest regions in the country are also places where agriculture is least developed and land and water resources are poorly managed. Because of poor irrigation farmers typically grow one crop and skip other crop opportunity. Irrigation potential in Orissa has not optimally tapped and there is evidence of poor utilization of the potential created. Agriculture in Orissa is mainly rain-fed, as only 35% of the cultivated area of 6.4 million hectares has irrigation facilities. Orissa is crisscrossed by many rivers and rivulets. Rivers of the State are rain-fed and they receive their runoff mainly from the monsoon rainfall. Only 20% of the total cultivable land in Orissa is covered under irrigation projects and the rest 80% depend on the vagaries of rainfall. Water harvesting practices is grossly neglected in the state resulting in augmentation of water problem.

Table 3
Natural Calamities in Orissa

1991	Flood
1992	Flood and Drought
1993	–
1994	Flood
1995	Flood
1996	Severe Drought
1997	–
1998	Drought
1999	Super cyclone, Flood
2000	Drought
2001	Severe Flood
2002	Severe Drought

Source: Board of Revenue, Cuttack; and Revenue Department, Govt of Orissa, Bhubaneswar.

Every year there is a need of about 1300 mm. to 1600 mm. of rain for *Kharif* crop. Though Orissa witnesses more than 1300 mm rain fall but due to its irregular and erratic nature the farmers are facing drought like situations almost every year. The data given in the table shows that there is no substantial increase in the irrigation potential created during recent years. It may be noted that there is decline in use of gross irrigation potential created between 1993-94 and 2000-01. This is really a matter of concern and need immediate policy actions.

Table 4

Net Irrigation Potential Created in Orissa by Different Sources

(area in 000 ha)

Years	*Major & Medium*	*Minor (Flow)*	*Minor (Lift)*	*Other sources**	*Total*
1990-91	937.1	378.9	264.3	447.1	2027.4
1994-95	992.7	409.9	294.5	518.5	2215.6
2000-01	1177.9	450.4	336.0	557.9	2522.2
2001-02	1190.5	455.6	338.7	557.9	2542.8
2002-03	1220.7	471.2	347.4	570.5	2609.7
2003-04	1234.9	497.5	358.1	572.6	2663.1
2004-05(P)	1236.5	503.9	376.3	579.2	2696.0

Source: Economic Survey, 2005-06, Orissa,
*includes water harvesting structures, Million Well Scheme & Schemes of PR Dept. etc.

The state is rich in natural resources like water, forests, mines and vegetation, if exploited properly the state can achieve a respectable level of growth. The state enjoys good rainfall, has perennial rivers and fairly good ground water regime. Though the rainfall is erratic in its behaviour, resulting in frequent droughts, while on the other hand the exploitation of the available water supply is very low (about 10 – 35 percent cultivated area is irrigated against the potential of 70 percent +). Consequently, the agriculture is subjected to wide fluctuations in production and productivity, resulting in wider fluctuations in incomes and employment of people. The short-term deprivations of people for food, drinking water, incomes and employment have had cumulative effect on the economy, ecology and human welfare, leading to concentration of poverty, unfavorable land distribution and unfavorable agrarian relations. Therefore, suitable water policy in the state can have lasting impact on agriculture, economic growth and human welfare. It appears that inappropriate water use in the state has resulted in frequent drought situation and it is proved to be a major constraint to poverty reduction and sustainable economic growth.

The State had 49.29 lakh ha. of land under net area sown, out of which 11.29 lakh ha.(23%) was irrigated through various sources of irrigation like canals, wells, tube-wells, tanks and others, which was higher comparing to 21% of 1990-91. The canal forms the most important source of irrigation having 67% share followed by tanks with 8% share only in terms of percentage of area irrigated by source to net irrigated area. The details of irrigation status are presented in the table below.

Table 5
Minor Irrigation Potential Created and Utilized in Orissa

(area in lakh ha)

SchemesGross Irrigation Potential Created (MI)Gross Irrigation Potential Utilized

	1993-94	2000-01	% change	1993-94	2000-01	% change
1. Dug Well	2.06	1.91	-7.40	0.92	0.77	-16.1
2. Shallow Tube Well	1.31	1.53	16.4	0.54	0.66	22.3
3. Deep Well	0.82	0.94	14.7	0.14	0.11	-20.6
4. Surface Well	6.98	6.90	-1.3	4.31	3.64	-15.5
5. Surface Lift	2.83	3.03	7.1	1.13	1.03	-8.5

Source: Economic Survey, 2005-06, Orissa, 2nd and 3rd Minor Irrigation Cencus

Table 6
Distribution of Net Irrigated Area and Net Area Sown in Orissa by Size Classes, 1995-96

(Area in Ha.)

Size Classes	Net Area Sown	Net Irrigated Area	Irrigated area as % to Net area sown
1. Marginal	1043891	267261	25.60
2. Small	1475515	346461	23.48
3. Semi Medium	1386023	304726	21.99
4. Medium	805369	166752	20.71
5. Large	218478	43825	20.06
All Size Classes	4929276	1129025	22.90

Table 7
Irrigation Status in Orissa

Year	Net Area Sown	Irrigated Area	Non-Irrigated Area
1970-71	56,10,236	7,26,374	48,83,862
1976-77	54,80,297	8,21,574	46,58,723
1980-81	52,25,500	8,01,400	44,24,100
1985-86	52,44,700	8,85,500	43,59,200
1990-91	52,71,423	10,00,524	42,71,899
1995-96	49,29,276	11,29,025	38,00,251
2000			

Source: Agriculture Census, Board of Revenue, Cuttack, Orissa

STATE INTERVENTIONS

Managing water resources is crucial in improving land and water productivities. The water policies of the states have three components: (1) drought relief works, (2) drought prone area programmes, now depending on watershed development approach and (3) agricultural policy which includes water management; irrigation development; research and development; dissemination of research, extension and training etc. Somehow these policies have non-left very favourable impact on water related issues. Evidence shows that the intensity and frequency of droughts and vulnerability have increased in the state over the years. The severe drought of Orissa in 2002 is a good example of this.

In short, the basic objective of water policy is to assimilate rainfed (drought prone) areas into mainstream of economic growth. This assimilation is essential at present because–

- Agricultural growth in the country can be sustained only through rainfed areas now, as the benefits from irrigated regions are relatively saturated.
- For sustained economic growth we need market participation of rainfed areas as producers and as consumers.
- Equity considerations as well as considerations for poverty reduction require that these areas are assimilated into the mainstream of economic growth.

Drought Relief Work

Efforts made by the state to fight droughts can be broadly divided into two type of works-irrigation works for protection of crops and employment. The state government undertakes drought relief works to ensure that affected people get employment/wages to buy food grains available through PDS shops. Drought relief works are also expected to ensure (1) supply of fodder for animals, (2) drinking water, (3) cash doles for the disabled and (4) tagari loans for farmers for the next seasons.

Water scarcity or droughts are declared at the block level on the basis of agricultural production (Annavari System) - if the crop production is 25 percent or less. Funds are allotted on the basis of the areas and people affected by it. State Relief Commissioner allocates the funds to the relevant departments as per their respective demands which are generally not worked out on the basis of any long term plans. This intervention is more like a firefighting exercise undertaken in the form of relief measures. The decisions about scarcity works are taken by the Panchayat bodies, supported by local administration. The relief works usually include construction of roads, digging ponds, deepening of tanks etc. Experiences revealed that these relief works are unable to address the gravity of situation nor to meet the needs of the people.

Provisions for immediate relief measures can be undertaken by the concerned authorities include from Sarpanch to Collector. In cases of starvation, prolonged illness or fight for survival, these officers can issue immediate relief works. The relief code has been revised recently to make it more amenable to relief. District level relief action plans for water, fodder and employment are often prepared in consultation with public representatives. Pani Panchayat, Biju Krushak Yojana etc some participatory approach has been formulated to provide a dependable and sustainable source of water. The state government has formulated some guidelines clarifying what kind of works can be undertaken under drought relief, with a priority:

- Water harvesting works, tanks, well and traditional structure for retaining water without digging.
- Earth work relating to water harvesting, construction and repair of roads and water and soil conservation
- Lift irrigation: small lift irrigation works to draw ground/ surface water. However, this has been a failure in the state, largely due to the unreliability of electricity supply and other operational and management issues.
- Forest related activities

Unfortunately, though such works are taken up under drought relief, their selection and implementation hardly matches to the local condition. Repairing of minor irrigation, soil conservation, construction of minor irrigation works, deepening of ponds and construction of check dams are some examples.

Our discussions and visits as well as review of studies have revealed that the state government is not able to ensure adequate relief to people due to (a) shortage of funds, (b) shortage of staff, (c) corruption in administration and (d) the general lack of commitment to help affected people. Scarcity relief works are increasingly expected to be linked with drought proofing, but this has been done, at best, in a highly haphazard manner. In fact there is lack of long term planning, shortage of fund, lack of departmental coordination and planning and delays in release of fund are major constraints.

IRRIGATION WORKS

Since the nature of droughts differs from year to year, depending on the duration and pattern of rainfall farmers can adjust by changing the sowing period, transplantation period or the variety of crops if rainfall is late. If the dry spells follow the rainfall in the beginning, farmers need some life saving irrigation to save crops. Farmers are suggested timely to change their farming practices and input uses to adjust to the changing situation. There is thus a need for regular drought prediction and forecasting. Weekly weather monitoring, recording information about water situation, prediction on the rate and pattern of rainfall in the coming days and the implications of these in saving crops can help farmers to cope with droughts. This calls for a good MIS to predict and monitor drought situation, and good system of extension and training to help farmers in using the management information system. The drought of 2002 has helped the state to make a small beginning in this area, but the state has a long way to go to acquire technical expertise and to acquire the administrative ability to reach the poor at the bottom.

DPAP and Watershed Development

The state policy is gradually shifting to soil moisture conservation, watershed development, development of smaller irrigation projects as well as to improve water use efficiency and agriculture practices. As directed by the government of India, watershed development has been accepted as the main approach for the development of DPAP areas in the state. At present DPAP covers 8 districts and 47 blocks of the state. In the first batch of DPAP, 192 watershed projects were sanctioned in the state in 1995-96. However, only 30 of these projects have been completed so far. By 2003 April, 684 projects have been initiated in the state so far. These projects have been taken up in all the DPAP districts, namely, Bargadh, Bolangir, Boudh, Dhenkanal, Kalahandi, Naupada, Phulbani and Sonepur districts. A Long Term Action Plan for watershed development has been prepared for this belt by the

Central Government for the period 2001 onwards. The annual fund allotted to these districts varies between Rs. 8-10 Crores.

Table 8
DPAP Area in Orissa

Name of districts	*Total Blocks*	*No. of DPAP Blocks*	*Percentage of area covered under DPAP*	*Percentage of population covered under DPAP*
1. Boudh	2	2	34.44	62.3
2. Kandhama (Phulbani)	12	12	80.02	93.47
3. Kalahandi	13	10	50.46	77.34
4. Nuapada	5	5	72.24	94.51
5. Bolangir	14	8	52.10	53.57
6. Sonepur	6	2	29.53	28.05
7. Bargarh	12	6	43.63	43.75
8. Dhenkanal	8	2	1.58	24.78
Total	73	47	24.04	67.26

Source: Govt. of Orissa, Department of Agriculture, Bhubaneswar.

Table 9
Crop Loss of 50% and More Due to Drought During Different Years

Unit	*1996-97*	*1997-98*	*1998-99*	*2000-01*
1. No. of Districts affected	28 (93.33)	15 (50.00)	27 (90.00)	29 (96.67)
2. No. of Blocks affected	263 (83.76)	86 (27.39)	163 (51.91)	216 (68.79)
3. No. of GPs affected	3762 (71.51)	849 (16.14)	1686 (32.03)	2511 (47.74)
4. No. of Villages affected	28837 (55.96)	4688 (9.10)	11431 (22.18)	16219 (31.72)

Note: (1) Figures in parentheses are % share.
(2) The Government had declared districts as drought affected (having crop loss of 50% or more) Source: Govt. of Orissa, Revenue Department, Various years.

WATERSHED DEVELOPMENT

Watershed development projects have emerged as a major programme for water scarcity and irrigation management and development and as a strategy for protecting the livelihoods of people inhabiting the fragile ecosystems experiencing soil erosion and moisture stress. The twofold aims of the programme are (1) to ensure availability of drinking water, fuel wood and fodder and (2) to raise incomes and employment of farmers and the landless through increased agricultural production and productivity. Some of the important features of this approach are:

- Simultaneous development of land, water and biomass resources in the light of symbiotic relationship among them
- Integrated farming system approach through water management
- Meeting food, fodder, fuel needs of human being and livestock
- Ensuring environmental sustainability along with water use efficiency.
- Improving land productivity by promoting better agronomic practices and input use
- Development of local institutions for water management and maintenance through participatory approach

However, the impact of watershed programme on reducing water scarcity and drought impact has been observed to be limited due to (1) poor focus on local cropping pattern, common lands and other natural resources, (2) low attention given to sharing water and cost of operation and maintenance, (3) low priority to drinking water, fodder and fuel wood for local people and (4) the lack of clear focus on long term drought proofing. An important implication of WSD programme that is missing is the fair participation and benefit the weaker sections.

The new version of WSD programme, i.e. Hariyali, has been introduced by the Government of India (2003) with emphasis that the programme will now be implemented through Panchayati Raj bodies and it cover all DPAP, DDP and IWDP areas.

Since the state has initiated watershed development programmes recently much is not known about its impact. A systematic evaluation of these programmes in the state is not yet available. The number of programmes and the area covered are very small as compared to the need. There is a need to expand the coverage of these programmes and allocate larger funds for the purpose. There is also a need to plan watershed development systematically, taking care of technology, activities, institutions etc, and also plan for watershed "plus" activities, to translate watershed development into agricultural growth.

To sum up, scarcity relief works are increasingly expected to include works that lead to drought proofing and watershed development programmes are expected to develop drought prone areas to achieve drought proofing along with agricultural growth. However scarcity works are still carried out independently by the Revenue Department as disaster management and the area development programmes do not focus enough on the critical needs for drought proofing. In the final analysis therefore, the major limitation of the water policy today is the lack of integration between scarcity works and drought proofing activities for promoting agricultural growth.

PANI PANCHAYAT

Orissa has initiated some reform measure in water sector. Commercialization of water participatory water management was initiated by the state few years before. "Pani Panchayat" programme is a part of this effort but has not become success in addressing the key water issues in the state. Equitable and timely water supply, water management, to create an awareness of common water use among farmers, to convince farmers for crop diversification and to share responsibility of operation and maintenance of the water infrastructure are the major objectives of this programme. However, prevalence of government managed irrigation systems in an uncertain condition has made farmers reluctant to adopt participatory approach. Heterogeneous background and apprehension of poor farmers about water charges and operation and maintenance charges discourage to participate voluntarily.

Decentralization of water use in the absence of effective dispute solving mechanism also lead to social divide among the users groups - dominance of a particular group ("Panidar" class). Some studies show that access to water by marginal groups has not improved (Sainath, 2000). The experience of pani panchayat model in Orissa seems limited with 'little pani, less panchayat' than addressing the issue of water in the states. A powerful landlords and contractors are in control of what was once a public resource.

There may be a rise in number of membership or the rise in common deposits but it does not reflect a growth of prosperity in the pani panchayat region. Much of the deposit often comes from outside of agriculture just it is all for the benefit of the better-off people. The small farmers are not equally benefited due to no rabi crops and increase in cost of cultivation and other expenditures. Big farmers who anyway dominate the village grab control of the created resource. There is lack of fairness in pani panchayat system all people do not participate in the scheme. Equity in sharing of water is missing and it becomes a threat to success of the programme. Under this condition the idea of Privatisation of irrigation system under current situation may not attract many users/farmers. Operation and maintenance cost of irrigation projects need to be supported by the State Government. In water sector NGOs may play vital role in managing water but there is need for an effective mechanism to monitor their activities.

CRITICAL ISSUES IN WATER POLICY IN ORISSA

The above discussion on the water management and drought intervention in Orissa provide understanding on the content and working of these programmes and throws light on the inadequacies of water policies that would lead to reduction of drought impact and sustainable agricultural growth. This section discusses the critical issues relating to these policies. We discuss below the critical policy issues.

Lack of Integrated Approach in Water Management and Irrigation Development

Absence of a comprehensive planning, lack of integrated water and land management policy and poor coordination among the line departments largely responsible for sub optimal use of water in the state. Short period drought relief works, watershed development without input supply, poor maintenance and operation of irrigation projects, long pending other projects, absence of effective water dispute mechanism, inadequate mass participation, equity and access to water ownership etc are key policy issues in the state and not yet addressed adequately.

Flaws in Scarcity/Drought Relief Works

The present form of water scarcity or drought relief works have an inherent disadvantage in their root in drought relief manual on the one hand and in the implementation agency, the revenue department on the other hand. In spite of the modifications introduced over the years, the manual is still focused on "relief" and not any integrated long term approach. Our field experience reveled that these interventions have not been very successful because of the following reasons:

- Lengthy procedure: of declaring an area as a drought affected area as well as initiating of relief works
- Huge gaps between what are people's needs and provisions for relief–employment, timely wage payment, drinking water, fodder for animals–and what they receive is too small due to unexplained delay and funds fall short.
- Low income: wage payment is made on the basis of the quantum of work done not on the minimum wages and the work is not available for a longer period.
- Malpractices and corruption: Our field experience shows that community coordinators hired from market for organizing SHGs often ask groups for a cut in the sanctioned fund.

Scarcity works need overhauling and reorientation: Firstly, these works should be components of a long term plan for drought proofing. Area specific long-term drought proofing plans should be formulated. Alternatively, scarcity works should continue through out the year and should go on till the specific drought-proofing plan is implemented. Need for changes in relief manuals in the context of the present development planning.

Research & Development, Extension and Training in Water Use

One major problem observed in the states was regarding the adequacy of the support for R & D, extension and training for water management, irrigation and crop practices. Even though some research has been developed at OUAT to meet local areas problem in water use and crop production but due to poor coordination among other departments and lack of extension services these technologies remained limited to few govt. farms only.

The major research support needed for water management in both water scarcity and water logging areas. It includes (a) crop sector-developing new crops and new varieties of existing crops i.e. short duration crops and drought resistant crops, mixed cropping, crop rotation etc, (b) agronomy–weed control, thinning of crops, reducing evaporation and loss of moisture through chemicals, new tools and equipments etc. and (c) other agricultural

Table 10
Status of Pani Panchayat (by 15-Sept-2006)

Name of the Circle	*Detailed Programme*		*Registered/ Formed*		*Handedover*		*Delineation*		*P. P. formed by conducting election as per P.P. Act & Rules as on 15.09.06*		*Programme for 06-07 (by Dec.' 06)*	
	No. of PPs	*Area (Ha)*	*No. of PPs*	*Area (Ha)*	*No. of PPs*	*Area (Ha)*	*No. of PPs*	*Area (Ha)*	*No. of PPs*	*Area (Ha)*	*No. of PPs*	*Area (Ha)*
Sub Total (Major & 294531 Medium)	2567	1117300	1427	623658	1123	494236	2331	1033529	302	109259	731	
Sub Total (Minor Flow)	1883	328291	1005	192294	775	149596	1883	328291	466	55019	468	69090
Sub Total (OLIC)	11033	243449	11033	243449	10377	229768	2472	51734	0	0	4305	86300
Grand Total 449921	15483	1689040	13465	1059401	12275	873600	6686	1413554	768	164278	5504	

Source: Dept of Water Resources, Govt. of Orissa

practices like spreading agricultural operations over time (for example staggered planting), spreading input use over time or giving spilt dozes of inputs, improving water use efficiency, mulching etc. Research inputs are also needed for promoting alternative land use patterns for development of grass lands, horticulture, forestry, medicinal plantation etc. It may be noted that some new varieties of rice suitable for dry areas has been developed by CRRI, Cuttack and OUAT, Bhubaneswar but their adoption in the state is negligible due to several factors.

Our discussions with experts and officials revealed that dry farming technology for water management has not been a priority area of research till recently, as irrigated agriculture always received a high priority. On the other hand, there is acute problem of fund, infrastructure and equipments, expertise in new areas like Remote Sensing, GIS, biotechnology, faced by state research organizations. However, R & D institutions including agricultural universities do not always respond to the needs of farmers at the field level. Similarly, delay in seed certification and lack of timely availability of certified seeds and other extension services remain major constraints of wider adoption by the farmers. On the farmers' side, their access to new research in drought proofing seems to be limited due to their lack of knowledge about the arrangements, poverty and indebtedness, unfavourable agrarian relations, and poor communication. In short, the process of water management from research to extension and training is not a smooth process. There are many hurdles, which need to be over come!

POLICY SUGGESTION

Water management and irrigation development in the state need urgent policy attentions. Water shortage does not only cause just crop failures, but also affects soil moisture, soil fertility, cropping pattern, crop health and crop planning. There is a need to plan suitable interventions in these areas to minimize the crop risks and to promote agricultural growth. There is a need to shift to less water consuming crops. Since paddy is the main crop, the demand for water will remain high unless there is a shift towards less water consuming crops. Also, water efficient irrigation systems need to be promoted to reduce water consumption. Expansion of irrigation in remote areas through suitable methods and land leveling/land development will help considerably. DPAP and particularly watershed development projects can contribute significantly in these areas. Therefore the scale of activities and the funds allotted to these activities suggest that the state has a long way to go to achieve this goal.

Mechanism for ensuring water use efficiency is grossly undermined in the state especially by ignoring the contribution of OLIC in augmenting food production, generation of employment, less capital investment and reduction wastage of water. Water policy in the state needs to be reoriented and reorganized with this new focus. This has implications for major overhauling of programmes, institutions and organizations on the one hand and diversification of land use and economic activities in affected areas on the other hand.

The irrigation system in the state is inefficient largely due to lack of coordination and planning, poor maintenance and operation and constraint of funds. The irrigation system needs more strategic policy actions in order to ensure water use efficiency and effective public-private partnership and commercially managed Water Service Agency (WSA).

Farmers sometimes also try to reduce costs by spending less on inputs till there is good rainfall. These strategies can be strengthened by scientific research and by supportive credit and other infrastructure facilities. Over exploitation of common property resources for fodder, wood, vegetation, fruits during dry year also common across region. Temporary migration seems to be a very common strategy adopted by many households.

In spite of its high level of water resources, the potential creation and utilization of water in the state have not contributed in economic growth, agricultural development, poverty reduction and human development as compared to the other Indian states. There is immediate need of suitable policy actions in managing water and production risks in the state.

The state is facing two major challenges at present: (1) How to link drought relief works to drought proofing more systematically and (2) How to integrate DPAP and watershed development with the overall agricultural development in the state. Long-term Water Policy: Long term water policy is expected to achieve drought proofing is necessary to minimize the uncertainties and risks associated with droughts. The main components of drought proofing are minimizing the fluctuations in agriculture, i.e.

stabilizing agriculture and promoting agricultural growth in drought prone areas by proper management of land and water resources. Drought proofing requires interventions in terms of (a) balanced management of land and water resources, i.e. watershed development, (b) research and development in developing new crops, new agricultural practices etc, (c) extension and training, (d) proper organizational and infrastructural support to carry out the above activities, and (e) institution building that involves and empowers communities to undertake these tasks. This comprehensive approach aims at ensuring sustainable livelihoods to people.

Water policy can be seen as a process of risk diversification for minimizing risk and uncertainty of droughts. Research and development activities, backed by extension and training, can promote (a) horizontal risk diversification by farmers by providing new crops, new varieties of old crops, mixed cropping possibilities, suitable crop rotation etc., (b) vertical risk diversification by enabling them to spread agricultural operations over time, like staggered planting, spreading inputs use over a period of time or planting many seeds per ha. etc.

Agricultural Policy: The problem of the state is not of shortage of water resources but it is of utilization and efficient management. Development of irrigation facilities, proper distribution of water in command areas through participative irrigation management, can improve water use efficiency and need more attention.

Adequate research and development efforts suitable to the local conditions, cropping patterns cultivation practices should be focused to increase water use efficiencies and to reduce fluctuations in agricultural production. Another neglected area is the extension and training activities which are required for dissemination of research and new methods.

Diversification of Agriculture and Non Agricultural Activities: The risk of droughts can also be reduced by promoting alternative uses of land to minimize the risk. Allotting land under horticulture, medicinal plants, grasslands will help farmers in diversifying the risk and minimizing income fluctuations. Similarly, diversification into non-agricultural activities will also reduce vulnerability of people to weather fluctuations. Promotion of handicrafts, small or large industries can help immensely in this context.

The central water policy needs to be viewed in the context of the above discussion. Since the state level policies are formulated under the broad framework of the Central Policy, such a discussion will be relevant and pertinent.

CONCLUSION

Orissa has been subjected to frequent, and severe water scarcity leading to drought in spite of its rich status in water resources. The impacts of water scarcity, namely, temporary sufferings, specific regional/area based constraints and isolation from the mainstream economy are addressed through three broad interventions: Drought interventions (DPAP), watershed development and agricultural policy. Though attempts have been made to integrate these three strategies into one comprehensive strategy, these have not been much successful. There is a need to establish explicit linkages between the three strategies so as to manage water and other resources effectively for agricultural and economic development.

REFERENCES

Bhatia, B.M., (1967), *Famines in India: A Study in Some Aspects of the Economic History of India*, Asia Publishing House, New Delhi.

Brahmananda, P.R., Undated, *Macro-drought: Where Policy Lags?*, www.rural nic.in.

Chen, M., (1991), '*Coping with Seasonality and Drought*' New Delhi Sage Publications.

Chetty, V.K., and D.K. Ratha, (1987), '*The Unprecedented Drought and Government Policies*', Mimeo, Indian Statistical Institute, New Delhi.

Deshai, B.M., (2003), *Drought Impact and Vision for Proofing*, Economic and Political Weekly, May 24.

Dreze, Jean, (1988), '*Famine Prevention in India*' Discussion Paper Series No. 3, Development Economics Programme, London School of Economics.

Government of Orissa, (1996), *Orissa Relief Manual*, Department of Revenue.

Government of India, (1994), *Report of the Technical Committee on Drought Prone Area Programme and Desert Development Programme*, Ministry of Rural Development, New Delhi.

Government of India, (2003), *Guidelines for Hariyali*, Department of Land Resources, Ministry of Rural Development, New Delhi.

Hirway, Indira, (2001), *'Vicious Circle of Droughts and Scarcity Works: Why not Break it?'* Indian Journal of Agricultural Economics. Vol. 56, No. 4

Jodha, N.S., (1980), *'The Process of Desertification and the Choice of Interventions'*, Economic and Political Weekly, Vol. XV No. 32.

Jodha, N.S., and A.C. Mascarenhas, (1985), *'Adjustment in Self-provisioning Societies' in Kates R.W., et al. (eds.),* Climate Impact Assessment, Jhon Willey & Sons.

Joy, K.J. and Suhas Paranjape, Undated, *Watershed Development Review: Issues and Prospects,* CISED, Bangalore.

Kalavalli Sashi and Jon Kerv., (2002), *Mainstreaming Participatory Watershed Development,* Economic and Political Weekly, January 19.

Mathur, Kuldeep and Nirajia Jayal., (1993), *Drought, Policy and Politics: The Need for a Long Term Perspective,* Sage Publications India Pvt. Ltd., New Delhi.

Parsuraman, S., and P.V. Unnikrishnana, (1999), *India Disaster Report: Towards a Policy Initiatives,* New Delhi Oxford University Press.

Pradhan Jagdish, (1993), 'Drought in Kalahandi', *Economic and Political Weekly,* May 29.

Rao, C. H. Hanumantha, S.K. Ray and K. Subbarao, (1988), *'Unstable Agriculture and Drought'*, New Delhi, Vikas Publishing House Pvt. Ltd.

Rao, C.H. Hanumantha., (2000), *Watershed Development in India: Recent Experiences and Emerging Issues,* Economic and Political Weekly, November 04.

Sainath, P., (1996), *Why Everybody Loves a Good Drought?,* Penguin, India.

Wilhite, D. A., (ed.) (1993), *Drought Assessment, Management and Planning: Theory and Case Studies,* Boston, Kluwer Academic Publishers.

Scott, C. (2000), "Mixed Fortunes: A Study of Poverty and Mobility Amongst Small Farm Households in Chile, 1968-86", Journal of Development Studies, Vol. 36.

World Bank (2001), Risk Management in South Asia: A Poverty Focused Approach,Washington DC: Report No. 23509-SAS.

Raghav G. and K. Imai, (2003), 'Vulnerability, Shocks and Persistence of Poverty-Estimates for Semi-Arid Rural South India', FMS, University of Delhi.

8

WATER RESOURCES PLANNING THROUGH NON-CONVENTIONAL METHODS: (A STUDY ON INTER BASIN TRANSFERS IN PENINSULAR INDIA)

*D. Rudrappan**

INTRODUCTION

The age old saying "not to waste money like water" need now to be updated by the saying "use water like money". Consequently conservation of water is the mainstay of all water related strategies. It has been predicted that if at all there is a third world war that may be due to the sharing of water. Therefore, water is undoubtedly one of the most important natural resources and is vital for all living organism, major eco-systems and sustainable development. While water is more or less a renewable source, its availability is largely determined by climatic conditions and technology that permits its exploitation, and also by the efficiency with which water is consumed and used.

* Dept. of Post-Graduate and Research Studies in Economics, Presidency College, Chennai

The vast variation in geographical space and time regarding the availability of water in different regions of the country has created what is called as flood-drought-flood syndrome, with some areas facing acute water shortage, while others suffering from flood damages. The concept of inter-linking of rivers through inter-basin water transfer from surplus river basins to water deficit basins was conceived for optimum development and utilization of water resources. In view of the looming water crisis, Interlinking of Peninsular Rivers would prove to be a blessing to the water starved areas and of course a relief to the excess water areas in preventing devastation caused by flood.

SUPPLY AND DEMAND FOR WATER

India, with its geographical area of 32.87 lakh sq. km has 2.45 % of the earth's land mass, supporting 17% of the world's population with barely 4 % of its water resources[1]. Though India is considered as one of the oldest civilizations with a kaleidoscopic variety and a rich cultural heritage, an average Indian has hardly one–sixth of land and one–fourth of water as compared to the world average. In view of the severe disparity in regard to water and land, optimal use there of, is of utmost necessity to ensure comfortable living to the people of India.

India is one of the few countries where the total precipitation is estimated at around 400 million-hectare meters[2] (mhm). Based on the study carried out by the Central Water Commission, the Irrigation Commission has worked out the supply of water resources of the country at 178 mhm which includes ground water runoff at about 45 mhm, while the rest is lost in evaporation or escapes into sea during the floods. Since there is temporal variation in rainfall and floods, mismanagement in some parts and population pressure have made India to be classified under the categories of countries having water scarcity. The average annual normal rainfall of the country is 1170 millimeter. However, there is a large variation in rainfall over space and time with Chirepungi in Assam receiving more than 1100 centimeters and parts of Rajasthan accounting for barely 100 mm per year. In view of the above, floods are recurring features in Brahamaputra and Gangetic regions in which almost 60 % of the river flows of our country occur. Further, whenever there is less rainfall, droughts are a common feature in Western India as well as parts of Andhra and Tamil Nadu. According to the practice followed by the Indian Meteorological Department, drought is taken to have occurred over an area where the annual rainfall is less than 75 per cent of the normal. When the annual rainfall is less than 50 per cent of the normal, it is called severe drought[3]. By 2050, it is expected that the population of India would increase to about 1640 million and that would require about 450 million tonnes of foodgrains[4]. To attain this target, it would be necessary to increase the irrigation potential, to 260 million hectares from the present 140 million hectares.

Further, it is calculated that by the year 2050, per capita availability of water in Brahamaputra region would still be around 9000 kilo liter (Kl), 800 Kl in the Ganges basin and 400 Kl in Cauvery and Pennar basin as against the annual minimum requirement of 900 Kl per person[5]. The need of the nation is at present large storages of flood water to tide over the water crisis during the lean seasons, through interlinking of rivers and inter basin transfers[6].

EQUITABLE DISTRIBUTION

The concept of larger good of larger number of the society should be accorded a due place while taking decisions like linking of rivers. Any lopsided approach is bound to create tension, friction, needless delay in the implementation of the projects and, in short, the whole exercise tends to become counter- productive. Hence, the linking of rivers with judicious handling of the 'dams component' is a most significant factor, in the context of the development of Indian economy. Now, more than ever, equitable distribution of water is the need of the hour. According to ancient literature, Bhagiratha, the legendry hero, is perhaps the first to have ever attempted a river-water project in the world. It was this great man, who along with a select team of young men undertook an expedition to the Himalayas and could successfully accomplished the incredible task by cutting the edge of the divine lake Manasasarovar, in such a way that the flow of water was diverted towards India, giving birth to the River Ganga or Bhagirathi, the holiest of Holy Rivers. Now, according to the interlinking project, the River Ganga has to cross the Vindhyas and reach the south to join the River Cauveri. This unfinished agenda of interlinking of rivers of our ancestors would be our endeavor at present.

PROPOSALS ON ILR–AN HISTORICAL PERSPECTIVE

To secure a bright future for the country, several interlinking proposals were mooted. Sir Arthur Cotton recommended to the British Council interlinking of Indian rivers facilitating navigation facility even as early as 150 years, for marketing of British goods. However, British Council rejected his proposal by pointing out that colonial market could be developed much faster with the laying of railway lines throughout the country. Captain Dastur, an Indian Airlines pilot proposed a Garland Scheme linking North Indian rivers and South Indian rivers with a canal ranging upto 2500 meters width at an outlay of more than Rs.1.20 lakh Crores (at the 1970 price level). As the Garland Scheme involved huge capital outlay, policy makers have not evinced much interest in its implementation.

The former Irrigation Minister, Dr. K.L. Rao, envisaged in his "Ganga–Cauvery" link, the transfer of about 65,000 cusecs of flood flow from the Ganga near Patna down to the Cauvery through a network of canals joining the various rivers enroute[7]. As it involved huge consumption of electricity for lifting of water over Vindhya Mountains, this proposal also has been dropped.

The severe drought that affected India in July 2002 and continued thereafter in several states, revived the demand for the "Ganga–Cauvery Link" by several political leaders and even the President of India, Dr. A.P.J. Abdul Kalam asked for the implementation of the inter–basin transfer proposal. At the same time, the Supreme Court also, as a result of public interest litigation asked the Government to implement the inter-linking project within ten years.

Accordingly, a Task Force on Inter-Linking of Rivers both under Himalayan and Peninsular components was set up with former Union Energy Minister, Shri Suresh Prabhu as its Chairman. The National Water Development Agency had estimated that the cost of inter-linking of Indian rivers would be a total of Rs.5,60,000 Crore. This is not the final figure which will be known only after detailed project reports of each of the thirty links. However, the new Government at the center has disbanded the Task Force set up by the previous Government.

INTERLINKING OF PENINSULAR RIVERS

The plans and estimates for Interlinking of Peninsular Rivers as part of a National Perspective Plan were prepared by the National Water Development Agency under the Ministry of Water Resources in 1980. The proposal contemplates linking of Mahanadhi, Godavari, Krishna, North Pennar, Palar, South Pennar, Cauvery, Vaigai and Gunder with 15 link canals. Peninsular river grid region is located from the Tabi river basin to Kanyakumari in the West and from the Mahanadi to Kanyakumari in the East. The link from Mahanadi to Godavari will be along the East Coast and will not involve any lift. The link between Godavari and Krishna will be partly by gravity and partly in the ultimate stage, by lifts of the order of 120 meters.

The Peninsular grid begins from a place called Manibhadra in Cuttack district in Orissa where a dam is proposed to be built for storage and diversion of excess water in the Mahanadi, during the flood season. From this place a 900 km long canal is to be constructed for taking surplus waters to the Godavari at Dowlaiswaram barrage near Rajamundry in Andhra Pradesh. Water for irrigation enroute, the canal, will also be provided for.

From Dowlaiswaram, the excess water of both the Mahanadi and the Godavari will be taken to the Prakasham Barrage across the Krishna river near Vijayawada in Andhra Pradesh. Provision has also been made to transfer Godavari water to the Krishna from the proposed Inchampally Project upstream of Rajamundry. There is also a proposal to take Godavari waters from Polavaram on its bank to the Krishna. From Krishna, a canal, augmented by transfer of surplus waters from Mahanadi and Godavari rivers, to the North Pennar river. Another canal from the Almati dam across the Krishna in Karnataka to the Somasila dam across South Pennar has also been proposed. From South Pennar, a canal will take water to the Grand Anaicut in the Cauvery. However, this is not the ultimate destination of the waters from the Mahanadi and the Godavari. From the Cauvery, waters will flow to Gundar and the Vaigai further South.

FRESH WATER WASTED TO SEA

The estimated average annual surface water flow of both east flowing rivers from Mahanadi to Gundaru (15 links) and west flowing rivers from Tabi to Kanyakumari (39 links) is estimated at

501.78 bcm[8], i.e. 17,720 tmc (one bcm is equal to 35 tmc approximately). While the actual quantity utilized upto 1998 for irrigation, domestic, industrial power generation is estimated at 257.20 bcm (9083 tmc), quantity of water wasted to the sea per annum is calculated at 244.58 bcm (8637 tmc) which accounts for 48.8%[9]. The enormous waste of water in the west-flowing rivers could be avoided by diverting them to the eastern side of the Western Ghats, as was done in the case of Periyar river 100 years ago. By adopting water management and water saving techniques, it is possible to irrigate 4000 hectares for one tmc ft of water in a fertile soil and it could produce 5.5 Tonnes per ha. which costs Rs.5,500/- The above yield has been worked out on the basis of the average productivity per ha. in Tamil Nadu which stood at 5.5 Tonnes during the year 2000. Out of the total unutilized 8,637 tmc ft of water, only 7,500 tmc ft is considered for irrigation leaving the remaining 1,137 tmc for drinking and other uses. The above 7,500 tmc ft of water would produce Rs.16,500 crores to 90,750 cr. worth of food grains annually.

COST– BENEFIT RATIO

Based on the area irrigated and quantity of paddy produced, its cost, man-days generated, the cost benefit ratio has been worked out for the unutilized surface water resource of the Peninsular region. It is estimated that the interlinking of Peninsular rivers would require about Rs.50,000 crores as per the 1996 cost calculation. At the ratio of 50 farm labourers per hectare, it is likely to generate employment about 150 million man-days to 750 million maydays per year. Taking into account the indirect employment opportunities generated, no other industry would give livelihood security for so many people as the agricultural industry. The cost benefit ratio works out to 1:2.4 for paddy alone[10]. If mixed crop is raised, this ratio will be more, providing "incalculable benefits".

FINANCE FOR THE PENINSULAR PROJECT

As inter basin transfer of river water is a National Programme, the project should be financed by the Central Government. Even an amount equal to 10 % cut in the total subsidy is sufficient to implement the programme. Considering the long-term benefit, the scheme could also be carried out as a self – financing scheme, a people's project without any financial allotment from the Central or State Governments. A statutory body in the name of Peninsular River Development Authority could be formed to borrow funds from the commercial Banks, beneficiaries of the project, or the World Bank and implement the Scheme on Build, Operate and Transfer (BOT) method, and handing it over to the respective Farmers Associations for maintenance.

PROBLEMS AND PROSPECTS IN PENINSULAR PROJECT

Lessons from History: For a number of years, many mega projects such as Mekong Delta Project, Three Gorges/Gangtze river links project of China, Tennessee Valley Authority of USA, man-made lakes for drinking water supply to New York city have been performing well, providing benefits without any hitch. Therefore, linking of rivers proposal is a feasible one and would not pose any major problem. As nature has bestowed river flows through certain regions/states, interlinking project alone would ensure water for everyone. Also, this demonstrates the principle that water is a common property resources. Some State Governments such as Kerala and Punjab are not yet convinced about the utility of the project. It is felt that those who possess water in excess of what is required need to think of sparing for the sufferers, rather than allowing it to be drained to sea.

Water as a National Asset: The most important aspect is the fact that the Constitution of India under Seventh Schedule, List 11, Entry 17 has allocated water as a subject to the States. As such, the States have more or less the final say in matters connected with irrigation, dam construction or hydro – electricity generation. Taking the extreme step of enacting a law in Parliament, under entry 56 making water, a national asset, could reverse this.

Satellite Surveys: Further, to facilitate early completion of feasibility reports of interlinking of rivers, 3-D radar satellite imagery could also be used to save time for Topo surveys and in evaluating alternative links of rivers.

Demand and Supply Management: Another fact to be remembered is that 85 % to 90 % of the river run-offs occur during the four monsoon months of a year, where as conditions of drought occur in over 80 % of the country's land area, even if there is a shortfall in rains of only 25% from the national annual average. The proposed interlinking project would both mitigate the sufferings of people in the flood havoc regions of the Mahanadi basin as well as the hardships of people in the drought- ravaged regions of southern and western parts of India.

Displacement of People: Obviously, these links, the dams and the reservoirs will result in displacement of a large numbers of people. In any case, the interlinking proposal will take full care of the project affected people by evolving a suitable rehabilitation package acceptable to the displaced people.

People's Participation: As people's participation is essential for the success of the colossal river linking project, Panchayats should be involved in all the stages of its implementation so that costs would be minimized and benefits would be maximized.

Easy Implementation: As the Himalayan component requires consent and co-operation of neighbouring countries, it is easier and less complicated to take up first, the implementation of Peninsular Rivers Project.

National Waterways and Employment Generation: Incidentally, Interlinking of Rivers also facilitates inland navigable waterways to the tune of 14,500 K.M. besides providing employment opportunities to the teeming millions. It is also estimated that mechanized crafts with a tonnage capacity of 18 M. Tonnes could be used in a length of 3,700 K.M. waterways.

Strategic Security: Much more than this, National River Waterways contributes to the most important strategic security to the northern and western parts of India.

Expansion of Agricultural Production: Being a mega project, Interlinking of Rivers not only provides access to drinking water to everyone living in the nook and corner of the country, but also leads to doubling of agricultural production of India to 450 million Tonnes in the year 2025.

Environmental Issues: Projects relating to interlinking of rivers should be so conceived as to minimize the adverse environmental impact and enhance the positive impacts such as regional development, equity, employment generation and pollution control. State of the art management techniques should be adopted to reduce adverse environmental aspects such as reduction of seepage in water conveyance, distribution and application.

Besides, interlinking of Peninsular rivers should be taken up early as it helps constant flow of water which is essential in order to look after plant, animal and micro – organism life downstream of dams. Further, interlinking facilitates uninterrupted water flows helping to clean the cities, towns and villages. Moreover, slow release of stored water will not only produce hydro-electricity, but also helps large–scale afforestation.

CONCLUSION

As stated above, besides providing immense benefits, interlinking proposal of Peninsular rivers is conceived to optimize the availability of water by promoting techno-economically feasible and environmentally sustainable development through storage and diversion of water from water rich to water-shortage and drought-prone areas. Irrespective of political boundaries, India needs to be looked as one single country. Leaving aside regional, cultural, linguistic differences and state boundaries, and with the active support of our President of India, Supreme Court, Central and State Governments, intelligentsia and the beneficiary community at large, it is hoped that the proposal would take concrete shape for implementation sooner than later. To sum up, the mega project of linking of Peninsular Rivers is clearly a vision that puts India on a higher growth trajectory and is rightly viewed as a welcome panacea for the water woes, besides forging not only national integration but also emotional integration.

REFERENCES

Government of India, (2002), India 2002, A Reference Annual, Publications Division, GOI, New Delhi, p.1.

Rudrappan, D., (2003), *Ground Water Management, Environment Security an Agricultural Development,* Southern Economist, Vol. 42, No.5, p. 5.

Indian Council of Agricultural Research, (2002), Hand Book of Agriculture, New Delhi. p.8.

Govt. of India, (2003), Interbasin Water Transfer Proposals, Task Force on Interlinking of Rivers, Ministry of Water Resources, Govt. of India, New Delhi, p.11.

Ibid, p.5.

Ibid, p.6.

Arabinda Ghose, (2003), Inter-linking of Rivers, *Kurukshetra*, Oct., 2003 p.70.

The Hindu, (2001), *Peninsular River Development Authority Suggested,* The Hindu, dt. 20.3.2001. p.4.

The Hindu, (2001), *Wasted to the sea,* The Hindu, dt. 13.3.2001. p.16.

Ibid, p.16.

9

DECENTRALISED MANAGEMENT OF WATER RESOURCES STRATEGIES FOR SUSTAINABLE DEVELOPMENT IN N E REGION

*N. Upadhyay & T.K. Giri**

INTRODUCTION

Water is a prime natural resource, a basic human need and a precious national asset. Optimum development and efficient utilization of water resources, therefore, assumes great significance. The total surface area of the earth is 510 million Sq. Kms. Almost 70% of the surface is covered by seas and oceans, which represents 1400 million cubic Kms. of water, which is only 0.5% of the earth's volume. The total volume of the water consists 97% of saline water and about 3% surface water. The distribution pattern of the mega reservoir needs a holistic view of the Hydrosphere, which raises high alarm as fresh water accounts for even less then 1% stock in Hydrosphere. Hydrosphere, the most dynamic and eventful part of the planet earth comprised of all the water resources of the earth. Our source for drinking water is

* Prof. Niranjan Upadhyay, Director & Dr. Tapas Kumar Giri, Asst. Professor, NIRD, NERC, Guwahati–22

mainly from surface and ground water. Indian rural folk and livestock are mainly dependent on water from wells, hand pumps, streams and public distribution systems. Ironically, these sources are considered safe even with manifold oddities.

Alarming Situation and Competing Priorities

It is now estimated that by 1995, 0.45 billion people in 31 countries experienced water-stress. By the year 2025, 48 countries with 2.8 billion people would experience water-stress condition and almost 4 billion people are expected to be living in water-stress condition in 54 countries worldwide by the year 2050. Although water seems to be a common commodity, of nearly 1370 million cubic kilometers of earth's total water quantity, only about two per cent of it is fresh out of which again nearly 97 per cent is tied up in polar caps in the form of permanent snow. The remaining three per cent is renewable through cycle of precipitation and evaporation from oceans and the surface sources on earth. The renewable fresh water falling on the continents and islands is only about 410,000 cubic kilometers, which is hardly 0.03 per cent of total quantity of water on earth.

Agriculture uses largest share, 70% of water and not efficient as half of the water used for the irrigation disappears in seepages and evaporation. In developing countries 80% of the fresh water is used for irrigation and they use twice per hector as compared to developed countries and produce only 1/3rd of the crop yield. Yet the full security for all is a distant dream to many living in poor and developing countries. To make provision for their food may need another 20% more water for irrigation. Since 1900 A.D. population has tripled, global economy has grown 20 fold, consumption of fossil fuel 30 fold and industrial activity 50 fold. In 1990 the global irrigated area was estimated 255 million hectare of which 200 million hectare were in low or middle-income countries (Patwari 2003).

MAN AND WATER- AN EMERGING THREAT

Approximately, 60 million people in our country, till today, lack access to clean drinking water. More than 200 towns in India, excluding the class-I cities, get less than 70 lpcd (liters per capita per day) of clean water. Compare this to 262 lpcd, average water consumption by US citizens. According to an estimate by Union

Ministry of Water Resources water demand of the top 35 cities throughout the country is expected to double to 12,906 million cubic meters in 2021 as the population of these cities shoots up to 202 million from the present level of 107 million.

The fresh water resources distributed unevenly throughout the region like rest of the areas of the country. Even though the country has four per cent of world's fresh water resources, most of it lies along Brahmaputra basin covering Assam and adjoining areas with a per capita availability of 14,000 cubic meters. Sabarmati basin covering Gujarat region has less than 500 cubic meters. While parts of Meghalaya receive more than 11,000 mm of rainfall during the year, arid and semi-arid regions of Rajasthan receive less than ten centimeters. Per capita availability of fresh water resources has also gone down significantly. According to the data released by the Ministry of Water Resources, at 1250 cubic meters of water per person per year, we are, already a water-stressed nation. This has also been noticed in information released *by Gardner-Outlaw & Engel man in 1997.*

Six out of ten bore wells dug in north Gujarat and Saurashtra-Kutch region yield no water even at a depth of 1200 ft. In Loej village of Gujarat, each of the 200 families has one member suffering from kidney problem. In Rajasthan, ten towns get water once in three days, and 31 once in two days. Twelve towns get it from 200 km away. Sixty-six million people in 200 districts risk fluoride contamination and 15 million risk arsenic poisoning. *(India Today, June 9, 2003).* The problems are same in the North Eastern India where most of the water resources are located at Brahmaputra and Barak valley basin and the people living in the areas are facing lot of problems and challenges regularly in managing these resources. Thus in the present paper a critical review of the water resources and its management practices are attempted and based on the analysis few suggestions are forwarded for considerations.

PHYSIOGRAPHY OF NE REGION

Water and land are the two most important and abundant bio-productive resources of North East India. The Northeastern Region of the country consists of eight states–Arunachal Pradesh, Assam, Manipur, Meghalaya, Mizoram, Nagaland, Sikkim, and Tripura. It covers 255158 Sq Km i.e 7.7 % of the geographical area and about

3.7 % (i.e. 39 million as per 2001 census) of countries population. It has 98 % international border with China and Bhutan in north, Myanmar in the east and Bangladesh in south and west. The region has wide variation of physical features. NE hills and basins accounts for 65 % area, Brahmaputra valley 22 % and Meghalaya nearly 13 %.Tribal population is about 60 %, however Assam accounts for only 13 %, these belong to Aryan, Dravidian, Indo-Burma, Indo-Tibetan stock, This apart many small tribes are also inhibiting in the remote hilly areas fortunately for whom the modern civilisation could not make inroad. Status of rainfall is quite comfortable as compared to many parts of the country.

Climatic Features

The climate of the region varies from sub-tropical to extreme alpine type. The normal mean minimum and maximum temperatures vary between 18°C to 32°C in summer and 0°C to 22°C in winter. Heavy fog is a common feature all over the mountain area throughout the year. The temperature in the snow-clad mountains in the north is well below zero degrees. These hill states receive the maximum rainfall in the country and mean annual rainfall varies from 1,400 mm to as high as 6,000 mm in Arunachal Pradesh. Cheerapunji (Meghalaya) is one of the wettest places in the world, which gets annual rainfall of more than 11,000 mm.

Rainfall

Rainfall is quite comfortable as compared to many parts of the country. It is high rainfall intensity zone of the country with annual rainfall of above 2,000 mm (Maximum up to 11000 mm). Annual rainfall occurs almost throughout the year in NE Region. It varies from 200 cm to 400 cm, the higher being in Khasi-Jaintia Hill region, where rainfall exceeds 1100 cm. Hence let us see the distribution pattern of data for the period 1987 to 1993 giving average variation from normal. (Table 1).

Variation in Rainfall

Rainfall varies significantly throughout the year as well as seasonally in the region. The seasonal and annual variation in rainfall is presented in Table 2. Prolonged duration of rainfall occurs during the month of March and April whereas steady rain occurs during full monsoon starting from early May continue upto

Table 1

Rainfall during Monsoon Season in Brahmaputra Basin

Sl. No.	*Metrological Sub Divisions*		*For the Period 1st June to 30th September.*						
			1987	*1988*	*1989*	*1990*	*1991*	*1992*	*1993*
1.	Arunachal Pradesh	A	2460	3644	2258	2507	1681	1116	2436
		N	2144	2527	1849	2052	1823	1824	1812
2.	Assam & Meghalaya	A	1935	2199	2184	1845	1687	1517	2066
		N	1918	1876	1857	1791	1651	1843	1835
3.	Nagaland, Manipur, Mizoram & Tripura	A	881	1309	1148	1294	1166	1017	1381
		N	1287	1119	1231	1254	1236	1239	1262
4.	Sub-Himalayan West Bengal & Sikkim	A	—	2194	1985	2160	2429	1674	1925
		N	—	1095	2050	2062	2058	2059	2058

A-Actual, N-Normal

Source: Seetapathi, Patra & Dutta: Hydrological aspects of Watershed Management in NE region p-35

first half of October in general. Highest Average rainfall (11,406 mm) in the world recorded at Mawsynram village in Meghalaya state.

Table 2
Seasonal and Annual Rainfall (in mm) in NE region

Stations	*Season*				
	Winter	*Pre-Monsoon*	*Monsoon*	*Post-Monsoon*	*Annual*
1. Dhubri	29.7	623.8	1774.4	159.0	2586.9
2. Guwahati	35.6	455.4	1050.9	92.8	1634.1
3. Tezpur	40.9	477.0	1192.1	137.2	1847.2
4. Sibsagar	80.3	668.0	1613.7	177.3	2539.3
5. Dibrugarh	96.3	657.1	1841.3	201.7	1796.4
6. Haflong	65.9	708.2	1269.6	243.1	2277.8
7. Silchar	69.4	946.4	2082.3	249.6	3347.7
8. Lumding	45.0	303.5	819.2	150.3	1318.0
9. Imphal	51.6	341.6	867.4	151.9	1412.4
10. Kohima	45.2	336.3	1375.7	162.8	1920.0
11. Agartala	40.1	556.0	1237.5	190.8	2024.4
12. Passighat	151.4	876.9	3147.8	318.0	4494.2
13. Tura	82.0	701.8	2289.1	270.1	3293.1
14. Cherrapunji	71.6	2223.3	8016.5	558.8	10869.4
15. Shillong	43.2	497.9	1479.6	232.6	2253.3

Source: Seetapathi, Patra & Dutta: Hydrological aspects of Watershed Management in NE region p-35.

WATER RESOURCES-NE PERSPECTIVES

The NE states have two-principle drainage system, Brahmaputra and Barak valley, both combining in Bangladesh to from Meghna Basin. The region has abundant water resources along with a substantial unutilised ground water resource base. The region's water resources potential accounts for 31% of country's total surface water potential. These rivers constitute India's National Waterway II and the basins contain seasonally flooded wetlands that sustain a broad range of biodiversity. There is an estimated 60,000 MW of economically viable hydropower potential (roughly about 33% of country's potential), of which only about 65 MW is developed or under construction. It is also clear that the abundant surface water resource imposes severe distress and costs on the region through frequent flooding and that this needs to be managed to improve economic development.

Water management challenges in North-East

Every mistake made in development process resulted in storm drain overflow, mine drainage, nutrient loading, over irrigation, sewage overflow, excessive withdrawal of ground water or top soil erosion from deforestation have caused a variety of problems including pollution to our water resources. Deepor Beel, 40 sq km, one of the only two in Northeast India, is encroached upon by agencies *viz.* the Government Departments (BSF camp and railways), nursing homes, brick kilns – and is swamped by water hyacinth. Apart from a storm cushion, it is a major migratory center and breeding ground for birds.

Flood and Soil Erosion Problem

The immense surface water resource of NER causes extensive flood damage and drainage problems in Assam as the rivers descend from hilly regions into the plains. Annual Flood problem is a grim reality of the region, particularly in the Assam Valley of the Brahmaputra River basin, the Cachar Valley in the Barak River basin, and in the lower reaches of the Teesta River in South Sikkim and the adjoining North Bengal region. The average annual flood damage is of the order of Rs.1720 million per year and this is despite an extensive network of embankments, river training works, drainage channels, etc., on all the rivers of the region.

There is no basin or even sub-basin wise planning for composite water use for providing drinking water, irrigation, generation of hydel power & industrial and with regard to ecology & environment. Devastating flood in the Northeast has become a regular phenomenon. In some years there's a worse flood than in others. Millions, as we know, are displaced. Last year nearly 350 embankments collapsed. In Assam, all 27 districts were hit by the floods including the hill districts, where landslides and rushing water snapped communications and disrupted life. More than 12 million people were displaced or otherwise affected. The efficacy of embankments in NE has to be re-looked and there is a clear felt need to find alternatives. The flood damage was estimated at some Rs. 6,500 Crore.

Besides floods, it is erosion that is gnawing away at the resources of people, their lands, their homes, and their hopes. India has some 15,675 km of embankments. Of this, 5,027 km are in

Assam alone, that is 32 per cent of the country's total, and the government still talks of more embankments.

Institutional Challenges

There are many challenges before the public institutions that operate in the North East These are:

- An unusually fragmented and often uncoordinated administrative structure–by issue and by location–in which there is often unwillingness to accept responsibility and a lack of awareness of actions taking place elsewhere.
- The relative weakness of the private sector in a region where the public sector has laid the dominant role in development. Coupled with the dependency syndrome, this means that the government acts with little involvement of either the private sector or communities.
- Several central, undressed problems arising from the mismatch between: a single water resource system in the Brahmaputra and Barak basin on the one hand, and the (relatively recent) division of the Northeast into eight states, several small, on the other hand; and a large asymmetry in the assignment of benefits and costs between upper and lower riparian.

For example, Arunachal Pradesh's concerns are development of hydropower and avoidance of resettlement, whereas Assam's concerns are primarily flood management. Arunachal's preference is for single-purpose run-of-the-river projects, whereas Assam would benefit most from largely multipurpose reservoirs. There is currently little systematic exploration of the tradeoffs and mechanisms for a development program that would take full account of the multiple benefits and costs.

DECENTRALIZED MANAGEMENT APPROACHES

In India's Northeast, notably, there is a range of states where tribal communities have traditionally been very strong in managing their forests. This coincides with the general call in the natural resources literature for providing management responsibilities to those actors "at the lowest appropriate level", whereby resources are managed by those local actors who can do so most effectively, rather than by distant bureaucrats. This is particularly the case for river basin management approaches, where decentralization of decision-making is strongly advocated in order to provide incentives to those stakeholders who are the closest to the resource to manage it better. In the case of the Northeast, this would certainly lead to an examination of the history of managing the Brahmaputra river basin through a central agency, such as the Brahmaputra Board, compared with an organization that would be more grounded in the northeastern states and would operate with their full and active participation.

A further central concept in water and natural resources management is benefit sharing. Infrastructure development – particularly in the water sector – creates both winners and losers. The benefits of dams and other hydrological infrastructure typically accrue downstream whereas the costs of inundation and construction activities are borne upstream. Attention to be paid to the distribution losses too. Appropriate water and forest development and management could provide benefits in the form of hydropower, agriculture, inland water transport, biodiversity conservation, reduced flood damage and erosion, longer dam-reservoir life, forestry and ecotourism.

GOVERNMENT INITIATIVES IN DECENTRALISATION

Since eighth five year plan period, Government of India has taken number of initiatives to manage the country's water resources and to provide safe drinking water to all the inhabitants including North East; a few of them are summarized in nutshell here below.

Safe Drinking Water Programme

Safe drinking water means, it is free from pathogenic microorganism, poisonous substances, excessive amount of minerals and organic materials, which is unfit for human use for drinking purpose. It is reported by WHO that 15 million children die every year because of water borne diseases. This is why Govt. of India has laid prime emphasis on the availability of water in rural areas. For sustainable and healthy village development through the Mandate of Safe Drinking Water by 2004-01-12, two major interventions Programme was launched during tenth plan period *viz.* 1. Accelerated Rural Water Supply (ARWSP) and 2. PM's Gramodaya Yojana- Rural Drinking Water (PMGY-RDW). The same was transferred to Planning Commission of GOI w.e.f. 01.04.2002. Under Rural Water Supply programme of GOI the

Accelerated Rural Water Supply Programme (ARWSP) was launched with missionary zeal:

- To ensure coverage of all rural habitation with access to safe drinking water; by 2004.
- To ensure source sustainability of drinking water system.
- To tackle the problem of water quality in affected habitation.
- To institutionalise the reform initiative in rural drinking water supply sector.
- To infuse cooperative attitude of the people.

Under this programme the major emphasis was laid on Rain Water Harvesting; Sustainability of sources; and Community Participation in water resources management.

Being state subject, states have been taking initiatives to negotiate the problem from their own resources. However Central Government come in a big way to support the states' programme. Unique funding patterns with 50% -90% share of Central Government were decided through a number of different schemes for the states.

Coverage in North East States

Under this scheme coverage habitation were taken into account rather then house hold concept. performance of the projects was examined with three indicators i.e. fully covered, partially covered and not covered. Central Government achieved considerable success in meeting rural drinking water needs with a composite investment of more than Rs. 40000 Crore and at the end of March 2003 nearly 91 % rural habitations were covered and rest by now in the verge completion. The details has been given in Table 3.

Sectoral Reform Approach

In this approach 67 districts in 27 States in the country has been identified and also interventions made through these programmes. for total project cost of Rs. 2060.45 out of which central share is 1922.85 Crore Under sector reform project 11 District of NE region.The details has been placed at Table 4. The hardware support is supplemented by other support service such as: Capacity Building, Social Mobilisation, Information & Education of Communication (IEC), Human Resource Development (HRD) and Management Information Service (MIS). In this approach govt. plays the role of facilitator. Awareness through IEC is created for

Table 3
Coverage of Habitation Under Rural Water Supply in NE Region

(Up to January, 2003)

States	Number of Habitats				% of Habitats			
	Fully Covered	*Partially Covered*	*Not Covered*	*Total*	*Fully Covered*	*Partially Covered*	*Not Covered*	*Total*
1. Arunachal Pradesh	3093	880	325	4298	80.15	19.85	7.56	100
2. Assam	52643	17585	441	70669	74.49	24.88	0.62	100
3. Manipur	2604	183	4	2791	93.30	6.56	0.14	100
4. Meghalaya	7524	788	327	8639	87.09	9.12	3.79	100
5. Mizoram	457	454	0	911	50.06	49.84	0.0	100
6. Nagaland	607	569	349	1525	39.80	37.31	22.89	100
7. Sikkim	1389	290	0	1679	82.73	17.27	0.00	100
8. Tripura	7299	20	93	7412	98.48	0.27	1.25	100
NE Total	75616	20769	1539	89924	606.10	165.10	36.25	100
All India	1295504	112804	14356	1422664	91.06	7.93	1.01	100

Source: Annual Report, MoRD 2002-03. P. 201

effective participation so that the community should be willing to be involved in the implementation process of schemes. This would also help in creating feeling of ownership of the assets created.

Table 4
Access To Safe Drinking Water To Households In 1991

	State	Rural	Urban	Total
1.	AP	66.87	88.20	70.02
2.	Assam	43.28	64.07	45.86
3.	Manipur	33.72	52.10	38.72
4.	Meghalaya	26.82	75.42	36.16
5.	Mizoram	12.89	19.88	16.21
6.	Nagaland	55.60	45.47	53.37
7.	Sikkim	70.98	92.95	73.19
8.	Tripura	30.60	71.12	37.18
	NE States	42.60	63.65	46.33
	All India	55.54	81.38	62.30

Source: 10th Five Year Plan, 2002-07, Volume-III, State Plans Trends, concerns and Strategies, Planning Commission, Govt. of India, New Delhi, P. 81.

Swajaldhara

The key elements of the programme are: Demand driven & community participation approach; Panchayats/communities to plan, implement, maintain & manage all schemes; Cost sharing; Full ownership of project assets with GP and Full operation and maintenance by users/panchayats. The funding pattern of the scheme is in the proportion of 10% by community, 5% for SC/ST community and 90% by the Government of India. Under this programme 3791 projects were sanctioned to 123 districts in 16 States with an outlay of Rs. 24143.4 lakh. Only two district (Cachar and Karimganj) of Assam State got 38 projects for an amount of Rs. 465.12 Lakh. The study was basically done on the basis of household survey. The safe concept and application today thrives on the water security in the village/habitat. The study was also unique in its covered of urban as well as rural household.

CONCLUSION AND RECOMMENDATION

From the statistics available, one would presume that water is abundant in the region. Being nearly 2/3 part hilly area region suffers from water distribution pattern, low quantity in hilly areas and perennial flood in the valley and plain areas but in reality the problem is quality. Claims that the almost hundred percent achievement needs an in-depth review. The problem therefore comes who would bell the cat? The states, the implementing agencies or by some independent evaluators. This may again lead to a centre-state conflict. Should we go for it? I feel that same should be avoided, which can be effectively done by making stake holders stronger to achieve sustainability and therefore recommend few interventions. Some of these are:

- Making an inventory of the NE states' fresh water resources.
- Identifying potential hot-spots and grey areas,
- Assessing key issues on quality, and allocation to meet human needs and not greed.
- Hilly areas face acute shortage of water both in quantity and quality which need special focus.
- Developing innovative economic, legal and institutional instruments for sustainable fresh water use through awareness, education and participatory approach.
- Implementing pilot projects in the key regions under the local users group to meet the need.

Therefore, to achieve these tages one has to finally understand user groups of the region with diverse culture, dynamics, and unique symbiotic life style with nature. The people are easily accessible once they are convinced with the ethos of well being and this can be achieved through the following measures:

- Effective co-ordination between various User groups
- Bringing stakeholders together
- Involvement of stakeholders for management of distribution system
- Education and training to strengthen local organizations especially for PRIs
- Users to pay and polluters to pay principle

Therefore we should demonstrate to relate our dependency on the environment and some of its parts, in a manner so that the rivers, rivulets and streams flow across mountain villages, urban settlements to see every one smiling.

REFERENCES

Planning Commission Report Vol.I, II, III, 2004-05.

Workshop Proceeding on Water Quality Assurance held at Itanagar (18-19 Apr 98).

Upadhyay.N: Participatory Water Shed Management: A policy by Choice Presented in the Seminar on Hydrological Aspects of Watershed Management in North Eastern Region Pp. 120-128.

Patwari. BC: Water Resource Problem of NE Region: A Vision Element 2050, Presented in Brain Storming Session on Water Resource Problems of NE Region, May 20. 2000 Guwahati.

Patwari. BC: Hydrological Aspects of NE Region: Future needs and Prospects P.14-15.

Thaplyal B.K: Evaluation Report of MWS in Assam 1998 : NIRD Study.

Venkatdri Evaluation Report of MWS in Nagaland.

Key Note Address: Palat Mohandas, Secretary, Dept of Grinking Water: Address Delivered at the IIPA New Delhi on 4th Aug 2003.

Statistical Handbook of Assam-1996.

NEC Basic Statistics-2000.

NEC Basic Statistics-2002.

Water for Health- taking Charge, WHO, 2001.

Rural Drinking Water Supply Programme - RGNDWM, Department of Drinking Water Supply, MRD, Government of India, 2000.

Guideline for Implementation of RWS Programme–RGNDWM, Department of Drinking Water Supply, MRD, Government of India, 2000.

Surveillance of Drinking Water Quality in Cities and Towns–A. Majumder and D. Kohali, National Workshop on Drinking Water Quality Surveillance, Advanced Technology in Water and Waste Water Management and Public Private Partnership Initiatives, Bangalore, November 6-9, 2001.

10

SUSTAINABLE APPROACHES TO WATER RESOURCES MANAGEMENT

*Ra. Murugesan**

INTRODUCTION

Water resources are an essential component of the earth's hydrosphere and an indispensable part of all terrestrial ecosystems. The water environment is characterised by the hydrological cycle, including floods and droughts, which in some regions have become more extreme and dramatic in their consequences. Global climatic change and atmospheric pollution could also have an impact on fresh water resources and their availability through sea level rise.

Water is essential for survival and needed for all aspects of life. The general objective is to make certain that adequate supplies of water of good quality are maintained for the entire population of this planet, while preserving the hydrological, biological and chemical functions of ecosystems, adapting human activities within the capacity limits of nature and combating vectors of water related diseases. Innovative technologies, including improvement

* Assist. Prof., NIRD, Hyderabad.

of indigenous technologies, and indigenous knowledge systems are needed to fully utilize limited water resources and to safeguard those resources against pollution.

The widespread scarcity, gradual destruction and aggravated pollution of fresh water resources in many world regions, along with the progressive encroachment of incompatible activities, demand sustainable water resource planning and management. Such integration must cover all types of interrelated freshwater bodies, including both surface water and ground water, and duly consider water quantity and quality aspects. The multi-sectoral nature of water resource development in the context of socio-economic development must be recognized, as well as the multi-interest utilization of water resources for water supply and sanitation, agriculture, industry, urban development, hydro-power generation, inland fisheries, transportation, recreation, low and flat lands management and other activities. Rational utilization schemes for the development of surface and under ground water supply sources and other potential sources have to be supported by concurrent water conservation and water waste minimization measures. However, priority must be accorded to flood prevention and control measures, as well as sedimentation control, where required.

With this background the authors have made an attempt to examine the need for (i) integrated water resources development and management; (ii) water resources assessment; (iii) protection of water resources, water quality and aquatic ecosystems; (iv) drinking water supply and sanitation; (v) water and sustainable urban development; (vi) water for sustainable food production and rural development and (vii) impact of climate change on water resources. Further, this paper has also focused the status of water and water quality management strategies in India.

WATER: AREAS OF CONCERN

Water is not a new development priority; the search for clean water and improved sanitation is as old as human civilization. What is new is that the issue has suddenly, after years of relative neglect, been thrust to the top of the global agenda. World Summit on Sustainable Development in Johannesburg, has endorsed that water is one of five WEHAB priorities–the others being Energy, Health Agriculture and Bio-diversity and recommitted to the

Millennium Development Goal target of halving the proportion people without access to safe drinking water by 2015. The reasons for all the attention towards water are:

- Of the more than six billion people in the world, at least 1.2 billion do not have access to safe drinking water;
- More than 2.4 billion people do not have proper sanitation facilities;
- Dirty water is deadly. More than 2.2 million people, primarily in developing countries, die each year from diseases caused by polluted water and filthy sanitary conditions;
- Every day, diarrheal diseases cause some 6000 deaths, mostly among children under five;
- More than 200 million tons of human waste are dumped daily into the world's rivers;
- While water covers 70.0 percent of the earth's surface, 97.5 percent is salt water; although the remaining water is fresh water, three quarters of that is frozen in ice caps;
- By 2025, nearly two-thirds of the world population will live in water stressed region;
- In the next two decades, water use by human will increase by 40.0 per cent and 17.0 per cent more water will be needed to feed increasing populations in developing countries;
- Over the last 30 years, the share of rural families with access to safe water rose from 10.0 per cent to about 60.0 percent; and
- Presently, 70.0 per cent of all available fresh water is used for agriculture, but because of inefficient irrigation systems, 60.0 per cent of this water is lost.

These alarming situations, underscores the urgent need to develop an integrated approach to the development and management of water resources. In addition, maintaining a balance between the availability and requirements, across basins region and between sections of the population adds another dimension challenge.

INTEGRATED WATER RESOURCE DEVELOPMENT AND MANAGEMENT

Integrated water resource management is based on the perception of water as an integral part of the eco-system, a natural resource and a social and economic good, whose quantity and quality determine the nature of its utilization. To this end, water resources have to be protected, in order to satisfy and reconcile needs for water in human activities. In developing and using waters resources, priority has to be given to the satisfaction of basic needs and safeguarding of ecosystem. Beyond these requirements water users should be charged appropriately. Further integrated water resources management should be carried out at the end level of catchment basin or sub-basin, for which four principle objectives should be pursued and they are:

- To promote a dynamic, interactive, iterative and multi-sectoral approach to water resource management that integrates technological, socio-economic, environmental and human health considerations;
- To plan for the sustainable and rational utilization, protection, conservation and management of water resources based on community needs and priorities within the frame work of national economic development policy;
- To design, implement and evaluate projects and programmes that are both economically efficient and socially appropriate within clearly defined strategies based on an approach of full public participation, including women, youth, indigenous people, local communities, in water management policy making and decision making;
- To identify and strengthen the appropriate institutional, legal and financial mechanisms to ensure that water policy and it implementation are a catalyst for sustainable social progress and economic group.

WATER RESOURCE ASSESSMENT

Water resources assessment comprises the continuing determination of sources, extent, dependability and quality of water resources and of the human activities that affect those resources. Such assessment constitutes the practical basis for their sustainable management and pre-requisite for evaluation of the possibilities for their development. In this context, reliable information is need about water resources, especially information on surface water and ground water and water quality. Establishment of national data bases is vital to water resource assessment and to mitigation of the effect of floods, droughts, desertification and pollution. Five specific objectives have been set for water resource assessment and they are:

- To make available to all water resources assessment technology, including methods for the impact assessment of climate change on freshwater;
- To allocate to water resource assessment financial resources in line with the economic and social needs for water resource data;
- To ensure that the assessment information is fully utilised in the development water management policies;
- To establish the institutional arrangement needed to ensure the efficient collection, processing, storage, retrieval and dissemination users of information about the quality and quality of available water resources; and
- To have sufficient numbers of qualified staff recruited and retained by the water resource assessment agencies and provided with the training and retraining they need to carryout their responsibilities successfully.

Protection of Water Resources, Water Quality and Aquatic Ecosystem

Fresh water is a unitary resource. Long-term development of fresh water requires holistic management of resources and a recognition of the interconnectedness of the elements related to freshwater and freshwater quality. Major problems affecting the water quality of river and lake arise, from inadequately treated domestic sewage, inadequate controls on the discharge of industrial waste waters, loss and destruction of catchments area, ill-considered sitting of industrial plants, deforestation, uncontrolled shifting cultivation and poor agricultural practices. This gives rise to the leaching of nutrients and pesticides. Aquatic ecosystems are disturbed and living freshwater resources are threatened. Thus, a preventive approach is crucial to protect water resources water quality and aquatic ecosystems. Under these circumstances, three objectives will have to he pursued concurrently to integrate water-quality elements into water resources management.

- Maintenance of ecosystem's integrity, according to the management principles of preserving aquatic ecosystems, including living resources and of effectively protecting them from any form of degradation on a drainage basin basis; and
- Human resources development, a key to capacity building and a pre-requisite for implementing water quality management.

Based on the experiences of many countries of the world, all governments could set the following objectives in order to protect water resources:

- To identify the surface and ground water resources that could be developed for use on a sustainable basis and other major developable water-dependent resources and simultaneously to initiate programmes for the protection conservation and rational use of these resources on a sustainable basis;
- To identify all potential sources of water supply and prepared outlines for their protection, conservation and rational use;
- The initiate effective water pollution prevention and control programmes based on an appropriate mixture of pollution reduction-at–source strategies, environmental impact assessments and enforceable standards for major point-source discharges and high risk non point sources, commensurate with their socio-economic development;
- To participate, as far as appropriate, water quality monitoring and management programmes;
- To reduce the prevalence of water associated diseases;
- To establish, according to capacities and needs, biological, health physical and chemical quality criteria for all water bodies (surface and ground waters) with a view to an on going improvement of water quality;
- To adopt an integrated approach to environmentally sustainable management of water resources, including the protection of aquatic ecosystems and fresh water living resources;
- To put in place strategies for environmentally sound management of freshwaters and related coastal ecosystems, including conservation of fisheries, aquaculture, animal grazing, agricultural activities and bio diversity.

The following activities should he undertaken and implemented by the governments, according to their capacity and available resources for protection of water resources:

- Water resources protection and conservation;
- Water pollution prevention and control;
- Development and application of clean technology;
- Ground water protection;

- Protection of freshwater living resources;
- Monitoring and surveillance of water resources and water receiving wastes; and
- Development of national and international legal instruments that may be required to required to protect as quality of water resources.

DRINKING WATER SUPPLY AND SANITATION

Safe water supplies and environmental sanitation are vital for protecting the environment, improving health and alleviating poverty. Safe water is also crucial to many traditional and cultural activities. An estimated 80.0 per cent of all diseases and over one third of deaths in developing countries are caused by the consumption of contaminated water and on average as much as one tenth of each person's productive time is sacrificed to water related diseases. In this contest, global institutions aimed to have "all people's, whatever their stage of development and their social and economic conditions, have the right to have access to drinking water in quantities and of a quality equal to their basic needs". One in three people in the developing world still lacks these two most basic requirements for health and dignity. It is also recognised that human excreta and sewage are important causes of the deterioration of water quality in developing counties and the introduction of available technologies, including appropriate technologies, and the construction of sewage treatment facilities could bring significant improvement, Hence, there is a need to provide safe drinking water in sufficient quantities and proper sanitation to all for which the essential guiding principles are;

- Protection of the environment and safeguarding of health through the integrated management of water resources and liquid and solid wastes;
- Institutional reforms promoting on integrated approach and full participation of women at all levels;
- Community management of services, backed by measures to strengthen local institutions in implementing and sustaining water and sanitation programmes; and
- Sound financial practices and widespread use of appropriate technologies.

WATER AND SUSTAINABLE URBAN DEVELOPMENT

Rapid urban population growth and industrialization are causing stress on the water resources and environmental protection capabilities of many cities. Special attention needs to be given to the growing effects of urbanization on water demands and usage and to the critical role played by local and municipal authorities in managing the supply, use and overall treatment of water, particularly in developing countries. Scarcity of fresh water resources and the escalating costs of developing new resources have a considerable impact on industrial, agricultural and human settlement development and economic growth. Better management of urban water resources, including the elimination of unsustainable consumption patterns, can make a substantial contribution to the alleviation of poverty and improvement of the health and quality of life of the urban and rural poor. It is well known fact that a high proportion of large urban agglomerations are located around estuaries and in coastal zones. This leads to pollution from municipal and industrial discharges, combined with over exploitation of available water resources and threatens the marine environment and the supply of fresh water resources. The strategy for water and sustainable urban development can be achieved with the following objectives:

- To support local and governmental efforts and capacities to sustain national development and productivity through environmentally sound management of water resources for urban use;
- To identify and implement sustainable strategies and actions so as to ensure the continued supply of affordable water for present and future needs;
- To reverse current trends of resources degradation and depletion;

Water for Sustainable Food Production and Rural Development

Sustainability of food production depends on efficient water use and conservation practices consisting primarily of irrigation development and management, including water management in rain-fed areas, livestock water-supply, inland fisheries and agro-forestry. Food security is a high priority in many countries and agriculture must not only provide food for rising population, but also save water for one uses. The challenge is to develop and apply water saving technology and management methods, through capacity building institutions and incentives for the rural population to adopt new approaches for both rain fed and irrigated agriculture. Water scarcity for food and livestock production are significant limiting factors that affect both human and animaldom. It is estimated that the current global livestock drinking water requirement is about 60 billion liters per day and based on the livestock population growth estimates, this daily requirement is predicted to increase by 0.4 billion liters per annum in the foreseeable future. Fresh water fisheries in lakes and streams are an important source of food and protein. Fisheries of inland waters should be managed so as to maximise the yield of aquatic food organism in an environmentally sound manner. This requires the conservation of water quality and quantity, as well functional morphology of the aquatic environment.

The key strategic principles for integrated sound management of water for sustainable food production and rural development are:

- Water should be regarded as a finite resource having an economic values with significant social and economic implications reflecting the importance of meeting basic needs;
- Local communities must participate in all phases of water management, ensuring the full involvement of women in view of their crucial role in the practical day to day supply, management and use of water;
- Water resource management must be developed within a set of policies for (i) human health; (ii) food production, preservation and distribution; (iii) disaster mitigation plans; (iv) environmental protection and conservation of the natural base;
- It is necessary to recognise and actively support the role of rural populations, with particular emphasis on women.

IMPACT OF CLIMATE CHANGE ON WATER RESOURCES

Higher temperatures and decreased precipitation would lead to decreased water supplies and increased water demands; they might cause deterioration in the quality of freshwater bodies, putting strains on the already fragile balance between supply and demand in many countries. Even where precipitation might increase, there is no guarantee that it would occur at the time of year when it could be used; in addition, there might be a likelihood of increased floods. Any rise in sea level will often cause the intrusion of salt water into estuaries, small islands and coastal aquifers and the flooding of low-lying coastal areas; this puts low-lying countries at great risk. This calls for more information about and greater understanding of the impacts of climatic change on water resources for which the need of the hour today is;

- To understand and quantify the threats of the impact of climate change on fresh water resources;
- To facilitate the implementation of effective counter measures as and when the threatening impact is seen; and
- To study the potential impacts of impacts change on areas prone to drought, and floods.

INDIA'S INITIATIVES IN THE WATER SECTOR

As per the Schedule VII of the Constitution of India, water is state and concurrent subject empowering the states to enact laws, frame policies or fix prices of related services. Planning and implementation of water development project, is handled both at the center and state levels there are numerous institutions dealing with water resources management both at the central and state level. India receives nearly 4000 cubic kilometers of water annually through the seasonal monsoon rains and rivers that flow down from the snow capped Himalaya and other mountain ranges. The distribution and availability of water is not uniform across the country or through the year. Surface sources often run dry in the summer and groundwater availability varies from the rich aquifers of the Indo-Gangetic – Brahmaputra plains to on comparatively low yielding hard rock regions of peninsular India.

The rapid increase in the country's population from 343 million at the time of Independence to over 1000 million in 2000, accompanied by growth of agriculture, rapid urbanization, economic growth and improved access to basic services has resulted in an increase in the demand for water. Most of the country's water is used to irrigate agricultural land (84.0 per cent). A small proportion of it is consumed in people's home (40.0 percent), while industry and other users claim about 12.0 per cent of water. The country's total water requirement is projected to grow to 1180 billion cubic meters by the year 2050 as against 629 billion cubic meters in 1997-98. The average consumption per person is only about 680 cubic meters per year for all uses, but it is projected to increase. Over exploitation of ground water is very serious problem and the water table has steadily been falling in many parts of the country. There are more than 4.79 million electric and 3.7 million diesel pumps withdrawing ground water across the country. The widening gap between demand and supply has led to a substantial increase in the share of ground water consumption by the urban, agricultural and domestic sectors. Ground water is a shared common property resources and there is a need to evolve appropriate institutional mechanism for its management. Moreover, the quality of water resource is threatened because of inadequate provision for the treatment of waste water.

Even after enactment of Water (Prevention and Control of Pollution) Act as early as in 1974, water quality continues to deteriorate in the country. Rivers, streams and ground water are severely polluted. When quality deteriorates water loses its economic value. Pollution also creates water security in region which otherwise have abundant water resources.

WATER QUALITY MANAGEMENT STRATEGY

Clean water is not a luxury. It is a necessity. With sensible policies, water bodies can be cleaned, saving crores of rupees in health costs and protecting our land and water resources for future generation. A comprehensive multi pronged water quality management strategy includes strict implementation of pollution control laws, promotion of Clean Technology Mechanism, fiscal incentives and economic instruments of appropriate prices, taxes and property rights. Inter-ministerial co-ordination is also essential for effective implementation. An aware, active and vigilant citizenry is critical for success. In general, the aim of water quality management strategy is to ensure water security for preserve and future generation, present water quality and assure that water is available equality to all members in society.

CONCLUSION

Water is still treated as a free good, to be used liberally and to be sullied with impunity. Pricing water to reflect its scarcity value can encourage users to be more prudent in its use. The price of water should also include opportunity costs and environmental impacts of its use. Such kind of price reforms will encourage resource conservation on the one hand and provide additional support for the fund starved municipal service providers or the other. Better pipeline management can curtail water losses. Domestic wastewater can be reused and recycled for agricultural and industrial purposes. Subsidies on rune electricity and agricultural water supply that encourage wasteful use of water must be reassessed.

The polluter is rarely held accountable for the damage, and continues to pollute with impunity. It is essential that "Polluters Pays Principles" is made operational. Large investment in treatment facilities are essential and these must he encouraged through economic incentives. Environment friendly practices like vermiculture, use of organic manures and integrated pest management practices should be vigorously encouraged so as to curtail over use of fertilizer and pesticides Above all, what we need today is community participation in management of maintenance of water projects at the local, regional, national and international levels. Vigilant stakeholders with strong and technically equipped institutional support can play a very important role in managing the water resources.

REFERENCES

Andrew S., Goudie and David J., Cuff (Eds): Encyclopedia of Global Change-Environmental Change and Human Society, Vo.2 (Oxford University Press, New York, 2002).

Government of India: Agenda 21-An Assessment (Ministry of Environment and Forests, New Delhi, 2002).

Government of India: Empowering People for Sustainable (Ministry of Environment and Forests, New Delhi, 2002).

Government of India: National Human Development Report 2001 (Planning Commission, New Delhi, March 2002).

Jauhari V.P.: Sustainable Development of Water Resources (Mittal Publications, New Delhi, 2002).

Ramasamy R., Iyer: Water–Perspectives, Issues, Concerns (Sage, New Delhi, 2003).

United Nations Development Programme: " Clean Water- An Agent of Change" Choices Vol. 12, No. 1, March 2003.

11

CONFLICT RESOLUTION AND INSTITUTIONAL ARRANGEMENTS FOR TRANS BOARDER RIVERS: A CASE OF KOSI BASIN

*N.M.P. Verma**

INTRODUCTION

Trans boundary conflict is one of the chronic riparian problems in Indo-Nepal region of South Asia. The construction of a barrage and canal system for irrigation in Nepal and India was initiated in 1953. Also afflux bund was executed in upstream of barrage. Indian Government has also initiated eastern and western canal system along with embankments and powerhouse. The construction of embankment reduced loss of livelihood, enhanced employment and earnings. This also gave sense of eco-security in the region. Major irrigation has been provided in both India and Nepal. In addition, flood prone area has also been protected. Despite this, there has drainage congestion, water logging behind the embankment, severe flood problem in unprotected area, rising riverbed and recurring maintenance problem in the region. These

Professor & Head, Department of Economics, Babasaheb Bhimrao Ambedkar University, Lucknow-226025, India, *E-mail:* nmpverma@rediffmail.com

are causes of concern for both the Government of India and Nepal. There is huge resource crisis for both the countries because of which promotional and protective measures are not taken to the desired extent. Even large reservoir could not be constructed although it was planned by the Indian Government a long back. There is lack of coordination between the two countries. This has constrained agricultural development in the basin. It has also enhanced economic insecurity in the region of upstream and down stream. The vulnerability of the people has also gone up. There are several contentious issues related to the above problems which need to be resolved. It also includes land dispute, flood planning, water discharge and water management. The aim of the present paper is to examine the development in Kosi basin in North India. The paper aims at is to highlight riparian conflicts in Kosi basin of Indo-Nepal region.

Since many conflicts become chronic and reduces the growth and diversification of agrarian economy, the paper focuses on characteristics of the basin, intensity of conflict, conflict minimization process, areas of joint venture and concludes that in such circumstances of inter-country conflicts institutional reform for minimum common governance (MCG) may be a sustainable solution. The paper is based on secondary data and other related information of trans-boundary problem. The paper, on the basis of secondary information, also proposes the modalities of MCG and its modus operandi. In the end, the paper concludes with common minimum governance in the Kosi basin of Indo-Nepal region which is a viable option for sustainable development of South Asia.

The need of joint minimum governance in the basin is being felt. In this framework both the countries will chalk out institutional reforms in several sectors and policies that will be of the nature of joint management. It may also include plan appraisal, ex post evaluation, monitoring and resource sharing. It realised that planning by a single country may not solved the disastrous chronic problem in this region of two South Asia countries.

If such policy is executed then it may bring higher agricultural growth and other related changes. The livelihood losses of the people of concerned countries can be reduced forever which will ultimately lead to sustained eco-security, food security, employment and income guarantee The river Kosi originates form the Himalayas. The total length of the river is about 74030 sq. km. The upper catchment of the river system falls in the region of Tibet and Nepal. This is 85% of total length. The remaining 15% lies in India where it mainly passes through north of Bihar state. Because of severe floods through sudden overflow during monsoon, lateral shifting tendency up to 20 km a year and having uncertain flood plain, the river is termed as ' the river of great sorrow' in South Asia. It brings large loss of livelihood in terms of loss of crops, life, property employment and earnings in the region. It carries enormous silt loaded discharge during flood and spills into the plains of Indo-Nepal border. About 1295 sq. km. of land in Nepal and 7770 sq.km. of land in India have become unused because of deposition of sand. It may highlight the nature and extent of common trans-boundary problems for India and Nepal.

Trans-boundary water conflicts settlement in river basin have several implications for agricultural development. Sometimes it leads to sudden underflow of water and some times it leads to overflow of water in the catchments of either of the two or more countries. As a consequence it leads to shortage of water for agriculture, industry and household sectors. Often it leads to overflow of water in the rivers during monsoon. Both ways various sectoral production and services are seriously affected which causes immense damage to crops, property and lives. If these situations become chronic and persist for a long time then the agricultural prospects of the region is severely affected in the concerned region. This may also affect increase in gross cropped area, cropping pattern, increase in production and productivity and it may decrease production There are many instances of conflicts of inter country nature in and around India.

The Indo-Bangladesh dispute over the Ganga water could emerge because of Farakka barrage project in West Bengal state of India. Similarly, projects on the Kosi and Gandak generated riparian conflicts between India and Nepal that was further accentuated by Tanakpur Barrage Project. Later on it was resolved slightly through Mahanadi Treaty, 1996. Some issues are still awaiting to be resolved relating to Pancheswar Project. The causes of conflicts may be many like misunderstanding among the parties based on principles, sensitivity ignorance, indifference to share information relating to water disasters and politicization for electoral benefits of the governments.

Since many conflicts become chronic and reduce the growth of the economy, therefore, the paper focuses on characteristics of the basin, intensity of conflict, conflict minimization process, areas of joint venture and concludes that in such circumstances of inter-country conflicts minimum common governance (MCG) may be a sustainable solution. The paper also proposes the modalities of MCG and its modus operandi. If it is properly executed then livelihood losses of the people of concerned eastern Indian states can be reduced forever which will ultimately lead not only to agricultural diversification but also to sustained eco-security, food security, employment and income guarantee. Ultimately there can be a better eco quality and reduction in socio economic miseries. This issue is very relevant. The South Asia Technical Authority Committee (SASTAC) of the Global water partnership in its document: Water for the 21st Century: Vision to action in South Asia emphasized that "poverty will be eradicated and living conditions of all the people will be uplifted to sustainable levels of health and well being, inter alia, through coordinated and integrated development and management of water resolution in the region." With this background the issue of discussing trans-boundary conflicts is very important especially in the context of flood management. It may be noticed that South Asia contains about 20% of world population that may go up to 25% in 2025. It is emerging as the poorest, the most illiterate, the most malnourished or undernourished and lagging behind in agricultural diversification and development. According to World Bank estimates over 500 million population is surviving below the poverty line that is 40% of the world is poor. Its share in global income is just 1.3 per cent.

CHARACTERISTICS OF KOSI BASIN

River Kosi is known as "river of immense sorrow" in India. It originates at an altitude of over 7000 Mts. above MSL in the Himalayas. The upper catchments is 62620 Sq. Km. (85% of total length) which lies in Tibet and Nepal. Remaining 11410 Sq. Km. falls in India and mainly passes through North of Bihar state. It originates in the Tibet, flows through India and Nepal. Flood is the chronic disaster in the Kosi zone of North Bihar, India and Nepal. As a result, there is loss of life, damage to agricultural crops and property, employment and earnings and enormous miseries to people in the region. It also carries enormous silt load discharge during monsoon and starts spilling into the plains of Indo-Nepal border. The meandering flow of the Kosi has rendered about 1295 Sq. Km of land useless in Nepal and 7770 Sq. Km. in India because of deposition of sand. The river is especially known for lateral migration and has shifted to West. The river has shifted about 20 Km. in a single year. As a result the river has ravaged lands to the tune of around 3000 to 15000 Sq. Km in North Bihar and 800 to 1000 Sq. Km in Nepal. It has also generated huge scattered swamps. There are at least three factors responsible for the Kosi led flood in Indo-Nepal region. The river hardly passes a well-defined flood plain. There is huge discharge. There is also excessive fluctuation in daily discharge during flood season. The range is 5 thousand cusec to 26 thousand cusecs. The riverbed has silted up considerably over the years. At several points ground level is lower than river bank. The problem becomes severe when the Ganga and other rivers also start overflowing. In 1950, it was decided to construct a barrage and canal system to irrigate1.65 million hectares in Nepal and Bihar state in India. However, after 1953 a barrage across Kosi in Bhimnagar could be initiated. Also afflux bounds were executed in upstream of barrage in Nepal. India has also initiated eastern and western canal system along with enlargement and powerhouse. Then construction of 468 km embankment reduced losses of livelihood and enhanced rural income and worked as a surge of security at India-Nepal border. The assured irrigation has been provided to 1.30 M ha in both India and Nepal and further 0.35 M ha is in process. Also 1.015 M ha flood prone area has been protected.

At the same time there has been drainage and water logging behind the embankment. Even the flood problem has enhanced in unprotected area. Embankment also needs recurring maintenance as riverbed has gone up.

In India, about 7% of disasters are because of floods. The trend of average area affected by flood hardly shows significant fluctuation during 1959-2001. However there is an increase in population affected and damages of agricultural crops. The mean area affected has been around 7.56 million hectare during the last forty years. Flood prone area in India is 40 million hectares. This constitutes about 25% of cultivable land. This magnitude is much higher in frequently flood prone states like North Bihar, Assam,

West Bengal and Coastal Orissa and Andhra Pradesh Flood water conservation is quite essential for irrigation and other uses. Flood water can be very useful during irrigation of rabi crop. Normally flood comes during rainy season that is followed by rabi season. If flood water is conserved in medium and big dams, water sheds, channels etc it can be used in various ways including aquaculture. But in order to store and consume flood water public private partnership is quite essential.

In several Asian countries flood control is under state control. The intended beneficiaries are the local community. They suffer a lot because of crop loss, property loss, losses to public utilities, animals and diseases of endemic nature. Some of the key statistics are given below providing damages in Bihar during 2002.

Damages Due to Heavy Rains and Floods During South West Monsoon in Bihar in 2002

Total Districts	38
Affected Districts	25
Taluks/blocks Affected	205
Villages affected	8208
Area affected (Lakh Hectares)	18.45
Population Affected (Lakh)	158.18
Human Lives Lost	434.00
Cattle Lives Lost	1380.00
Damages to Cropped Area (Lakh Hectares)	8.10
Estimated Value of Crops (Rs. In Crores)	467.44
Damages to Houses (No. in Lakh)	3.96
Estimated Value of Damages to Houses(Rs. In Crores)	451.98
Estimated Value of Losses to Public Properties (Rs. In Crores)	296.21

Flood disasters are very common in India, China, Sri Lanka, Bangladesh, Nepal, Pakistan and several other Asian countries. The Rio-Earth Summit (1992) mooted the idea of social participation into the process of sustainable development. There is problem of sustainable livelihood of people in flood prone area. In India the material damages estimated was to the tune of Rs 5846 Crore in 2000. It is the public department that takes care of various activities such as forecasting, rescue, rehabilitation and relief operations. Flood water conservation has hardly attracted the attention of the people. Dams conservation may be quite useful for flood water conservation. This will stop the overflowing of flood water in Bay of Bengal, Indian Ocean and other seas.

INTENSITY OF CONFLICT

There are conceptual and divergent opinions for planning, design, construction and operations of joint projects on trans-boundary river. Both the countries want to maximize their benefits and hence there is a clash of interest. In one such case, a water treaty has been signed by both India and Nepal. Both set up a joint project for detailed database. The conflict could be resolved through a free and frank dialogue where both the countries talk with a compromising attitude. There is always a lack of mutually agreeable regulatory machinery that generates disputes between these two nations. Rigidities, strengthens disputes, whereas flexible dialogue reduces it. Therefore, as has happened in the case of the Indus river basin between India and Pakistan and the Ganga basin between India and Bangladesh, a proposal of similar footing is required for India and Nepal. According to the views expressed by the local people of the countries, flood is not only a natural process. Often people cut the embankment in fear of flooding of their own region. As a consequence, the protection of people is diluted. Therefore the bone of contention is safety of the local people if the flooding is naturally done or it is a manmade process. Nepal is on upstream and Bihar state of India is in downstream. It is essential to take pre-flood measures, post flood measures and structural and non-structural measures. These measures may have long term consequences for the people of both the countries. It may reduce damages of various types and it may provide production friendly environment in the region. It may also reduce the depth and duration of the flood. Even water logging can also be reduced in the major part of the area. The focus groups of local people intend to have joint authority in the area for their welfare.

CONFLICT MINIMIZATION

Kosi being a large river, passes through Nepal and India (mainly Bihar), therefore, from the viewpoint of hydrology, the political and geographical boundaries of the two countries may be ignored. From the viewpoint of basin management the whole basin or sub basins may be treated as integrated unit. The administration over this basin and sub basin has to govern in isolation to the governance of other geographical nature. The basin may be treated as a regional unit for water management and other planning issues. A regional authority may be created for this purpose. In this authority the technical and other professional members of both the countries should jointly plan for the development of the area linked to the basin. The country having larger basin can have chairman and the country having smaller basin can have the position of deputy chairman. The other professionals, technical, managerial and administrative personnel may be recruited on the basis of share of basin. The finance for these purposes may be generated through proportionate contribution of the concerned countries. It may have its own budget and may function as an autonomous body. Its annual report will be presented before both the governments.

The justification for common minimum governance is simply to strengthen regional, planning of basin. The development of basin area depends on local circumstances where both the countries may cooperate for success of a regional planning. The planning of basin needs comprehensive but flexible approach, it needs a practical friendly approach rather than a theory based planning. The planning at the regional basin level may plan formula for water sharing, agricultural production including horticulture and animal husbandry, industrial growth, growth of services including tourism and electricity. This may be slightly flexible depending on circumstances of each year. The regional authority of the basin will do proper justice in sharing of basic amenities also. The authority, although exercises the power over water, will also give due importance to the problems of the country of upper riparian (Nepal) and lower riparian (India). This way such environment will be created which may ultimately minimize the existing conflicts and discourage further conflicts to grow.

Nepal and India both like other South Asia countries depend on monsoon for fresh water. The monsoon is of about 90 days. Nearly 80% of available water appears in river basin. The total average rainfall in Nepal is about 200 BCM out of which 75% occurs during monsoon i.e. June to September. In India average precipitation including snowfall is around 4000 BCM. During monsoon all the major rivers like the Ganga, Brahmaputra, Kosi and Meghna overflow which leads to flood disaster. Agriculture provides above 40% to GDP in Nepal and provides employment to 80% of working population. In India agriculture provides around 30% to GDP and employs above 65% of population. There is wide spread poverty in the region. Food security is also quite less. The

Kosi basin is basically a mono crop region. During monsoon because of flood Kharif crop is hardly grown. Therefore, it is felt that river basin may be treated as one of the area of planning and management of water resources. In this process both the countries should come together and plan jointly. So far, it has not been possible to setup effective measures for managing water resources of Kosi basin. It is suggested to take institutional mechanism for flood water conflict resolutions in these countries. This may bring prosperity in the region in a sustainable manner.

The trans-boundary treaty may focus on such provisions which can help in developing congenial relationships between the two countries. There may be enough scope for consultations, convergence, mediation and judication. It may function like an autonomous International agency. The memorandum of understating may be also very transparent so that nothing is pending for future resolution. This may also facilitate taking up growth friendly agenda for future. Normally, it has been seen that riparian conflicts arise because of conflicts between the country of upper catchments and downstream, between those having loss of land, crops, property, livelihood and those who enjoys the benefits like irrigable water, hydroelectricity and also between the head reach farmers and tail-enders in sub basin. Such conflicts can be alleviated on humanitarian grounds in addition to cost benefit analysis. The down stream country like India must be at least not worse because of launching a new project.

The Indus water treaty between India and Pakistan for sharing the water of the Indus river has set up a permanent Indus Commission. Both the countries may appoint an executive Head of the Commission. Despite political discords between these two nations it has been found that the commission has been working satisfactorily since last 30 years. Similarly India and Bangladesh have reached at long-term treaty on sharing of the Ganga water in 1996. General storage at down sites on tributaries of the Ganga are in the political territory of Nepal. It has enough infrastructure for hydropower generation, irrigation extension, flood management and navigation. The development of both India and Nepal may be accelerated in order to have sustainable livelihood. India and Nepal have already jointly undertaken the construction of the Pancheshwar project on the river Mahakali which is their common border. The common minimum cooperation needs to be strengthened further in order to make the Kosi basin developed as well.

AREAS OF JOINT VENTURE

There are several areas where both the countries may come forward for joint management of flood in general and improvement in the basin in particular. Construction of a major dam may be one option. This may be explored further for construction. The construction of the dam may be designated in such a manner that the benefits outweigh the costs. Other ecological impacts may also be considered for variable planning and management. Dam may also be constructed in the light of the chronic flood disaster. This may generate enough hydropower that may be used for economic activities like agriculture, industry and households. This may be a good substitute of thermal power. The third area where both the countries can participate relates to surface irrigation. A major canal network may be constructed in such a manner that both the countries can develop agriculture, horticulture, fishiculture, tourism and encourage several other activities. This may also regulate drinking water supply. The fourth area may be social forestry. This may reduce soil erosion during floods, generate livelihood for tribes and animals, develop tourism, reduce the change in river course and provide as a good source of earnings as forest products. If necessary the theory of community approach may be applied because the local people are more familiar with regional planning. Their participation may strengthen the success rate of social forestry and increase the survival rate. The next area may be increase in bio-drainage system. Since water logging is a big problem for vast area during post-flood situation, therefore, plantation of such trees may be promoted which can absorb maximum water. This system may be helpful along the railway track, canal and the catchments.

All the activities involve heavy investment in water sector. In addition to this there may be financial requirement for research and development. It may also be thought to establish a water Research and Development Council Jointly. At present there are a few institutions to continue this research on local field issue especially of collaborative nature. South Asia is especially poor to carry out research in this sector. There is also need of inter disciplinary regional apex institution leaving research thrust on water sector. Thus the issue of effective governance is quite essential. The governance may develop common progressive issues, chalk out minimum program for execution and avoid differences. All these things need flexible institutional structure for both the countries. In addition to this, many economic reforms and law reforms are required in the existing governance set up. Development with equity and social justice may hardly be possible without transparent joint working of the two countries. Although there are a few inter-country set up in the region like joint river commissions for joint management of water including flood water amongst the neighboring nations but horizontal and vertical linkages are very narrow and weak.

CONCLUSION

Flood disaster is a chronic problem for the Kosi basin of India and Nepal region. It is a problem of trans boundary water conflict as-well. Both the countries are trying to resolve the contentious issues like pre-flood management. preparedness, post-flood management, structural measures and non-structural measures. But still it has not been possible to manage flood disaster in the region. As a consequence, economic vulnerability is further deepening, losses of various types are on increase every year and water saving devices are not available. As a result, flood passes through Nepal and Bihar state of India. It ultimately fails in the Bay of Bengal and thus of no use for agriculture. After highlighting this chronic problem in the region the present paper concludes that (1) an institutional change is direly required for resolving the Trans-boundary conflicts. (2) a joint authority may be setup. It may be the function of the authority to take care of most contentious issues, evolve a minimum common program effectively, take the assistance of local beneficiaries, professionals and non-professionals and resolve the most contentious issues first. In the second stage the authority may also chalk out other developmental water related plans that may be framed in the light of regional environment. It is hoped that both the countries of South Asia can resolve the issues in an eco friendly manner.

REFERENCES

Biswas, Ajit, K. (1993 ed.), Management of International Water Resources: Some Recent Development in International Water of the MIddle East, OUP.

Chauhan, B.R. (1992), Settlement of International and Inter-State Water Disputes in India, ILI, Delhi.

Cosgrove, William J. (2000), Water Resource Sector in the Coming Decades-Global Perspectives in Kamta.

Herath, Srikantha et.al. (1984), Sustainable Water Resource Management, Proceedings of a Conference, Vol. 1, United Nations University, Tokyo, Japan.

Iyer, Ramaswamy, R. (2000), Water: A Note Regarding Conflicts, Proceeding of an International Conference, IRMED, Delhi.

Mohile A.D. et al (1995), Indian Experience in River Basin Planning and Management, Indo-French Workshop on River Basin Management, New Delhi.

Parsuraman, S. (1999), Indian Disasters Report, OUP, New Delhi.

Prasad, Kamta (2003), Water Resources and Sustainable Development, Shipra, Delhi.

Ray,Isha (2005), Get the Price Right: Water Prices and Irrigation Efficiency, Economic and Political Weekly, Vol. XL, No. 33, August 13, 2005.

Verma, NMP (1993), Irrigation in India: Themes on Development, Planning, Performance and Management, MD Publications, New Delhi.

Verma NMP (2004), Sustainable Management of Flood Water through Community Participation in India in Herath et.al (ed.) Sustainable Water Resource Management, Proceeding of a Conference, Vol. 1, United Nations University, Tokyo, Japan.

12

LEGAL REGIME OF GROUND WATER IN INDIA

Sujith Koonan[*]

INTRODUCTION

One of the peculiar facts in the history of water law is the separate development of doctrine governing surface water sources-such as lakes and rivers- and that governing ground water. While land ownership is critical to both surface water and ground water, riparian rights in surface water are usually subject to the doctrine of reasonable use, whereas the ground water has always been governed by 'freewheeling rule of capture'. Ground water was considered as a 'chattel' with no separate existence apart from the land. Hence common law approved the land owner's absolute right to use the ground water to any extent. It was mainly due to the ignorance about the structure and flow of ground water and the lack of sophisticated devices to extract ground water at that time.

Since independence, India also carried forth the colonial legacy of the common law tradition. Legal principles to deal with the surface water have been developed according to the change in situation. But ground water regime continued to be controlled by

* Law Researcher, International Environmental Law Research Centre (IELRC), #75, Sidharth Enclave, New Delhi-14.

the old common law principle of absolute proprietorship. In the post independence period food security was one of the major aims of the government. Hence government has supported the expansion of cultivable area through irrigation. In the beginning canal irrigation was the main source. The uncertainty in the availability of surface water has gradually prompted the farmers to rely on ground water. Now ground water contributes nearly 60 percent of total cultivable area and nearly 80 percent of water for domestic purpose.[1] Government has encouraged the ground water use by providing subsidies in the form of free power to agricultural pump sets, low interest loans for deepening existing wells or constructing new wells, for purchasing pumps and other equipments etc. At the same time little attention was paid to maintain the already existing traditional irrigation sources such as canals and spring channels, which was a good source of ground water recharging. This extensive increase in ground water extraction for irrigation coupled with the corresponding increase in use of ground water for domestic and commercial purposes resulted in deterioration of ground water quality and quantity, which tilted the ecological and economical balance.

This alarming situation necessitated legal intervention. It prompted the central government to come out with a model Bill in 1970 to guide the state governments to enact ground water legislation. Literally all the states have ignored the model bill. The Central government has circulated the recent versions of Model Ground water Bill in 1992 and 1996. The latest Bill has been introduced in 2005.[2] Few states have so far responded to this attempt by the central government, namely Kerala, Andhra Pradesh, Delhi and West Bengal. Since the development of ground water laws is in the beginning stage, there is a need to examine the legislative framework that is being adopted.

Apart from the statutory framework, the Constitution also contains certain schemes which are applicable to ground water. The 73rd and 74th amendment of the Indian Constitution provides for local self government. As a result water management to a limited extent is under the purview of local bodies like Gram Panchayat, Municipalities etc. It is the duty of the state governments to frame laws for devolution of power to local bodies. Panchayat Raj Act has been enacted to bring the 73rd and 74th amendment in force. Though the local bodies are invested with power to manage ground water, it lacks financial and technical capacity. Therefore the decentralisation principle has not been implemented in its true spirit.

Judiciary has also played an important role in interpreting Article 21 of the Constitution to include right to clean water as a derivative fundamental right. The judiciary has used this weapon to control the over exploitation of ground water by private parties in few occasions. Plachimada is a classic example in this series to discuss the principle guiding ground water regime in India. Hence the statutory, constitutional and judicial initiatives altogether can be considered as the legal regime of ground water in India. This is the focus of this paper.

The paper explains in its paper part the constitutional scheme applicable to the ground water resources. This analysis is relevant because modern resource management principles strongly argue for decentralised participatory management. The second part of the paper deals with the judicial activism in ground water regime. There are very few cases which directly addresses the ground water rights. The Plachimada case is the one of the recent cases which specifically discusses the principles governing ground water regime. Discussion of this case is the focus of the second part. The third part analyses the statutory framework in India. Since water is a state subject, states have the power to enact upon the subject. There has been some initiative from the part of central government to guide the state governments. But few states have come forward to enact laws on the basis of center's guidelines. Hence this section of the paper analyses the legal frame work suggested by the center and few state legislations for regulating the ground water use. The selection of the state statutes has been done on the basis of year of enactment and the nature of the statute, i.e., statutes enacted after 1995 (past the reforms phase) and statutes dealing either exclusively or in an innovative manner combined with other subject matters such as the general environment have been selected. The fourth part is an analysis of the ground water legal regime in India as a whole. It also includes a comparative analysis and appreciation of the parts discussed in paper.

CONSTITUTIONAL SCHEME

The Constitution of India does not address the ground water problem directly or expressly. But there are few provisions in the

Constitution which can be applied to ground water regime. The major provisions applicable to ground water are the 73rd and 74th amendments and Article 21 of the Constitution. The 73rd and 74th amendments provide for decentralisation and local self governance and as a result local bodies are empowered to manage water resources. Hence the Constitutional scheme envisages planning and implementation at the lower level. This can be appreciated as a positive trickle down of power which is a more acceptable and desirable management approach in the water sector.

Part III of the Constitution also does not address ground water expressly. But the right of community over the ground water resource has been recognised by the judiciary on more than one occasion. Hence the ground water right is derivative fundamental right and its violation by anyone through over drawing amounts to violation of the right to life under Article 21 of the Constitution. The judiciary has further evolved principles like public trust doctrine to restrict the right of the government and the individual over the natural resources like ground water. This Constitutional scheme is important because it provides the individuals to the right to approach the court for enforcing their ground water rights and thereby provides the jurisprudential scope for a separate legal regime for ground water.

Decentralisation principle

In the present Constitutional scheme the legislative power, in relation to water, is divided between the union and the states. The union has been empowered to legislate in matters relating to the regulation and development of inter state rivers and river valleys.[3] The states on the other hand, have been empowered to legislate in respect of water supplies, irrigation and canals, drainage and embankments, water storage and water power.[4] Hence it is clear that states have the exclusive legislative jurisdiction over intra-state water resource.

The Indian constitution has only used the term 'water resources' in general. However there is no specific reference to ground water. It is very difficult to assume that Constitutional makers have intended to include all water resources. The framers of the constitution seem to have been thinking mainly of river waters. Nonetheless ground water can be read into the broad framework of the term 'water resources' and it would, therefore, be logical to conclude that ground water falls within the domain of the state's legislative competence.

The Constitutional scheme envisages that water resources are to be managed at the local regional level. The Constitutional approach can be interpreted in such a way that management of water resource should be done by considering the factors like hydrological nature of the area, natural water storage capacity, annual rainfall and local needs. Hence planning and implementation of water resource management can effectively be done at regional or local level. This approach has been strengthened after 73rd and 74th Constitutional amendment in 1992 by envisaging water resource management at grass root level.[5]

The 73rd and 74th Constitutional amendment cast duty upon the state government to constitute panchayats in rural areas and municipalities in urban area as the basic unit of local self government.[6] It further provides that the state may, by law, enable the local bodies to function as institutions of self government. The object of the amendment is the decentralisation through the devolution of powers and responsibilities upon the local bodies.[7] The decentralisation is more evident in the fact that the Constitution prescribes selection of members of the local bodies through direct election.[8]

As a result of the incorporation of the decentralisation principle, the management of ground water comes within the purview of local bodies. The subject 'minor irrigation, water management and water shed development' and 'drinking water' has been included in the schedule of powers and functions of the panchayat.[9] The subject 'water supply for domestic, industrial and commercial purpose' has been included under the responsibility of municipality.[10] The amendment further provides the Constitutional support for the state to establish their regulatory power over the natural resources including ground water.

The decentralisation principle as reflected in the Constitutional scheme does not envisage a geographically compartmentalised approach in natural resource management. Such an approach is inappropriate especially in the case of ground water. Because unsustainable use of ground water in one area is most likely to have an adverse impact upon the quality and quantity of ground water in nearby areas. Hence a co-operation and consultation between local bodies, the state and the central unit are necessary.

Supreme Court in M.C.Mehta case supported this view and directed the setting up of the Central Ground Water Board for the purpose of regulation and control of ground water management. Even though the constitutional mandate lies with state and local bodies, the underlying spirit is to evolve a culture of co-operative federal management of water resources in the country.[11]

The decentralisation principle as envisaged in the Constitution empowers the state to extend its 'eminent domain' power to preserve and protect the natural resources including ground water in the larger interest of public. Any reasonable restriction through law upon the individual rights as consequence of the exercise of this power is justified as per the Constitutional scheme. Moreover a wider interpretation of the Constitutional scheme reflects the participatory management of natural resources. But the local bodies should be empowered, financially, legally and technologically, to achieve the real spirit of this Constitutional scheme. The local bodies should be supported by the Union and State institutions and at the same time planning and implementation need to be done by the local bodies.

Ground Water as a Constitutional Right

Part III of the Indian Constitution speaks about the fundamental right of individuals. The rights guaranteed under this part are of negative nature. It demands the State not to interfere with the enjoyment of the rights by individual unreasonably. Generally, Part III of the Constitution does not demand affirmative action on the part of the State for realisation of the rights. The Constitution does not guarantee right to ground water expressly. But the judiciary by playing its interpretative role has made it a derivative fundamental right through case laws.

The Supreme Court in Subhash Kumar's case held that right to life under Article 21 includes right of enjoyment of pollution free water and air. It was held as necessary for the full enjoyment of the right to life.[12] The consequence of such an interpretation is that if anything endangers or impairs this right, the individual shall have the right to recourse to constitutional remedy under Article 32 of the Constitution.[13] The court reasoned its decision on the ground that pollution free water and air is a basic determining factor of quality of life, which is a guaranteed right under the Constitution. Thus the right to pollution free environment was declared as a justiciable derivative fundamental right. This right has been used in the past on more than one occasion to preserve environment, viz, to prevent tanneries from discharging effluent into river[14], to prevent construction in wetlands[15], to prevent state from dissipating river to private individuals for constructing hotels,[16] etc.

The interpretation of Article 21 has further been expanded to include ground water within its purview. It was held that, "...Large scale defoulment of the quality of water so as to make it unusable by others or as a result where of the water is contaminated and becomes unpotable would be violative of Article 21 of the Constitution".[17] The court reasoned its decision on the ground that over extraction of the ground water by an individual is likely to have adverse impacts upon the life of others. Hence in the larger interest of society, individual behaviour towards ground water is to be regulated.[18]

There is no doubt that in the present stage of pollution and declining ground water level, its use needs to be regulated. But how far can an individual rely on Article 21 of the Constitution to prevent an individual from extracting ground water from his land? The nature of the fundamental right under Part III of the Constitution provides that individual is entitled to use it as a weapon to restrain the State from interfering with their guaranteed rights. This weapon can only be used against the State. Hence the question arises: can a private individual be brought within the purview of 'State'? This question has never been addressed by the judiciary, though the court had issued directions to private parties under the umbrella of fundamental rights. Hence this approach is very unstable due to the weak jurisprudential basis.

The absence of a comprehensive statutory framework is the reason which prompted the judiciary to go for the extremes. Though the decision was jurisprudentially weak, such an intervention was necessary in the public interest. Therefore a legislative intervention is more appropriate than building a weak jurisprudence on the basis of the Constitutional right. The recent Plachimada case, which is discussed in the next part, also shows that judicial solutions are not proper and stable always. A comprehensive legislation and its effective implementation is the workable solution for the present ground water crisis.

PLACHIMADA

In the discussion of ground water rights and ground water legal regime in India, 'plachimada' is an indispensable part. Plachimada appeared on the Indian legal map after Hindustan Coca Cola Beverages Private Limited (herein after referred as 'the company') started a soft drink manufacturing factory in plachimada. It became an "opportunity" to discuss the ground water legal regime in India when the local people of plachimada started an agitation against the company by alleging that the over extraction of ground water by the company had resulted in deterioration of ground water quality and quantity in plachimada.

The question was addressed by both a single bench and a division bench of the Kerala High Court. Plachimada addressed the core question of the nature and extent of the individual's right over the ground water in his land and the State's power to regulate its use. Before going into the details of the plachimada case a brief description of the factual background may be useful for an easy understanding of the case.

The company set up a soft drink manufacturing factory in plachimada in the year 2000. Plachimada of Perumatty Panchayat in Chittoor Taluk is a little hamlet of Palakkad District, Kerala State, popularly known as the "rice bowl of Kerala". This part of Kerala State is almost exclusively agricultural land and depends heavily on canal irrigation and ground water and the villagers are predominantly landless agricultural workers. The land which was said to be reclaimed by the company was previously poonthal padam (a traditional slushy paddy land which dries up in summer with a unique and different ecosystem).[19] The potential for ground water recharge in the district is good due to the undulated terrain. However the area is characterized as being in the rain shadow area of the Western Ghats and the statistics show how the monsoon have been falling during a number of consequent years.

The local people started their agitation against the company within one year of the implantation of the company. Their protest was on the basis of the allegation that ground water has declined and wells in the locality were polluted due to the activity of the company. This has also alleged to have resulted in unemployment due to decrease in agricultural production. The people of plachimada argue that the water crisis is the reason for decline in agricultural production. As a result of the growing agitation,

Perumatty grama panchayat rejected the application of the company to renew their license on the ground that the over extraction of ground water by the company has resulted in acute drinking water scarcity in plachimada.[20] This action of the panchayat has actually dragged the issue to Kerala High Court by challenging the power of panchayat to regulate the right of the company to extract water from its land.[21]

The plachimada case deals with the legal controversy between common law right of the land owner to extract ground water from his land to any extent and the principles evolved through modern environmental jurisprudence which demands State to ensure sustainable use of natural resources. At the first instance, the Kerala High Court took an express pro-environmental approach and supported the State's power to regulate individual's right to extract ground water from his land. In the appeal, the division bench reversed the single bench's decision and took a decision in favor of private property right of individuals.

The single bench of Kerala High Court upheld the power of panchayat to regulate the extraction of ground water by the company.[22] Court made it very clear that: "The panchayat and the State are bound to prevent it. The duty of the panchayat can be correlated with its mandatory function No. 3 under the third schedule to Panchayat Raj Act namely, 'maintenance of traditional drinking water resources' and that of the state of Article 21 of the Constitution of India".[23] The Court further held that though the ground water has not been mentioned expressly, it can be read with section 218 of the Panchayat Raj Act which makes the panchayat 'the custodian of all natural water resources'.[24] Hence it can be said that the single bench decision is in tune with the decentralisation principle as reflected in the Constitutional scheme.

The single bench has relied upon the public trust doctrine as a theoretical support for its decision. The Public trust doctrine provides that State is the trustee of all natural resources which are by nature meant for public use and enjoyment. The public at large is the beneficiary of the sea shore, running waters, air, forests and ecologically fragile lands. The State as a trustee is under a legal duty to protect the natural resources. These resources meant for public use cannot be converted into private ownership.[25] Public trust doctrine was originated under Roman law to ensure public access to certain common properties like rivers, sea shore etc. The application of the doctrine has been extended to all natural resources and took shape for protecting environment at large as a result of the judicial interpretation of American courts and influential academic literature like that of Joseph L. Sax.[26]

By relying on the public trust doctrine it was further held that the under ground water belongs to the public. The State is duty bound to protect ground water against excessive exploitation. The court has linked the trust duty with right to life under Article 21 of the Constitution and said that, 'the inaction of the state in this regard will tantamount to infringement of the right to life under Article 21 of the Constitution of India'.[27] The outcome of the decision is that the individual is not free to extract ground water to any extent from his land. The land owners right is subject to public interest and the State as a trustee can regulate the right of land owner to protect the public interest. By arriving at this conclusion, the single bench has rejected the traditionally existing common law right of absolute proprietorship.

With regard to the major question, it was held that, "the order of the Panchayat to close down the unit on the finding of excessive extraction of ground water is unauthorized. The Panchayat can at best, say, no more extraction of ground water will be permitted and ask the company to find out alternative source for its water requirements".[28] Therefore the court allowed the company to extract the quantity of water that a land owner would extract for his domestic and agricultural purposes.[29] Hence the Court upheld the power of the State (Panchayat) to put reasonable restriction upon the individual right to extract ground water from his land even in the absence of a specific law on this matter and further held that underlying ground water does not belong to the land owner absolutely.[30]

The assertion of single bench was reversed by the division bench of the Kerala High Court in Perumatty Gram Panchayat v. State of Kerala.[31] The Division bench has disapproved the single bench's findings by asserting that, "ordinarily a person has right to draw water in reasonable limit without waiting for permission from the panchayat and the government. This alone could be the rule and the restriction an exception".[32] The Court further held that reliance upon the public trust doctrine by the single bench is inappropriate in the present case. The right of the land owner to extract ground water from his land was approved as a 'basic right' and held that such basic rights can only be curbed by valid legislation.[33]

The Division bench relied upon the property rights of the individuals. The Court said, "Even with reference to mandatory referred to in the third schedule of the Panchayat Raj Act, namely 'traditional drinking water resources' could not have been envisaged as preventing an owner of a well from extracting water there from as he wishes. The Panchayat had no ownership about private water resources, in effect denying the property rights of the occupier and the position of law laid down by the learned judge (single bench) is too wide and unqualified acceptance".[34]

The division bench decision seems to be based on the common law rule of absolute right of the land owner to extract water from his land. But certain propositions of the division bench are contradictory to the common law principle, that is, the proposition of 'reasonable limit'.[35] And further the proposition made by the court that the "company should actively involve in the community development programmes for the people residing in the locality, especially in the matter of health and drinking water supply at the supervision of Panchayat", also is not in tune with the legal proposition on which the decision is seems to have been rendered.[36] The contradiction in the judgment makes it possible different interpretations of the decision and therefore the decision is of poor precedential value.

Thus the possibility of laying down a new rule of law in tune with the requirements of modern time has been tilted by the Division Bench. This approach allowed the private owner to go ahead with his exploitation of underground water with out any hindrance. Development of a new legal principle in tune with the felt needs of the society was possible in this case. It is unfortunate that property law and jurisprudence was considered by the court as sacrosanct and refused to touch, though very strong need was there to do so. Thus the court lost an opportunity to mould the law to suit the needs of the present day society. It would have been more proper for the court to mould law in tune with the decentralisation principle as envisaged in the Constitution and the new evolving environmental jurisprudence.

STATUTORY FRAMEWORK

The Constitution of India empowers the State to make legislations on the subject 'water resources'. In spite of the alarming need for a legal framework to control and regulate ground water use, State governments have not come forward to respond on the issue. Hence the central government introduced a model bill in 1970 and circulated among the states. Literally all states have ignored the model bill. The States started to enact for the regulation of ground water only in the beginning of 21st century. Almost all ground water legislations came into existence after 2000. The Central government has circulated the recent versions of Model Ground Water Bill in 1992 and 1996. The recent version of the model bill has been introduced in 2005. Since statutes are the important part of legal framework, this part of the paper analyses the schemes envisaged in model bill and few State legislations (Kerala, Andhra Pradesh, West Bengal and Delhi) for ensuring sustainable use of ground water. Hence the observation of statutory framework in this paper means analysis of the above mentioned four statutes and the schemes envisaged in Model Bill, 2005.

The state legislations to regulate ground water use in India adopted more or less the same scheme as envisaged in the model bill 2005. Hence an analysis of the scheme in the model bill gives an overall outlook of statutory legal regime of ground water in India. The basic object of the model bill is the constitution of an authority to control the use of ground water. The Scheme envisaged in the model bill is that the authority is empowered to 'notify' certain areas.[37] The control over the ground water use is subject to the notification. The regulations envisaged in the statutes are only applicable to the ground water users of notified areas.

The regulations are in the form of permit system for the new wells and registration for the existing wells in the notified areas. The model bill does not make any difference between competing users. The definition of the term "user of ground water" includes domestic users also.[38] But proviso to section 6(1) provides that the wells which is proposed to be fitted with hand pumps or water is proposed to be drawn by manual devices need not seek permit. Hence every user in the notified area needs to seek permit from the authority to sink new well. The decision of the authority in granting or denying the permits are based on a number of factors which include technical factors such as availability of ground water, the quantity and quality of water to be drawn and the spacing between the ground water structures.[39] The same procedure is prescribed for the existing user to register the wells.[40]

The permit granted under the statute is not for ever. The authority is empowered to cancel the permit/certificate of registration on reasonable ground. If the authority is satisfied that the ground water situation in a particular area warrants limitation or restriction of use, the statute empowers the authority to alter the permit/certificate of registration.[41] The authority can also close-down the extraction of ground water if the user of ground water violated any provisions of the statute.[42] Apart from this the statute confers several powers to the authority to carry out the object of the Act.[43]

The conservation idea is also reflected in model bill. It has been assumed that the over exploitation of ground water due to ever increasing population and other developmental activities have led to fall in ground water table, drying up of wells etc. Therefore model bill suggests rain water harvesting for ground water resource augmentation. It has also been suggested to make rain water harvesting as a stipulated condition for granting approval for the building constructions.[44] The model bill envisages water shed management to facilitate ground water recharge in rural areas.[45] Altogether the scheme of the bill is to promote sustainable use of ground water by restricting the ground water use to the extent of natural replenishment. Wherever, there is mismatch, steps will be taken to ensure augmentation of ground water resources in addition to regulatory measures.

Although the center had repeatedly circulated the revised versions of the model bill, majority of the States are yet to enact ground water legislation.[46] Out of the state legislations examined, Delhi and Kerala Acts followed the 'permit/registration system' as discussed in the model bill.[47] At the same time the idea of participatory management and conservation have been discarded in both the Acts. Though the statutes give special emphasis on public drinking water resources, the priorities have not been properly streamlined in the statute. It would have been proper to follow the priorities as reflected in the National Water Policy, 2002.[48] Setting up of priorities is very important in the present conditions of declining ground water quality and increasing competitive uses.

The West Bengal Act went a step ahead by incorporating some additional principles. The Act expressly gives much importance to the conservation and augmentation of the ground water resource. The Act put a duty upon the authority to initiate policy to conserve ground water resource by way of recharging, replenishing, recycling or reusing in a coordinated manner.[49] The Act envisages participatory management of ground water resource. The interaction between the modern technology and age old practices of ground water conservation have been recognised and promoted under the Act.[50] The Act promotes the traditional rain harvesting system to reduce the over dependence upon the ground water. The idea of conservation and people's participation in planning and management of ground water resources as envisaged in the Act is in tune with the modern environmental jurisprudence and therefore worth appreciating.

The authority (district level authority) is further duty bound to prepare district profile on ground water resources in every five years. The Act makes it the duty of the authority to keep a regular vigil on the quality and quantity of the ground water and to initiate action upon any sudden deterioration or contamination of ground water resources.[51] This is a very progressive approach. The base data preparation and its regular revision are necessary for the authority to take proper and timely action to preserve the ground water resource. Absence of the base data regarding the quantity and quality of the ground water was pointed as the difficulty in plachimada by Central and Kerala State Ground Water Board.

The Andhra Act more expressly provides some stringent measures. As per the Act the authority is empowered to prohibit the extraction of ground water if in its view such water pumping in such area is likely to cause damage to the level of ground water or deterioration for a period of not more than six months which after review may be extended for a further period of not more than six months at a time.[52] The authority is also empowered to declare certain area as over exploited areas.[53] No one is allowed to sink wells in such 'over exploited areas'.[54] But the Act gives much emphasis to drinking water. The wells sunk for public drinking water purposes or hand pump for public or private drinking water purposes are exempted from the blanket prohibition applicable to 'over exploited areas'.[55]

Further no one is allowed to dig a well within a distance of two hundred and fifty meters from a public drinking water resource.[56] Any person who intends to sink a well within this stipulated space shall seek permission from the authority. The regulation is more stringent if such a well is intended to be used with power driven pump. In such cases the concerned person shall also obtain permission from concerned authority related to electricity.[57] The authority is further empowered to prohibit the extraction of water for commercial, industrial or other purposes if such use is likely to have adverse impact on the public drinking water source. This power is wider that the authority can close or seal off wells, and if necessary the authority can take over private wells on the consideration of availability of drinking water.[58]

The AP Act is more comprehensive because it deals with the conservation of ground water through rain water harvesting and the idea of incentives for recycling and reuse of waste water by industrial, commercial users etc.[59] The Act prescribes rain water harvesting as a mandatory provision to be implemented by the local bodies before giving approval for constructing buildings.[60] The principle of public participation in natural resource management is also reflected in the Act.[61] The AP Act went a step ahead by including the principle of polluter pays principle. The Act provides that in addition to the penalty, "the cost of its repairs or remedying the same shall be recovered as an arrears of land revenue".[62] By including the principles of modern environmental jurisprudence and more comprehensive provision to regulate ground water use, it can be concluded that the AP Act is far ahead in complying with the object of the ground water laws sought to be achieved.

A brief analysis of the four statutes gives an impression that though some statutes like Andhra Act have comparative advantage over others, the statutory framework need to be improved further by incorporating regulatory measures and conservation strategy in a more workable manner. The conservation measures as prescribed in the statute like the Andhra Act is not mandatory. A 'command approach' is not feasible for conservation strategy. Participatory approach is preferable to promote conservation of all natural resources. The local bodies and the user's associations can play more vital role in this regard.

AN ANALYSIS OF LEGAL REGIME OF GROUND WATER IN INDIA

Historically, there has not been specific principles or laws developed for ground water. Common law considered it as a part of individual's right to enjoy his property (land). There was no limitation as to how much quantity of ground water a particular land owner may draw.[63] The common law approach of absolute proprietorship developed because of lack of knowledge about the ground water hydrology at that time. Moreover the necessity of regulating ground water might not have been felt at that time. In the present day context of increasing dependence on the ground water resource for various competing uses, legal intervention is necessary. The idea of justice and equity also demands such a legal intervention to ensure accessibility and availability to all. It can be assumed that these notions are the basis of legal regime of ground water in India also.

The Constitutional scheme, evolution of principles by the judiciary and the recently originated statutes form the legal regime of ground water in India. Like legal intervention in any other area, ground water legal regime also places restrictions upon individual activities. The State assumes its right or power over the ground water resources, though the resource is in private land. The State is gradually establishing its eminent domain power over the natural resources (ground water). The reasonable restrictions upon the individual right is justified on the ground of 'public interest'. No individual can defend and argue for his property right on the basis of traditional common law property right. Hence, as a result of the new evolving legal regime, ground water has been or is being shifted from individual to governmental control.[64]

The absolute right of the land owner has also been disapproved in the plachimada case. Though the division bench decision of Kerala High Court appears to be relying on common law rule of absolute proprietorship, a close reading gives different result. The court has not allowed the company to extract ground water to any extent. The right of the company was recognised but it was held that the right is subject to restriction. The court has used the term "reasonable limit" to denote the nature of restriction. The available rainfall has been recognised as the determining factor of the "reasonable limit". Hence the plachimada case has not recognised the uncontrolled right of land owner to extract ground water, even though the court was lenient towards the property rights.

Judiciary has recognised ground water as a derivative fundamental right by making it as a part of Article 21 of the Constitution of India. Few jurisprudential problem need to be solved or clarified before considering ground water as a

Constitutional right. The nature of the Constitutional rights includes the claim of individuals against the State only. Hence there exists a theoretical problem for the judiciary to issue orders against the private individual on the basis of the Constitutional rights. At best, the court can order the Government to take appropriate action by relying upon the 'state inaction' rule. However the Supreme Court of India entertained number of litigations against the private parties on the basis of the Constitutional rights. Unfortunately the court has not addressed this issue properly so far.

Moreover competing uses of ground water is very wide. It include basic necessity like drinking purpose, developmental needs like industrial use, commercial use etc. It is not viable to include ground water right with all these contents as part of the constitutional right. It is also very difficult to set limits of these rights and set priorities among the competing uses through the Constitutional rights. The realisation of the ground water rights through the Constitutional means would create legal confusion and stability cannot always be expected from the judiciary though the rules of precedents are there. The judicial approach is worth appreciating because the court has tried to evolve a legal regime of ground water in the absence of specific and express statutory framework.

The Constitutional scheme envisages local bodies as the appropriate authority for planning and implementation of water management. The decentralisation principle has not yet been implemented in its true spirit. Local bodies do not have autonomy in financial and technical level. The role of local bodies in implementing the principle of community participation is very important. The present statutory framework has not considered this factor. An improvement in this regard needs to be made in the ground water legislations to ensure the role of local bodies in planning and implementation of resource management programmes.

The analysis of ground water legislation makes it clear that the regulation methodology is based upon the 'permit/registration system'. All legislation provides for the authority to implement the regulations. Except West Bengal Act, all others prescribe constitution of authority in State level. West Bengal Act provides for District and Corporation level authorities also. The authority is entrusted with powers to regulate the extraction of ground water mainly through issuing of permit and certificate of registration. The authority has the power to stop extraction from any well if the situation warrants such extreme actions. At the initial stage of development of ground water legal regime this can be considered as a good beginning to establish the governmental control over traditionally existing property rights of individuals. But the statutes need improvements to achieve the long term object of sustainable use of ground water.

In one way or the other all the legislations give preference to drinking water resources. The priorities after the drinking/domestic purpose have not been set in any of the legislations. The law is dangerously silent about the situation which demands regulation or prohibition of any competing use of ground water for protecting the drinking water resource. The legal regime has neglected the priorities prescribed in the National Water Policy, 2002. Another often highlighted criticism against the ground water laws is the absence of provisions for community participation. Thought the concept of public participation in decision making is practically difficult, the presence of such a provision can at least empower some enthusiastic individual or organization to challenge erroneous decisions or plans of government. It can also very well serve an educative purpose for the present to achieve the real results in future.

The ground water laws concentrate more on the regulatory schemes and give little emphasis for conservation of resources, though conservation is one of the established objects of the legislation. Both these objectives should have been given equal importance. It is preferable to include mandatory provisions for rain water harvesting, recycling and re use of used water for suitable purposes etc. The conservation programmes should be implemented by the local bodies with the assistance, technical and financial, of state and central governments. In drawing up the conservation scheme, it is also important to distinguish the different uses to which the groundwater is being put, so that the rules and regulations are appropriately framed.

The lessons like Plachimada reveal the importance of mandatory provisions for preparation and regular monitoring of ground water quantity and quality. It must be one of the duties of the authority to prepare and keep such base data of its jurisdiction. An environment impact assessment of ground water should be made mandatory for large scale users likely to affect the ground water. The preparation of base data and mandatory impact assessments will help the authority to take timely actions. This can also be a good legal weapon in the hands of the public to protect their natural resources through courts. Had such a provision been there in Kerala, the State institution would not have been showed themselves as helpless and leave the plachimada people to suffer miseries.

Polluter pays principle is one of the essential principles in modern environmental law. The Supreme Court of India has made this principle a part of Indian law in Vellore Citizen's Welfare Forum case.[65] This principle can be an effective part of penalty, but the legislations in India have neglected it.[66] This principle is very effective in social as well as environmental point of view. Hence the ground water legal regime need to be improved by giving equal importance to conservation measures with regulatory measures and by incorporating the principles of modern environmental law, like polluter pays principle. It is equally important to envisage within the statutory regulatory framework, an effective role for community participation. There must also be an effective implementation of legislative provisions without which these legislations are meaningless.

CONCLUSION

The development of ground water legal regime shows the shift occurred in the concept of property concept. The land and all resources in the land were considered as the absolute right of the land owner. Now the recent ground water legal regime supports the idea that property rights are no longer a sacrosanct right. Land rights and all other rights in connection with the enjoyment of land rights like right over the natural resources are subject to restriction on the ground of public interest. The results of the analysis reveal the expansion of state's police power to invisible natural resources in private properties. The shift from private to governmental control can be used to establish the community right, to a limited extent, in all natural resources. The recognition of such community right contains an inbuilt restriction on the individual rights.

Though the ground water laws recognised the expansion of state's police powers and to some extent the community right in ground water resource, it has not gone to the extreme of "public property" concept. The regulatory provisions in the statutes give necessary respect to the rights of the land owner. The provision of compensation to land owners in the case of take over of wells in Andhra Act clearly conveys this message. Hence, it can be concluded that the ground water legal regime in India has taken a via media between extreme private property concept and public property concept.

NOTES

1. S. Janakarajan & Marcus Moench, "Are wells a Potential Threat to Farmer's Well Being? Case of Deteriorating Ground Water Irrigation in Tamil Nadu", XLI Economic and Political Weekly 3977 (September 16-22, 2006).
2. Model Bill to Regulate and Control the Development and Management of Ground Water in 1970. [herein after referred as Model Bill].
3. Constitution of India, Schedule VII, List 1, Entry 56.
4. Constitution of India, Schedule VII, List II, Entry 17.
5. Herein after referred as 'The Amendment'.
6. The Constitution of India, Article 243B, 243Q.
7. Id., Article 243G.
8. Id., Article 243C (2), 243R (1).
9. Id., Article 243G, Eleventh Schedule, Entry 3 & 11.
10. Id., Article 243W, Twelfth Schedule, Entry 5.
11. M.C.Mehta v. Union of India, (1997) 11 SCC 312.
12. Subhash Kumar v. State of Bihar, AIR 1991 SC 420, 424.
13. Ibid.
14. Vellore Citizen's Welfare Forum v. Union of India, (1996) 5 SCC 647.
15. Jagannath v. Union of India, AIR 1997 SC 811.
16. M.C.Mehta v. Kamalnath, (1997) 1 SCC 388.
17. M.P.Rambabu v. District Forest Officer, A.I.R 2002 A.P. 256.
18. Venkatagiriyappa v. karnatakan Electricity Board, 1999 (4) Kar.L.J 482; see also Attakoya Thangal v. Union of India, 1990 (1) KLT 580; F.K.Hussain v. Union of India, AIR 1990 Ker. 321.
19. Jananeethi, Report On the Amplitude of environmental and Human Rights Ramifications by the HCCBPL at Plachimada", July 2002;

Yuvajanvedi, Report on the environmental and social problems raised due to Coca Cola and Pepsi in Palakkad District, November, 2002, p. 8.

20. Resolution passed by the Perumatty Grama Panchayat on 07-04-2003.
21. Apart from the ground water extraction issue, there are other problems like pollution of water, pollution due to solid wastes etc. Since this paper discusses only ground water legal regime, these problems have not been discussed here. Kerala Pollution Control Board has ordered the closure of the company on the ground that the company has not set up treatment plant for hazardous wastes as required under the Hazardous Wastes (Handling and Management) Rules, 1986.
22. Perumatty Grama Panchayat v. State of Kerala and Others, W.P. (C) No. 34292 of 2003 decided on 16-12-2003 by Justice K.Balakrishnana Nair.
23. Id., para. 13.
24. Ibid.
25. For detailed discussion on public trust doctrine see M.C. Mehta v. Kamalnath, (1997) 1 SCC 388.
26. The origin, development and expansion of public trust doctrine have been discussed elaborately in Joseph L. Sax, The Public Trust Doctrine in Natural Resource Law: Effective Judicial Intervention, 68 Michigan Law Review, 471.
27. Supra Note 49, Para. 13.
28. Id., Para. 12.
29. Id., Para. 15.
30. Id., Para. 13.
31. W.A. No. 17600 of 2004 decided on 07-04-2005 by Justice M.Ramachandran and Justice K.P. Balachandran.
32. Id., Para. 43.
33. Id., Para. 35.
34. Id., Para. 35.
35. Id., Para. 43.
36. Id., Para. 54.
37. Model Bill, 2005, S. 5(2).
38. Id., S. 2 (10).
39. Id., S. 6 (5).
40. Id., S. 7(5).

41. Id., S. 11(c).
42. Id., S. 12 (1) (h).
43. Id., S. 12.
44. Id., S. 19.
45. Ibid.
46. Fourteenth Report of the Standing Committee on Rural Development to the Lok Sabha, 2005.
47. The Delhi Water Board (Amendment) Bill, 2002 which is still under consideration; Kerala Ground Water (Control and Regulation) Act, 2002.
48. National Water Policy, 2002, See S. 5.
49. Id., S.6 (2).
50. Ibid.
51. Id., S.9
52. Andhra Pradesh Water, Land and Trees Act, 2002, S. 9 (1).
53. Id., S. 11(1).
54. Id., S. 11(3).
55. Ibid.
56. Id., S. 10(1).
57. Id., S. 10(2).
58. Id., S. 12.
59. Id., S. 18.
60. Id., S. 17.
61. Id., S. 6(f).
62. Id., S. 35(2).
63. Chhatapati Singh, Water Rights and Principles of Water Resource Management, N.M. Tripathi Pvt. Ltd., (1991, Bombay), P. 39.
64. Philippe Cullet, "Water Law Reforms- An Analysis of Recent Developments", 48 Journal of Indian Law Institute 206, 218 (2006).
65. Vellore Citizen's Welfare Forum v. Union of India, (1996) 5 SCC 647.
66. To some extent the principle is reflected in Andhra Act, Supra Note 89.

13

PUBLIC-PRIVATE PARTNERSHIPS IN WATER AND SEWERAGE AN AFRICAN PERSPECTIVE

*Reto Thoenen**

FROM PUBLIC GOODS TO PUBLIC-PRIVATE PARTNERSHIPS

Basic infrastructure such as the provision of water and sewerage were long seen as typical cases of natural monopolies and public goods. In infrastructure, economies of scale exist due to large initial investments. The cost of every additional connection to the water and sewerage system is-compared to the initial investment of setting up the network-comparatively low. Economic theory states that the provision of such infrastructure and the services linked to it can be more cost effective when it is done by one single provider that takes advantage of the economies of scale. This situation is referred to as natural monopoly which is frequently used to justify government intervention.

The provision of infrastructure and linked services share another economic characteristic that justifies government intervention: externalities. Improvements in a person's access to water and sewerage tend not only to improve his or her personal

* Economic Commission for Africa (ECA)rthoenen@uneca.org

situation and well-being but at the same time increase the overall economic and social outcome in the economy. Further, public goods-such as public health-share the characteristic that private provision without government intervention leads to a demand that is below the optimum for the economy.

Provision of basic infrastructure has therefore for a long time been considered as the exclusive responsibility of the public sector. But the public sector by itself, for a variety of reasons, has not been able to meet infrastructure requirements arising in Sub-Sahara Africa (SSA). The monopoly granted to public entities in charge of providing basic infrastructure often led to under-provision. Frequently, the delivery of infrastructure was rationed to a limited part of the population and associated costs of production were high. Further, the ability to raise capital for financing new projects is constrained in the context of poor credit ratings of African States and macroeconomic stability programmes (Estache et al. 2005, Nellis 2005).

The importance of delivering quality infrastructure, however, has been underlined by the United Nations Declaration of the Millennium Development Goals (MDGs). Research finds that investing in water and sewerage positively have impacts on the livelihoods of people. Boosting the provision of those services has a great potential in helping to reach the MDGs as a whole (ECA 2005b, ECA 2005c). In this context new ways of approaching the provision of basic infrastructure involving the private sector have been explored in developing regions and Africa.

The private sector can participate in the provision of basic infrastructure in various ways. Private sector participation ranges from rendering specific contracted services (like management services, construction, etc) to full privatisation. One form of private sector participation often talked about are public-private partnerships (PPP). PPP emphasise the collaborative element between the public and the private sector for achieving a particular goal. They have the potential of combining the concerns of the public sectors for equity and universal service delivery with competencies and strengths of the private sector such as efficiency, cost-effectiveness and responsiveness to consumers' needs (ECA 2005c, Labuschagne 1998).

Public-private partnerships are often perceived as a middle ground between full privatisation and state ownership. They allow governments to tap into the resources, financial and human, of the private sector while still hanging on to ownership and control over the asset. However, governments should be aware that public-private partnerships are not a "light" version of privatisation. Public-private partnerships involve capacities and commitment from all sides as much as do privatisation (Mitchell-Weaver et al. 1992). PPP also bear significant risks for the public sector and for public service delivery.

PUBLIC-PRIVATE PARTNERSHIPS ON THE RISE IN DEVELOPING COUNTRIES AND IN AFRICA

Public-private partnerships have become more widespread in the developing world. Since 1990 a large number of developing countries have tried public-private partnerships in different sectors and in different forms. Investment in water and sewerage with private sector participation increased from zero in 1990 to reach a US$ 10151 million in 1997 (World Bank, PPI Database). This boom was partly caused by the acknowledgment of states of their huge infrastructure needs while facing at the same time fiscal constraints in their attempts to maintain macroeconomic stability. As a consequence, many governments opened up their infrastructure sectors and invited private investors (Thomsen 2005).

Figure 1
Investment in Water and Sewerage with Private Sector Participation by Region, 1990-2005 (US$ millions)

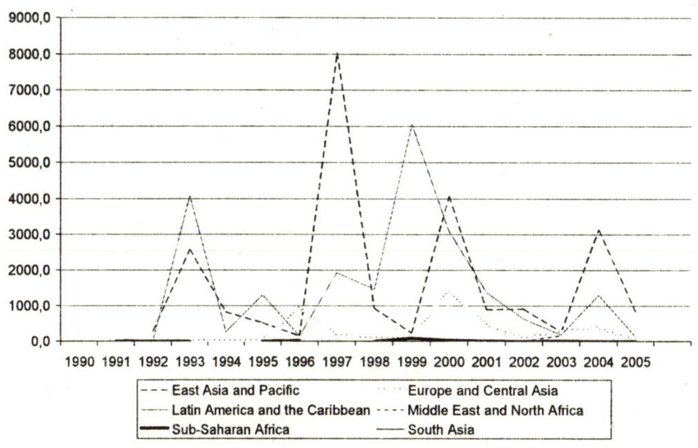

Source: World Bank, PPI Database

After 1997 investment in water and sewerage with private sector participation dropped and at its lowest point in 2003 (US$ 912 million) was less than on tenth of the 1997 level (World Bank, PPI Database). The lower levels of investment in infrastructure with private sector participation in the late 1990s and the new millennium were caused by a variety of reasons including the Asian financial crises and financial difficulties experienced by international companies in their home countries that reduced their appetite for risky investments in developing countries (Thomsen 2005).

The experience of private participation in Sub-Saharan Africa differed from the overall situation for developing countries. Figure 1 depicts the evolution of private sector participation in water and sewerage by region. Investment in water and sewerage in SSA only peaked in 1999 and has dropped since than. In international comparison Sub-Saharan Africa only represents a small percentage of investment in water and sewerage with private sector participation (0.3 per cent), less than the Middle East and North Africa region wit 1.3 per cent (see Figure 2). The biggest share goes to East Asia and the Pacific with 46.7 per cent followed by Latin America and the Caribbean with 43.5 per cent.

Figure 2
Share of Investment in Water and Sewerage with Private Sector Participation by Region, Averages 1990 - 2005

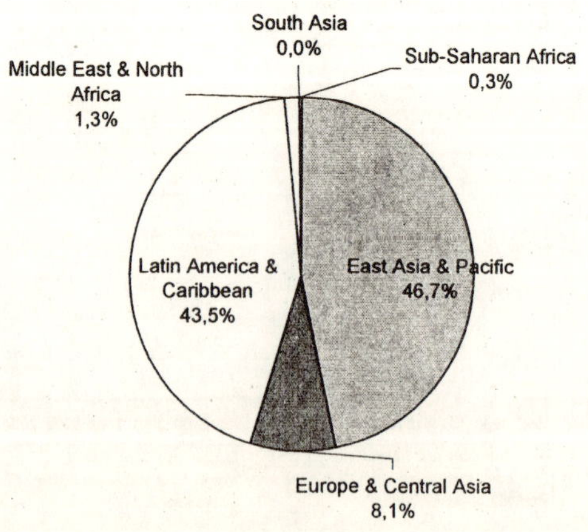

Source: World Bank, PPI Database

Consequently there is a marked difference in the presence of the private sector in the delivery of infrastructure in different regions. Particularly, in developed countries the share of the private sector in these industries is significantly higher than in developing countries. Table 1 shows the share of countries that have some sort of private sector involvement (including privatisations).

Table 1
Presence of the Private sector in the Delivery of Infrastructure (% of countries)

Developed Countries	80%
Developing Countries	35%
Sub-Saharan Africa	20%
East Asia	64%
Eastern Europe	62%
Latin America	41%
Middle East	18%
South Asia	13%

Source: Estache et al. (2005)

The gap between Sub-Saharan Africa and other developing regions is substantial in the water and sewerage sector. In 20 per cent of Sub-Saharan African countries the private sector is involved in the delivery of these services. In comparison, more than 60 per cent of East Asian and Eastern European countries involve their private sector; in Latin America it is just above 40 per cent.

The above table also suggests that there might still be room for increased private sector participation in Sub-Saharan Africa. The potential to involve the private sector to expand service delivery seems even more promising considering the sheer size of infrastructure investment requirements (ECA 2002). To meet its growing infrastructure needs, Africa has to spend an additional US$ 20 billion each year from 2005 to 2015. Only then the 7 per cent growth rate in GDP needed to halve poverty is attainable (Commission for Africa 2005).

A FRAMEWORK FOR PUBLIC-PRIVATE PARTNERSHIPS

Public-private partnerships can take a variety of contractual forms. Accordingly the government hands over more or less managerial competences and control over assets. Risks associated to the project are shared as well. Adapted from Thomsen (2005), the most common forms of public-private partnerships in the provision of infrastructure can be summarised in three categories: 1) service contracts; 2) delegated management contracts; and 3) construction support contracts (see Table 2).

The overall operational responsibility remains with the public sector in a service contract. The contracted private sector firm provides a number of specific services. The service contract has two main forms: the management support contract and the operation and management (O & M) contract. Management support provides technical and human resources against a fee. Ownership and some management responsibilities remain within the sphere of the public authority. Operation and management places more responsibility in the private hands. The private contractor is responsible for operating and maintaining the facility. It is paid according to the achievement of agreed performance criteria.

In delegated management contracts the public entity preserves ownership of the asset but hands over the responsibility for the management to a private entity. The duration of the contract is longer than for service contracts. The lease agreement is one form such a contract can take. The private firm is responsible for directly invoicing the end-user and it takes charge of the existing assets and all personnel. However, the responsibility for financing and building new investment remains with the public sector. Another contractual form is the concession. Under this type of contract the private firm takes on all responsibilities for management and investment for a longer (e.g. 20 years) period. The private firm invoices the end-user directly. The public authorities retain control over service terms and decisions on tariffs and enlargement targets. The affermage contract is more frequently used in Francophone Africa. The private sector entity bills all consumers and collects the revenue at the tariff set by the government. However, the company receives a fixed fee for each unit sold (covering costs and a regulated profit) that might differ from the tariff. Affermage contracts can incorporate investment requirements as well as incentives in the fee structure to meet targets on leakage, bill collection and connections of poor households.

In construction support contracts the private sector is involved in the design, construction, and operation of a new investment. The private sector firm bears some of the risks involved. The build

Table 2
Different forms of Public-private Partnerships

	Form of contract	*Operation & maintenance*	*Ownership*	*Investment*	*Commercial risk*	*Duration in years*
Service contracts	Management support	Public and private	Public	Public	Public	1-2
	Operation & management	Private	Public	Public	Public	3-5
Delegated management contracts	Lease	Private	Public	Public	Semi-private	8-15
	Affermage	Private	Public	Public	Semi-private	8-15
	Concession	Private	Public	Public/private	Public and Private	20-30
Construction support contracts	BDO	Private	Public	Public	Private	20-30
	BOT, BOO	Private	Public/private	Private	Private	20-30

Source: Adapted from Thomsen (2005)

design operate (BDO) contract delegates the design, construction, and operation of a new facility to a private sector operator. The property remains with the public sector. The private operator assumes the risks linked to design and management of the facility and is remunerated by an agreed fee. In the build operate transfer (BOT) contract ownership remains with the public sector. The private firm is given the responsibility of designing, financing, and managing the facility over a period long enough to allow the private firm to recover the costs it incurred. The build own operate (BOO) arrangement essentially differs from BOT in that the private sector firm retains ownership over the facility. Other arrangements are the build own operate transfer (BOOT) and the build operate train transfer (BOTT) contracts. Both are similar to the BOT contract, with variation in initial ownership and emphasis given to training before transferring the asset to the public entity.

Circumstances under which public-private partnerships in water and sewerage are concluded vary largely between countries and sectors. The public sector entity has to evaluate separately for each project whether a public-private partnership is appropriate and which contractual arrangement (concession, lease, BOT, etc.) to chose. The appropriate contract depends on the size of the project, the sector, the political environment, and the financial situation, only to mention a few factors (ECA 2002, ECA 2005c, Khalifa et al. 2003, WEF 2005). In the water and sewerage sector concessions were used most frequently accounting for 69.0 per cent of all investment and 41.5 per cent of all projects in developing countries (Sub-Saharan Africa: 52.4 per cent and 11.8 per cent respectively) from 1990 to 2005 (World Bank, PPI Database).

In order to isolate the determinants of the performance of public-private partnerships it is necessary to look at both, successful and failed projects. The Data of the Private Participation in Infrastructure (PPI) Database of the World Bank provides a first insight of the proportion of failure and success. In developing countries as a whole, cancelled or distressed investment in water and sewerage with private sector participation accounted for 11.0 per cent of projects and 33.7 per cent of investment from 1990 to 2005. In Sub-Saharan Africa 17.6 per cent of projects were cancelled or were distressed but they only represent 5.8 per cent of total investment (see Figure 3). The high global failure rates as percentage of investment were mainly driven by the two regions with the largest share in investment in infrastructure with private sector participation: East Asia and the Pacific and Latin America and the Caribbean (World Bank, PPI Database). This implies that on an average large projects failed less in Sub-Saharan Africa than in Latin America, the Caribbean, East Asia and the Pacific.

Figure 3
Cancelled or Distressed Private Sector Participation in Sub-Saharan Africa and the World, % of Total, Averages 1990 - 2005

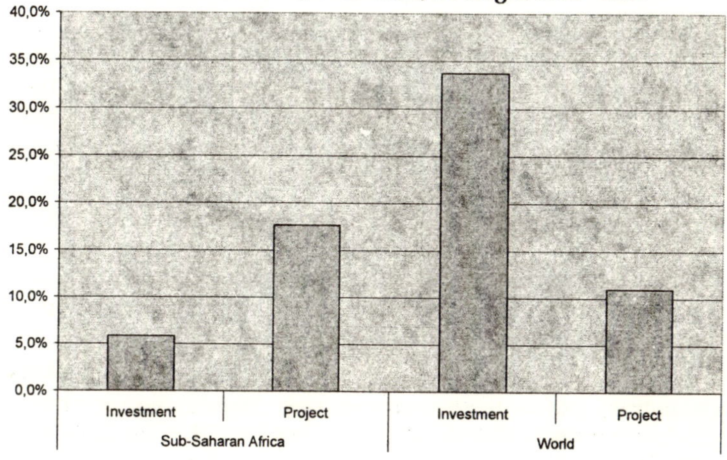

Source: World Bank, PPI Database

In general public-private partnerships have been more successful in areas such as touris projects and telecommunication compared to water and sewerage. The reason for this lies much in the nature of these industries. In clearly defined and limited projects, risk allocation is more obvious. Thus return on investment can be assessed with greater accuracy. Water and sewerage lack a priori these clear definitions. However, with proper regulation and strong commitment from both sides, public-private partnerships can succeed in these sectors as well (Farlam 2005). The following section looks at examples of public-private partnerships in water and sewerage.

EXAMPLES OF PUBLIC-PRIVATE PARTNERSHIPS

The case of Gabon and Société d'Energie et d'Eau du Gabon (SEEG) which combines the delivery of water and energy is an interesting one in many respects. SEEG is mostly foreign owned company but has been present in the water and electricity sectors in Gabon for a while.

Preparation for the public-private partnerships started well ahead. Since the early 1990s reforms in the water and electricity sectors saw prices increase to cost covering levels albeit reducing employment. The private sector participated in service delivery in the two major cities, Libreville and Port-Gentil. This long initial phase allowed government and the private firm to build capacities and to establish the base for mutual trust.

In 1997, the Government of Gabon and SEEG signed a 20 year concession contract for the provision of water and electricity. The contract specified investment obligations, fixing specific targets for the water and the electricity sectors. Incentive mechanisms were included that reward timely service expansion to more remote regions and to poorly connected neighbourhoods. The combination of water and electricity allows for cross-subsidies in the provision of infrastructure. Even though water only made up 15 per cent of SEEG's turnover, 60 per cent of investment went to rehabilitating and extending the water network (Farlam 2005).

Public-private partnerships can involve big (multinational) companies or small local enterprises. Mauritania developed a scheme of small scale independent water providers for small towns. In 1993 the government decided to decentralise the water supply management system in small towns. The new rules give local governments the possibility to engage in public-private partnerships for the provision of water. Municipalities still can assume the responsibility for water provision but in slightly more than 70 per cent of small towns water is provided in the framework of a public-private partnerships.

The private operators are usually given a one year concession if the groundwater is pumped by a diesel pump or a one month concession if it is pumped by electricity generated by solar panels. The contracts governing the concession are often loose, relying more on informal checks and balances between the community and the provider. The small size of the concessions allows for this more informal way of contracting.

Network maintenance should be provided by the National Water Department. But as it is often short of capacities, spare-parts, and means of transport, many private concessionaries have taken over some of the maintenance responsibilities. The private concessionary is in principle responsible for the extension of the network. However, as funds are short and concessionaries face difficulties in borrowing money from the formal banking system, extensions are often paid for by the community, rich individuals living in the community, remittances from the diasporas, or associations such as farmers' cooperatives (IRC 2004).

In Senegal the government and Sénégalaise des Eaux (SDE), a private firm with the French company SAUR as the main shareholder, entered a public-private partnership for the provision of water and sewerage in 1995. Since then the quantity of water supplied has increased by 20 per cent. The number of connections increased by 35 per cent. The collection of water bills has improved and water losses have been reduced significantly. The private operator was one of the first water companies in Africa to receive ISO 9001:2000 certification.

The preparation and negotiation were a long process. International donors helped the government build capacity in the involved ministries. The type of contract chosen was that of an affermage. The ownership and tariff setting power remain with the public entity. The private firm is responsible for the collection of tariffs, water treatment, and new connections. It is paid a fee that incorporates incentive structures for tariff collection, leakage reduction, and expansion of service.

The contract was chosen to avoid political backlashes that are frequently associated with private tariff setting and ownership. Further, tariffs were set in a manner to supply the first ten cubic meters at a lower rate. This scheme should render water provision more pro-poor. It allows poor households to cover their needs at a lower cost while recovering some of the costs from bigger (and presumably richer) users. This kind of tariff structure assumes connectivity of poor households. However, it has certain drawbacks. For example, if a couple of poor households share one connection, the tariff scheme will not work in their favour (Kerf 1999, Thomsen 2005).

A key factor of success was the flexibility on both sides. At the beginning, when SDE realised that the initial investment would be larger than planned because the quality of the existing installations was not as good as assessed, the government and the private operator entered a renegotiation process which allowed the private operator to carry out the necessary refurbishments. During the first two years the private operator was making losses but subsequently was able to capitalise on its investments.

LESSONS LEARNT

The advantages of small, local contracts involving the private sector are numerous as the example from Mauritania shows. Local involvement creates a sense of ownership among the local community. This translates into willingness to participate in the cost of network extension. Local concessions are a key driver of local economic empowerment as they permit local capacity and capital to build and accumulate (Berthélemy et al. 2004).

Contracts can be less formal as they are small and of short duration. Local contracts make constant communication and adoption of services easier, as both parties come from the same community and share a broader a priori understanding of the goals and expectations of a concession. The small size also facilitates an informal monitoring of achievements.

On the other hand, a large number of small contracts increases the work load of the central regulatory agency (if existent) as it needs to monitor and negotiate with a large number of concessionaries. These concessionaries most often do not have subsidiaries in the capital which makes communication between the central agency and the concessionary more costly.

The examples from Gabon and Senegal highlight several important issues. First, it is crucial to plan sufficient time for the preparation, planning, and contracting phase. Public-private partnerships that will satisfy both sides typically take time to be formed. Objectives and information need to be shared, compromises elaborated, and goals clearly specified. Second, public entities involved in the process should be very concerned about building the capacities required for public-private partnerships with their employees. Third, flexibility is key. Once trust has been established between the two partners, it is easier to renegotiate the contract along the way. In complex projects as water delivery, there will often be contingencies which will make it necessary for one side or the other to ask for renegotiation. Both sides need to be sufficiently flexible to allow for balancing interests (Kerf 1999, Thomsen 2005).

In the case of water and sewerage, ownership is less important for achieving efficiency than in other sectors such as transport or telecommunication. The difference in efficiency for water and sewerage between public and private providers is in fact statistically not significant (Estache et al. 2005).

It is far more important that the operators deliver their services in a competitive environment. When incentives are set in a way that they remunerate enhanced outcomes, the operators will improve their performance. Here, an interesting difference between public and private operators surfaces. In general, private operators are quicker to improve efficiency and lower cost than public operators. However in the long run, private operators will not be more efficient than public operators (Estache et al. 2005).

Further, a strong and capable regulatory body will improve performance in the water and sewerage sectors independently of ownership. Other variables influencing the outcome are the general quality of institutions and the level of corruption. In a setting of feeble institutions and/or generalised corruption, handing over the provision of water and sewerage to a private operator is not likely to improve the service or efficiency at all (Estache et al. 2005, Nellis 2003).

POLICY RECOMMENDATIONS AND CONCLUSIONS

Preparation

Even public-private partnerships might produce faster results than pure public sector projects, it is important that the initial phase is undertaken carefully. The risk of renegotiation increases with a less careful planning phase (South African National Treasury 2004, Thomsen 2005). Starting off with small projects involving the private sector and gradually moving to bigger and more complex ones helps building capacities within the government over time. For example, Botswana started off with relatively small contracts in constructing and maintaining government offices.

Donors or central governments can play a crucial role in the initial phase of a project by providing technical assistance and capacity building (e.g. PPIAF or Public Sector Comparator model developed by the Government of South Africa, see below) and informing the government as well as the private bidders about the needs and priorities of the respective other party. Helping to adjust unrealistic expectations on both sides and playing the role of an "honest broker" they can help to put public-private partnerships on a sound basis.

One way to assess the adequacy of a public-private partnership is the Public Sector Comparator as recommended by the South African National Treasury. The Public Sector Comparator compares costs of the government providing the service as opposed to a public-private partnership. To obtain the risk-adjusted Public Sector Comparator model, risks have to be identified, their respective costs assessed, and the likelihood of the risk estimated. The procedure is the following:

Identifying risks associated to each step in the project.

Identifying the impacts of each risk and estimating the costs associated to it.

Estimating the likelihood of occurrence for each risk.

Obtaining the expected cost of each risk by multiplying the cost and the likelihood of it occurring.

Identifying and developing appropriate strategies for minimising the likelihood of a particular risk occurring or mitigating the cost associated to the risk.

Allocating risks between the contractual parties. It is advisable to be explicit about which risks are born exclusively by which party and which risks are shared and in what manner they are shared.

Constructing a risk matrix (which consolidates all identified project risks, their impacts and their associated costs).

Building the risk-adjusted Public Sector Comparator model

This process helps building consciousness of the real cost of the project. Comparing the risk-adjusted Public Sector Comparator model with the institution's medium term budget frame will give an indication of the project's affordability (Adapted from South African National Treasury 2004).

Information

Governments at the local, provincial and national level should try as much as possible to inform the public about plans and possibilities for public-private partnerships. Political and popular resistance to public-private partnerships is a serious limiting factor for its success. When public acceptance is low, leading a fruitful and for both sides satisfying partnership will be hard to achieve. Government must inform well in advance about plans involving the private sector. They must inform the public about their reasons for asking for private sector involvement. Advantages and drawbacks should be discussed openly. Acceptance is generally higher when the process is transparent; this is particularly true for the final contract awarding. Rumours of corruption, true or false, can damage a public-private partnership crucially (Farlam 2005).

Participation

All stakeholders should be able to participate in the design of a public-private partnership. Active participation increases acceptance as concerned groups and individuals feel that their needs are heard and taken into account. Participation can start at the very beginning while assessing demand. In the case of water and sewerage, it might not be possible to connect all households directly to the main water supply. Alternative solutions (such as central fountains, flexible plastic hoses, etc.) could meet the needs of consumers better.

Community involvement allows for a design that responds, from the beginning, to local needs. Further, local business can be involved as subcontractors. One such approached is pursued in South Africa with the Black Economic Empowerment (BEE) policies where public-private partnerships must include a certain amount of local firms owned by individuals of previously disadvantaged communities. These measures aim at sharing the "profit" among the local community, while giving those firms and individuals an opportunity to build capacity and improve their services over time (Farlam 2005).

Transformation

In the context of free or under-priced provision, a gradual transformation is key. Increasing tariffs in a single big step often proves to be politically unsustainable. Planning for a sufficiently long period of time for price adjustments, enforced bill collection, and labour adjustment is likely to increase public acceptance. This is particularly the case if improvements in the service delivery, extension of connections, and better reliability are observed at the same time.

Political Commitment

For public-private partnerships to be successful, strong support form the political side is crucial. Most successful public-private partnerships in Africa took place in an environment of strong political commitment. A political champion, a person on the political side that advocates private sector participation, can make a huge difference (Farlam 2005, WEF 2005).

Contracting out Regulatory Functions

In the short run, contracting out regulatory functions can bridge the capacity gap. It gives regulatory bodies the time to build up their capacity while already being functional (Nellis 2003). A world-wide survey among regulatory bodies found that three quarters contracted out certain tasks. These tasks included tariff reviews, compliance monitoring, and dispute settlement. The findings of the external party were binding in only 15 per cent of the agencies. The agencies saw the main advantages of out-contracting in improving competencies, building trust between stakeholders, and ensuring independence (Thomsen 2005).

Mitigating Private Risk

Risks are attributed in the contract to one party or the other. Political and currency risks are often interlinked. Private firms might not be willing to assume those risk or only with a large risk premium. In this case donors can provide schemes for mitigating those risks, making the private sector more likely to be interested in a project and lowering the risk premium considerably (Nellis 2003).

Currency Risk

Sudden and substantial devaluation poses a problem to private operators that had to raise capital in foreign currency. Rapid adjustments of tariffs often are not possible for political reasons. One solution proposed and implemented in a public-private partnership in Brazil is a sort of liquidity backstopping facility. Tariffs are indexed to local inflation, the liquidity backstopping facility provides, in case of sudden devaluation, a loan in foreign currency that is used to service the debt. The private operator can repay the loan to the liquidity backstopping facility as local inflation catches up with the devaluation (Thomsen 2005).

The Public-Private Infrastructure Advisory Facility

The Public-Private Infrastructure Advisory Facility (PPIAF) is an initiative supported by 11 donor governments and three multilateral organisations. Its goal is to help developing countries to improve their infrastructure by involving the private sector. It provides technical assistance to the government in the following areas: regulatory and institutional reforms; infrastructure development strategies; consensus building; and capacity building, among others (PPIAF 2005).

Synergies

One way of advancing service provision is through creating synergies for private initiative and infrastructure provision. In Tanzania, the Barrick Gold Corporation planned to construct a water pipe for operational needs. They aligned the track of the water pipe with the District Development Plan for water. This coordinated effort facilitated the provision of water to 35,000 people in the region (WEF 2005).

Another domain where improvement to the conditions of livelihood can easily be made is in tourism. The approach of pro-poor tourism tries to maximise the welfare impact of tourism on local communities. One aspect of this is that infrastructure provided for the tourist facilities can be extended to include local communities, often at low cost. These infrastructures include roads, communication, healthcare, electrification, water and sewerage. Integrating local communities and sharing economic prosperity with them has obvious benefits for the tourism industry as well. It is key that these issues are addressed in the initial stage of a project, in order to avoid backlashes (see PPT 2005).

Public-private partnerships are not a "light" version of privatisation or a way around capable institutions and sound regulation. In order to successfully involve the private sector, the public sector must have strong capacities within its institutions in order to negotiate a fair and satisfying deal. Public-private partnerships have the potential to combine concerns of the public sectors for equity and universal service delivery with competencies and strengths of the private sector such as efficiency, cost-effectiveness and responsiveness to consumers' needs. If implemented poorly, though, they can bring out less desirable aspects of each sectors. Hence, public-private partnerships demand careful project assessment, project implementation and specific measures to make it profitable for the poor (pro-poor).

REFERENCES

Berthélemy, Jean-Claude, Céline Kauffmann, Marie-Anne Valfort, and Lucia Wegner (Editors), 2004. "Privatisation in sub-Saharan Africa". Development Centre Studies, OECD, Paris.

Brugger, Fritz, Richard Franceys, Peter Gleick, Jim Lamb, Meena Panaliappan, Dieter Rothenberger, Neigel Scott, Luis Uzin, Gary Wolff, 2005. "Public-Private Partnership for Water Supply and Sanitation: Implementation Guidelines Manual for Sustainable Municipal Water Services", Public Release Version 1, as of April 2005. www.partnershipsforwater.net, accessed August 31, 2005.

Capital for Development CDC, (2005), Commission for Africa, Business Contact Group, "Infrastructure", London, 1st March 2005.

Commission for Africa (2005), "Our Common Interest: Report of the Commission for Africa", March 2005.

Economic Commission for Africa (ECA), (2002), "Improving the Policy Framework for Private Public Partnership", Ad Hoc Expert Group Meeting on Improving the Policy Framework for Private, Public Partnership, Addis Ababa, 29-30 November 2002.

Economic Commission for Africa (ECA), (2004a), "Economic Report on Africa 2004: Unlocking Africa's Trade Potential". Addis Ababa.

Economic Commission for Africa (ECA), (2004b), "Scoring African Leadership for Better Health." Addis Ababa.

Economic Commission for Africa (ECA), (2005a), "Economic and Social Conditions in Southern Africa 2003: The Challenge of Private Sector Development in Southern Africa". ECA Subregional Office for Southern Africa, Addis Ababa.

Economic Commission for Africa (ECA), (2005b), "The Millennium Development Goals in Africa: Progress and Challenges", Addis Ababa.

Economic Commission for Africa (ECA), (2005c), "Public-Private Partnership for Service Delivery: Water and Sanitation", Third meeting of the committee on Human Development and Civil Society, Addis Ababa, 4-6 May 2005.

Estache, Antonio, Sergio Perelman, Lourdes Trujillo, (2005), "Infrastructure Performance and Reform in Developing and Transition Economies: Evidence from a Survey of Productivity Measures". World Bank Policy Research Working Paper 3514, February 2005.

Farlam, Peter, (2005), "Working Together: Assessing Public-Private Partnership in Africa". The South African Institute of International Affairs Nepad Policy Focus Report No. 2.

Global Environment Facility (GEF), (2005), "About the Global Environment Facility (GEF)". www.thegef.org, accessed August 30, 2005.

International Telecommunication Union (ITU) (2004), "African Telecommunication Indicators 2004". ITU, Geneva.

IRC, (2004), "Mauritania", Water and Sanitation Program - West and Central Africa, Thursday 15 January 2004. http://www.irc.nl/page/7605, accessed August 7, 2005.

Kerf, Michel, (1999), "Do State Holding Companies Facilitate Private Participation in the Water Sector?: Evidence from Côte D'Ivoire, The Gambia, Guinea, and Senegal". World Bank Policy Research Working Paper No. 2513, November 30, 1999, Infrastructure Department, World Bank.

Khalifa, Nawal, Driss Essaouabi, (2003), "Public-Private Partnership: Which Strategy for the Drinking Water and Sanitation Sector in Morocco?". Water Resources Development, Vol. 19, No. 2, June 2003.

Labuschagne, Jean-Pierre, (1998), "Development debate and practice: Public-private partnership in the health industry". Development Southern Africa, Vol. 15, No. 1, Autumn 1998

Mitchell-Weaver, Clyde and Brenda Manning, (1992), "Public-Private Partnerships in Third World Development: A conceptual Overview". Studies in Comparative International Development, Winter 1991-92, Vol. 26, No. 4, 45-67.

Nellis, John, (2003), " Privatization in Africa: What Has Happened? What is to be Done?". Center for Global Development, Working Paper No. 25, February 2003.

Nellis, John, (2005), "The Evolution of Enterprise Reform in Africa: From State-owned Enterprises to Private Participation in Infrastructure - and Back?". Fondazione Eni Enrico Mattei, Nota di Lavoro 117.2005, September 2005.

Pro-Poor Tourism (PPT), (2005), www.propoortourism.org.uk/, accessed August 31, 2005.

Public-Private Infrastructure Advisory Facility (PPIAF), (2005), www.ppiaf.org/, accessed October 12, 2005.

South African National Treasury, PPP Unit, (2004), "Public-Private Partnership Manual". http://www.ppp.gov.za/ accessed August 30, 2005.

Thomsen, Stephen (external consultant to the OECD/DAF), (2005), "Encouraging Public-Private Partnerships in the Utilities Sector: The Role of Development Assistance". Overview study prepared for the NEPAD/OECD Investment Initiative conference on "Investment in

African Development: Making it Happen", Imperial Resort Beach Hotel Kama Hal, Entebbe, Uganda, 25 - 27 May 2005.

World Bank, (1998), "Public Policy for the Private Sector", Note No. 147, August 1998. World Bank Group, Finance, Private Sector and Infrastructure Network.

World Bank, PPI database. http://ppi.worldbank.org/reports/customQuery Aggregate.asp, accessed September 24, 2006.

World Economic Forum (WEF), (2005), "Building on the Monetary Consensus: The Growing Role of Public-Private Partnerships in Mobilizing Resources for Development". United Nations High-level Plenary Meeting On Financing for Development, September 2005. Geneva.

PART II

MANAGEMENT OF IRRIGATION

14

DEVELOPMENT AND MANAGEMENT OF IRRIGATION WATER

*S.K. Rau**

WATER RESOURCE DEVELOPMENT (WRD)

A lament echoes from the past days that "we have excellent water project construction engineers but not enough irrigation engineers". The raison d'etre for this rather facetious remark lies in the vicissitudes of the after ministration of the enormous quantum of water created by them: It is felt that it has not been applied to irrigate the project estimated ayacut, on the farmers fields, in time, as per a dependable schedule.

But the scenario has changed much in the concept of water, methodologies in its usage and the need for it in greater quantities. A planned forecast of the requirements including irrigation is made by Government of India and this perceptive plan is being implemented. The National Commission for Integrated Water Resource Development 1999 (NCSWRD) and the National Water Policy (April 2002) of the Government of India provides estimates of the current availability and the future requirements of irrigation

* Former Secretary to Govt. of India and Director General, NIRD, Hyderabad.

among other needs. The projection of irrigation requirements basing at 534 km as on 1997-98 is estimated to be at 557 km in the year 2010 and 611 km in 2025. Another way of looking at these has been 688 bcm in 2010 and 910 bcm in 2025. The Central Water Commission has some variations based on the figures of potential created till the end of the VIII plan and the utilization. So have SK Gupta, PS Minas et al in a paper at the International Conference on "Managing Natural Resources for the 21st Century" February 2000. In the later two cases, it is also pointed out that the potential created has not been fully utilized by the States. The estimates or actuals of water available or utilized cover all sources as flow water, underground accumulation and rainwater harvested and stored. The balancing of these sources for long term use is a part of the plan for Water Resources Development (WRD). The effort to increase the availability of water for irrigation is an active process.

There is sometimes a preference and weightage given to one or the other of these sources and methods of extraction. There are some who do not prefer high dams and big storage provision. Their view is also anchored on an International Committee too, on this mode. Their objections broadly based on : (a) high costs without optimal utilization of the water as also the time factor; (b) the human problems of settlement and rehabilitation of project affected people; (c) the damage to ecology; (d) problems of water logging, salinity and sedimentation, and (e) the likely differences within the States that could arise on riparian rights. Hence the preference of no dam protagonists is for smaller water collection devices. They prefer low cost minor irrigation works, tanks, reservoirs, wells, lift irrigation projects, tapping streams through pipes, rainwater storage in situ, ground water charging, watershed treatment and the like.

It is true that at one stage of our history, small-scale water tapping and preservation measures were taken up. Technology was limited but the innovation wide. A remarkable measure frequently quoted is the anikut created in 2nd century AD by Karikala Chola across the Kaveri River. It was built entirely with stone (Kal Anai). No mortar was used. Many reservoirs and tanks were the vital source for irrigation and drinking water. At one time (1997) there were 3.92 lakh tanks in Tamil Nadu (the terrain suited tanks) and over 1.65 lakh sizeable wells. The interesting documentation is in the inscriptions (stone and copper plate) where provisions were made for their upkeep. Andhra seems to have only 11,274 big tanks but plenty of wells for drinking of water. North India seems to have favoured wells in earlier times. Irfan Habib states that "wells were probably the major source of artificial irrigation in most areas of northern Indian during the Sultanate". Bore well irrigation has become more common now. It is pointed out that this process has "shifted the determinants of water access away from the communities and into the hands of individuals". A greater danger pointed out the water resources are under severe threat from several sources.

But with the growth of the enormous demand for water, for different needs including irrigation, and with huge sources being available and also with the dawn of new technologies, some of the old order had to yield place to the new. A case like the Three Gorges Dawn in China (2006) has changed the geography of China in its coverage; years of "sorrow" of people has changed into comfort. Brahmaputra has a catchment of 1,94,413 Sq km and has water potential of 436312 kms. Godavari River has a catchment of 48357 kms, and water potential as striking; 80% of its water, in seasons, causes floods, or flows away into sea. All this water is direly needed to the different needs of the growing population. This water has hardly be fully tapped through minor works of engineering and storage, such water cropping innovations may follow the canals constructed for the flow irrigation and have validity in a limited way for the villages. Situations sanctify the salience of big dams too. It is not either, or always. Fortunately Government of India is encouraging both paths in the Plan for WRD.

Another facet in the national policy is that a National Water Grid (as in Power Grid) is also visualized and discussed. In the seventies, the Ganga-Kaveri linking was propounded by Dr. K.L. Rao. It connected the northern rivers to the Southern with a wide canal length of about 2640 kms. A Garland Canal Project of linking Himalayan Rivers with those of the Peninsula, in stages was also examined by Dasturs. The ideas achieved publicity and a Task Force was also appointed (2002) by Government of India. A comity of interlinked water flows obviates inter-state disparities. This is also conducive to a more predictable agricultural production not conditioned by the vagarious of precipitation in different parts of the country, since the gain in one zone can be shared by a deficit of another. The task can possibly be handled technically too. But a caveat raised sometimes, is the cost-benefit ratios. Another one, which is more portentous and scary, is the totally provincial outlook of the States on Water problems. Even if the flows and shares between two or more states are streamlined, the local pressure and political compulsions would be a weak impost on the entire structure due to the strong demands and riparian claims in scanty seasons. The pressures of the powerful and articulate lobby of farmers can hardly be resisted by the persons in power. These demands and obstinacies become avalanches and glissades of mass movements. The riots that took place in 1991 in Karnataka after the Kaveri water Tribunal interim verdict are an example of the well documented water riots. Around 50 thousand Tamils fled Karnataka; apart from deaths, the material loss was estimated at Rs. 17 crores. It has been the experience in other river water disputes too that there are no permanent agreements or disagreements where water spirit is funneled through popular mass uprisings. There could be an overspill of democracy sometimes.

One possible step is to declare some rivers that necessarily come under water transfers, as 'national' for governance by the central government. If necessary this has to be done by appropriate amendments to the constitution. There is a proposal to bring water sources under a concurrent list. If not directly controlled, water will be iniquitously appropriated or structures elected unilaterally even on inter-state rivers. Kaveri, Krishna, Godavari and others recently produced ambiguous facile assertions by concerned States in their own favour. Amartya Sen may have to now write about the 'Equivocations Indian". In a way, irrigation water has to be nationalised for the common good.

WATERSHED

A step in the process of water Resource Development is encouraging the seepage of rain water on to the soil and further into the ground. This effort can be regarded as inhering in the watershed development for dry land agriculture. But these operations should not be limited to dry land rehabilitation but as a step for water conservation and underground water development also. There is much consideration of the cultural practices, people's participation, technologies, land shaping, postoperative maintenance and the like on watersheds. The essential objective is to capture the rainwater for retention on the field and for percolation into the ground. It is not easy to immediately quantify the water that developed underground except for the effect on the aquifer known later. Of course, through the diverse rainwater preserving methods and a suitable cropping pattern dry land agricultural production may increase, as it should; just as the water productivity in irrigated areas. Initially the water nutrient is fed to the arid land and by nourishing further through long term engineering structures under ground water becomes robust. This is a slow process with the gradation of retentivity constructions "watershed is not a technology but a concept which integrates conservation, management and budgeting of rain water". The concept should conceive of different technological devices for the soil moisture retention and seepage for groundwater improvement. In hilly areas where the slopes are prominent, water flows from top downwards. Cultural practices are adopted only to temporarily retain the water on the cross bonded terraces. The Angamies are one of the best terrace cultivators in the north-east of India.

MANAGEMENT

In the wake of the success of WRD as per the plan, the management of all the water available and that which could accrue becomes the real task "Management" now-a-days is "a many splendoured word". In the present context it covers many animations. It is not limited to structures of control and supervision or arrangements for monitoring such livers. The government has to balance allocation of the available water among the other needs covering the irrigation of necessities, industrial demands, human and bovine requisites. Much admitness is needed in meeting the respective demands, lest the water could become an inflammable commodity!

Irrigation water distribution by the government department is different from the management of the conjoint use of all sources, while also sourcing them, and both are different from Participatory Irrigation Management (PIM). In the use of water for several needs and the varied sources have to be carefully husbanded. The production increase of these without hurting the sources, the equilibrium in their use are as important as encouraging the conjoint use of the flow, underground and rainwater while in others there is scarcity. The distribution of the stored flow water by the governmental department is sometimes known for small adjustment in crop needs, in timings, in adequacy and in equitable reach to all.

The ground water exploitation is causing concern all over. With absolute individual rights over land with all the under ground wealth (a frequent compulsory clause of rights on treasure trove and other underground wealth in all land sale deeds), there is enthusiasm to extract this wealth through tube wells. This is frequent in dry lands with the aquifer getting more low. In areas where such exploitation is taking place, it is estimated that groundwater tables are falling a meter or more per year. It was also calculated that out of 5711 block (Mandals, Talukas) in the country 310 are over exploited and that no water can be extracted in the near future. The figures are growing and 4.5 lakh sq km show this fast declining trend (Ground Water Board calculations).

With the liberal usage of chemicals for crops, alkalinity and salinity are also noticed in ground water (Garrett Hardim's well known conceptualization of the "tragedy of the commons" is evident). However where the tube wells are near a canal, there would be extraction of water spared through seepage. Dhawan estimated that half of the crop output originating from tube well irrigated lands in Punjab is from ground water that is of canal origin. This is indirectly a pilferage of canal water!

But the situation arising out of tube well extraction is getting distressing. Even those who could not have otherwise afforded the installation of a deep tube well and run it for a water demanding crop are now entering into the numbers. One of the causes for this tube well cupidity is located as electricity. "A major cause of wiser spread ground water depletion in India is the perverse incentives created by energy subsidies". In some states power is also given free for some categories. This gives scope for them to run pump sets without costs or time constraints and in some cases spare it to the non-categorized sections.

The government is aware of the rapid decrease of ground water levels and the water availability. Apart from the exportations on the prudent and far-sighted use of the ground water, a Central Ground Water Authority (Act 1986) was constituted covering environmental angles also. The Board implements several regulatory and advisory measures. So do some of the States in Legislative Enactments. Attempts are to enforce the law. But it would appear "well nigh impossible to enforce on 20 odd million pump owners, a member that is growing at a rate of 0.8 million an year". The Government, farmers, Voluntary Agencies, Panchayati Raj Institutions all have to concentrate on ground water, by contriving the rain water seepage under the ground. Several measures have been often repeated: Earth bunds, erecting stonewalls, trenching in the contours, digging percolation tanks and other minor engineering works. This aspect requires as much attention as the flow water under the total Water Resource Development (WRD). After all the canal irrigation is also from rain water. It does not rain in Spain only!

There is somewhat of a piercing discussion on the origin of the people's water distribution bodies and irrigation societies. It is averred that the "Principal proponents involved in the debate are the bureaucrats, academicians of the country and the International Agencies like the USAID and the World Bank. Surprisingly a farmer is yet to make a debut on the debating scene". The writer further mentions that the relegation of the water distribution to a society is to lighten the department responsibility and the reduction of their operating costs. This is a critical but probably a germane comment. It should also be remembered that the democratic governments and the intellectual analysts felt that rural development can be resuscitated only when people have access to their natural resources and can utilize them by themselves. This has been the bit motif of many of the people's organizations encouraged by all including the stakeholders. Simultaneously a vast government programme of water development was on way. Undoubtedly it is a great burden monetarily and manpower wise. It was therefore an ideology and a necessity of cost sharing synchronized. Some economy does arise in the people distributing the water among themselves and also maintain the system. In a way, such saving of government expenditure has been the trend elsewhere too. After attending a conference on "Irrigation Management Transfer" in Wuhan (China) Dr. Vaidyanathan, while reviewing the papers and ruminating on the impressions observed "A common motive for attempted transfer in most countries seems to have been the inability of the State to cover the O & M costs from water charges, resulting in the growing strain on the government finances". Randolph Barker of the "International Water Management Institute (Sri Lanka) also shares this view: "The interest in the transfer of responsibilities to a User Group is two fold (i) to help reduce government responsibility (ii) to increase productivity". In view of the pronounced reason of cost and manpower saving by the government department after Water Users Association started functioning, it would be appropriate to study the actual results. This would cover government budgets and the work by WUA in taking over part of the irrigation responsibility.

The Water Users Associations (WUA) has experienced many criticalities in their formulations and functioning. But the lack of cost consciousness of water and its monetary value seems to be still endemic in some. Competitive claims for more water, excessive use for their particular crops, wasteful discharge of water into their neighboring nullah after flooding their own fields show inadequate understanding of the costs in producing the water and those of its distribution process on to individual fields.

Among the many estimates, depending on the location and distances, one estimate is that a "farmer enjoying surface irrigation (flow from major or medium projects) pays 0.2 p. per 1000 litres of water. The same in the government owned Panchayat managed tube wells is estimated at 0.19 p. as against 0.25 p. in a cooperative lift irrigation project." It is well known that the very high costs of major, medium and to an extent minor irrigation are entirely borne by government. The farmers however pay negligible costs for the water that is supplied to them on their field. A modicum of cess (or levy) is paid in some states based on the area basis. The water for irrigation is virtually free for all categories of farmers.

This unconscionable subsidized system has been recognized by many and mentioned often. But there is always a hesitation in political circles to alter the situation. There is a great solicitude towards the farming community as they are the anna givers (Anna data) and also due to the great stress on agricultural production, as also the priority being given to rural areas. Supply of irrigation water is the prime concern and not the costs. The farmers lobby is also a powerful political force.

But WUA itself is a farmer's organization distributing water not as largesse but having secured it at a cost from government. This arrangement of government charging for its bulk supply will introduce the cost consideration and a market orientation to the distribution process: Purchase and sell; the sale proceeds go to meet the cost of the bulk supply. Part remains with the WUA.

Government can be as considerate as possible in fixing its sale price. One of the WUA models in Philippines is that "the Association assists in the delivery of the water to the various turnouts, prepare the list of the irrigated and planted areas, distributed the Irrigation Service Fee Bill (ISF) prepared by the National Irrigation Administration (NIA), collect the payment of ISFs from farmers and remit the collections to NIA. For this activity they are entitled to a share of the amount collected. Similar functions were performed in India by an irrigation oversees, a Revenue Inspector or a similar village paid functionary. A type of decentralization to the new water institution has to be contemplated. The only suggestion is that the water supplied by government be charged at a reasonable commercial rate. Business with water is becoming a feature when tube well water is sold to the neighbor at an agreed price. When there is no over exploitation of water out of cupidity this is a benefit to the giver as to the receiver. The financial transactions suggested to be extended are for the flow water. Land tax, where it still exists now is not much of a revenue to the state. Similar is the negligible part of the water "cess". The situation is sought to be altered for the WUA. They pay to government a reasonable rate for the water and charge among themselves the costs of distribution. Any difference remains for the common benefit of the constituents of the Association. The Agricultural Prices Commission takes into account the commercial costs to the farmers in fixing the prices of the products.

Where farmers associations exist as registered companies the legal position and the operations are opaque and straight. The company taken up other commercial activities as purchase of inputs their sale, handling of the farmers produce for the market. They can handle water also as a commodity transacted between the Government and the farmer. This will be a schematic business procedure. The ultimate gain is for the shareholders.

To an extent, the process of costing water, presupposes the supply on a calculated basis in estimating the discharge accurately. At present time is the calculator. This will have to alter with meters and other scientific control devices. Water has to be calculated on volumetric basis and not by area basis. "In the Jingru PIM project, the Bank funds are used to install meters at each of the WUA so that the farmers could save in water fees, if they reduced water use. Maintaining the meters distributing the water and collecting the water fee and turning it over to the local bureau was all that the WISA did".

On a different model from the WUA, the Pani Panchayats are formed in some states (eg. Maharashtra and Orissa). They are in a way like PRIs in their constitution. They handle the irrigation water distribution, by common understanding without the prescriptive rights over water. But if they are supposed to own water and trade in it, this likely to distorted benefits to some segments of the society and iniquitous market conditions. P. Sainath has been crusading against this privatization of water handling, and felt that "the so called Pani Panchayats of Angul in Orissa (was) one of the biggest frauds in water sector of this country". He quotes the situation in many Latin American countries where water barons have sprung up. He also felt that the World Bank is harming the needy. But when it is said that the Mexican reforms required that the government to "create a new legalframe work and a system of private property rights over water" one need not read too much into it other than the creation of autonomous and semi autonomous water institutions. The fair play or lack of it in an elected Panchayat is another story not linked with privatization.

The Pani Panchayats which are democratically organized institutions and are handling the distribution of water to the farmers are being encouraged by Government of India too. They have announced subsidizing some such Pani Panchayats by Rs.50,000 and also training one male and a female in each Panchayat as "Water Masters". They will be organizing and steering entities. Government of India proposes to provide financial support to this scheme.

PARTNERSHIP

"Participatory development is a new socio-economic force aiming for sustained development at the village level". This is the new credo frequently chanted all over. In some ways this is revivalism of the past. Nostalgic statements about inter-dependence of the past, mutual help, community action, voluntariness are recorded. In most Asian countries these were locally organized bodies called by different indigenous names.

Such social or situational alliances are now being resurrected to suit the present time, in the shape of democratic institutions. In irrigation, government is supplying water to the farmers fields at their own expense; or improving the dry areas through soil conservation departments. Such activities being now handed over to peoples committees: Water users associations or watershed committees.

Governmental initiatives have become necessary. In the 20th century individualism is a force against formatted societal mores and group cohesion. Heterorganic actions do gain the place. Caste, self-interest, kinship, political attachments or internecine rivalries in the villages make the constitution of partnership units, such as WWAs or WSCs, difficult. Where there may be a common production interest, or trade fraternity or where there is already a group formation (induced or voluntary) such as a commune (China) Kibutz or Moses (Israel) Edijo (Mexico) it is easy to launch a collective action on common interests. Fortunately irrigation is imperious to the inbound agriculturist. Common action is predictable.

For the climate of united action a demonstrated success is an incentive success stories of shared waters encourage emulation. In considering subsidies and responsiveness in sharing costs Amita Shah mentioned, "one of the important achievements in this regard is the approach initiated by MYRADA and subsequently adopted by KAWADA". It has been demonstrated that if the farmers are convinced of the economic benefits (in terms of increased returns) they can be made to share as much as 50 to 60% of the costs. This is a sign of partnership with government. Since water is vital, the prospect of allotting it among themselves as per individual consents is bound to be very attractive particularly when they see a fruitful effort elsewhere. The palate has to be wetted for the appetite to grow.

While enthusiasm for united action grows with a perceived benefit, problems did arise in starting the project with peoples volition. Felt need, leadership and proximity help. Small projects for water sharing such as tube wells, mini-irrigation have been more hospitable. Smallness of size made inter-personal relationships easier.

Rakesh Hooja feels that "awareness generation, capacity building, attitudinal orientation" are extremely important. But he favoured "wherever water is to be distributed among 30 or more farmers the PIM and WUA approach should be mandatory". Unfortunately voluntariness cannot be packaged nor ordained.

Most State Governments have passed Acts to enable the formation of irrigation or water saving elective institutions. The Andhra Pradesh Farmers Management of Irrigation Act (1997) provides a "legal backing for implementing measures to institutionalize the participation of irrigation users". A formatted organization with a legal status is created since individuals cannot ply the network of water distribution in any given area. The government manages the irrigation system (PIM) through the instrumentality of the institutions installed. "Participatory process, even at its best, needs an organizational and formalized frame for its stability and procedure". Bodies constituted by the electoral process, with "entitled" members, discharge the functions of government in a decentralized irrigation system.

The structured institutions constituted by the electoral process of single membership seem to be developing occasional hiccups in functioning. The usual election politics seem to cloud the voluntary spirit. In view of the limited light of prominence, some prospective candidates shy away from contesting in WUAs. The proviso that "office bearers" of WWA are disqualified for Panchayat electoral posts has also become a discouraging factor in the participation in WUA Committee. As on October 2006, in Andhra there are about 2065 vacancies out of which 317 are those of Chairman, 298 are those of Dy. Chairman and 1450 are members. The process of participation in Participatory Irrigation Management seems to be getting stylized, attracted more to power than partnership and only as a hapless alternative. But thousands of WUAs are functioning, probably as instruments.

COORDINATION

The line departments of Government, other organizations which are statutory, semi-governmental or voluntary, all do apply themselves to water problem in the rural areas. The coordination among them has become a nettlesome problem. In the studies done in 1982 and 1985, Purandare and Jaiswal noticed the absence of inter-sectoral coordination among irrigation assess groups. Above all, the Panchayati Raj Institution is the powerful decentralized replica of the entire state government. Its functions, contemplated under the 11th Schedule of the Constitution (Article 243 G) extend to 23 departments of the state government. Item 2 of the Schedule is "Land Improvement and Soil Conservation". Item 3 is "Minor

Irrigation and Watershed Development". Items 11 and 29 contemplate the powers of drinking water and maintenance of community assets. Powers under Item 2 and 3 coincide with the work of the watershed and that of the Water Users Association. The legal jurisdiction of PRIs and that of these two specialized institutions clash. The geographical jurisdiction may not be coterminous with that of a Panchayat, and may cover that of more than one Panchayat. Thus legally two sets of Institutions created by government and financed to an extent by them, clash. Dual membership of the villagers may be embarrassing to them when any difference of opinion between the institutions arises. Coordination among these institutions is problematic. Many suggestions for this have been made. One of them is the creation of a new village institution comprising of the younger educated members of the WSC and WUA as a subsidiary of the Panchayat on the lines of the Village Development Board of Nagaland. Segregation of membership on youth and education is difficult due to different standards, time and locations. Youth vs. the aged can raise another problem of coordination. Some suggest that the specialized functions committees (WSC Committee and WU Committee) should be only sub-committees of the Panchayat set up. A collegium of the "Heads" of all the local institutions for coordination is another suggestion. "Signing a Memorandum of Understanding between the Project Authority of WSC and WUA" is mentioned. Also "formal and informal linkages should be established between watershed associates and the line departments including the PRIs to integrate various developmental schemes". Rather simplistic solutions that emerge in workshops!

This lack of collaborative action at the ground level organizations and the cutting edge activities seem to be a problem elsewhere too and fodder for the theoreticians. Talking of USA situation Mark Lubel et.al, felt that the "Agencies throughout the federal government increasingly favour the growth of new governing institutions that encourage cooperation between local actors with conflicting interests, divergent geographic bases, overlapping administrative jurisdictions to solve continuing disputes over resource management". This is exactly what the Govt. of India did when all the developmental activities at the local level (23 of them) are brought under the decentralized authority of the State - the Gram Panchayat. Unless this comprehensive survey of all programmes of different local rural institutions is definitely vested in the Gram Panchayat the coordination problems will sometimes to be problematic.

Managing different facets involved in exemplary irrigation practices is a labyrinthine travel. Bringing harmony among heterogeneous rural population (in India) who could have divergent ideas with conflicting interests, also ridden with caste, creed and localism is an ambition in our epics too. To convince them of the advantage of a conjoint utilization of the flow, underground and rainwater frugally and pecuniary needs considerable persuasion and fiat. To ground group action and participatory spirit is becoming an exasperating task in both watershed and water distribution schemes. To achieve coordination among different government departments and local natural organizations has been the cogitation of the ideologues and the endeavor of administrators for a long time. To handle these strains successfully a new management orchestrisation may have to be heralded. In the dichotomous situation between tradition and American management force, Japan is said to have evolved a Confucian-Shinto-Nippon style of management. We have to evolve an Indian version.

REFERENCES

These GOI Projections have been taken from the Indian Journal of Public Administration - Vol. XLIX - No.3, Page 284. Quoted by Ramaswamy R Iyer.

Ibid S K Das, Page 266.

Irfan Habib and Tapan Ray Chaudhury. The Cambridge History of India Vol I, Cambridge 1982, P. 43.

"Are Wells a Potential Threat to Farmers Well Being", S.Janakiranjan and Marcus Moench. Economic and Political Weekly (EPW), September 16, 2006, P. 3977.

"Conflicts over Water and Land in South Indian Agriculture" Steen Folkes. Economic and Political Weekly (EPW) February 14, 1998, pp.343-345.

Ratna Reddy et al "Watershed Development and Livelihood Security", Centre for Economic and Social Studies, Hyderabad 2001, p.12.

Dhawan BD (1993), "Indian Water Resource Development for Irrigation. Issues, Critiques and Reviews" Commonwealth Publications, New Delhi.

"Irrigation Institutions in a Dynamic Economy". Tushar Shah, Mark Giordono and Jinxia Wang, EPW, July 31, 2004, p.3455.

Ibid, page 3656.

S. Satish, WAMANA, Oct 1987, p.10.

EPW Nov 19, 1994, p.2966.

"Organisational Change for Participatory Irrigation Management", APO. Tokyo, 2002, p.24.

Ibid, p.10

"Irrigation Institutions in a Dynamic Economy", Tushar Shah et.al, EPW, July 31, 2004, p. 3452.

P. Sainath 'Hindu', March 22, 2006.

"Water Sector Reforms in Mexico", Tushar Shah et.al, EPW, Jan 24, 2004, p. 361.

Hindu, 4-10-06.

"Integrated Watershed Management", Rajesh Rajoria, Rawat Publications, 1998, p. 169.

"Economic Rationale, Subsidy and Cost Sharing in Watershed Projects", EPW, June 25, 2005, p. 2667.

Revisiting Participatory Irrigation Management (PIM) in India, The Indian Journal of Public Administration, Vol. XLIX, No.3, p.330.

"Water Users Associations in Andhra Pradesh", Jarween Jarath, CESS, Hyderabad 2001, p.16.

Journal of Rural Development, NIRD, Hyderabad, p.288.

Eenadu, Oct 11, 2006, p. 3.

"Watershed Development in India", NIRD, Hyderabad 1995, p.23,27,28.

NIRD Bimonthly April 2005, p. 3.

"Kandi Watershed Development Project" A Critical Evaluation, EPW, Dec 1993, p. A 127.

"Integrating Equity Concerns in Watershed Projects" B. P. Maithani, *Political Economic Journal*, Vol. 9, p. 59.

NIRD, Bi Monthly, April (2005), p.3, Workshop on Project Management of Watersheds under Different Programmes.

"Watershed Partnerships and Emergence of Collective Action Institutions" *American Journal of Political Science*, No.46, Jan 2002, p. 148.

POINTS

A. Water Resource Development

1. Perspectives of Water Resource Development
2. Alternatives of Big Banks and Minor works
3. National Water Grid. Nationalize water
4. Watershed programme should also encourage groundwater accumulation

B. Management

1. Conjoint use of all water sources
2. Prevent overexploitation of ground water
3. Commercialize water use
4. Partnership
5. Coordination
6. Different management techniques needed

15

DEMOCRATISING IRRIGATION MANAGEMENT AND REFORMING WATER PRICING SYSTEM IN INDIA

*Mamata Swain**

INTRODUCTION

In the wake of liberalisation, privatisation and globalisation, there has been a dramatic shift in irrigation management paradigm with respect to both institutional pattern and financing mechanism. As farmers are the end users of water and ultimate beneficiaries of irrigation system, it is now emphasized that they should actively participate in the planning, development and management of irrigation infrastructure. Some responsibilities which had been traditionally performed by the irrigation agency such as operation, maintenance and repair of physical structures, water allocation among water users and collection of water rates are now being transferred to Water Users' Associations (WUAs) (Brewer et al., 1999; Hooja et al., 2002; Jairath,2001; Svendsen and Gulati, 1995). It is presumed that the farmers will take extreme care in maintaining the irrigation structures and proper allocation and utilisation of water as their livelihood crucially depends on use of

* Professor & Head, Department of Economics, North Orissa University, Baripada, Orissa, India-757003, *E-mail: mama_swain@hotmail.com*

water in crop production. In fact most of the states in India are now implementing Participatory Irrigation Management (PIM) through Irrigation Management Transfer (IMT) to farmers in a mission mode with financial support from international donor agencies like World Bank, Department for International Development (UK), European Commission, Japan Bank for International Co-operation and under state initiative as a part of overall economic restructuring programme.

While, it is observed that the institutional restructuring by devolution of irrigation management responsibility to farmers is undertaken with a high spirit, the financial reforms of irrigation has not been considered with the same vigor. Financial devolution should be a pre-condition for sustainability of WUAs once they are formed. Different sources of income of WUAs are membership fees, share capital, collection of water rate, operation and maintenance grants from government, income from commercial operations etc. However, of all the sources of income, water rate stands out as the single most important source on which the financial strength of a WUA crucially depends. The WUAs can undertake the devolved responsibility of Operation and Maintenance (O & M) and repair of the irrigation system only if they have adequate financial resources. Mere transfer of responsibility will not yield desired results, unless fund is made available for the purpose by proper resource mobilisation mechanism. Therefore, while bringing about institutional change in governance structure of irrigation by democratising and decentralising its management, it is important that devolution of financial power to farmers' organisations should also be ensured to carry out the handed over responsibility. In many states in India, in the case of minor flow and lift irrigation projects, the WUAs are now empowered to fix and collect water rates and they operate and maintain the systems with the revenue collected from water users. In the case of major and medium canal irrigation projects in few projects such financial autonomy has been granted.

If we analyse the practice of irrigation financing in India, it is clearly visible that due to capital intensive nature of canal irrigation projects and long payback period coupled with difficulty to recover project cost from innumerable small poor farmers, the irrigation projects (major and medium) have remained mostly state owned, state funded and are departmentally managed in a traditional top-down approach. It is well known that the water rates are fixed on the basis of political considerations rather than on economic criteria. Leave aside the capital cost of the projects, the revenue receipts from sale of water hardly cover the recurring expenditure on operation and maintenance of irrigation structures.

However, during the last two decades, because of resource crunch and financial constraint, the budgetary support to irrigation sector has been drastically reduced and emphasis is placed on internal generation of resources by increasing water rates. In 1992, the International Conference on Water and the Environment in Dublin, issued four guiding principles, one of which was that "water has an economic value in all its competing uses and should be recognised as an economic good." A consensus has emerged that irrigation water should be priced based on economic principles and the undue high subsidy given to irrigation sector should be withdrawn in phases. The financial performance of irrigation sector should be improved to ensure sustainability of the irrigation system (Gulati et al. 1994; Swain,1998; Vaidyanathan, 1992; Winpenny, 2003). In recent years, reforms in irrigation water pricing has become necessary due to the following reasons :

(i) High subsidy in provision of irrigation service and the consequent financial burden on state exchequer has become unsustainable due to financial crisis faced by the government;

(ii) Realisation of scarcity value of water and increasing demand for water for other uses like domestic use, industry and environmental needs as a result of increase in population, urbanization and industrialisation;

(iii) To encourage prudent use of water without waste through demand management;

(iv) Internal generation of resources to improve quality of irrigation service;

(v) To take into account environmental externality effects like water pollution and water logging and water salinity and to reduce water use to tackle such effects;

(vi) Implementation of general economic reforms towards reduced public involvement in commercial activities.

Sustainable financing for irrigation systems will require increase in water rates, removal of subsidy, improved cost recovery from water users and increased management efficiency through farmers' participation. Therefore, in the context of institutional restructuring and irrigation management transfer to water users, there are several intricate issues relating to water pricing reforms, which need to be addressed with much care and precaution. The pertinent issues are as mentioned below:

(i) Who has to fix the water rate? Whether state government should fix a uniform water rate for all projects or the concerned WUA should be given the power to determine the rate?

(ii) What should be the norms or criteria to fix water rates? Who will fix the norms?

(iii) What should be the method of water pricing? Whether water is to be priced according to area irrigated or actual volume of water used or output achieved from the use of water?

(iv) Is there any need to regulate water rate fixation to ensure rational pricing of irrigation water?

(v) Whether WUAs need to be entrusted with the responsibility of water rate collection? Will it improve water rate collection? How proceeds from water rate collection should be distributed between the irrigation agency and the WUAs?

(vi) If the farmers do not pay water rates to WUAs, what punitive measures can be taken against them by the WUA?

As irrigation comes under the domain of state being included in the state list of Indian Constitution; the rules, regulations and Acts pertaining to irrigation differ from state to state. Also, water rates are fixed by the concerned state government and thus vary from state to state. In this paper we are addressing the above issues in Indian context with special reference to Orissa, a pioneering state in implementing IMT programme under state patronage with strong political will of the party in power.

WATER RATE STRUCTURE

In most of the states in India, water rates are charged on the basis of area irrigated differentiated by season and crops grown. With some exception In Assam and North Eastern states no irrigation charges are levied directly or indirectly. In some states like Andhra Pradesh and Tamil Nadu water charges are integrated with land revenue by charging wet lands at a higher rate than dry lands. Water rates for paddy and wheat in flow irrigation projects in major producing states are shown in Table 1. An inter-state comparison of water rates reveals in case of paddy it is the lowest in West Bengal and highest in Gujarat. For wheat the water rate is the lowest in Manipur and highest in Assam.

Table 1
Water Rates for Paddy and Wheat in Major Producing States Utilising Flow Irrigation in India (May 2004) (Rs/ha)

States/Uts	*Year Enforced*	*Paddy* *Maximum*	*Minimum*	*Year Enforced*	*Wheat* *Maximum*	*Minimum*
Andhra Pradesh	1996	494.00	247.00	–	–	–
Assam	2000	751.00	281.24	2000	562.50	562.50
Bihar	–	247.00	108.68	2001	185.25	138.20
Gujarat	2001	825.00#	701.00#	2001	240.00##	200.00##
Haryana	2000	148.20	148.20	2000	123.50	111.15
Jharkhand	2001	217.36	108.68	2001	185.25	138.32
Jammu & Kashmir	2000	49.40	49.40	2000	24.70	24.70
Karnataka	2000	247.10	247.10	2000	148.25	148.25
Kerala	1974	99.00	37.00	–	–	–
Madhya Pradesh	1999	494.00	200.00	1999	200.07	200.07
Maharashtra	2001	180.00##	180.00##	2001	360.00##	360.00##
Manipur	1977	75.00	37.50	1977	37.50	37.50
Rajasthan	1999	197.60	49.40	1999	148.20	64.22
Tamil Nadu	1987	49.42	37.07	–	–	–
Uttar Pradesh	1995	287.00	40.00	1995	287.00	128.00
Uttarakhand	1995	143.00	143.00	1995	143.00	143.00
West Bengal	1977	37.05				

Note: # : Increase in Water Rates @ 25% per Annum.
: Increase in Water Rates @ 15% per Annum.
Source: Central Water Commission

In case of Orissa water rate is assessed on the basis of area irrigated and types of crops grown. The present practice is to levy a water rate on a per hectare basis for the kharif crop and rabi crops. Kharif crop is levied with a compulsory basic water rate on the basis of class of allocation (the total depth of water) that each project is designed to supply (Table 2). The basic water rate has to be paid whether the water is used or not. The rabi rate is not compulsory and it is paid only if a crop is irrigated.

For the staple cereal crop like kharif paddy, the water charge for class-I irrigation has been substantially enhanced (nearly two and half times) from Rs.39.54 per ha in 1981 to Rs.100 per ha in 1998 (Table 2). The water charges for different rabi crops as revised from time to time are shown in Table 3. The water rates are fixed according to the water requirement of crops. Water rate is very high for water intensive crops like ganja, betel leaf and saru. The per hectare water rate varies from the minimum of Rs. 30 for low water consumptive crops like mustard and *til* to the maximum of Rs. 465 for *ganja*, which is a highly water intensive crop.

Table 2
Compulsory Basic Water Rate in Orissa

Sl. No.	*Class of Irrigation*	*Depth of Supply (in inches)*	*Irrigation Rate for Flow Irrigation in Rs. Per* 61-62	68-69	75-76	81-82	98-99 Onwards
1.	Class I	28"	19.77	9.88	19.77	39.54	100.00
2.	Class II	23"	14.83	7.41	14.83	29.65	75.00
3.	Class III	18"	9.88	4.94	9.88	19.77	50.00
4.	Class IV	9"	4.94	2.47	4.94	9.88	25.00

Source: Department of Water Resources, Govt. of Orissa.

Table 3
Rabi Water Rates in Orissa

(in Rs. Per Hectare)

Sl. No.	*Crop*	*1968-69*	*73-74*	*74-75*	*81-82*	*98-99 onwards*
1.	Paddy	19.77	39.54	59.30	88.96	225.00
2.	Tobacco	37.07	37.07	55.60	83.40	210.00
3.	Potato	24.71	24.71	37.07	55.60	140.00
4.	Vegetables	19.77	19.77	29.65	44.48	115.00
5.	Onion	24.71	24.71	37.07	54.36	140.00
6.	Wheat	4.94	14.83	22.24	32.12	85.00
7.	Maize	12.36	12.36	18.53	27.80	70.00
8.	Mung	2.47	2.47	3.71	5.56	14.00
9.	Groundnut	12.36	12.36	18.53	27.80	85.00
10.	Orchards	29.65	29.65	44.48	66.72	167.00
11.	Sugarcane	34.59	44.48	66.72	100.08	250.00
12.	Jute	7.41	7.41	11.12	16.68	42.00
13.	Fodder	12.36	12.36	18.53	27.80	85.00
14.	Pulses	4.94	4.94	7.41	11.12	30.00

Contd....

Sl. No.	Crop	1968-69	73-74	74-75	81-82	98-99 onwards
15.	Cotton	24.71	24.71	37.07	55.60	140.00
16.	Til (Oilseeds)	4.94	4.94	7.41	11.12	30.00
17.	Betel Leaf	74.13	74.13	111.20	166.79	420.00
18.	Arhar	12.36	12.36	18.53	27.80	85.00
19.	Sunhemp	17.30	17.30	25.95	38.91	100.00
20.	Chilly	12.36	12.36	18.53	27.80	85.00
21.	Saru	74.13	74.13	111.20	166.79	420.00
22.	Ragi	6.18	6.18	9.27	13.99	35.00
23.	Mustard	2.47	4.94	7.41	11.12	30.00
24.	Ganja		123.55	123.55	185.33	465.00

Source: Department of Water Resources, Govt. of Orissa.

In most of the Indian states, the revision of water rate has been infrequent, hesitant and quite modest in comparison to the increase in cost of maintaining irrigation service and also increase in irrigation benefits due to substantial increase in support prices for crops. The Irrigation Commission (1972) has recommended for reviewing and adjusting water rates every 5 years. However, no states in India have followed the practice of revising water rates every five years (Table 4). As raising of water rate is a sensitive matter and will most likely invite the wrath of the rural electorate, the water rates are determined on the basis of political considerations rather than from the viewpoint of economic necessity.

Table 4
Duration between Current and Previous Water Rates in Selected States in India (March 2004)

	States	*Duration between Current and Previous Water Rates in Selected States in India(In Years)*
1.	Andhra Pradesh	10
2.	Assam	8
3.	Bihar	12
4.	Gujarat	20
5.	Jammu & Kashmir	24
6.	Karnataka	15
7.	Rajasthan	17
8.	Uttar Pradesh	12
9.	Madhya Pradesh	7

Source: Central Water Commission

The point to be noted is that as water rate is assessed on the basis of area irrigated, the payment that a farmer has to make has no relationship to the actual quantity of water used. Therefore, the farmers have no incentives to economise in water use, which results in excess use and wastage of water. Moreover, low price of water does not reflect its scarcity value to the water users and hence not encourages conservation of water.

ASSESSMENT AND COLLECTION OF WATER CHARGES

Assessment and collection of water charges in Orissa is undertaken by the officers of the Revenue Department on area basis. The Tahsildar is responsible for the annual assessment and collection of water charges and works under the control of the Collector for the district. The Tahsildar after jointly agreeing with the Water Resources Department on certification of the area irrigated collects water charges accordingly. The amount of water charges collected from the water users by the Revenue Department goes to the general revenue or state treasury. Thus, water is supplied by the Water Resources Department, water charges are collected by the Revenue Department and whatever is collected goes to the state treasury. Therefore, the irrigation agency does not feel itself accountable to the farmers, as it is not receiving the payments for providing the service. Further, the revenue yields from water charges are not earmarked for irrigation development. It is advocated that the whole or substantial part of the receipts from each irrigation system should be earmarked for operation and maintenance of that system. There should be partial financial autonomy.

It is observed that the revenue receipts from sale of water is far less than the cost of providing irrigation service. In Orissa, it is estimated that the total annual cost of irrigation projects, including interest is around Rs.686 million (GOO 1993). But the matter of concern is that the income from irrigation projects in terms of annual revenue receipts from sale of water comes to a meager sum i.e. around Rs.50 million. Thus, the drain on the state treasury because of the present subsidised water rate policy is to the tune of Rs.636 million per year. Such financial burden on the state exchequer is not actually sustainable in the long run.

Moreover, there is a wide gap between the amount of water charges assessed or demand raised and the actual amount collected. About more than 50 per cent of the assessments issued is remaining as arrears (Table 5). Most important reasons for non-payment of water rates usually adduced by the farmers are inability of the farmers to pay, unwillingness to pay and faulty assessment (Pant, 1981). Also, it is observed that mostly the large and influential farmers evade payment on different pretexts. The collection process of water charges from water users is beset with difficulties due to several reasons as follows (GOO 1993):

Table 5
Demand Raised and Actual Collection of Water Charges in Orissa (1986-87 to 1991-92)

Year	*Demand Raised (Rs. In Lakh)*	*Actual Collection (Rs. In Lakh)*	*Collection as % of Demand Raised*
1986-87	925	383	41.4
1987-88	952	415	43.6
1988-89	960	467	48.6
1989-90	920	367	39.9
1990-91	921	417	45.3
1991-92	956	492	51.5

Source: Vaidyanathan Committee Report (1992)

- The person collecting water charges hails from a different department and is not certain whether an area was supplied with water or not;
- Problems in water supply with regard to timeliness, adequacy, reliability and predictability resulting in gross discrepancy between the standards of service provided and that promised;
- Little perceived benefit by farmers especially in tail ends;
- Arguments with landowners over whether basic water rate should apply when water is received indirectly from uncontrolled canal leakages etc;
- Absence of any direct link between the charges, the quantity of water received and the entitlement to receive water;
- Irrigation causing adverse effects like water logging, salinity and crop loss.

The sad picture of collection of water charges falling short of the assessments made and the resultant piling up of arrears is also visible in other states of India (Table 6). Moreover the high cost of collecting water rates from thousands of farmers is a matter for concern. An extreme case is Bihar, where the cost of collection exceeds the amount of water charges collected. Further, in all the states, the amount of water charges collected hardly covers the operation and maintenance expenses of irrigation projects.

Table 6
Demand Raised, Actual Collection, Accumulated Arrears of Irrigation Charges and O&M Costs of Major and Medium Projects in Some Indian States

(Area in Lakh Hectare)
(Amount in Lakh Rupees)

States	*Area Irrigated*	*Demand Raised*	*Actual Collection*	*Annual Arrears*	*Cumulative Arrears*	*O&M Cost*
1. Bihar	21.49	475	217	371	3482	4156
2. Gujarat	6.98	1110	765	346	3264	2400
3. Haryana	20.23	1179	1298	-118	1150	3067
4. Madhya Pradesh	15.97	2792	1264	1528	8065	1451
5. Maharashtra	N.A.	1953	1228	735	7910	6060
6. Orissa	13.59	920	367	553	N.A.	N.A.
7. Punjab	28.59	1077	1121	Nil	Nil	3986
8. Uttar Pradesh	56.63	6429	6041	388	582	10158
9. West Bengal	15.58	234	80	154	N.A.	2246

Source: Vaidyanathan Committee Report (1992).

IMT AND PRICING REFORMS

In most of the countries where Irrigation Management Transfer has been implemented like Portugal, Turkey and Philippines, the water rates are fixed and collected by the WUAs. Farmers are charged a two-part tariff. The first part is meant to recover the O&M costs of irrigation schemes and is based on individual farm acreage receiving irrigation service. The second component is meant to reimburse the State over a fifty year period for the capital costs invested in projects. The first part is retained by the WUAs for incurring O & M expenditures and the second portion is paid to the parent irrigation investment agency towards capital cost.

In Orissa, in the case of major and medium canal irrigation projects, the power to fix water rate has not so far been vested in the WUAs/Pani panchayats. While delineating the functions of Pani Panchayats (PPs), the Orissa Pani Panchayat Act, 2002 envisages that in the case of major and medium canal irrigation projects the Pani Panchayat will assist the Revenue department in the preparation of demand and collection of water rates. However if the government prescribes and the farmers' organisation decides, the Pani Panchayats can levy and collect water rate. In Orissa as on April 2005, irrigation management has been turned over to 845 Pani Panchayats covering 3.82 lakh ha in major and medium canal irrigation projects. But, they have not been given the power to fix and collect water rate.

But in the case of Minor Irrigation (Flow) projects and lift irrigation projects the farmers' organisation/Pani Panchayat is empowered to fix and collect water rate and retain it for O&M expenses. The farmers' organisation shall fix the water rate, which may cover energy charges and maintenance charges of the project.

If the individual Pani Panchayat is empowered to fix water rate, the water rate will vary depending on climate, water scarcity, cropping pattern, output prices, ability to pay and affordability and many other factors. Also, it is observed that in case of minor irrigation projects where the Pani Panchayats are fixing and collecting water rates, many problems are cropping up in assessment, billing, collection and utilisation of funds raised from water rates. This is happening more frequently where the office bearers of Pani Panchayats are corrupt and enjoy unchallenged power belonging to rural elite class. In many turned over lift irrigation systems, it is observed that the farmers are reluctant to use water as they apprehend that water rate will be increased by the Pani Panchayats. Therefore, regulation of water rates is necessary by putting enough checks and balances in place. It is suggested that a regulatory body like Orissa Water Rate Regulatory Commission should be established to provide guidelines, norms, criteria to fix water rates as has been done for power sector by setting up Orissa Electricity Regulatory Commission. On the basis of the criteria prescribed by the regulatory body the Pani Panchayats will be required to fix/revise their water rates from time to time. There should be hearing from water users if they have any grievances.

WATER RATE DETERMINATION

While fixing water rates, factors affecting both demand for and supply of irrigation water should be taken into consideration. The costs incurred in supplying water is as important from the point of view of the supplier of the service/irrigation agency, the benefits derived from the use of irrigation water is equally a matter of prime concern for users of water/farmers. Now there is increasing emphasis on cost recovery from water users to provide improved and sustainable irrigation service. The irrigation project cost has two components: fixed or capital cost and variable or O&M cost. The annual cost of providing irrigation consists of three elements: (a) operation and maintenance expenses (O&M); (b) depreciation and (c) interest on capital invested. There is diversity of opinion as regards which components should be taken into account to fix the water rates.

The National Water Policy adopted in 1987 envisaged that water rates should cover annual maintenance and operation charges and part of the fixed cost. Recent National Water Policy 2002 emphasizes the physical as well as financial sustainability of existing irrigation systems. It stipulates that water charges should cover at least operation and maintenance costs of providing the service initially and a part of the capital costs subsequently. However, it prescribes subsidy to the disadvantaged and poorer sections of society, which should be transparent and reach the targeted population.

The Vaidyanathan Committee (1992) appointed by the Planning Commission is of the opinion that full-cost recovery cannot be sought without improving quality of irrigation. Therefore, in the first phase the aim should be to recover O&M costs and one per cent interest on capital employed. On an India-wide basis this would mean a five-fold increase in water charges and account for six per cent of the gross value of irrigated production. The Vaidyanathan Committee recommendations were further studied by the Group of Officers set up by the Planning Commission in October, 1992. The Group in its report in December, 1994 recommended that the irrigation water rates should cover the full annual operation and maintenance cost in phases in next 5 years period. Irrigation being the State subject, the recommendations of the Group were communicated to the states by February 1995 for further action. The States have started taking action and those who have revised their water rates are: Andhra Pradesh (January 1997), Bihar (November 1995), Haryana (September 1995), and Maharashtra (July 1994) and Orissa (1998).

There is difference of opinion with regard to the costing principle to be followed in fixing water rate. Whether water rate should be equated with average cost (AC) or marginal cost (MC)?

Champati and Patnaik (1985) have determined the average cost and marginal cost of irrigation water in the case of Salandi irrigation project in Orissa and have indicated that there is significant gap between water rate and the costs of providing irrigation water. They are apprehensive of fixing water price at MC/AC, as farmers will be hesitant to use water at such high prices. Though it is theoretically sound, it is practically unworkable due to the low income and inability of the farmers especially marginal and small farmers to pay the cost.

Now coming to the demand for irrigation water, from the point of view of water users it is not the cost of providing irrigation service but the benefits that accrue from use of irrigation water need to be taken into consideration to determine water rates. In this regard the Irrigation Commission (1972) envisages that water rates should be fixed at around 5 per cent of gross income for food crops and 12 per cent for cash crops. At present, the actual gross receipts per hectare of area irrigated by major and medium projects is barely 2 per cent of the estimated gross output per hectare of irrigated area. In Orissa farmers are paying only 0.4% to 0.7% of value of production. The incremental benefits due to use of irrigation water is considerably greater than the payments that the farmers make to get irrigation service. Thus farmers enjoy significant economic rent, which need to be tapped for improving irrigation service or extending irrigation facility. Thus there is enough scope to increase water rates without causing any discouragement to the farmers.

SUBSIDY IN WATER CHARGES

Many economists argue in favour of high subsidy in irrigation charges. The premise of such argument is that irrigation has substantial positive externality effects on the rest of the economy through enhanced crop production, increased food supply, lowered food prices, increased food security and also many other forward and backward linkage effects causing overall prosperity in the whole economy. Therefore, it is argued that in the interest of the entire populace, irrigation water should be subsidised and the cost of provision of irrigation should be met from state revenue from general tax payers' money. Again, it is argued that the beneficiaries of irrigation projects are mostly the poor peasants and if water is charged on the basis of cost of provision, the poor farmers will not be able to pay high water charges. Many also argue that the cost of irrigation facility is covered in the land price while purchasing the land.

However, the above arguments are not sound as the water users are not only the poor farmers. If we subsidise water rates, given the poor state of government finances it will not be possible to extend the irrigation facility to the poorest farmers in the dry land areas and also the operation and upkeep of the canal will suffer due to paucity of funds resulting in poor irrigation system performance and lower production. Therefore, the farmers who are already benefiting from public investments in irrigation should bear the cost of the services that they receive.

If water is made available at a highly subsidised rate, the farmers will not be aware of the scarcity value of water and will misuse it resulting in wastage water. If farmers are charged economic price of water, they will be conscious of using water efficiently by maximizing yield per unit volume of water. Thus, an upward revision in the level and structure of water rates is necessary from the point of view of both efficiency and equity.

Though costing irrigation is a critical and sensitive issue, as it has political implications, this problem, however can be resolved by improving quality of irrigation service and effecting increase in water rates. The water users will be willing to pay more for irrigation services (Vaidyanathan, 1992), provided-

(a) They are assured of a better quality of service in terms of timeliness, adequacy, reliability and predictability in water delivery and the rates are linked to this;
(b) They are convinced that the allocation rules/procedures are fair and enforced in a non-discriminatory manner irrespective of the farm size or farmer class and location of the farm in high reach, middle reach or tail end of the canal;
(c) They are not asked to bear the burden of high costs resulting from inefficiency and waste in the government; and
(d) The systems demonstrate a greater concern for keeping costs down.

In the light of the above considerations, while thinking of curtailing subsidy and increasing water rates, the Government should take adequate measures to improve irrigation service and to keep a check on costs. Moreover, government may think of cross subsidizing the use of water for irrigation while charging higher water prices for industrial uses, where returns/profits are also higher.

Furthermore, while increasing water rates, its possible repercussions on water use need to be assessed carefully. Farmers may respond to price increases in various ways, which include: (i) changes in cropping patterns;(ii) reductions in the amount of irrigated land; (iii) improvements in on-farm water management practices; (iv) changes in irrigation technologies; and (v) abandonment of irrigation altogether. While increasing water rate, all the possible effects on water use efficiency and farm income should be taken into account. However if wastage of water is taking place due to lack of supply control, water price as an instrument to control water use and manage water demand is of less significance.

Now there is a world-wide consensus based on experience and experimentation that the quality of irrigation service can be improved by democratising and decentralising irrigation management through turning over the irrigation management responsibility to WUAs at appropriate level. This will create a sense of belongingness among water users and instead of thinking irrigation structures as government property, they will consider it as a common pool resource and will be concerned about its proper upkeep and sustainability. In fact, if the system of water rate fixation becomes transparent and it is fixed in consultation with water users in a democratic process, the farmers will be interested to pay the high water rates, provided there is improvement in irrigation service. In Australia an open and transparent public process and a high level of public involvement have been important in enabling the water agency to increase water prices towards full cost recovery.

In Japan irrigation development is demand driven and based on beneficiary pays principle. The proposal of constructing an irrigation project comes from the water users through Land Improvement Districts (LIDs) with a clear commitment by the prospective/potential water users that they are willing to pay the water rates as decided for getting irrigation service. Therefore, before commencing of the project at the appraisal stage the financial viability of the project is ensured by obtaining signatures of the prospective water users that they would pay the estimated water rate. Also, subsidy and loan components are made available to the LIDs from the government and financing institutions, but farmers' contributions are clearly specified. In India it is the Government's irrigation project and farmers participate, but in Japan the Government participates in Farmers' project. This should be practised in India. Irrigation development should be need based as perceived by the farmers and their willingness to pay towards cost of provision of irrigation service should be ensured by obtaining their signatures before construction of the project.

A view that is gaining ground is that there should be provision of smart subsidies i.e. grants/subsidies should be linked to performance and local resource mobilisation capacity of WUAs (INPIM, 2004). Moreover, the WUAs should be accepted as credit worthy by financial institutions and loans may be advanced to them for construction and repair work. The WUAs may create reserve fund to meet future needs. They should prioritize the repair works and gradually take up the works in a phased manner by mobilising resources locally. Now the WUAs are given back a certain percentage of water rates collected by the government to carry out O&M work. The resource mobilisation capacity of WUAs should be strengthened by appropriate devolution of financial power to them to collect water rate and retain part of it for irrigation improvement.

PRINCIPLES OF PRICING WATER

In pricing water, the internal generation of resources through payments by water users should not be considered as an end in itself, but is only important in so far as it improves irrigation performance. Therefore, certain principles are to be followed while fixing water rates in a rationalised way. Widely accepted principles of pricing irrigation are (OECD, 1999)–

(i) Allocative efficiency
(ii) Equity
(iii) Financial Requirements
(iv) Environmental Sustainability
(v) Consumer Acceptability
(vi) Administrative Costs

Water should be priced in such a manner so as to facilitate efficient utilisation of water, equitable distribution of income, recovery of costs of irrigation, sustainable use of water resources, consumer acceptability and reduction of administrative costs of assessing, billing and collecting water charges from water users. However, these objectives of water pricing should be judiciously combined by according due importance to each objective for devising a rationalised water price structure (Swain;1998, 1999). The high water rates will not be acceptable to the poor farmers. Therefore there should be subsidy for poor farmers. However, subsidies should be targeted, transparent and, where they are intended to ease the transition should be tapering.

METHODS OF PRICING WATER

Based on above mentioned principles of water pricing, an appropriate method of irrigation pricing needs to be adopted that would facilitate efficient, equitable and sustainable use of water resources. Water users are no more considered as beneficiaries of a public irrigation system, rather they are the clients or customers of irrigation agency and in order to receive the service, they are required to pay user charges. Different types of user charges for irrigation service can mainly be grouped into three categories. These are: (i) area based fees, (ii) quantity based prices, and (iii) output based fees (Small and Carruthers, 1991).

Area Based Fees

Area based fees are assessed on the basis of area irrigated and fees per unit area may be differentiated with respect to the cropping season and the types of crop grown or both. Area based fees are observed in many developing countries like India, Pakistan and Philippines. In this case as water rate is assessed on the basis of area irrigated, throughout the cropping season, fee for water is a fixed cost of production regardless of actual quantity of water used by the farmer. As a result, water charges will have no effect on quantity of water that farmers use. This will result in excess use of water and wastage of a scarce resource.

Quantity Based Prices

In the case of quantity based water prices farmers are required to pay water fees on the basis of actual quantity or volume of water used by them. Proxies for volume such as duration of time water is delivered, number of watering and the size offtake or depth over a weir may be considered. In this pricing method, the amount of water charges to be paid by the farmer depends on his decision about the amount of water to use. This causes the cost of water to become a variable rather than a fixed cost of production, thereby creating a financial incentive for the individual farmer to use less water. From the viewpoint of equity, increasing block rate pricing is sometimes adopted to discourage excess use of water, as a farmer has to pay at a higher water rate, if he uses more water.

Output-based Fees

In this system payments are based on production achieved at the end of the crop season. In regions where the entire irrigated area is devoted to a single crop, payments could be based on a percentage of total production. This type of fee structure is sometimes found in private irrigation systems. This approach has the advantage of giving the irrigation agency an incentive to provide high quality irrigation services. This method of pricing irrigation has been reported in Viet Nam.

An assessment of different direct pricing methods reveal that irrigation water pricing is the most efficient method of charging irrigation water, as it provides incentives to economise in water use. But in case of India in general and Orissa in particular, in surface irrigation systems serving large number of small farms pricing of irrigation water volumetrically is likely to be prohibitively expensive. In particular the cost of implementing farm level water pricing for systems in which rice is the dominant crop, with water frequently flowing more or less continuously to many farm turnouts is likely to be very high. Capital cost of measuring equipments and the cost of its installation and maintenance is significant. Further, the protection of measuring equipments will be difficult. Also there is additional cost in administration, reporting, billing and collection procedures. Thus in Orissa, the establishment of reliable system of water measurement to individual farm is likely to prove technically difficult, administratively unmanageable, and economically costly. Therefore, it is questionable whether the economic gains from introduction of volumetric water pricing will exceed the economic costs. The feasibility of introducing volumetric water delivery and collection of water charges quantitatively is to be assessed carefully.

It is felt that water pricing is likely to be feasible only if groups of farmers forming Water Users' Association (WUA) at the level of tertiary canals can be supplied water volumetrically and served with a single consolidated bill. There will be bulk sale or 'wholesaling' of water by the irrigation agency operating the main system to the WUA, with the latter responsible for the subsequent distribution and "retailing" of the water to the individual farmers. It will be the responsibility of the WUA to collect water rates from the farmers. Thus in the imposition and collection of water charges there must be administrative decentralisation by delegating powers to farmers' association.

China has reportedly experimented with water wholesaling method. In Mexico, Philippines, Indonesia, Sri Lanka, South Korea, Portugal and Turkey, the farmers within minor/sub-minor of a canal are forming Water Users' Organisation and taking up the responsibility of operation, maintenance of the system; distribution of water to member farmers and collection of water charges. In India, states like Maharashtra, Gujarat, Andhra Pradesh, Tamil Nadu, Orissa and Bihar, are trying to institutionalise Pani Panchayats or Water Users' Associations for handing over the downstream parts of irrigation systems to them for their operation and maintenance. Recently Karnataka, Andhra Pradesh and Orissa have enacted Pani Panchayat Acts for farmers' management of irrigation systems. It is expected that community pressure and the fear of social stigma of being a defaulter will encourage the water users to make payment of water charges in time. According the Orissa Pani Panchayat Act, 2002 a defaulter of water charges ceases to be a member of Pani Panchayat and thereby he is deprived of the benefits of participation. With persuasion and penalty measures the WUA can improve water rate collection by way of granting discounts for payment of water charges before due date and imposing fines for late payment.

SUGGESTIONS

In the light of the above discussion the following suggestions are offered for restructuring water charging system in India.

1. Water pricing reforms should supplement and complement the institutional restructuring process in irrigation management. It is emphasized that an integrated and holistic approach is necessary to reform the irrigation

sector. The financial reforms and crafting of new institutions like WUAs or Pani Panchayats for irrigation management at farmers level will ensure efficient, equitable and sustainable use of water resources.

2. In charging irrigation water the `user pays' principle should be followed. Water rate should be fixed at a level so as to cover at least O&M expenses of irrigation. From the point of view of both efficiency and equity the subsidy should be phased out and rational water rates should be introduced, taking into account the acceptability and ability to pay by the farmers. In addition to variable water rates covering O & M costs, a fixed sum towards capital cost of the project to be recovered in 100 years from water users may also be considered. The first part of the water rate may be reimbursed to the WUA and the second part may go to the government.
3. As regards methodology of water pricing, theoretically charging irrigation water volumetrically according to the actual quantity of water used is considered to be the best. But in the institutional context, particularly in the case of gravity irrigation system, volumetric irrigation pricing is technically infeasible, administratively unmanageable and economically expensive. Therefore, the current practice of charging water according to area or time for irrigation, differentiated by crops grown and season may be used.
4. The principles followed in fixing water rates, the allocative efficiency objective should be accorded the utmost importance. The pricing policy should improve irrigation performance by encouraging optimum utilisation of water. The equity objective may be taken care of by extending subsidy to the poor and the disadvantaged farmers, whose income is below poverty line. The subsidy component should be transparent and well targeted to the resource-poor farmers only.
5. For improving performance irrigation system a necessary precondition is that irrigation agency should have sufficient financial autonomy. This will result in greater accountability of the irrigation agency to the water users and will also encourage water users' involvement and participation in irrigation water management.

6. For assessment, billing and collection of water charges, there must be administrative decentralisation. The Water Resources Department will make bulk sale of water to the WUAs. The WUAs will distribute water among the members according to their entitlements. The responsibility of collection of water charges from individual water users will rest on the WUA. However, there should be an independent and competent regulating authority to oversee water rate fixation procedure and to provide guidelines and norms on the basis of which water rates will be determined.
7. A public information and awareness campaign on costs and benefits of irrigation needs to be taken up. Publication of disaggregated project wise data on irrigation service costs and its different components and benefits of irrigation should be published. The information should be shared with the general public and customers, particularly farmers of the command area. The WUAs or Pani Panchayats should be involved in decisions on determination of irrigation water charges and other water services management and expenditures. This will help in diffusing the farmers' strong resistance and opposition to increase water rates.

REFERENCES

Asian Development Bank (1986), Irrigation Service Fees, Proceedings of the Regional Seminar on Irrigation Service Fees, International Irrigation Management Institute.

Brewer, J.S., S. Kolavalli, A.H. Kalro, G. Naik, S. Ramanarayan, K.V. Raju and R. Sakthivadivel (1999), Irrigation Management Transfer in India: Policies, Processes and Performance. New Delhi: Oxford and IBH publishing House.

Champati, M. and S.C.Patnaik (1985), 'A Functional Approach to Fixation of Water Rate under the Principles of Marginal and Average Costs: A Case Study of Salandi Irrigation Project of Orissa.' *The Indian Journal of Economics,* Vol.46, Part I, No.260, pp. 21-37.

Dalua, A.K. (1991), Irrigation in Orissa, Water and Land Management Institute, Orissa.

Government of Orissa (1993), Report on Water Charges for Major and Medium Irrigation Projects, Water Resources Consolidation Project, Orissa Irrigation Department.

Government of India (1987), National Water Policy, Ministry of Water Resources.

Government of India (2002), National Water Policy, Ministry of Water Resources.

Gulati, A. and et al. (1994), 'Major and Medium Irrigation Schemes: Towards Better Financial Performance' Economic and Political Weekly, Vol.29, No.26, June 25, pp. A72-A79.

Hooja Rakesh, G. Panagare and K.V. Raju, (eds.) (2002), Users in Water Management. Jaipur: Rawat.

INPIM (2004), PIM and Financing.

Jairath Jasveen (2001), Water User Associations in Andhra Pradesh, Center for Economic and Social Studies, Hyderabad and New Delhi: Concept.

Merrett Stephen (1997), Introduction to the Economics of Water Resources: An International Perspective, London:UCL.

Organisation for Economic Cooperation and Development (1987), Pricing of Water Services, Delhi: Oxford.

OECD (1999), Agricultural Water Pricing in OECD Countries, OECD Working Paper No. 33.

Pant,N. (1981), Some Aspects of Irrigation Administration, Calcutta: Naya Prokash. Chapter VI.

Rivera Daniel (1996), Private Sector Participation in the Water Supply and Wastewater Sector: Lessons from Six Developing Countries, Washington D.C: The World Bank.

Small, L.E. and Ian Carruthers (1991), Farmer-financed Irrigation: The Economics of Reform, Cambridge: University Press.

Subramnian, Ashok, N. Vijay Jagannathan and Ruth Meinzen-Dick. (1997), User Organizations for Sustainable Water Services, World Bank Technical Paper No. 354.

Svendsen Mark and Ashok Gulati. (eds) (1995), Strategic Changes in Indian Irrigation, New Delhi; ICAR, Washington DC: IFPRI.

Swain, Mamata (1998), 'Water Rate Fixation in Major and Medium Irrigation Projects in Orissa: Issues and Problems.' Water and Energy International, Vol. 55, No. 2.

Swain, Mamata and D.K. Das (eds.) (1999), Emerging Trends and Reforms in Irrigation in India: A Perspective of Orissa, Delhi: M.D. Publishers.

Swain, Mamata and G.C. Kar. (eds.) (2000), Farmer and Local Participation in Irrigation Management and System Turnover, New Delhi: Commonwealth Publishers.

Swain, Mamata (2003), 'Demand Driven Rural Drinking Water Supply with Cost Recovery Component: Applied Policy Issues', Man & Development, Vol. 25, No. 1, pp. 85-94.

Vaidyanathan, A. (1992), Report of the Committee on Pricing of Irrigation Water, Planning Commission.Government of India, Indian Journal of Agricultural Economics, Vol. 49, No. 1, pp. 107-23, 1994.

Winpenny, James (2003), Financing Water Infrastructure, World Water Council.

16

PARTICIPATORY IRRIGATION MANAGEMENT IN INDIA POLICY AND PRACTICE

*Deepak Kumar Das**

INTRODUCTION

In India, starting from the very inception of planning and in subsequent five-year plans a major chunk of plan outlay has been meant for irrigation development for increasing agricultural productivity and improving the standard of living of the rural mass. It is well known that irrigation development holds the key to agricultural growth as availability of irrigation water triggers the use of other yield stimulating inputs like high yielding varieties of seeds, chemical fertilizer, farmyard manure and encouraging farmers to adopt advanced agricultural techniques and improved agronomic practices. Irrigation increases agricultural production and farm income by enhancing the crop yield, diversifying the cropping pattern in favour of remunerative cash crops and increasing the cropping intensity (Bharadwaj 1974,1990; Dhawan 1987,1988). Construction of irrigation infrastructure is capital intensive in nature and the pay back period is too long. Therefore,

* Reader, Water Resources Engineering, Water and Land Management Institute, Cuttack, Orissa, India-753051, *E-mail: deepak_das62@yahoo.com*

major and medium irrigation projects in India have remained fully state owned, state funded and departmentally managed by a hierarchical bureaucracy in a centralised and top down approach.

There is no denying the fact that during 1960s, irrigation coupled with high yielding varieties of seeds and application of fertilizers played the key role in ushering in green revolution and made India self-sufficient in food grain production. Nevertheless a scrutiny of the performance of irrigation sector in India reveals that the overall performance of irrigation systems is sub-optimal, inefficient and inequitable falling far short of expectations. Irrigation sector is confronted with multi-dimensional problems of deteriorating physical structure in the absence of proper maintenance, low cost recovery, under-utilisation of created potential, tail-end water deprivation, uncontrolled water delivery, water logging, soil salinity and poor quality of irrigation service. In spite of massive investments made in irrigation projects, the physical as well as financial performance of irrigation sector is observed to be quite discouraging. The agricultural productivity from irrigation commands has not increased appreciably. The yield increasing potential of irrigation is rarely achieved. The estimated benefit-cost ratio and the internal rate or return are never realised. As regards the financial performance of irrigation projects, leave aside the capital costs of irrigation projects, even revenue receipts from sale of water hardly cover the recurrent operation and maintenance (O&M) expenses. The political economy of water pricing stands on the way of rationalising water rate structure. Therefore water rates are highly subsidised to serenade the rural electorate. Irrigation sector is, thus, unable to generate resources internally required for the operation and maintenance of irrigation structures.

In most of the government managed irrigation projects a substantial portion of budgetary allocation for O&M (70 to 80 per cent) is for salary bills of the employees and establishment expenditure leaving an insignificant amount for works component of repair and maintenance of the physical structures. This leads to situation of improper maintenance of irrigation structures allowing them to deteriorate and causing system inefficiency. The poor performance of irrigation systems is now a matter of grave concern for irrigation planners and executives.

In the wake of liberalisation, privatisation and globalisation, during the last decade many state governments in India have reduced subsidy in water charges and increased the water rates. But the water users, who are accustomed to get water at a very nominal price, are reluctant to pay higher water charges unless there is improvement in quality of irrigation service. To improve the irrigation system performance the government has to increase the water rate to generate resources for the upkeep of irrigation systems, but the farmers are reluctant to pay the increased water rate. Thus, irrigation sector is confronted with a vicious circle and a deadlock situation (Jairath 2001; Swain 2002). As an escape from this impasse and to make the vicious circle a virtuous one, during the last one decade most of the state governments in India have undertaken profound institutional and financial reform measures in irrigation sector.

Prior to 1980s, irrigation has been mainly considered as a public sector technical enterprise aiming at construction of hardware like dams, reservoirs, weirs, barrages and canals. It is well known that in a bureaucratically managed canal irrigation system, it is extremely difficult on the part of irrigation agencies to provide dependable irrigation service catering to the needs of numerous marginal and small farmers having fragmented land holdings scattered over extensive irrigation commands. The irrigation executives do not have adequate information on local resources and socio-economic conditions of farmers. Moreover, due to rigid procedures followed in a bureaucratically managed system and interference of influential oligarchs, the irrigation engineers are unable to take appropriate and timely decisions based on field conditions. In the absence of responsibility fixation and downward accountability, the irrigation executives are under no pressure to accomplish their job catering to farmer's need.

The impact of productivity on irrigation is critically dependent on the way water is applied and utilised. The quality of irrigation service in terms of adequacy, timeliness, equity, dependability and convenience in its supply affects the yield from irrigation commands to a great extent. For obtaining optimum yield water should be provided in time and in adequate quantity according to the crop water requirement of plants at its various growth stages, as water stress affects crop yield adversely. Irrespective of the location and size of the farm, water should be allocated equitably among the head reach farmers and tail reach farmers, large and small farmers. For proper crop planning, farmers need to know in advance the timing and quantity of water supply and system need to be reliable. Quality of irrigation service should be maintained over time in a sustainable manner. In the new agricultural technology, proper water management holds the key to increase agricultural productivity.

To establish such an improved water delivery system and for optimal utilisation of scarce water, new trends advocate much more active participation of the water users in all aspects of irrigation water resources development and management, which include planning, design, construction, operation, maintenance, on-farm development, rehabilitation, modernisation, water distribution, financing, resource mobilisation and collection of water rates (Balland and Platteau, 1996; Chambers, 1988; Meinzen-Dick et al., 1997; Ostrom et al., 1993, Vaidyanathan, 1994; Wade, 1987; and others). It is stressed that the water users have an important role to play in water resources management more importantly at the tertiary or downstream part of the canal system. Specifically the water users can better manage the operation and maintenance of canals, distribution of water among farmers and collection of water rates.

Internationally also many countries all over the world are broad basing their PIM programme. Pioneers in PIM are The Philippines, Mexico, China, Japan, Egypt, Turkey, Sri Lanka, Chile, Colombia, Morocco (Johnson et al., 2002; Meinzen-Dick et al,,1997). In these countries the key background conditions that paved the way for initiation of PIM programmes leading to the turnover of management of irrigation systems to WUAs include: national financial crisis, political will of the ruling party to place the irrigation sector on a sound economic footing, progressive deterioration of irrigation infrastructure due to deferred maintenance (Johnson et al., 2002). In Mexico, the Irrigation Management Transfer (IMT) programme was implemented due to political will at the highest level of the government. In 1989, with change in administration, comprehensive water management was recognised on top priority basis. The National Water Commission (CNA) was set up and a national policy on privatisation of irrigation was formulated. In Colombia it was the National Planning Department who in 1991 submitted the Land

Reclamation Programme 1991-2000. In Turkey, a budgetary crisis led to a squeeze on financial allocations to the state water resources agency and provided the initial impetus. In Argentina, a programme to modernise the entire economy, which began in 1990 with the privatisation of large electricity utilities, led to the turnover of water management to the provinces. In the Philippines the National Irrigation Administration (NIA) launched an ambitious programme in 1974 to increase rice production. There was a provision to phase out subsidies for the O&M of irrigation over a five year period, at which time NIA would be directly dependent on collections from farmers for its O&M expenses.

Factually Participatory Irrigation Management through Irrigation Management Transfer to farmers is now regarded as the most appropriate mechanism to bring about efficient utilisation of irrigation water, its equitable distribution and sustainable irrigation service. During the last decade most of the states of India like Andhra Pradesh, Orissa, Gujarat, Maharashtra, Karnataka, Tamil Nadu, Bihar and West Bengal have initiated reforms in irrigation sector under the economic restructuring programme mostly under the mandate of international funding agencies. Under these Programmes systematic institutional and organisational changes have been undertaken to increase farmers' participation in irrigation management through formation of Water Users' Association (WUA) or Pani Panchayats either under externally assisted economic restructuring programme like Water Resources Consolidation Project (World Bank), system rehabilitation programme (European Commission), irrigation infrastructure development (Japan Bank for International Cooperation), PIM in lift irrigation projects (Department for International Development, UK) or through state government initiative. These states are putting emphasis on farmers' participation in irrigation management and encouraging the farmers to form Water Users' Association (WUA) or Pani Panchayat (PP). Some state governments like Andhra Pradesh and Orissa are implementing the Participatory Irrigation Management (PIM) Programme with high priority and state patronage.

BENEFITS AND COSTS OF PIM

Benefits and costs of farmers' participation to two major stake holders i.e. farmers and irrigation agency have been indicated by several authors (Maloney and Raju,1994; Singh,1991; Svendsen, et al.,1997), which are as follows:

Benefits to Farmers

(i) Better water management at the tertiary level
(ii) Farmers' flexibility in use of water, choice of crops and land use
(iii) More crop production and more income
(iv) Optimal use of water in agriculture
(v) Ensuring equity in water allocation
(vi) Helping resolve disputes on water distribution
(vii) Encouraging community responsibility for the management of assets
(viii) Forum for facilitating effective communication between farmers and government departments
(ix) Better collection of irrigation fees
(x) More economic use of water, less wastage
(xi) Better maintenance-reduction in the cost of irrigation
(xii) Encouraging collective management of agricultural input supply and marketing of agricultural produce
(xiii) Less water logging because of careful use of water
(xiv) Less opportunity for corruption
(xv) Better mutual trust and understanding between farmers and officers
(xvi) Sense of ownership
(xvii) Scope for securing water right for agriculture sector

Costs to Farmers

(i) Higher costs, increased water rates
(ii) More time and effort required to manage
(iii) No assured rehabilitation assistance
(iv) Less disaster assistance

Benefits to Irrigation Agency

(i) Improved relations with client farmers; less mistrust
(ii) Reduced complaints on inequitable distribution of water between head reach and tail end
(iii) Increase in irrigation efficiency, job satisfaction
(iv) Sufficient time to attend technical matters and using their expertise

(v) Less bothering about unauthorised outlets, obstructions and maintenance problems
(vi) Improvement of credibility of the irrigation agency and irrigation officers
(vii) Better collection of water rates and saving on maintenance cost

Costs to Irrigation Agency

(i) Reduced bureaucratic and political influence
(ii) Uncertainty over agency role
(iii) Reduced opportunity for rent seeking
(iv) Reduced control over water resources
(v) Reduced O&M staff levels

Evidently, the major stakeholders i.e. water users and irrigation agency the benefits of PIM outweigh the costs. Therefore, it can be construed that PIM is a win-win strategy for the farmers as well as the irrigation agency and should be encouraged to improve quality of irrigation service.

EVOLUTION OF PIM IN INDIA

The concept of PIM has evolved gradually through three distinct phases (Maloney and Raju, 1994:16). In 1980s the concept was in its nascent stage limiting to farmers' participation through their representatives. It was felt then that in the decision making process of irrigation development, the views of farmers should be taken into account and they should be consulted in planning, development and management of the system. However, mere farmers' representation in scheme level committees could not yield much result. In the latter part of 1980s, it was realised that farmers couldn't have many stakes in irrigation management without a formal structure/forum to express their views. Therefore, the catchword became farmers' organisation.

In some states like Andhra Pradesh, Tamil Nadu and Maharashtra thousands of outlet associations/chak committees had been formed only in pen and paper but practically became on dysfunctional after a short period. By this time the concept of PIM received due importance under the Water Resources Management and Training (WRM&T) project of Government of India supported by United States Agency for International Development (USAID). Motivating farmers to foster Water Users' Associations was an essential feature of action research programme taken up by Water and Land Management Institutes (WALMIs) in eleven states of India (INCID, 2003). The publication of the Ministry of Water Resources, Union titled 'Farmers' Participation in Irrigation Water Management' in 1991 recognised the status of the erstwhile existing WUAs. By early part of 1990s it became apparent that the concept of farmers' participation in few activities is not sufficient. Therefore, a radical concept of Irrigation Management Transfer (IMT) and Farmers' Organisation and System Turnover (FOT) has evolved in which it is envisaged to entrust the WUAs with the responsibility of operation and maintenance of minor/distributary, allocation of water among farmers and collection of water charges from water users.

During late 1990s most of the state governments in India (Andhra Pradesh, Gujarat, Maharashtra, Madhya Pradesh, Orissa and others) have taken policy decision to introduce Participatory Irrigation Management and turning over the management of tertiary segment of the canals like minor/sub-minors/distributaries to Water Users' Associations (Maloney and Raju, 1994; Jairath, 2001; Hooja et al.,2002, Brewer et al.,1999). Thousands of WUAs have been formed and they are entrusted with the responsibilities of operation and maintenance of the tertiary units, distribution of water among water users and collection of water rates. It is contemplated that the irrigation agency will make bulk sale of water volumetrically to WUA at distributary /minor head and retailing of water to farmers will be the responsibility of WUA. In most of the states the responsibility is limited to crop planning, water distribution and tertiary system maintenance and participation in water rate fixation.

WATER POLICY AND LEGAL ENVIRONMENT

Water is a prime natural resource, a basic human need and a precious national asset. Therefore, it is stressed that planning, development and management of water resources should be governed by national perspectives. In India the first National Water Policy was adopted in 1987. As regards PIM, this policy envisages that

> 'Efforts should be made to involve farmers progressively in various aspects of management of irrigation systems, particularly in water distribution and collection of water rates. Assistance of

voluntary agencies should be enlisted in educating the farmers in efficient use of water and water management.'

Recently the National Water Policy, 2002 has been enunciated. The new policy is more explicit in emphasising the need for farmers participation in irrigation management and broadly outlines a Participatory Approach to Water Resources Management as follows:

> 'Management of the water resources for diverse uses should incorporate a participatory approach: by involving not only the various government agencies but also the users and other stakeholders, in an effective and decisive manner, in various aspects of planning, design, development and management of the water resources schemes. Necessary legal and institutional changes should be made at various levels for the purpose, duly ensuring appropriate role for women. Water Users' Associations and the local bodies such as municipalities and gram panchayats should particularly be involved in the operation, maintenance and management of water infrastructures/facilities at appropriate levels progressively, with a view to eventually transfer the management of such facilities to the user groups/local bodies.'

As irrigation is included in the state list of Indian constitution, irrigation policy, law, act, rules and regulations differ from state to state. In this paper we are discussing the Indian scenario with special reference to Orissa, which is a pioneering state in implementing Participatory Irrigation Management in a mission mode.

In Orissa, the first State Water Policy was adopted in 1994. The State Water Policy of Orissa follows the principles enunciated in the National Water Policy, 1987 and aims at promoting citizen participation in all aspects of water planning and management and handing over of operation and maintenance of irrigation systems to the users in due course. Orissa Water Resources Policy is implemented by the Department of water Resources. Recently subsequent to the declaration of the National Water Policy in 2002, a draft Orissa state water policy has been formulated, which has not yet been finalised.

ENACTMENT OF LEGISLATION FOR PIM

Recognising the need for sound legal framework for PIM in the country, the Ministry brought out a model act to be adopted by the state legislatures for enacting new irrigation acts/amending the existing irrigation acts for facilitating PIM. In accordance with the model act ten State Governments, Namely, Andhra Pradesh, Goa, Karnataka, Madhya Pradesh, Orissa, Rajasthan, Tamil Nadu, Kerala, Bihar and Maharashtra have legislated new acts or amended the existing acts. Details of the Acts/Rules enacted by different states are given in Table 1.

Table 1
State-wise Position of Enactment of New Act/Amendment of Existing Irrigation Act

Name of State	*Position of new Act/Amendment of Irrigation Act*
1. Andhra Pradesh	Andhra Pradesh Farmers' Management of Irrigation Systems Act, March, 1997
2. Goa	Goa Command Area Development Act 1997 (Goa Act 27 of 1997)
3. Karnataka	Ordinance on 7ln June 2000 for amendment of the existing Karnataka irrigation Act 1957
4. Madhya Pradesh	Madhya Pradesh Sinchai Prabandhan Me Krishkon Ki Bhagidari Adhiniyam, September 1999.
5. Orissa	The Orissa Pani Panchayat Act, 2002
6. Rajasthan	Rajasthan Sinchai Pranali Ke Prabandh Me Krishkon Ki Sahabhagita Adhiniyam, 2000
7. Tamil Nadu	Tamil Nadu Farmers' Management of Irrigation Systems Act, 2000
8. Kerala	The Kerala Irrigation and Water Conservation Act 2003
9. Bihar	The Bihar Irrigation, Flood Management and Drainage Rules, 2003 under the Bihar irrigation Act, 1997
10. Maharashtra	The Maharashtra Management of Irrigation Systems by Farmers Act, 2005

Source: Ministry of Water Resources, Government of India.

Andhra Pradesh has legislated Farmers' Management of Irrigation Systems Act in 1997. Government of Orissa has enacted Orissa Pani Panchayat Act in 2002. Madhya Pradesh and Rajasthan have promulgated laws based on Andhra Pradesh model. In Uttar Pradesh and Maharashtra membership of WUA is mandatory for water users. In Andhra Pradesh and Orissa more than ten thousand Pani Panchayats have been formed due to the political will of the party in power. The target is to cover the entire irrigated command area under Pani Panchayat scheme within a couple of years. Some

States like Gujarat had experimented with the idea of farmers' co-operative movement in irrigation management. Subsequent to the enactment of laws on PIM, various states of India have taken steps to form WUAs. The state-wise details of WUAs formed are given in the Table 2.

Table 2
State-wise Number of WUAs Formed and Area Covered

	Name of State	*Number of WUAs Formed*	*Area covered ('000 ha)*
1.	Andhra Pradesh	10790	4800.00
2.	Arunachal Pradesh	2	1.47
3.	Assam	37	24.09
4.	Bihar	37	105.80
5.	Chhatisgarh	945	N.A.
6.	Goa	42	5.00
7.	Gujarat	576	96.68
8.	Haryana	2800	200.00
9.	Himachal Pradesh	875	35.00
10.	J&K	1	1.00
11.	Karnataka	2284	1062.10
12.	Kerala	3930	148.48
13.	Madhya Pradesh	1470	1501.45
14.	Maharashtra	1299	444.00
15.	Manipur	62	49.27
16.	Meghalaya	99	N.A.
17.	Nagaland	25	N.A.
18.	Orissa	12688	995.00
19.	Punjab	957	116.95
20.	Rajasthan	506	219.65
21.	Tamil Nadu	7725	474.28
22.	Uttar Pradesh	24	10.55
23.	West Bengal	10000	37.00
	Total	57174	10,327.77

Source: Ministry of Water Resources, Government of India.

The table 2 indicates that Andhra Pradesh has covered highest area under PIM programme followed by Madhya Pradesh, Karnataka and Orissa.

APPROACHES TO PIM

PIM programmes differ according to strategies of implementation. In some countries the programme is executed in a gradual, incremental way (such as in the Philippines, Indonesia and Orissa, India). In some other places the PIM is implemented rapidly for several thousands or even millions of hectares (such as in Mexico, Turkey and Andhra Pradesh, India). The latter is referred to as the 'big bang' approach. In some cases government mandates transfer of all targeted systems (as in Turkey or Indonesia). In other cases, the government negotiates on a case-by-case basis and systems are only transferred if water users agree (such as in the Philippines). Many advocate the 'big bang' route as the impact is rapid, deep, and sustainable (Groenfeldt, 2000; Society for Promoting Participative Ecosystem Management, India).

Since the eighties, the concept of PIM has been introduced in almost all the states in India under USAID assisted Water Resources Management and Training (WRM&T) Project. In later part of nineties Andhra Pradesh, Madhya Pradesh, and Karnataka enacted legislation making PIM a statutory requirement to access irrigation water. Implementation of PIM seems to have two approaches - the legislative and the motivational. Andhra Pradesh and Madhya Pradesh first enacted legislation and went in for fast and extensive introduction of PIM i.e. going in for a top down approach. As against this, Maharashtra and Gujarat adopted the motivational strategy i.e. a bottom-up approach. On the other hand Orissa has adopted a in between process. To start with, in the implementation of PIM, for a certain period a bottom up approach with motivational strategy was adopted. Later on when the movement attained a certain level of momentum, the state adopted somewhat a top down strategy legislating the Orissa Pani Panchayat Act, 2002. The strategy adopted by state of Orissa is proving to be more effective. There is a steady progress in achieving the goals of PIM.

In Andhra Pradesh, where a big-bang approach was followed by forming nearly ten thousand WUAs within a year of implementation of the programme by legislating Farmers Management of Irrigation System Act in 1997, the performance of the WUAs has been poor with a lot of problems (Jairath 2003). In fact, the irrigation water management has been subjected to gross and unwarranted political interference and vested interests all over India. It is a major challenge to achieve the ultimate goal of PIM.

The motivational approach, on the contrary, so far adopted in Maharashtra and Gujarat may not have achieved quick spread all over the state but the motivated water users groups showed spectacular success in few of the irrigation projects of Maharashtra and Gujarat. From the above examples it can be seen that the strategy that has been adopted by Orissa seems to be more effective.

PANI PANCHAYAT PROGRAMME

In this section the salient features of the Participatory Irrigation Management programme or the Pani Panchayat programme have been discussed which have been implemented in the sate of Orissa since 1996. Recognising the need for systematic involvement and participation of farmers in irrigation management, initially in 33 irrigation projects covered under the Orissa Water Resources Consolidation Project (OWRCP) funded by World Bank, PIM process was launched christened as Farmers' Organisation and Turnover (FOT) programme. In these project areas steps have been taken to motivate farmers to form Water Users' Associations and to turn over the operation and maintenance of the downstream part of the canal system to WUAs (Swain and Kar, 2000). The major functions of the WUAs are to operate and maintain the distributary/minor canals, to ensure equitable water distribution among the WUA members, to advise the department on main system operation and ultimately with gaining experience to collect irrigation fees in due course. In OWRCP area the implementation of FOT programme has been entrusted to Water and Land Management Institute, Command Area Development Authority, Nabakrushna Choudhury Centre for Development Studies (an Indian Council of Social Science Research Center) and some local Non-Governmental Organisations. Since 2000 Pani Panchayat scheme has been implemented by the personnel of Department of Water Resources in non-OWRCP areas as well. In Orissa the Pani Panchayat is a three tier organisation for medium irrigation project and four tiers for major irrigation projects as indicated below:

(i) WUA/Pani Panchayat at primary level consisting of several chak or outlet committees.

(ii) Distributary committee at secondary level (major projects) is a federation of all the WUAs / Pani Panchayats under the distributary.

(iii) A project committee at project level is a federation of all distributary committees for major Projects. Similarly for medium irrigation projects, a project committee at project level is a federation all the WUAs/Pani Panchayats.

(iv) A state level committee is constituted by the Government with presidents of the project committees not exceeding ten.

A Pani Panchayat/Water Users Association is an association of all persons owning land within a hydrologically delineated portion of the command area ranging in size approximately from 300-600 ha in case of major/medium/minor irrigation project. It may be in respect of minor or sub-minor or direct outlets from the main or branch distributary of the project. In case of minor flow or lift irrigation the area is limited to project command area when the project command area is less than 300 ha. At the lowest level, outlet (chak) committee is formed electing three farmers, one from head reach, one from middle reach and one from tail reach of the ayacut of an outlet. An elected representative from among the outlet committee members (called chak leader) of each chak committee is a member of executive body of Pani Panchayat. The president, secretary and treasurer of the Pani Panchayat are elected out of the executive body of concerned Pani Panchayat. It may be mentioned that all the water users are members of general body of the Pani Panchayat. The details of organisational structure of farmer organisation is given in the following figure.

As envisaged in the Orissa Pani Panchayat Act, the principal objectives shall be 'to promote and secure distribution of water among its users, adequate maintenance of the irrigation system, efficient and economical utilisation of water to optimize agricultural production, to protect the environment, and to ensure ecological balance by involving the farmers, inculcating the sense of ownership of the irrigation system in accordance with the water budget and the operational plan.' The major functions of Pani Panchayats, therefore will be as follows:

- To prepare a cropping programme suitable for the soil and agro-climatic condition with due regard to diversification;
- To prepare a plan for the maintenance of irrigation system in the area of its operation at the end of each crop season and carry out the maintenance works with the funds of the Pani Panchayat;
- To regulate the use of water among the various pipe outlets under its area of operation according to the warabandi schedule;

- To promote economy in the use of water allocated;
- To assist the revenue department in the preparation of demand and collection of water rates;
- To resolve disputes between the water users;
- To raise resources;
- To conduct regular water budgeting;
- To conduct annual audit of its accounts and periodical social audit, as may be prescribed.

Structure of Pani Panchayat in Orissa

Pani Panchayat

Delineated by S.E of DOWR Under Section 4 of Pani Panchayat Act ' 02

General Body
(All Water Users within the PP Boundary)
Executive Body
(All Chak Leaders within the PP Boundary)
(Non Voting Members: One officer from DOWR and one officer from DOA nominated by the Government

Distributary Committee

Delineated by C.E. of DOWR Under Section 5 of Pani Panchayat Act ' 02

General Body
(All Presidents, Secretaries & Treasurers of PP)
Executive Body
(Maximum 9 persons elected from GB of D.C)
(Non voting members: Officer not below the rank of Executive Engineer of DOWR and DAO of DOA are permanent invitees to the executive body meetings nominated by the Government)

Project Committee

Delineated by the Government

Under Section 7 of Pani Panchayat Act ' 02

General Body
(All Presidents, Secretaries & Treasurers of DC)
Executive Body
(Maximum 9 persons elected from GB of P.C.)
(Non voting members: Officer not below the rank of Executive Engineer of DOWR and DAO of DOA are permanent invitees to the executive body meetings nominated by the Government

STATE LEVEL COMMITTEE

Constituted by the Government Under Section 9 of Pani Panchayat Act'02

Constituted by the Government with such number of Presidents of Project Committee considered necessary but not exceeding ten and Equal no. of Government Officials and Professionals not exceeding ten nominated by the Government

Structure of Farmers Organisation in Major Irrigation Projects

For successful implementation of PP Programme, the implementation process is divided into several phases viz. (i) Preparation, (ii) Assessment, (iii) Organising, (iv) Joint management and (v) Turnover. The sequencing of various Pani Panchayat activities is as mentioned below:

1. Delineation of Pani Panchayat Boundary,
2. Motivation of farmers regarding advantages of PIM
3. Identification of villages, collection of village maps
4. Collection of Irrigation Record of Rights
5. Preparation of beneficiary list and beneficiary-wise land holding
6. Selection of contact farmer
7. Conducting Participatory Walk Through
8. Socio-economic baseline survey
9. Collection of membership fees
10. Opening of bank account
11. Conduct of election
12. Training of accounts procedure to office bearers of WUA
13. Hydraulic testing
14. Discussion on Memorandum of Understanding
15. Joint management
16. Motivation of farmers on On Farm Development Work
17. Handing over of the system for final turnover

Table 3
Progress of Pani Panchayat Programme in Orissa as on June 2006

Types of Irrigation Project	*Irrigation Potential Created* '000 ha.*	*No. of Pani-Panchayats Formed*	*Area '000 ha.*	*Irrigation management transferred*	
				No. of PPs	*Area '000 ha.*
1. Major and Medium	1234	1426	623	1122	494
2. Minor					
3. (Surface)	497	920	180	663	127
4. Minor					
5. (Surface					
6. EC Projects)		56	9	56	9
7. Minor					
8. Lift	384	11033	243	10377	230
Total	2115	13435	1055	12218	860

* Irrigation potential from other sources are not included

Table 3 provides data on progress of PIM in Orissa. By June 2006, 13435 PPs have been formed covering an area of 10.55 lakh ha. Irrigation management has been transferred to 12218 PPs covering 8.60 lakh ha, out of the total command area of 21.15 lakh ha.

CONSTRAINTS IN OPERATIONALISING PIM

Crafting a new institution in the management of irrigation infrastructure is obviously a difficult task. Changing the mindset and ingrained attitude of farmers and officials of water agency is not so easy. Therefore, while motivating the farmers to form WUAs and implementing various activities of FOT programme many problems are encountered in the field, which hinder the formation of Pani Panchayat and their sustainability in the long-run. Some major constraints in organising farmers as experienced by the author in implementing Pani Panchayat programme some irrigation projects in Orissa during 2002-2004 are discussed in the following sections.

Class, Caste and Community Conflict

Peasant agriculture is most often confronted with the problems of a heterogeneous and hierarchical society having several caste and class cleavages. The more homogeneous the water users with respect to their socio-economic status, the easier it is to organise them to form WUAs. It is very difficult to form WUAs in an inegalitarian society having caste antagonism, class differential and unequal production relations. If the inter-village and intra- village socio-economic and cultural conflicts are severe, it is most difficult task to foster Pani Panchayat. The conflicts of interests and lack of collective efforts on the part of water users are major hindrances in implementing Pani Panchayat activities. There are also political differences. The farmers having different political affiliations naturally have differences of opinion. Therefore, the less the differences within the farmer community, the easier it is to form farmers' association.

It is a fact, that it takes a long time to change peoples' attitude and to make people realise the importance of forming a unified association for better water management. Implementing personnel must have the patience and perseverance to overcome such systemic and endemic problems in organising farmers. Learning by doing approach should be followed to determine the model and modalities of organising Pani Panchayat depending on the socio-economic conditions, psychology and cultural heritage of the farmers in the locality.

As the boundaries of Pani Panchayat/WUAs are determined on hydraulic basis i.e. command areas of minor/sub-minor, most of the Pani Panchayats cover more than one village. If historically the villages in a single Pani Panchayat having hostile relationship for some reasons or other, this may pose as a serious setback in forming WUAs.

Further, the farmers who have their holdings at the head of the canal tend to appropriate more water than required, whereas the farmers at the tail end often fail to get their apportioned share of water. Head-enders, therefore, have vested interest in continuing the existing arrangements. The tail-enders are showing interest to form WUAs as water supply in such areas remains inadequate and erratic and they become apprehensive that the situation may not improve if an association is formed. These differences in perceptions and conflicts of interest inhibit the coming together of head end and tail end farmers.

Election of Farmers' Representatives

At the time of election of leaders of chak committee, executive committee of Pani Panchayat and of project committee, many problems crop up. Water users having different political affiliations reveal their differences in interest and opinion. Political differences result in formation of separate groups within a Pani Panchayat/WUA. Though the elections are supposed to be non - political, the unwanted political interference does spoil the very purpose of the programme. In fact, the degree of political interference has an important bearing with the successor failure of Pani Panchayats.

Success of any democratic institution depends on an enlightened and informed electorate. Therefore, the water users should be made aware of the details of the Pani Panchayat programme like the objectives and benefits of WUA, its structure, the organising procedure, functions, rights, duties and responsibilities of different stakeholders, incentives for Pani Panchayat, etc. The different clauses of Pani Panchayat Act 2002 and the Memorandum of Understanding between the Department and the Pani Panchayat/WUA for handing over of the irrigation system to the Pani Panchayat/WUA should be explained to the farmers in simple terms. If they have any doubt in any clause, it should be clarified immediately. The farmers should be apprised of the need for forming Pani Panchayat/WUA for better operation and maintenance of canal minors/sub-minors, equitable water distribution between head and tail enders, adoption of common cropping pattern and other items of common interest to ensure sustainable quality irrigation service.

One the important constraints that affects the functioning and sustainability of Pani Panchayat/WUAs is lack of credibility of office bearers of Pani Panchayat and/or local leaders after holding the post. Unless the office bearers and leaders are honest, dynamic and farmer friendly, the very purposes of the programme may not be successful. Therefore, the local accountability mechanism should be strengthened. In the Pani Panchayat Act of Orissa, 2002 there is a provision to recall the president of a Pani Panchayat, if he is not responsive to farmers' needs and misuses his power and position. This is an exemplary feature of Orissa Act which need to be emulated by other states.

Physical System Inefficiency

Due to resource constraint, most of the irrigation projects suffer from several physical system inefficiencies. Many irrigation structures like head regulators, outlets, cross drainage works, village road bridges are in deteriorated and derelict conditions. Unless the faulty structures are renovated, rehabilitated and become fully operational, there is little hope that farmers will come forward to take up the responsibility of operation and maintenance of an inefficient system.

The older projects have many problems like deterioration of old control and measuring structures, leakages and seepage at various places, erosion of banks and beds, siltation and weed infestation. These are serious problems, discouraging farmers to take over the system management on technical and financial considerations. Hence it is imperative that irrigation structures be kept fully operational so as to facilitate smooth operation of irrigation system by the farmers.

Before handing over the minor/sub-minors projects to Pani Panchayat, the project authority should conduct a proper hydraulic testing of the irrigation system in the presence of Pani Panchayat executives and farmers. The water users of the concerned Pani Panchayat should be satisfied that the canal is discharging the designed volume of water and water is proportionately distributed in different reaches of the canal without any tail-end deprivation.

Information, Education, Training and Incentive

The water users should have thorough knowledge of various aspects of PIM such as the process of forming WUA, its structure, short-term and long-term benefits, functions, rights, duties and responsibilities of Water Resources Department and WUA. The water users need to understand and appreciate the goals and objectives of the programme. However, in some cases, farmers wrongly infer that it is a process of privatisation by which government wants to get rid of the problem of distribution and operation by simply handing over the system to the farmers. In some other cases, farmers feel that operation and maintenance of irrigation structures is the task of irrigation department which is shifting its responsibility on the farmers without any additional benefits. These types of numerous misunderstandings stand on the way of smooth formation and functioning of WUAs.

In Orissa, the Department of Water Resources provides many incentives for forming Pani Panchayat, which include annual maintenance grant and prizes for better functioning of Pani Panchayats. If 75 per cent of water users within the boundary of a Pani Panchayat become members of the Pani Panchayat, then grant-in-aid at the rate of Rs. 100 per hectare multiplied by the command area of the Pani Panchayat is allotted to the concerned Pani Panchayat every year. These incentives have induced the farmers to form PPs to avail the financial benefits. Similar approaches can be introduced by other states.

The WUAs will be able to operate and maintain the minors/ sub-minors provided they have personnel with technical know-how and do-how to manage the irrigation system. It is therefore necessary to provide appropriate managerial and technical training to office bearers of PPs and farmers. The main areas of capacity building for the office bearers of PPs are:

Technical: O&M, inventory and asset management, preparation of estimates, procurement, quality control, crop planning, water management, crop water requirement, preparation of canal operation plans, water scheduling and budgeting.

Legal: Law Awareness and legal provisions and procedures, dispute resolution.

Financial Management: Accounting, audit and social audit, raising resources, levy and assessment.

Administrative: Conducting of meetings, recording minutes/ proceedings, communication and negotiating skills, conflict resolution and conjunctive management with the irrigation agency/government.

Irrigation officials should also be trained from time to time to imbibe the right mindset, attitudinal change and technical expertise required for participatory irrigation management.

Monitoring & Evaluation

Systematic monitoring and periodic evaluation (M&E) with modern tools and techniques during the process of implementation of PIM helps in rectifying the mistakes and improves the functioning of WUAs (Das 2000). M&E procedure can be either goal oriented or multi-perspective. The M & E should cover the following aspects

Monitoring

1. *Performance Monitoring:* To track the use of project inputs and production of outputs and to identify delays and problems
2. *Financial Monitoring:* To monitor the correct use of funds, disbursements, and internal cash flow and assess cost-effectiveness
3. *Diagnostic Studies:* To understand why implementation and sustainability problems have occurred and propose solution
4. *Midterm Assessment of Project Performance:* To assess the overall progress of the project in order to identify key issues and required changes
5. *Monitoring Operations, Maintenance and Sustainability:* To assess the capacity of the project to continue delivering intended services and benefits throughout its planned economic life

Evaluation

For evaluating the impact of PIM, data need to be collected on technical, agricultural and socio-economic indicators of irrigation system performance in pre and post PIM situation with the objective to examine the following factors:

- Achievement of program goals and objectives
- Satisfaction of other participating agencies regarding their relations with the PP
- Community satisfaction with the performance of PP
- Effectiveness of coordination with other agencies
- Effectiveness and efficiency of monitoring and evaluation arrangements
- Financial administration and control
- Overall efficiency and effectiveness of the PP's management procedure
- Effectiveness of beneficiary participation in O&M of irrigation projects.

Apart from goal oriented activities, a multi-perspective M&E involving multiple stakeholders and multiple indicators should be carried out to identify the areas where the restructured institutions need strengthening. Number of methodology for multi-perspective M&E based on several stakeholders and multiple indicators as developed by Das, Mangaraj & Acharya (2003, 2004) has been applied for Orissa. Several indicators have been taken into account which include process indicators, outcome indicators and impact indicators, which may be expressed either in quantitative or qualitative terms. In this perspective collected opinion of various stakeholders regarding the performance of PPs are aggregated to obtain the composite information on performance of PPs. These studies and other evaluation studies in limited number of projects in Orissa unveil that though Pani Panchayat Programme has benefited farmers in some projects, still it is observed that additional strengthening of the institutions in the following aspects is required for ensuring sustainability.

- Marketing of produce
- Agricultural support services
- Continued support for Pani Panchayat, field personnel need to work directly with Pani Panchayat to ensure that they are effective.
- Resource mobilisation

Furthermore, detailed M&E studies based on Vermillion (2000) and Das, Mangaraj and Acharya (2003,2004) should be carried out in different types of irrigation projects for true understanding of the impediments in implementation of PIM and sustainability of PPs.

CONDITIONS FOR SUCCESS OF PIM

Though several state governments in India have taken policy decision to encourage farmers' participation in irrigation management and attempts are under way to motivate farmers to form WUAs, the farmers' response in this regard is not up to the satisfaction. The success of any organisation depends on some fundamental factors as mentioned below:

(i) Felt needs
(ii) Common interest
(iii) Collective effort
(iv) Effective leadership
(v) Bureaucratic commitment of the agency involved
(vi) Political will of the party in power
(vii) Financial viability
(viii) Enabling policy and legal environment
(ix) Effective role of change agents
(x) Agriculture is the major source of income and employment for the farmers
(xi) Land tenure security
(xii) Benefits due to participation are tangible, substantial and guaranteed
(xiii) Existence of indigenous community organisations

Some of the above factors are also highlighted in review of world-wide experience with water users' organisations (Johnson et al,2002; Meinzen-Dick et al.,1997) as critical factors that influence the likelihood of success of devolution programmes.

It is often argued that state sponsored programmes meet with failure due to its top down approach to solve localised grass roots problems. The local problems in natural resource management like water and land should be resolved through local initiatives and peoples' participation. In case of irrigation management the farmers should themselves feel the necessity of forming an organisation or association to improve the situation. There should not be disproportionate distribution of benefits among farmer groups. If the group of water users is small and homogeneous, conflicts are less. Capable, efficient, honest, trustworthy farmers should be elected as office bearers of Pani Panchayats.

The irrigation bureaucrats should be committed to the cause of organising farmers to form WUA and to decentralise and delegate power to them. The political party in power should provide all out support for implementing the programme successfully. In Andhra Pradesh and Orissa the success of PP is solely due to the initiative taken by the ruling party under the dynamic leadership of the concerned Chief Minister.

There should be well defined water rights of WUAs. Rights to water are detailed below:

(i) Rights to use the resource
(ii) Rights to derive income from a resource
(iii) Rights to change a resource
(iv) Rights to transfer the resource to others

Not all of these rights must be involved in a rights allocation and registration scheme. For example, rights to use and to derive income from a resource could be specified initially, while rights to change the resource and to transfer it could be kept in abeyance until a later stage. Moreover, it is not necessary to assign water rights to individuals, which can be traded in the water market.

Furthermore, the WUA should be financially viable by raising its own resources from different sources. The WUAs need to have reliable sources of income; otherwise they can hardly be sustainable. Different categories of income to WUAs are:

- Sale of water
- Sale/leasing out common property resources
- Fees and contributions
- Commercial operations
- Maintenance Contracts
- Subsidies and grants

The role of change agents in fostering WUAs is very important. Change agents are roles coming from outside the government-service provider-farmer chain of responsibility (Johnson et al., 2002). These include local and international consultants, international banks, bilateral funding organisation, UN agencies, local and international research organisations, NGOs, including WUAs. In playing the catalysing role the change agents bring in new ideas, evidence of successes and failures elsewhere, pragmatic action plan, robust research findings, and perspectives and values that may challenge local orthodoxy and inertia. To be effective, international change agents should collaborate and work with local change agents to provide technical support, new ideas, legitimacy, methodologies and financial assistance for strengthening the restructured institutions steering them in right direction.

Many factors must be considered in selecting an effective irrigation system management transfer method. Agronomic and socio-economic characteristics can affect the speed, success, and sustainability of a given method of irrigation transfer. Agro-economic factors include the size of land ownership holding, resource inequality, soil quality, scarcity of water, and the types of crops grown. The physical condition and effectiveness of the irrigation system, marketing and credit opportunities, and legal framework protecting farmers' titles and recognition of their organisation are also important. Socio-economic factors include the farmers' level of education, community solidarity, their knowledge and understanding of the participatory process, irrigation traditions, agricultural expertise, and familiarity with appropriate agricultural and business management practices.

Bardhan's (2000) empirical study of 48 indigenous irrigation communities in South India reveals that cooperative behaviour in an irrigation community is by and large negatively related to inequality of landholding, to urban or market connections, and positively to duration of access to water, monitoring by guards, and in some cases to social homogeneity, small group size, proportional cost-sharing rule, and collective hostile relation with other villages over water.

Sustainable WUAs require an enabling legal and administrative environment, strong political will with clarity of objects, accountable partnerships, and incentives. Empowerment of water users through capacity building, adequate resource mobilisation, appropriate incentives with timely monitoring and remedial actions can go a long way in strengthening the WUAs. The sustainability of a WUA clearly depends upon internal factors: farm size, location, education, social stratification and heterogeneity; and external factors such as the institutional environment, legal framework, financial and technical assistance, land reforms, agricultural policies and markets. To ensure effectiveness and sustainability of decentralised irrigation management institution, an integrated and comprehensive action plan is necessary.

SUGGESTIONS

Sustainability of WUAs depend on some fundamental factors. In order to have effective and sustainable WUAs, the following are suggested:

- As far as possible the formation of Pani Panchayats / WUAs should be spontaneous and demand driven, The WUAs should be endogenously created based on felt needs, common interest and collective effort. The existing social capital like social network, kinship ties, community solidarity should be used to foster WUAs. The approach should be voluntary, bottom-up and flexible.
- Successful implementation of the PIM programme requires involvement and cooperation of all stakeholders that include farmers, government, personnel of Department of Water Resources, Department of Agriculture, Department of Revenue and the implementing NGOs. Political will of the party in power along with cooperation of opposition party and bureaucratic commitment are highly essential for decentralising irrigation management. The role of change agents both external and internal is important in fastening the process.
- The dilapidated irrigation system should be improved and the canal should be rehabilitated and renovated to discharge the designed quantity of water and make proportionate distribution of water. There should be a proper hydraulic testing of the canal in the presence of water users. A fully operational and fault-free irrigation system should be turned over to WUAs for its upkeep.
- The organising efforts should be closely integrated with physical improvements to ensure that the farmers are committed to maintaining the improvements.
- Farmers' representatives should not misuse their power. Department of Water Resources must support the group decision in their internal affairs and refrain from interfering.
- Commitment and active interest of top policy makers and irrigation executives is essential in overcoming bureaucratic inertia and resistance to PIM programme.
- As the PIM process is very complex, no detailed blueprint solution or road map for achieving the changes can be laid down. In fact, the process requires experimentation,

negotiation, adaptation to local circumstances and time. It is emphasised that it may take a long period to achieve attitudinal changes needed at all levels for true success of PIM programme. Target oriented achievement should not be the objective of implementing the programme.

- It is observed that turnover to WUAs is partial and the irrigation agency continues to exercise control in water allocation and budget. This signals that the irrigation executives are reluctant to devolve power to WUAs and may cause management instability. The roles and responsibilities of various stake holders should be clearly defined and accordingly exercised without any overlapping or encroachment..
- WUAs are concerned about legal complexity due to emergence of free riders and anti-socials. Sufficient legal support should be given to WUAs for tackling such situations.
- The WUAs often demand a minimum period of warranty from Water Resources Department and working agencies for the physical work done in terms of workmanship and quality.
- Regular monitoring and evaluation of PIM program is necessary.
- Appropriate managerial, financial and technical training should be imparted to office bearers of WUAs/Pani Panchayats to build their capacity to manage the irrigation system efficiently. Irrigation officials as well as field staff should, also, be trained from time to time to imbibe the right mindset, attitudinal change and technical expertise required for PIM.
- There should be downward accountability of irrigation officials to client farmers.
- WUA should have proper legal status and clear water right. The water resource department should supply canal water as entered in the Memorandum of Understanding/Agreement between the WUA and the department.
- There should be a regulatory body to adjudicate any conflict or breach of contract between WUAs or between WUA and the department. Water courts may be established to render justice in time.
- The WUAs should be multifunctional. Along with the management of water, WUAs need to facilitate timely supply

of good quality agricultural inputs like improved seeds, fertilizer, farmyard manure and farm implements at reasonable prices. The WUAs may also take up other allied activities like transfer of technology through effective extension services, marketing of agricultural produce and agro-processing for value addition; so that agriculture can be a profitable and paying enterprise.

- Decentralisation of irrigation management should be accompanied with agricultural intensification programme to increase agricultural productivity. Land reform measures should be undertaken to recognise/record tenancy contracts, so that tenants who constitute an important segment of peasantry can become members of WUA and receive the benefits of farmer participation in management of water resources.
- The transfer process may be 'site-specific', and the approach to each region must be location specific keeping in view the geography, hydrology, ecology, agriculture and socio-cultural, psychological and economic conditions. Therefore, in general, a more gradual approach, balanced between the extremes of 'big bang' and 'gradualism' is probably most appropriate for a country like Indian sub-continent. Proceedings more gradually and adopting a cautious approach will allow appropriate response to various factors that will affect the sustainability and ultimate success of the transfer in the long run.

REFERENCES

Baland, Jean-Marie and J.P. Platteau (1996), Halting Degradation of Natural Resources: Is there a Role for Rural Communities? Oxford: FAO and Clarendon.

Bardhan, Pranab (2002), 'Decentralization of Governance and Development', Journal of Economic Perspectives 16(4):185-205.

Bardhan, P.K. (2000), 'Irrigation and Co-operation: An Empirical Analysis of 48 Irrigation Communities in South India', Economic Development and Cultural Change 48(4):847-865.

Bharadwaj, K. (1974), Production Conditions in Indian Agriculture: A study Based on Farm Management Surve. London: Cambridge University Press.

Bharadwaj Krishna (1990), Irrigation in India: Alternative Perspectives. ICSSR Research in Economics, Second Survey, Monograph 3.

Brewer, J.S., S. Kolavalli, A.H. Kalro, G.Naik, S.Ramanarayan, K.V. Raju and R. Sakthivadivel (1999) Irrigation Management Transfer in India: Policies, Processes and Performance. New Delhi: Oxford and IBH Publishing House.

Chambers Robert (1988), Managing Canal Irrigation: Practical Analysis from South Asia. New Delhi: Oxford.

Coward E. Walter. (1980), Irrigation and Agricultural Development in Asia: Perspectives from the Social Science. Ithaca: Cornell University Press.

Das, D.K. (2000), 'Participatory Irrigation Management with Special Reference to Orissa'; in M.Swain and G.C. Kar (eds.) Farmer and Local Participation in Irrigation Management, New Delhi: Commonwealth Publishers.

Das, D.K., B.K. Mangaraj, and R.C. Acharya (2003), 'Multi-Perspective Monitoring and Evaluation of PIM (Pani Panchayats) in Orissa: An Analytical Approach' Proceedings of Tenth National Water Convention, Ministry of Water Resources, Govt. of India, pp.339-348.

Das, D.K, B.K. Mangaraj, and R.C. Acharya. (2004), 'Performance Monitoring of Irrigation Management Reform (Pani Panchayat) in Orissa: A Multi-perspective Approach', 45th Technical Annual, Institution of Engineers (India), Orissa State Centre, pp-425-431.

Dhar Bonani (2001), UP Water Sector Restructuring Project: Social Assessment, World Bank Report.

Dhawan, B.D. (1988), Irrigation in India's Agricultural Development: Productivity, Stability and Equity. New Delhi: sage.

Dhawan, B.D. (1987), 'Productivity of Irrigated Farming: Survey Evidence from Three States.' Journal of Indian Water Resources Society, 17(3): 23-37.

Gooneratne, W. and S. Hirashima (1990), Irrigation and Water Management in Asia. New Delhi: Sterling.

Govt. of India, Ministry of Water Resources (1987), 'National Water Policy'.

Govt. of India, Ministry of Water Resources. (2002), 'National Water Policy'.

Govt. of Orissa (1994), State Water Policy of Orissa.

Groenfeldt, D. (2000), 'News Letter' International Network for Participatory Irrigation Management (INPIM).

Groenfeldt, D. and P. Sun (2004), 'Demand Management of Irrigation Systems Through Users' Participation', Mimeo.

Gulati, A. et al. (1994), 'Major and Medium Irrigation Schemes: Towards better financial Performance', Economic and Political Weekly 29(26): A72-A79.

Hardin, Garrett. (1968), 'The Tragedy of the Commons', Science, No.162, December.

Hooja Rakesh, G. Panagare and K.V. Raju, (eds.) (2002), Users in Water Management, Jaipur: Rawat.

INPIM. (2003), Report on Participatory Irrigation Management, New Delhi.

Jairath Jasveen (2001), Water User Associations in Andhra Pradesh, Center for Economic and Social Studies, Hyderabad and New Delhi: Concept.

Johnson Sam H., Mark Svendsen and Fernando Gonzalez (2002), Options for Institutional Reform in the Irrigation Sector, Presented at the International Seminar on Participatory Irrigation Management: 21-27 April, Beijing.

Maloney C. and K.V. Raju (1994), Managing Irrigation Together: Practice and Policy in India, New Delhi: Sage.

Meinzen-Dick Ruth, Meyra Mendoza, Loic Sadoulet, Ghada Abiad-Shields and Ashok Subramanian (1997) Sustainable Water User associations: Lessons from a Literature Review', in Subramanian,A., N. Vijay Jagannathan, Ruth Meinzen-Dick (eds.); User Organizations for Sustainable Water Services, World Bank Technical Paper No. 354.

Mitra, A.K. (1992), 'Joint Management of Irrigation Systems in India - Relevance of Japanese Experience', Economic and Political Weekly, 27(26): A-75, A-82.

Mitra, A.K. (1996), 'Irrigation Sector Reforms - Issues and Approaches', Economic and Political Weekly, 31(13): A-31-37.

Palacios, Enrique V. (1999), Benefits and Second Generation Problems of Irrigation Management Transfer in Mexico, EDI Participatory Irrigation Management Case Studies Series, World Bank.

The Orissa Pani Panchayat Act, (2002).

Sengupta, N. (1991), Managing Common Property: Irrigation in India and the Philippines, New Delhi: Sage.

Singh, K. (1994), Managing Common Pool Resources: Principles and Case Studies, Delhi: OUP.

Singh, K.K. (1991), (ed.) Farmers in the Management of Irrigation Systems. New Delhi: Sterling.

Svendsen Mark and Ashok Gulati (eds.) (1995), Strategic Changes in Indian Irrigation, New Delhi; ICAR, Washington DC: IFPRI.

Svendsen Mark, Jose Trava, and Sam H. Johnson III. (1997), Participatory Irrigation.

Management: Benefits and Second Generation Problems, Economic Development Institute, World Bank.

Swain Mamata (1998), 'Water Rate Fixation in Major and Medium Irrigation Projects in Orissa: Issues and Problems', Water and Energy International. 55(2): 63-72.

Swain Mamata and D.K. Das (1999), Emerging Trends and Reforms in Irrigation in India: A Perspective of Orissa. New Delhi: MD.

Swain Mamata and G.C. Kar. (2000), eds.. Farmer and Local Participation in Irrigation Management. New Delhi: Commonwealth.

Swain Mamata, (2002), 'Turning Over Irrigation Management to Farmers: Prospects and Constraints', Kurukhetra,.50(8): 4-10.

Swain Mamata and M. Swain (2003), 'Farmers' Organization and Turnover Programme: Some Lessons from Orissa Model', Kurukshetra, 51(12): 42-48.

Vaidyanathan, A. (1994), 'Better Resource Management for Poverty Alleviation', Economic and Political Weekly, 31(23):1397-1404.

Vaidyanathan, A. (1999), Water Resource Management: Institutions and Irrigation Development. New Delhi: Oxford.

Vermillion L. Douglas (2000), Guide to Monitoring and Evaluation of Irrigation Management Transfer; The Japanese Institute for Irrigation and Drainage (JIID), Tokyo.

Vermillion L. Douglas (2002), Current Issues and the Need for Monitoring Evaluation: Visit Note on PIM Programme in Andhra Pradesh; Users in Water Management, in Rakesh Hooja, Ganesh Pangare,.K.V.Raju (eds.), Rawat Publication, Jaipur.

Wade Robert (1987), Village Republics: Economic Conditions for Collective Action in South India. Cambridge: University Press.

17

MINOR IRRIGATION IN ANDHRA PRADESH: EQUITY AND OTHER ISSUES

*Rahul A. Shastri**

INTRODUCTION

The class divide between tank and groundwater irrigation forces choice on state policy. This paper suggests that their purported complementarity is highly overrated. Storage tanks are the least efficient form of charging groundwater among alternative water harvesting forms. They cannot be converted into percolation tanks. Further, if groundwater irrigation includes not just wells but also water harvesting structures, then more water for the wells means less water for tanks. There is a clear trade-off between tanks and wells.

Tank irrigation suffers from several technical deficiencies that groundwater does not. The collective nature of the asset aggravates these problems and leads to neglect. Groundwater being privately managed, has organisational resilience. Further, economic use of groundwater may ensue if electricity charges are metered. The paper also points to the equity inbuilt into surface lift schemes,

* Jt. Director, National Akademi of Development, 12-10-627, Warasiguda, Secunderabad-500 061.

and ways to bring deep tube wells into the reach of marginal farmers.

In minor irrigation, we have to choose between: (a) revival of tanks (b) new minor lift irrigation schemes, and (c) better and more groundwater use, or a mix of all three. Irrigation policy must be equitable in addition to being efficient. Equity in irrigation is understood here to mean some degree of parity in access to the form of irrigation between different sizes of farm holdings. Equity in water distribution is all the more important, because the basic agricultural input, land is very inequitably distributed.

DISTRIBUTION OF LAND, AND WATER FROM MINOR IRRIGATION

Land distribution in Andhra Pradesh is highly skewed and inequitable. Large farms (>10 ha.) are few in number (<1%), and operate a large part (nearly 10%) of the area. Medium farms (2-10 ha) are substantial in number (18%) and operate nearly half of the land, while 80% of the holdings are small and marginal, operating less than 40% of the land. Estimates of the distribution of land by size of holding are presented in Table 1 below.

Table 1
Estimated Distribution of Holdings and Land by Size in Andhra Pradesh

Items	*Marginal 0-1 Ha*	*Small 1-2 Ha*	*Medium 2-10 Ha*	*Big >10 Ha*
Share in Holdings	59.3%	21.4%	18.5%	0.8%
Share in Land Operated	17.5%	23.3%	49.7%	9.4%

*Source: NSSO, 1992.

Like land, distribution of and access to minor irrigation water is also inequitable. This is shown by Table 2. Row 1 of the Table shows that the share in number of units with access to various forms of minor irrigation was about 1/3rd each for marginal and small, about 1/4th for medium farms, and only 5% for large farms. Row 1a shows the ratio of these shares to the proportion of each category of holdings in the total number of holdings. It shows that the share in numbers of marginal farms was only 0.6 times their share in total number of holdings. On the other hand, the share of small and medium farms was about 1½ their share in total number of holdings. The proportion of large farms with access to various types of minor irrigation was more than 6 times their share in total number of holdings.

Row 2 brings out the share in land irrigated by MIFs by different types of holdings. As the Minor Irrigation Census does not provide this data, it has been estimated in this study on the assumption that the average irrigation capacity of different forms of minor irrigation is the same for all sizes of holdings. This assumption overestimates the share of marginal and small farms in irrigation because their size falls short of the irrigation capacity of most forms of minor irrigation. For instance, from the 3rd Minor Irrigation Census, it is found that even shallow tube wells have an average irrigation capacity of 1.5 ha., which is larger than the size of a marginal holding. Even so, Row 2 estimates that the share in land irrigated by all MIFs is nearly equal for all holding sizes. This despite the fact their share in number of holdings is widely unequal. Row 2a shows that by these estimates, the share of marginal farms in total land irrigated by MIFs is only 0.4 times their share in number of holdings. The share of small and medium farms in land irrigated by MIFs is about 1.25 times their share in number of holdings. Finally, the share of large farms in land irrigated by MIFs is 30 times their share in total holdings!

Inequality in distribution of water resources is closely associated with inequality in land holdings. Since water and land are strongly complementary, we can venture to suppose that inequalities in distribution of water resources reflect and perpetuate inequality in land ownership. This leads us to ask whether inequality afflicts the access to all forms of minor irrigation equally.

The 3rd Minor Irrigation Census enumerates five forms of Minor Irrigation. There are dug wells, shallow tube wells, deep tube wells, surface flow and minor surface lift projects. We make two changes in this classification. First, we shall refer to surface flow irrigation as Tanks. Surface flow irrigation consists mostly of tanks, since 90 percent of the surface flow units (74229/84223) are based on storage tanks, and another 4 percent are based one water conservation. Secondly, we shall, for the purpose of analysis lump together all three forms of groundwater irrigation. This may be justified in case of dug wells and shallow tubewells, which are similar in capacity and structure of distribution, but less so for deep tube wells (Table 3).

Table 2
Share of Different Categories of Holdings in Minor Irrigation*

Share of	Marginal 0-1 Ha	Small 1-2 Ha	Medium 2-10 Ha	Big >10 Ha
1. Number of Units with Access to all forms of Minor Irrigation	33.7%	35.3%	25.8%	5.1%
1a. Ratio to share in Total number of holdings	0.57	1.65	1.40	6.40
2. Share in land irrigated by all MIFs*	25.2%	27.2%	23.0%	24.7%
2a. Ratio to share in Total number of Holdings	0.42	1.27	1.24	30.82

*This is calculated on the assumption that the average area irrigated by each form of irrigation is the same for all holding sizes. This assumption overestimates the irrigation in marginal and small holdings because the holding size in these categories is much less than the irrigation capacity of different forms of irrigation. On the same grounds, irrigation in medium and large farms is underestimated.
Source: 3rd Minor Irrigation Census, 2004.

Table 3
Share of Different size of Holdings in five Minor Irrigation Forms

Share of → Number of units with access to	Marginal 0-1 Ha	Small 1-2 Ha	Medium 2-10 Ha	Big >10 Ha
II. 1. Dug wells	36.9%	36.7%	24.6%	1.8%
Ratio to share in Total holding	0.62	1.72	1.33	2.27
III. 2. Shallow Tubes	34.1%	36.8%	27.4%	1.7%
Ratio to share in Total holding	0.57	1.72	1.48	2.12

Deep tube wells, having a much higher capacity, tend to be out of the financial reach of marginal farmers, and favour medium and large farmers. We shall dwell upon this question later.

SHARE OF HOLDING SIZES IN DIFFERENT FORMS OF MINOR IRRIGATION

The share of different holding sizes in the three forms of irrigation: groundwater, tanks and others (surface lift) is reported in Table 4.

Cols. 2, 4, 6 of Table shows the share of holdings of different sizes in access to groundwater irrigation. Cols. 3, 5, 7 give the ratio of this share to the share of different holding sizes in the total number of agricultural holdings. Thus it is clear that there is considerable difference in the distribution of access between tanks and other forms of minor irrigation. Cols. 2 & 4 of the table indicate that in access to groundwater and other forms, the share of marginal, small and medium farmers varies between 30-35% approximately, while only 2-8% is the share of large holdings. In sharp contrast to this, large holdings are 82% of the units with access to Tank irrigation! All other farms put together are only 18%! Cols. 3,5,7 bring out the inequity in access to minor irrigation forms in relative terms.

Table 4
Access to three forms of Minor Irrigation by size of holding

Items	*Groundwater Irrigation*		*Others*		*Tanks*	
	Share in Access to	*Ratio to share in Total holding*	*Share in Access to*	*Ratio to share in Total holding*	*Share in Access to*	*Ratio to share in Total holding*
marginal 0-1 ha	35.1%	0.59	29.6%	0.50	4.3%	0.07
small 1-2 ha	36.7%	1.71	29.5%	1.38	4.2%	0.20
medium 2-10 ha	26.5%	1.43	32.8%	1.77	9.7%	0.52
big >10 ha	1.8%	2.25	8.1%	10.17	81.8%	102.23

Source: 3rd Minor Irrigation Census, ibid.

- It can be seen that the share of marginal farms is less their proportion in population for all forms. But where this is 0.5-0.6 times their population percentage for groundwater and surface lift, it is less than 0.1 times for Tanks.
- In the case of small and medium farms, for groundwater and other forms, their access is larger than their share in population, ranging between 1.4 to 1.8 times. However, for tank irrigation their share is only 0.2 to 0.5 times of their share in population, respectively.
- Finally, whereas the share of large farms in access to groundwater and other forms is only 2 to 10 times their share in population, their share in access to tanks is 100 times their share in the total population!

The above data brings out clearly that the difference between tank irrigation and other forms of minor irrigation is not a difference in degree, but a difference in kind. Access to tanks is highly inequitable, and is limited mostly to large farmers. The inequality in access to tanks means a much higher inequality in the distribution of water, since the average irrigation capacity of tanks is much larger (11.6 ha.), than the average irrigation capacity of groundwater forms (1-3 ha), and surface lift schemes (5.5 ha.).

SHARE OF MIFS IN IRRIGATION OF DIFFERENT HOLDING SIZES

We now consider the share of the three forms of irrigation in the total irrigation of each size of holding. Since the Minor Irrigation Census does not provide direct data on irrigation by size of holding, it is indirectly estimated by weighting the number of units with the average irrigation capacity of each form of minor irrigation. Table- 5 estimates the share of different forms of minor irrigation in the total irrigation of each size of holding. It can be seen that groundwater supports 90% of the irrigated area of marginal and small farmers, and over 80% to medium size farmers. In contrast, tanks are the source of 94% of the irrigated of large farmers. Tanks are also a little more important to medium farms, where it accounts for 12% of the irrigation.

Table 5
Share of different forms of Minor Irrigation in the Irrigation of each Holding size

Share of	*Marginal 0-1 Ha*	*Small 1-2 Ha*	*Medium 2-10 Ha*	*Big >10 Ha*
Groundwater Irrigation	91%	91%	82%	5%
Tanks	5%	4%	12%	94%
Surface lift	4%	4%	6%	1%

Computed from figures provided by the 3rd Minor Irrigation Census

POLITICAL ECONOMY OF MINOR IRRIGATION

The above analysis shows clearly that tanks are for large farmers. Large farmers get an estimated 94% of their irrigation from tanks, and large farms are 82% of the agricultural units with access to tanks. Tank irrigation is class irrigation: irrigation for the class of large farmers. In contrast to tank irrigation, groundwater and surface lift irrigation are mass irrigation: supplying irrigation to the masses. supply 95% of the irrigation need of marginal and small farmers, and over 80% of the irrigation of medium farmers. These three categories of farmers comprise 92-98% of the units that have access to groundwater and surface lift irrigation. The extreme class bias of tank irrigation raises two questions. The first is about the nature the state in Andhra Pradesh. The second relates to the choice between the two forms of minor irrigation.

There are wide differences in the assessment of the class nature of the Indian state. These range from the rule of intermediate classes[1], to national bourgeoisie, big-bourgeoisie cum landlord, feudal cum comprador bourgeoisie and so on[2]. Whatever the importance attached to non-agrarian classes in the analysis, most analysts seem to agree that at the regional level of state government, large landlords have an important influence. If so, it should reflect in state policy. Indeed, it can be seen from the debates of the AP legislative Assembly, that there were many demands to revive tanks from 1956 to 1980s. Nevertheless state neglect seems to have continued,[3] and tank irrigation continued to decline. Before the seventies, this neglected seems to have been accompanied with an emphasis on major irrigation in state policy. After the seventies, the state seems to have supported the borewell revolution through subsidies of various forms, while the neglect of tank irrigation continued.[4] After this, even the demands for revival of tank irrigation in the state assembly seem to have weakened.[5]

The above facts raise uncomfortable questions. If large farmers exercise state power, or share substantially in it, why is it that the A.P. Government so neglected tank maintenance and concentrated instead on major irrigation, and borewell irrigation thereafter? And this despite repeated demands in the State Assembly debates. Clearly, the political influence of landlords did not translate into state policy. Further, even in debates, tanks seem to have lost their importance in the state assembly in the recent decades. And this is inspite of the fact that they provide an estimated 95% of the irrigation water to large farmers! All this suggests that the influence and power of large farmers over state policy is somewhat overestimated.[6]

How can we explain the close relation between tank irrigation and large farms? One reason lies in history. Tanks are an ancient form of irrigation. As such they were constructed in all places where the soil conditions and surface flows were favourable to storage. Since they were constructed within the framework of the feudal system, it seems plausible that feudal/large landowners owned large parts of tank irrigated land. The pressures of development, and later land reforms, forced traditional landowners to reduce the size of their holdings. It seems plausible that in the process of alienation marginal and less productive lands would have greater priority. The more productive tank-irrigated lands would be retained by the traditional landowners. This could be the reason why large holdings account for 82% of the units with access to tanks.

It would appear that by retaining tank-irrigated holdings, traditional landowners held on to a disproportionate share of water. Land was alienated, but not water. Land distribution was quantitative not qualitative. Tank-irrigated lands were retained while the non-tank lands went to smaller, and non-traditional landowners. The new classes got more land but not proportionately more water.

Redistribution of water was only possible outside the frame of tank irrigation. Was major irrigation one way of redistributing water in favour of small farmers and non-traditional landowners? Whether this was true or not, wells are the primary means by which these classes can get water for their land. The declining costs of borewell irrigation has helped in this process.[7] Thus, the growth of groundwater irrigation at the expense of tank irrigation reflects a redistribution of water in agriculture that compensates for the qualitative inequities of land distribution. The class divide between tank and groundwater irrigation raises the question of choice between the two forms.

CHOICE BETWEEN TANK AND GROUNDWATER IRRIGATION

Many scholars deny the choice between tank and groundwater irrigation. They assert a high complementarity between tanks and groundwater. For instance, it has even been argued that the decline in tank irrigation may explain the low levels of groundwater in Telangana and Rayalseema, and that it is necessary to revive tank irrigation to improve the groundwater conditions.[8] It can be argued against this view that the presumed complementarity is highly over-rated, and that a choice between the two does exist.

Everyone agrees that groundwater has to be recharged through surface bodies. But the best surface bodies for recharging ground water are not storage tanks. Storage tanks are meant to store water not to lose it. It has been estimated that the inadvertent seepage of water into the ground from storage tanks is only 9-20% of the live storage capacity over the year.[9] As against this, seepage from percolation tanks is as high as 50%! This difference is due to the difference in the beds of the two types of tanks. In fact, storage tanks are the least efficient of the alternative surface forms of groundwater recharge. It has been reported that "infiltration rates in irrigation tank beds are much lower, only about one fourth to one-sixth of the rates in the stream beds and lower than the rates in cultivated and uncultivated (fallow) land"[10] (emphasis added). This is one reason why storage tanks are not an important water harvesting structure.

We cannot get round this problem by converting irrigation tanks into percolation tanks, as has been suggested.[11] That is because percolation tanks cannot be constructed everywhere. The soil type must favour percolation, and the location must favour recharging some aquifers. That is, the conditions must favour loss of water into the soil. This is precisely what the storage tanks are meant to avoid. That is why, they are likely to have been chosen at spots where percolation would be low rather than high. To recharge groundwater, we have to look to forms other than storage tanks.

Water harvesting increases the water in wells, but reduces the water in tanks. Thus, it is noted that "Water harvesting activities, when done on a substantial scale, imply a spatial reallocation of water availability and use in a catchment, while negatively influencing surface storage systems."[12] One recent study estimates that in some areas, the increase in water harvesting to recharge groundwater level has, over 12 years, reduced the inflows into tanks by 10-50%.[13]

This indicates a conflict, and not complementarity, between groundwater irrigation (inclusive of water harvesting) and tank irrigation. Active promotion of one form conflicts against the other form. Choice of forms is necessary, at least at the margin. It has already been shown that of groundwater irrigation is more equitable, while tank irrigation is predominantly for the rich. It can also be argued that storage tanks are less efficient both on the technical and economic considerations.

Tanks are a less efficient form of storing water than wells. This is because, tanks suffer from evaporation while groundwater does not. During summer, when water evaporation is twice as high as during monsoons, this can cause salination.[14] Siltation increases loss due to evaporation. That is because, it increases the water spread ratio. Typically, the surface to volume ratio of a water body should be between 0.23 and 0.3 to minimise loss due to evaporation.[15] Siltation reduces the volume of the tank and increases this ratio, increasing the loss due to evaporation.

The surface water distribution system of tanks also suffers from inefficiencies including lopsided water delivery (water logging), as well as loss of water due to evaporation and seepages. These can be minimised in groundwater irrigation which has a smaller irrigation capacity, and is privately managed. Tanks also suffer from all the ills of collective management. This results in Siltation, weed infestation, Closure of link/feeder between chains of tanks, encroachment into tanks, cultivation of tank beds etc. all of which are leading to the decline of tanks. Groundwater is privately managed, and as such, has an inbuilt organisational resilience. Finally, groundwater distribution depends upon electricity. As long as electricity charges are metered, they put a cost on the use of water that encourages the economic use of water. This is advantageous both at the micro and macro level. At the micro-level it avoids over exploitation of land and water. At the macro-level it helps because Andhra Pradesh is a water short state.

DEEP TUBES AND SURFACE LIFT IRRIGATION

The more capital intensive forms of irrigation are deep tubes and surface lift schemes. Surface lift schemes use pumps to lift flowing stream water into irrigation channels. Deep tube wells exceed 70 meters in depth and typically have a discharge that is 15 times the average of a shallow tubewell in normal conditions.[16] The comparison of shares of different size holdings in access to these forms of irrigation with wells and shallow tubes is presented in Table 6.

Table 6
Share of Different Categories of Farmers in access to Deep tubes and Surface Lift Schemes

Share of	*Marginal 0-1 Ha*	*Small 1-2 Ha*	*Medium 2-10 Ha*	*Big >10 Ha*
Wells and Shallow tubes	35.9%	36.8%	25.6%	1.8%
Deep Tubes	17.3%	35.4%	44.8%	2.5%
Surface lift	29.6%	29.5%	32.8%	8.1%

Source: 3rd Minor Irrigation Census, op. cit.

It can be seen from the table that the share of marginal farmers in deep tubes is substantially less than in dugwells and shallow tubes. On the other hand, their share in surface lift schemes is somewhat closer to their share in dugwells and shallow tubes. However, the share of medium and large farmers in surface lift schemes is somewhat higher than their share in dugwells and borewells. The smaller share of marginal farmers in deeptubes may be explained as follows. First, deep tubes are privately financed, and as such, are out of reach of the marginal farmer. Also, the small size of marginal farms may preclude their full utilisation. State subsidies and/or collective use of deep tube wells may obviate these problems.

As regards surface lift schemes, the greater share of medium and large farms (when compared to dugwells and shallow tubes) may be explained by the local initiative that propels these schemes from conception to delivery. It is to be expected that these classes of farmers will have a greater initiative and also a greater say in the scheme of things. This may explain their greater share. In spite of their larger share, marginal farms remain a fairly large proportion of those with access to these schemes. That is because their lands cannot be bypassed by irrigation channels distributing water for the schemes. Thus, geographical dispersal of non-tank lands among different categories of farmers ensures that equity is inbuilt in surface-lift schemes.

CONCLUSIONS

It is found in this study that land distribution in Andhra Pradesh is inequitable, as is access to minor irrigation forms, and the irrigation from them. It is found that the share of marginal farms in access to minor irrigation forms is only 0.6 times their share in total farms, while for other classes this rises from 1.65 to 6.4 times (small to large farms). If share in irrigated land is considered, this inequity appears to be much larger.It is found that Tank irrigation is class irrigation. Large farms are 82% of the units with access to tank irrigation, and an estimated 94% of their irrigation is from tanks. In contrast, groundwater and surface lift irrigation is mass irrigation. An estimated 91% of the irrigation of marginal and small farmers comes from groundwater, while for medium farmers this figure is 82%. Their share in holdings with access to these forms is closer to their share in holdings, with the ratio of the two ranging form 0.6 (marginal farms) to 2.25 (large farms).

The state neglect of tank irrigation and the active promotion of other forms of mass irrigation is difficult to explain on the premise of landlord control or influence on state policy. This throws doubt on many conventional views on the nature of the state in Andhra Pradesh.It is suggested that pressures of development, and later land reforms led to alienation of non-tank lands and concentration of tank irrigated lands in the hands of traditional landowners. This mode of land distribution distributed more land than water. Equitable redistribution of water was a result of non-tank modes of irrigation that were favoured by state policy after independence.

The class divide between tank and groundwater irrigation forces choice on state policy. This paper suggests that their purported complementarity is highly overrated. Storage tanks are the least efficient form of charging groundwater among alternative water harvesting forms. They cannot be converted into percolation tanks. Further, if groundwater irrigation includes not just wells but also water harvesting structures, then more water for the wells means less water for tanks. There is a clear trade-off between tanks and wells. Tank irrigation suffers from several technical deficiencies that ground water does not. The collective nature of the asset aggravates these problems and leads to neglect. Groundwater being privately managed, has organisational resilience. Further, economic use of groundwater may be ensured if electricity charges are metered. Deep tube wells however, seem to be out of reach of marginal farmers. Remedial measures such as state support and collective drilling and use of such wells may help.

NOTES

1. "One of the most original and striking of Kalecki's conceptions is the notion of 'intermediate regimes', in which the old feudal class has been dispossesed and the big business scarcely exists, so that political power descends to the lower middle class". In the Indian context this has been interpreted by K.N. Raj to mean the rule of a coalition of "rich peasants, small traders, educated professional and civil servants ..." p10 Joan Robinson, 1976.
2. Various branches of the communist movement.

3. It has been suggested that minor irrigation suffered as after independence, major projects were given more importance. For instance, see Venkateshwarlu and Srinivas, 2001.
4. cf. p 104, GoAP, 2003b.
5. "by 1980s and 1990s ... tanks began to take a low priority in the Irrigation debates of the A.P legislative Assembly" B. Prabhakar, 2005.
6. This is not to deny the power of large farmers over the state apparatus and functionaries, especially in rural areas. We distinguish here between control over policy and direction, and influence and control over the organisation of the state apparatus.
7. P. 172, V. Ratna Reddy, 2003.
8. P. 174, V. Ratna Reddy, ibid.
9. GoAP, 1984.
10. P. 142, K.R. Karanth, 1995.
11. "A better option would be to convert the existing tanks into percolation tanks so that groundwater potential would improve." P. 175, Ratna Reddy, op.cit.
12. P. 41, GoAP, 2003a op. cit.
13. The study was of five tanks located in Kalyandurg and Dhone, in Rayalseema, GoAP, 2003c.
14. K.R. Karanth, op. cit.
15. K.R. Karanth, op. cit.
16. P. 6, 3rd Minor Irrigation Census, op. cit.

REFERENCES

B. Prabhakar (2005), "Study of Tank Irrigation in Mid Godavari Basin", Unpublished.

GoAP (1984), Groundwater Estimation Committee Report.

GoAP (2003a), A.P. Water Vision, Vol. 1, Water Conservation Mission.

GoAP (2003b), AP Water Vision: Methods, Position Papers And District Reports, Vol. 2, Water Conservation Mission.

GoAP (2003c), Andhra Pradesh Rural Livelihoods Programme Water Audit, ed. Dr M.S. Rama Mohan Rao, Dr C.H. Batchelor, Dr A.J. James, Dr R. Nagaraja, Dr J. Seeley, Dr J.A. Butterworth, APRural Livelihoods Programme.

GoAP (2004), 3rd Minor Irrigation Census, 2000-2001 of Andhra Pradesh, Directorate of Economics and Statistics.

J. Robinson (1976), "Introduction", Essays on Developing Economics, Kalecki M. The Harvester Press, 1976.

K.R. Karanth (1995), "Role of Storage Reservoirs and Percolation Tanks in Meeting the Challenge of Diminishing Water Supplies", in Groundwater Management: The Supply Dominated Focus of Traditional NGO and Government Efforts, Vikram Sarabhai Center for Development Interaction, VIKSAT, Natural Heritage Institute.

NSSO (1992), Operational Land Holdings In India 1991-92: Salient Features, NSSO, Department of Statistics, Ministry of Planning & Programme Implementation Government of India

V.R. Reddy (2003), "Irrigation: Development and Reforms", in Andhra Pradesh Development: Economic Reforms and Challenges Ahead, Ed. C.H. Rao and S. Mahendra Dev, CESS.

Venkateshwarlu and Srinivas (2001), Troubled Waters – Status of tank Management Institutions in Rainfed Areas of Andhra Pradesh, SDC/ ICNGO (Swiss Agency for Development and Cooperation/Inter-cooperation Non Government Organisation) Programme.

18

ECONOMIC AND INSTITUTIONAL ASPECTS OF IRRIGATION WATER RATES IN ANDHRA PRADESH[1]

U. Hemantha Kumar[*]

INTRODUCTION

The approaches to irrigation pricing needs to be understood in relation to the objectives of a nation's irrigation development programme. The research on socio-economic and institutional issues involved in the development and distribution of investment on irrigation through public investment programme have started receiving serious attention in recent past (NCAER, 1959, p. 65; Prasad and Rao, 1985, p. 17). Many of the studies did not give much importance to water rates. Some studies had noted the benefits from irrigation, i.e. improving production and productivity levels Vaidhyanathan (1987), Dhawan (1983), and Nadkarni (1979).

1 This paper is based on authors' unpublished Ph.D. thesis on "Economics of Irrigation: with special reference to water rates in Andhra Pradesh, 1865-1991" University of Hyderabad, Hyderabad and Reddy, V. Ratna, Prudhvikar Reddy, P and Hemantha Kumar, U, (2004), Draft Report on "Formalising Irrigation Institutions: A Study of Water User Associations in Andhra Pradesh", Centre for Economic and Social Studies (CESS), Hyderabad.

* Assistant Professor, Center for Water and Land Management, NIRD, Hyderabad.

Others stressed on the investment allocation and financial obligations, i.e. allocation of funds and financial autonomy of irrigation systems GOI (1993), ADB and IIMI (1986), Nickum (1979), Kimura (1977) and Khan (1981). Some other studies dealt with the groundwater markets Shah and Raju (1988), Janakarajan (1993), and Raju and Rao (1991). The canal irrigation and water rates are not analysed at length. The economic analysis of irrigation pricing hardly attracted the attention of irrigation researchers (except Vaidhyanathan Committee on Pricing of Irrigation, 1992) in recent times especially in two areas. The first is to design a pricing mechanism that covers the appropriate costs, and the second is to achieve high collection rates through effective water management (Formation of Water User Associations- WUAs).

The irrigation policy of the British rulers was marked by an apparent change in the methods of fixing of water rates. This was mainly because of their strong commercial motivation. The need for rationalization of water rates arose with the commencement of the irrigation under major river deltas in the early 1850s in the Madras Presidency. The basis of charging for water in the British period, when supplied by Government works, was chiefly the area irrigated. But different systems of charging were in operation in different parts of the Presidency at different points of time. An attempt is made here to trace the origin, evolution and rationalization of water rates in the British period as well as in Independent India in general and Andhra Pradesh in particular.

METHOD AND ANALYTICAL FRAMEWORK

This is a empirical study as well as historical research. The sources of data include various publications, Government reports and records. At the empirical level, as a second step, the approach was to study the problem extensively as well as intensively. A sample of 222 WUAs were selected from 22 districts (10 WUAs in each district) in the state. A detailed WUA schedule was canvassed among these 222 WUAs in order to examine their structure and functioning (see Table 1).

This paper is organized in four sections. The first section discusses the water rates policy and economics of irrigation both in British and Independent India; section two deals with pricing mechanism in Andhra Pradesh at the macro level over a period of time. The last section presents the concluding remarks and policy

Table 1
Details of Sample Selection of the WUA Households

Region	*District*	*Mandal*	*Name of the Village/WUA*	*Source of Irrigation*	*No. of Households Selected by Location of their land*			*Total Sample Households Selected*
					Head	*Middle*	*Tail*	
Rayalseema	Chittoor	Nagula Palem	Rajula Kandriga	Big Tank	15	20	15	50
					(60)	(83)	(59)	(202)
		Nagari	Mangadu	Small Tank	10	17	23	50
					(16)	(26)	(36)	(78)
Telangana	Nalgonda	Anumala	Ibrahimpatnam	Head Reach (Canal)	26	8	16	50
					(198)	(61)	(121)	(380)
	Khammam	Pallipadu	Singaraya Palem	Tail End (Canal)	10	12	28	50
					(187)	(225)	(525)	(937)
Coastal Andhra	East Godavari	Katreni Kona	Pallemkuru	Canal	14	18	18	50
					(565)	(725)	(725)	(2015)
	Srikakulam	Laveru	Marupaka-II	Tank (Medium)	20	16	14	50
					(56)	(45)	(39)	(140)

Note: Figures in brackets are the actual number of households in the respective category. Selection of Households within the Head reach, Middle and Tail end was based on the proportion of number of farmers in the respective location i.e., Head, Middle and Tail end to the total number of farmers in the respective WUA in the Village.

implications of Pricing of Irrigation as well as Participatory Irrigation Management through strengthening of irrigation institutions.

WATER RATES POLICY

Water Rates in Madras Presidency

Various principles were laid down for determining the viability of individual irrigation projects. G. N. Taylor found in 1858 that the increased revenue collections of Government in specific areas were mainly due to the construction of the anicuts in the Godavari and Krishna districts (Government of Madras, 1858, pp. 19-20). The great objective was to ascertain what the land actually yielded when supplied with anicut water. By then, the need of legislation in order to clearly define the position of the state in respect of irrigation was recognized and enacted act in 1865 by the Madras Presidency. The 1865 Act says that "whenever water is supplied or used for purpose of irrigation from any river, stream, channel, tank, or work belonging to or constructed by Government, it shall be lawful for the Government to levy at pleasure on the land so irrigated a separate cess for the use of the water, which shall be additional to any land assessment that may be leviable on the said land as un-irrigated or punjah (Government of Madras, 1865, pp.130-131). The irrigated products were viewed under two classes. Those required a large quantity of water to bring the crop to maturity, and those for which a moderate supply was sufficient. The former include paddy, sugarcane, and coconut plantations; and latter included turmeric, onions, and all ordinary garden produce. Two scales of rates for the different soils of a district were thus fixed; one for lands enjoying a full and certain supply of water, the other for lands, the irrigation of which was imperfect. Thus, during the 19th century the test for economic soundness of an irrigation project in India was based chiefly on the direct returns on the investment.

Later on came into being the criterion of productivity of the projects. Irrigation projects were regarded as productive or unproductive accordingly as they yielded net revenue or below certain fixed percentage (4%) calculated at the sum-at-charge, ten years after the completion. The sum-at-charge constituted the total capital outlay and the arrears of interest (PWD, 1879, pp.III-XXI).

This principle was followed since then and the rate of return required, before an irrigation projects could be considered as financially productive and economically sound was fixed from time to time according to the prevailing conditions. Later, the Royal Commission on Irrigation (First Indian Irrigation Commission, 1901) stated that the indirect advantages of irrigation in India were so great and incontestable that the question of the direct financial returns which might be anticipated on the capital outlay was one of minor importance to which very little regard should be paid. But in the context of water rate on productive works, it should include a charge for the supply of water which would cover primarily the debt charges and overhead expenses (GOI, 1903, p.27). This was again observed by the Taxation Enquiry Committee (1924-25), that it was difficult to evaluate exactly the extra net income of land from irrigation and that this being so the simple thing to do would be to charge a fraction of gross produce of irrigated land as water rates (Government of Madras, 1926, pp. 99-110). But the committee held that the direct returns from irrigation project were maintained.

Except in the Godavari and Krishna Deltas and on the Kurnool-Cuddapah canal, the assessment on irrigated land was consolidated, and the shares due respectively to land and water could be only approximately determined. In the Godavari and Krishna Deltas and on the Kurnool-Cuddapah canal, the assessment on irrigated lands was not consolidated, but separate water cess was levied on the area irrigated. For the determination of the share due to irrigation resulting from the disintegration of consolidated assessments, elaborate rules were framed (Government of Madras, 1879, p.38). Where settlements had not been revised the average dry rate of the village, or where that was not available for the taluk, or where that could not be determined for the district, the rate was applied to the total area irrigated. The product was then assigned to land and the remainder to water. With regard to double crop lands, the charge for the two crops was consolidated- the charge for the second crop being essentially in the nature of a water rate. The rates were fixed according to the revenue settlements, and the difference in the water-rate both in Godavari, Krishna and the Kurnool-Cuddapah Deltas were due to the variations in duration and supply of water to the tracts, where both wet and dry crops were shown (see Table 2).

Table 2
Water Rates in the Godavari and Krishna Deltas and on the Kurnool-Cuddapah canal-1869 (Rs. Per acre)

Nature of Crop	*Godavari and Krishna Deltas*	*Kurnool-Cuddapah Canal*
1. Single Wet Crop	4	4
2. Second Wet Crop	4	3
3. If Compounded for both Crops for not less than five years	6.5	6
4. Single Dry Crop	2	1
5. Second Dry crop on irrigated lands if not Compounded	2	1
6. Garden Crop	8	6
7. Garden Crop if Compounded for not less than five years	6.5	–
8. Crops requiring frequent flooding	–	3

Source: Based on C.D. MacLean, 1987, p.411.

In tracts of the Godavari and Krishna Deltas, the assessments of irrigated lands were made up of dry lands with the addition of a separate water rate. In all districts of Madras Presidency that had been re-settled by the Settlement Department, the assessment of the same class and dry lands were determined accordingly. In other cases, were approximately made, the application of average dry rates to wet areas, the product representing the land assessment proper and the balance of the wet assessment was assigned to irrigation. The incidence of irrigation revenue per acre for each class of works was deduced from the same accounts on the works of Classes I and II (see Table 3).

Table 3
Incidents of Irrigation Revenue under each class of work 1869

Class of Work	*Average Irrigation Revenue per acre (Rs.)*
1. Productive Public Works	3.27
2. Public Works not classed as productive works for which neither capital nor revenue accounts are kept:	3.28
3. Works for which in a continuous records of expenditure is maintained individually	4.93

Source: Based on C.D. Maclean, 1987, p. 419.

This was a practice till 1869, and then it was revealed that the rates themselves were numerous and unfair (Government of Madras, 1869, para. 7), which led to the introduction of a system of fixed water- rates Rs. 4 per acre in 1874 being charged for a wet crop grown on dry lands and Rs. 2 for dry crop. A distinction between occasional and systematic irrigation was drawn in 1890 and crops systematically irrigated became chargeable as wet (Government of Madras, No. 272, 1890). Prior to 1890, the charges for water was regulated by the class of crop raised, but not by the number of waterings it received. There were some defects raised by collectors of several districts that it made a distinction between systematic and occasional irrigation, that it assumed only two classes of irrigation sources where as in most districts there were four or five or six classes, that it involved a classification of crops, a procedure inconsistent with the principles underlying the ryotwari settlement, and that it penalized the bad lands and let off the good lands too lightly (Government of Madras, 1923, pp. 34-35). Then the Madras Government replaced it with the differential system, which stated that, when a dry land was irrigated, the appropriate water rate was the difference between the wet assessment and the corresponding dry assessment for the same class and sort of land. The Government accordingly, drew up a scheme in 1906 for water rates in varying terms of the difference between the wet and dry assessments. The scheme was introduced first in Kurnool district.

Some rules for the levy of water-cess on dry lands irrigated from a Government source or work in lieu of the existing rules relating to standard scale of water rates, the Godavari, Krishna and Guntur districts, the Kurnool-Cuddapah canal, Kurnool districts, etc., were revised. Irrigation sources were divided into three groups for purposes of charging water rate (Government of Madras, 1923, p. 66). Irrigation sources which could ordinarily afford a supply of water for not less than eight months in the year were placed in the first group; irrigation sources which could afford supply of water from five to eight months were placed in the second group. Other irrigation sources were in the third group (see Table 4). Second and third wet crop would be half of the rate charged for a single wet crop with variation in the supply of water.

Table 4
Water Rates of Different Crops, 1923

Nature of Crop	*Group-I) (Rs.-Annas*	*(Group-II (Rs.- Annas)*	*Group-III (Rs. -Annas)*
For a single wet crop per acre	6-4	4-3	3-2
For a single dry crop per acre	3-2	2-1	1-9
For a double crop per acre	0-6	6-4	4-11

Source: Government of Madras, Revenue, No. 1089, dt. 19.07.1923 (Ms.), p. 66.

Thus, in the British administration there were several systems of charging irrigation water in different parts of the country. Hence, the British government invented a separate and uniform system of charging the irrigation water to suit the newly emerging irrigation environment. The uniform and fixed system prevailed in the initial stages of the British rule. Later the differential water rate system was introduced in 1909 and experimentation was carried out for some years. Since it proved to be very complicated in practice, again the fixed and uniform water rate system, with some modifications, was re-introduced in 1931.

Water Rates in Independent India

Water rates in Independent India vary not only among states but also among different projects of the same state. These variations and inconsistencies in water rates in India were attributable to the peculiarities of the historical forces which went into their formulation either the colonial past or from the erstwhile princely administration. They emerged as ad-hoc arrangements to be substituted by consistent policies later on. But this substitution was never made and as such, water rates in India have never been fixed on a scientific basis (Hemantha Kumar, U, 1992, pp. 69-70). There existed a wide variety of systems for the calculation of water rates. There was multiplicity both in principle and in the rates for assessment (Central Water Commission, 1992, p. 2). Sometimes, the value of the crop at maturity served as a rough guide in fixing irrigation charges. At other times, the cost of providing water has been the overriding consideration for the rate fixing authority while calculating water rates.

This issue of the economics of irrigation pricing was received the widest attention in the 20th century especially after

Independence. The first deviation from the traditional past was made by D.R. Gadgil. This study has referred to a number of economic benefits from irrigation to the society as a whole and made a strong plea for a change in the productivity criterion to test the economic viability of an irrigation project. It was mentioned that "the investment policy of the state could be determined from a point of view broader than that of the direct net return on investment (Gadgil, D.R. 1948, p. 173). The situation changed during the period of Five Year Plans. Though, productivity criterion is not given up fully, it was not adhered to strictly in practice. It has been realized that on the basis of existing rates of irrigation charges hardly any major work could be expected to be remunerative (GOI, 1952, pp. 354-356). The charges were on par with the investments in irrigation have risen both in monetary and real terms as a result of works of construction for the irrigation projects. The taxation enquiry committee had opined that broadly, the assessment of water rates may be regulated according to three different principles, viz., no loss and no profit (to cover maintenance charges), no profit and some loss (dry land irrigation development), and some profit and no loss (well-off area where irrigation already developed) (GOI, 1953-54, pp. 252-55). Later, the attention has shifted from the question of investment decision to that of recovery of cost. Thus, there have been two contradictory rules (GOI, 1957, p. 43). It is in this context they have invariably recommended the upward revision of water rates during successive Five Year Plans.

Another view was that what is relevant for fixation of water rates is not cost but benefits from irrigation (NCAER, 1959, pp. 68-74). The only sound and justifiable basis for the irrigation rate should be the additional net benefit available to the cultivators for the use of water supplied. The Maharashtra Irrigation Commission 1962, suggested irrigation facility is a service and that it is reasonable that it should be paid for by the persons who take advantage of it and recommended that irrigation works as a whole should give an annual income which is equivalent to the annual cost thereof and that the burden of providing irrigation should not fall, even in part, on the general tax payer (Government of Maharashtra, 1962, pp. 67-75). Further, the commission indicated the following as the facts to be taken into consideration in formulating water rates for individual crops: the cost of irrigation, water required by the crop, the level of average gross income obtained from the crop under the usual conditions of fertility, supply of water, level of efficiency of cultivation, and prices.

The Government of India appointed a committee to suggest ways and means of improving financial returns from irrigation projects under the Chairmanship of Nijalingappa (GOI, 1964, pp. 1-31). This was to be measured, according to the Nijalingappa Committee by the extent of net benefit after irrigation over the net benefit before irrigation. The committee recommended water rates ranging 25 to 40 per cent of the additional net benefit derived per acre production of crops produced before and after irrigation, and those regarding per acre cost of production before and after irrigation. It is obvious that this committee tried to compromise the recommendations of the NCAER and the Maharashtra Commission as the former has recommended water rate varying between 25 per cent to 40 per cent of the net benefit and the latter 6 per cent to 12 per cent of the gross produce. There were studies also available from demand side suggesting that, in order to wipe out the resulting deficit between the construction cost of the modern irrigation project and the net return, a considerable rise in rates become necessary (Ansari. N. 1968, p. vi). The problem of the relationship between the growth of irrigation facilities and cost recoup principle was also studied by the Second Irrigation Commission (GOI, 1972, p. 406). The Commission highlighted the issue that the water rates should be related to the benefit which irrigation confers rather than to the cost of irrigation projects (GOI, 1972, pp. 407-10).

The system of irrigation water rates in various States of the country is as diverse as the culture and people. Rates have been fixed considering a number of factors. Sometimes, without any rational and objective considerations, they are seldom revised. Major factors leading to variation of water rates are: geographical unit, system of irrigation and crops and season. In states like Uttar Pradesh, Punjab, Haryana, Rajasthan, Bihar, Gujarat and Tamilnadu water rates are charged for actual areas irrigated under different crops, the rates varying with crops, region and projects. In Orissa, there is a flat water rate per acre of land within the cultivable command area of an irrigation work payable to the state government for supply of water whether used or not from an irrigation work (Central Water Commission, 1992, p. 20). In Andhra

Pradesh and other states, consolidated wet assessment in levied on lands registered as wet. Water rates are not being realised in Assam and Himachal Pradesh. In Nagaland and Meghalaya, there are no irrigation rates at present. In Sikkim, irrigation rates have not yet been finalised. There is thus, no uniform basis for irrigation charges in the country.

WATER RATES IN ANDHRA PRADESH

Water Rates in Andhra and Rayalseema Areas

All cultivable lands in the Andhra area, are classified and registered in the land revenue accounts under three heads, viz,. wet, dry, and manavari lands. Irrigation cess or water rate is levied on drylands for use of water from a Government source and irrigation of a crop other than the first crop on registered single crop wet lands. In revenue parlance, water rates are levied on registered dry lands are called "Thiruva Jasthi" and the irrigation cess payable for irrigation of second crop in a single crop wet land is known as "Fasli Jasthi" (Government of Andhra Pradesh, 1973, p.171). A second wet crop raised on such lands is liable to an extra charge, which is ordinarily half the wet assessment. Broadly speaking, the two systems for levying separate water rate in the Andhra Area are the differential water rate system and the fixed water rate system.

Water Rates in Telangana Area

In Telangana, all cultivable lands were classified and registered in the land revenue accounts under three heads, viz., dry, wet and garden. Wet assessment was levied on registered wet lands. The rates were fixed as maximum wet rates for a group of villages having identical conditions for the first crop. Wet lands were classified as single first wet crop. (Ekfasala Abi), single second wet crop (Ekfasala Tabi), and double crop wet (Dufasala ae) (Government of Andhra Pradesh, 1973, p. 167). A differential water rate system more or less same as in the Andhra Area was in force in Telangana area also while charging a dry land for use of water from a source. Resettlements were abandoned in Andhra in the year 1939 and in Telangana in the year 1953 as a matter of state policy. Instead of revision of assessment on wet and dry lands periodically through resettlements, thereafter, the rates of assessment came to be revised from time to time. Thus, in Andhra Pradesh, prior to 1986 land revenue was being levied on both dry and wet lands at specific rates and the charge for use of water from a Government source of irrigation was levied in the form of consolidated wet assessment. The rates, which were standardized in 1956, were enhanced in the years, 1967, 1975, and 1986. Substantial modifications for the basis, rates and collection of water tax have been made with effect from 1st July 1996 through Andhra Pradesh Water Tax Act 1998. All lands were regarded as dry and separate land revenue was abolished. The recent Act 1996, contemplated recovery of water taxes in respect of lands irrigated by category I and category II sources only. For this purpose, all major and medium irrigation projects are to be regarded as category I while other government sources which supply water for not less than 5 months in a year belong to category II. Further, water tax is chargeable with reference to the nature of wet crop, irrigated dry and dufasal crop (see Table 5).

Table 5

Revised Water Rates for Different Crop Categories in Andhra Pradesh

Crop CategoryType of source wise water rates per acre in Rs.				
	Category I	*Category II*		
	Old	*New*	*Old*	*New*
1. First / Single wet crop	60	200	40	100
2. 2nd and 3rd wet crop	60	150	40	100
3. 1st crop irrigated dry	40	100	20	60
4. 2nd and 3rd irrigated dry crop	40	100	20	60
5. Two-season crops per year	120	350	80	350
6. Aquaculture per year	00	500	00	500

Note: old rates with effect from July 1986 and the new rates from July 1996.
Source: APERP Project Implémentation Plan, 1998.

Judging by the reported trend in revenues, the impact of these revisions on users of public irrigation system has not been significant nor have they led to any improvement in the financial performance of these systems. These changes are essential and important constituents of any effort to improve public finances generally and that of the State Government in particular. There were several systems of charging irrigation water in different parts of the country during the British administration. While the British tried to rationalise the high water rates by searching for principles of fair returns to investment by the government, the present Independent India does not seem to have any rational economic principles on which to determine low rates.

Table 6
Productivity of Paddy per–acre of some Irrigated Districts in the Madras Presidency and Independent India (in quintals)

Year	*Districts*					
	East Godavari	*West Godavari*	*Krishna*	*Guntur*	*Kurnool*	*Nizama-bad*
Original Settlements	2.00	2.00	1.00	4.00	3.50	–
Re-Settlements	4.60	4.60	4.60	4.70	4.40	–
1918-19	9.1	8.6	8.2	7.3	N	–
1946-47	8.7	7.8	7.6	7.0	6.3	–
1953-54	9.1	9.1	8.2	8.2	6.5	–
1955-56	6.6	6.0	6.7	6.0	5.6	4.5
1965-66	7.2	7.3	6.7	6.8	6.2	8.2
1975-76	7.3	8.3	6.0	6.2	6.3	7.7
1985-86	10.1	10.8	11.4	10.7	7.0	7.6
1996-97	11.9	12.0	11.2	12.0	10.3	12.1

Source: 1. Government of Andhra Pradesh, 1959, pp. 85 and 309.
2. Government of Madras, Season and Crop Reports, Office of the Economic advisor to Government of Madras, for different years.
3. Based on data given in the Season and Crop Reports of Andhra state and Andhra Pradesh and Statistical Abstracts of Hyderabad State, for different years, Directorate of Economics and Statistics (DES), Hyderabad.

Note: one lb=0.454g. 1 k.g=2.20 lb, 1 quintal=220 lbs, Collins: Gem English Dictionary (p.612), 1990.

PRODUCTIVITY, PRICES, COST OF CULTIVATION AND WATER RATES IN ANDHRA PRADESH

The supply of water to the farms is the chief service rendered by the government when it constructs an irrigation project. Where irrigation structure is profitable for water users, they have high stakes in the related activities and they would be willing to co-operate in the improvement of their management practices. Also, managers can make credible claim on government resources for systems operation and maintenance budgets, and regimes would be readily willing to invest in their operation and maintenance (Uphoff, N, et.al., 1991. p.90). There are two categories of direct farm benefits i.e., increased family incomes and increased cash allowances. Thus the direct economic returns to irrigation in

India can be indexed by analyzing the trends in the crop productivity, their prices and their water rates.

Because of lack of comparable data relating to the cost of cultivation of all crops and other details for the past years, it is felt that it would be convenient to compare the gross value of the yields of rice per acre measured by the annual average wholesale prices prevailing at the chief market centers. Thus, irrigated farming in the State is represented almost by a single crop, viz., rice. And finally, the four delta districts of East Godavari, West Godavari, Krishna and Guntur in Coastal Andhra, Kurnool in Rayalseema and Nizamabad in Telangana region have the largest proportion of their irrigated area under the canals.

Trends in Productivity

The yields per acre of rice generally showed an upward trend overtime in the sampled rice producing districts. The productivity of rice per acre (in quintals) in some selected districts of the Madras Presidency increased between the Years 1860- 1909 (see Table 6).

The yield per acre of rice increased steadily from quinquennium to quinquennium. Before the construction of the Nizamsagar Project there was no assured water supply and the yield of rice fluctuated year to year. After the Nizam Sagar canal came into operation the yield of rice increased constantly (see Table -7). Yields of almost all crops in Nizamabad district increased and got stabilized after Nizamsagar came into operation. Thus, the impact of irrigation on yields per acre particularly in case of rice was relevant.

Table 7

Yield per acre of rice in Nizamabad District During the Period 1925-26 to 1944-45

(In quintals)

Years	*District Nizamabad*
1925-26 to 1929-30	5.0
1930-31 to 1934-35	3.6
1935-36 to 39-40	3.7
1940-41 to 1944-45	4.3

Source: Agricultural Statistics, Estimates and Yield of Principal Crops 1925-26 to 1944-45 (3 issues), H.E.H., the Nizam's Dominions, Hyderabad.

Note: Figures calculated for five years average.

Trends in Prices

There had been also an enormous increase in the general price level in Andhra Pradesh as well as in the country as a whole. The old prices were, strictly speaking, not comparable with recent prices on account of the variety of the measures in terms of which the prices were quoted in the old days and the uncertainty as to their contents, but they nevertheless hives a fairly correct general idea of the direction of changes that occurred. Representing the average prices of the rice in the five years beginning with 1853 by 100, the average prices of paddy at the quinquennial periods had increased during the period of 1861-1865 164 points (see Table 8).

Prices rose rapidly after 1853 till they reached their culmination in the five years ending 1865, when they were two-and- a-half times what they were prior to 1853 and twice as high as in the early years of the century (Raghavaiyangar, S.S, 1893, pp. 58-59). The prices of Paddy in some irrigated districts of both in the Madras Presidency as well as in Independent India increased (see Table- 9). Thus, the prices and the yield per unit of irrigated area have increased substantially.The technological improvements in agriculture, and especially in the case of paddy grown under assured irrigation conditions were main reasons for increase of yield and prices.

Table 8
Average Price (indices) at the Quinquennial Periods-1853 to 1888 in Madras Presidency

Grain	*Average for five-years ending*			
Paddy	1853	1865	1874	1888
	100	264	216	234

Source: Based on Raghavaingar, S.S, 1893, P. 58.

Trends in Water Rates

There were several concepts involved in fixation of water rates (Raghavaingar, S.S, 1893, pp. 187-190). The first step was to divide the soils into certain main classes according to the mechanical composition and chemical properties of the land dealt with. The second step was to ascertain the grain out-turn of the lands irrigated and un-irrigated. The third step was to find out the money value of the grain out-turn. The fourth process was to ascertain, by enquiries, the expenses of cultivation for each kind of soil. Accordingly, water rates for Godavari delta was fixed at Rs. 3 an acre in 1861. It was raised to Rs, 4 in 1865 in G.O No. 1006 dt. 12.5.1865. This was again raised to Rs. 5 per acre in G.O. No. 562, Revenue in 1894. Since 1931, with the resettlement the rate has been increased to Rs. 6.25 per acre. Subsequently, there was no revision in the water rates till 1956-57. The water rate was increased in stages in later years and in 1986-87 a uniform water rate was fixed which was enhanced to Rs.60. The Government of Andhra Pradesh had further enhanced the water rate through ordinance in 1996 (Government Ordinance, dated 01.07.1996) (Table 10).

Table 9
Farm Gate Prices of Paddy and Average Wholesale Prices of Rice in some Districts of Andhra Pradesh (1914-15 to 1996-97)

(Rs. Per quintal)

Year	*Districts*					
	East Godavari	*West Godavari*	*Krishna*	*Guntur*	*Kurnool*	*Nizamabad*
1914-15	4.17	–	4.54	4.53	6.05	–
1920-21	7.52	–	7.73	7.94	8.48	–
1931-32	3.63	3.59	3.84	3.71	4.32	6.70
1944-45	8.74	8.48	8.59	9.20	9.78	4.25
1947-48	10.58	10.72	10.53	12.27	15.53	3.50
1969-70	72.00	76.90	68.08	74.52	61.00	59.08
1970-71	93.71	97.86	103.79	110.35	101.25	98.15
1979-80	190.23	185.50	184.60	186.80	190.60	190.30
1980-81	202.20	210.10	205.00	203.20	218.60	209.30
1996-97	461.80	466.60	410.00	552.80	520.10	472.50

Source: Based on Statistical Atlas of the Andhra State, Season and Crop Reports, DES, Hyderabad for different years.

There was differential water rate system operated in Kurnool district. The rate was enhanced by 18 ¾ percent in 1956-57 and again by 18 3/3 percent (G.O. No. 982. dt.27.05.1957), in 19657-58 and by 1962-63, the rate was doubled to Rs. 14-20, which is also the current average wet rate of assessment. (G.O. No. 168, dt. 31.01.1963). Further, it was enhanced to 10 per cent (i.e., Rs. 15), in the year 1965-66 and it was Rs. 30 in 1975-76 (G.O. No. 412, dt. 22.03.1975). During the year, 1986-87, a uniform water rate was fixed throughout the state (G.O. Act -11 of 1989, dt. 24. 08. 1989). Wet assessment was abolished and water rate was Rs. 60. It was further enhanced to Rs. 200 per acre in 1996-97. At the time of the formation of Andhra Pradesh the rates of land revenue were already considerably higher in the Telangana districts as compared to the remaining districts of the state and therefore, as part of the programme of rationalization of land revenue assessment in the state, only a small enhancement of about 30 per cent was effected in Nizamabad and other Telangana districts as against 100 per cent in wet assessment in the Andhra region (Government of Madras, 1931, p. 277). The average wet rate of assessment Nizamabad district was Rs. 10.60 till 1952-53 when it was enhanced by one anna in a rupee, to Rs. 11.

Table 10
Water Rates per acre in the Delta areas 1931-32 to 1996-97 (category-I)

Year	*Rs.*
Till 1931-32	5.00
1931-32	6.25
1956-57	7.50
1957-58	9.40
1962-63	12.50
1965-66	15.00
1975-76	30.00
1986-87	60.00
1996-97	200.00

Source: 1. Based on data in the Government Orders, Revenue Department, Government of Andhra Pradesh, in different years and Andhra Pradesh Irrigation commission Report, 1973.

Note: Old denomination of prices-192pies=Re.=1, New denomination of prices-12 paise=1 anna, 16 annas-Re.=1, 100 paise=Rs.1.

Thereafter the average rate was enhanced every year. Further in the year 1986-87 a uniform rate was fixed instead of consolidated assessment, and it was enhanced to Rs. 60. Further it was enhanced to Rs. 200/- per acre (see Table 11). Thus, there was a variation in water rates from project to project in different districts. In delta districts consolidated wet assessment was implemented. In Kurnool district, differential water rate system was implemented. Nizamabad, was under the domain of Nizam's till 1948. It was during the period 1986-87, onwards uniform rate implemented throughout the state of Andhra Pradesh.

Table 11
Average Wet Rate levied on the lands in Nizamabad District (per acre) 1952-53 to 1996-97 (category-I)

Year	*Rs. Ps.*
Till 1952-53	10.40
1952-53	11.00
1962-63	14.30
1966-67	15.00
1975-76	30.00
1986-87	60.00
1996-97	200.00

Source: 1. Based on data in the Government Orders, Revenue Department, Government of Andhra Pradesh, in different years and Andhra Pradesh Irrigation commission Report, 1973.

Note: Old denomination of prices-192pies=Re.=1, New denomination of prices-12 paise=1 anna, 16 annas-Re.=1, 100 paise=Rs.1.

Share of Water Rates in Gross value of Output

The proportion of Government assessment to the gross produce was estimated by the settlement calculations at 6.3 per cent, taking the value of the gross outturn at 50 crores of rupees, and land revenue at 3.16 crores in Madras Presidency. They have excluded from land revenue, 1.37 (2.9 per cent) crores as water charges and not forming part of land tax proper (Raghavaingar, S.S, 1893, pp. 112-113). Taking the ryotwari lands alone, the average rate of assessment for wet lands was Rs. 5/- per acre and for dry lands Rs. 1/- per acre and these rates are between one-fourth and one-fifth and one-fourth and sixth respectively of the gross outturn according to settlement calculations after deducting from the average outturn 16.66 per cent in the case of wet and 25 per cent in the case of dry lands as allowance for vicissitudes of season. The share of water rate to the gross value of output was around one-sixth in dry areas, while in wet areas it was one-fifth (see Table 12).

The gross value of output per acre of rice increased through out the period (see Table 13). In Nizamabad district, during the period, 1925-26 to 1944-45, it was decreased by 1 per cent due to severely effected successive famines, but later years, a significant increase in the productivity was observed in the Nizamsagar area (see Table 14). The low water rates in relation to gross value of output, therefore indicate the decreasing trend of water rates. Gross value of production and productivity increased, water rates too increased but their proportion in the gross output had declined over the time period of 1854 to 1999 (see Table 15).

Table 12
Average Yield, the cultivation expenses and the rent per acre in Madras Presidency-1861 to 1880

Item	*Dry*		*Wet*	
	RS.	*PS.*	*RS.*	*PS.*
1. Value of Produce	8	11	25	43
2. Cultivation expenses	3	28	9	57
3. Water Rates	1	20	5	15
4. Farmer's Profit	3	65	10	79
5. Ratio of water rate to the produce	14.87		20.08	

Source: Based on S.S. Raghavaingar, 1893, Memorandum on the Progress of the Madras Presidency: during the last fifty years, Appendix-Tables 3, p. cxlvii.

Table 13
Gross Value of Rice per acre (in Rs.) in some Districts of Madras Presidency and Independent India

Year	*Districts*					
	East Godavari	*West Godavari*	*Krishna*	*Guntur*	*Kurnool*	*Nizama-bad*
1918-19	53.33	–	46.25	47.30	53.73	–
1939-40	28.72	28.86	26.42	33.78	33.30	–
1947-48	90.98	90.10	77.92	90.80	102.71	–
1955-56	123.75	105.78	99.23	110.64	92.4	36.00
1965-66	288.00	313.00	268.00	277.00	331.00	331.0
1975-76	1329.3	1189.8	977.4	957.5	928.2	1213.10
1985-86	3154.2	3114.7	3381.2	3359.8	2412.9	2301.30
1996-97	5495.4	5599.2	4592.00	6633.60	5356.00	5717.30

Source: Computed.

Table 14
Value of Gross output of Rice Per acre and Water Rates as Percentage of Gross value of output of rice in Nizamabad District

Years	*Gross Value of Output (in Rs. Per acre)*	*Share of Water Rates in Gross value of output*
1925-26 to 1929-30	26.47	39.3
1930-31 to 1934-35	32.00	32.5
1935-36 to 39-40	30.40	34.2
1940-41 to 1944-45	25.37	41.0

Source: Computed.

Table 15
Share of Water Rates (%) to Gross Value of Output Rice per acre (in Rs.) in some Irrigated Districts of Madras Presidency and Independent India

Year	*Districts*					
	East Godavari	*West Godavari*	*Krishna*	*Guntur*	*Kurnool*	*Nizamabad*
1918-19	9.4	–	10.8	10.6	N	–
1939-40	21.8	21.7	23.7	18.5	20.7	–
1947-48	6.7	6.9	8.0	6.9	6.7	–
1955-56	5.0	5.9	6.3	5.6	8.5	30.6
1965-66	5.2	4.8	5.6	5.4	4.5	4.3
1975-76	2.3	2.5	3.1	3.1	3.2	2.5
1985-86	1.0	1.0	1.0	1.0	1.2	1.3
1996-97	3.6	3.6	4.4	3.0	3.7	3.5

Source: Computed.

Share of Water rates in Cost of Cultivation

The cost of cultivation of the crops varied from district to district depending upon local conditions prevailing and availability of irrigation facilities. In early times, the cost of cultivation consist of initial preparatory for cultivation, manures, seeds and sowing, costs of after cultivation, irrigation and harvesting expenses. (Government of Andhra Pradesh, 1959, pp.130-55). During the settlement scheme reports of original and resettlements i.e. 1860 and 1931, the cost of paddy was very low. The cost of paddy per acre in the year 1958 had increased tremendously (see Table 16).

A number of cost concepts such as cost A1, Cost A2, Cost B and Cost C have been followed in the analysis. The concept of total cost is used as synonymous with cost C and this includes the expenditure incurred on human labour (owned, exchanged and levied), bullock labour (owned, exchanged and levied), seed (owned, exchanged and levied), organic manure (owned and purchased), fertilizers, pesticides, land revenue cesses and taxes, irrigation charges, depreciation, interest on working capital, interest on fixed capital, rent paid for leased-in land, imputed rental value of owned land, imputed value of family labour and others miscellaneous expenditure. The cost of cultivation of paddy per acre has gone up and the share of water rates was very low (see Table 17).

Table 16
Cost of Cultivation of Paddy per acre (in Rs.) in some Irrigated Districts of Madras Presidency and Independent India

Year	*Districts*					
	East Godavari	*West Godavari*	*Krishna*	*Guntur*	*Kurnool*	*Nizama-bad*
1. Original Settlements	3.50	3.50	6.50	5.00	3.60	–
2. Resettlements	14.00	14.00	20.00	14.00	12.25	14.60
3. 1958 (DES)	175.00	151.00	120.00	150.00	260.00	210.00

Source:
1. Raghavaindhar, S.S, 1893, Appendixes, Table 3
2. Government of Andhra Pradesh, 1959,

Table 17
Water Rates as Percentage of Cost of Cultivation of Rice (per acre) in Andhra Pradesh

Year	*Total Cost of Cultivation (in Rs.)*	*Water rates (in Rs.)*	*Water rates as percentage of Cost of Cultivation*
1971-72	597.00	15.00	2.5
1981-82	1556.74	30.00	2.0
1989-90	3210.89	60.00	1.9

Source: Government of India, "Cost of Cultivation in principle crops", 1990, DES, New Delhi.
APAU, "Cost of Cultivation Cell, 1989-90,Hyderabad.

The present study has revealed that the gross receipts in general had fallen short of the working expenses and this wide gap resulted in the paucity of funds for operation and maintenance, and not to speak of any quantum of new investments in irrigation sector. The average cost on the cumulative irrigation potential created through major and medium schemes during this period and increased the per hectare/acre development cost. The study also found that the gross value of output per acre of rice increased throughout the period (since 1865), where as water rates increased marginally, leading to a declining proportion of the share of the water rates in the total output of the crops irrigated.

REFORMS IN ANDHRA PRADESH: FORMALIZATION OF WATER USER ASSOCIATIONS

It is often argued that the reason for the ills of irrigation management is the alienation of farmers/beneficiaries from the process of planning and implementation. This was as such during the British irrigation policy. Maintenance and management of irrigation systems through user societies and participatory process is expected to bring in efficient equal distribution of water resources. But such processes often remained at a micro level as good examples are often found to be difficult to replicate (Age old Practices of Kudimaramath and Cheruvumaramath-Management of Tank irrigation in South India). The state of Andhra Pradesh has, however, shown the way to scaling up of participatory irrigation management through formalising the irrigation institutions. The most interesting feature of these reforms is that they are 'top down' with a 'bottom up' approach. It has the advantage of greater reach (possible under 'top down') and intensity through involvement of the community (possible under 'bottom up'). The irrigation reforms in AP, however has problems. These problems can be identified at two levels: (a) conceptual or formulation level, and (b) implementation level.

At the formulation level

- Tank and canal systems are treated as similar in fund allocations. While the degenerated tanks needed more allocation for rehabilitation, the allocations under WUA were based on the command area. This has led to under funding and resulted in little improvement in the systems per se and availability of water. The study made it crystal clear that there is hardly any improvement in the tank systems. Official estimates show about 69 per cent of the tanks in the state are in disuse (Reddy, 2003), and farmers are asking for more funds/works pertaining to tank deepening and strengthening the bunds.
- Second problem in this regard is that groundwater resources, the single largest source of irrigation, are completely left out of the purview of the WUAs and irrigation reforms are neither comprehensive nor followed an integrated water resource management approach.

At the implementation level,

- An important eye opener fact of the study is the 'elite capture' of the WUAs. Most of the cases, forward castes and large farmers are occupying the position of presidents disproportionately to their actual populations. This is mainly due to the reason that they are loosing their positions in the constitutional bodies of PRIs due reservations. Equal distribution of water rights is seen as vital for sustaining water institutions (Deshpande and Reddy, 1990). One of the recent demands is to make women participation mandatory.

- One of the reason of initiating parallel institutions in AP, by passing PRIs is that these institutions would be apolitical and focus on delivery. But, the emphasis on elections has resulted in the fight between parties to capture the power, which makes people insensitive to common good. These institutions in their present form tend to dilute the social capital rather than strengthening it.

Impact of Formation of Water User Associations

Impact of WUAs is seen as vital source of funding and sharing pattern, functional aspects of WUAs and share of water rates in total cost of cultivation. Our field results show that the major sources of funding are from the government followed by water cess. The share of water cess is about 50 per cent in the case of canal head reaches while it is 30 percent in tail ends and only 19 percent in the middle reaches. In the case of tank WUAs government contributes about 90 percent and the share of water cess is less than 2 per cent (Table 18). However, the difference between government contribution and water cess is not clear as the sharing of water cess between the WUA and other institutions was initiated only recently. Hence, these two categories could be treated together as external funding. On the other hand, people's contribution is less than 4 per cent in the case of canal WUAs and less than 6 percent in the case of tank WUAs (Table 19). This indicates clearly that the mandatory 10 per cent contribution in the works is not adhered to in the entire sample of WUAs.

Table 18

Proposed Revenue Sharing Pattern among the PIM Structures (%)

Level	*Major*	*Medium*	*Minor*
1. Water User Association (WUA)	50	60	90
2. Distributary Committee (DC)	20	30	–
3. Project Committee (PC)	20	–	–
4. Local Government	10	10	10

Source: PIM, 1997 Act.

The only important activity in which the WUA has say in majority of the cases is regarding the works. The dominance of the department is clearly seen in the most important aspects like assessment, fund allocation, cess collection and water management, though the involvement of WUAs is better in the case of tank WUAs. Despite the promises the devolution of powers has not taken place even in the case of land assessment, let alone fund allocation and collection (see Table - 20). This is one of the main bottlenecks for the sustenance of the WUAs. On the other hand, the role of contractors appears to be marginal even in the case of works.

In line with the policy of transfer of management responsibilities to WUAs and making them financially sustainable, it was recognised that the water users have to pay the real o & m costs for the irrigation service. The results indicate that since 1997, farmers were paying only 3 percent of the actual cost of cultivation (2 to 1.5 percent in total gross value of output) in different locations viz., head, middle and tail end reaches of the canal system; while 1 to 1.5 percent of the actual cost of cultivation (1.5 percent in total gross value of output) in different locations of the tank areas (see table 21). It is also observed that there is a shift from cereal crops to cotton, castor, and other cash crops like papaya, banana and mango especially in different locations of the tank areas. However, the shift to higher value crops is driven by prices, and influence of other factors and not entirely due to the irrigation reforms. In general there is an increase in income from all sources and in all locations.

The irrigation reforms in ap, through external funding, have provided an opportunity to revive and strengthen the ailing irrigation systems. The improved systems are expected to enhance the credibility of the department in providing assured water supplies, a precondition for increased water tariff. The funding is expected to help in building the necessary social capital for sustaining the reform process. Strengthening the social capital is a prerequisite for the reforms that aim at transferring irrigation management to farmers. Reforms also focus on financial sustainability through price reforms.

Though water rates were increased initially by 3 times, they are still short of O&M expenditure. Moreover, increased prices are not translated into revenues due to weak institutional structures

Table 19

Funds (in Rs.) from Various Sources to WUA under Canal and Tank area by location in Andhra Pradesh During the period 1997 to 2002

Sources of Revenue	*Canal*						*Tank*			
	Head		*Middle*		*Tail-end*		*Irrigation Tank*		*Percolation Tank*	
	Per WUA	*Per acre*	*Per WUA*	*Per acre*	*Per WUA*	*Per acre*	*Per WUA*	*Per acre*	*Per WUA*	*Per acre*
1. Government	97592	199	16696	485	91650	272	38527	1437	28820	1140
	1	(48)	20	(76)	0	(46)	6	(90)	1	(90)
2. Water-cess	28306	208	42543	124	60187	179	9851	37	3999	16
	5	(50)	5	(19)	9	(30)		(02)		(01)
3. Contribution	23441	5	33904	10	69598	21	22051	82	20553	81
		(02)		(02)		(04)		(05)		(06)
4. Others	2804	0.60	61203	18	39752	118	12749	48	6432	25
				(03)	1	(20)		(03)		(03)
5. Total	12852	412.6	21901	637	19854	590	42992	1604	31918	1262
	31	(100)	62	(100)	98	(100)	7	(100)	5	(100)
Avg/year	21420	69	36502	106	33091	98	71655	267	53197	210
	5		7		6					

Note: Figures in brackets are respective percentages to total.

Source: Field Survey.

Table 20
Functional Aspects and Decision Makers of WUAs under Canal and Tank areas in Andhra Pradesh

Functional aspects	*WUA*	*Irri.Dept /MRO*	*Canal Con-tractor*	*Total*	*WUA*	*Irri.Dept / MRO*	*Tank Con-tractor*	*Total*
1. Organisation of meetings	92	22	0	114	152	31	1	184
	(80.7)	(19.3)		(100)	(82.6)	(16.8)	(0.6)	(100)
2. Assessment of Command Area	19	83	0	102	83	101	0	184
	(18.6)	(81.4)		(100)	(45.1)	(54.9)		(100)
3. Work Related	269	228	8	505	325	352	12	689
	(53.3)	(45.1)	(1.6)	(100)	(47.2)	(51.1)	(1.7)	(100)
4. Funds Allocation	18	75	0	93	12	129	2	143
	(19.4)	(80.6)		(100)	(8.4)	(90.2)	(1.4)	(100)
5. Water Management	73	155	1	229	256	102	1	359
	(31.9)	(67.7)	(0.4)	(100)	(71.3)	(28.4)	(0.3)	(100)
6. Collection of Water Charges	0	81	0	81	4	107	0	111
		(100)		(100)	(3.6)	(96.4)		(100)
7. Others	381	103	1	485	287	204	7	498
	(78.6)	(21.2)	(0.2)	(100)	(57.6)	(41.0)	(1.4)	(100)

Source: Field Survey.

Table 21
Cost of Cultivation and Returns per acre of Paddy of the Sample WUA House Holds by Size Class

Farm Size/Cost/Returns	Canal			Tank		
	Head	Middle	Tail	Head	Middle	Tail
Marginal Farmers						
1. Electric Charges (%) to total cost	0.2	0.4	0.1	6.1	3.6	4.6
2. Water Cess (%) to total cost	3.1	2.9	2.7	2.7	0.5	1.3
3. Total Cost (Rs.)	604	585	672	623	616	578
	(100)	(100)	(100)	(100)	(100)	(100)
4. Gross Returns (Rs.)	1122	1012	942	954	885	921
5. Net Returns (Rs.)	518	428	270	331	270	343
6. BC ratio of Gr.	1.9	1.7	1.4	1.5	1.4	1.6
7. BC ratio of Nr.	0.9	0.7	0.4	0.5	0.4	0.6
Small Farmers						
1. Electric Charges (%) to total cost	0	0	0	9.0	13.0	5.4
2. Water Cess (%) to total cost	3.1	2.9	2.7	2.7	0.5	1.3
3. Total Cost (Rs.)	742.2	724.6	585.2	659.1	523.0	654.7
	(100)	(100)	(100)	(100)	(100)	(100)
4. Gross Returns (Rs.)	830.2	1043.1	955	969.0	1161.3	849
5. Net Returns (Rs.)	87.9	318.5	370	310.3	638.3	194.2
6. BC ratio of Gr.	1.1	1.4	1.6	1.5	2.2	1.3
7. BC ratio of Nr.	0.1	0.4	0.6	0.5	1.2	0.3

Contd....

Farm Size/Cost/Returns	Canal			Tank		
	Head	Middle	Tail	Head	Middle	Tail
Medium Farmers						
1. Electric Charges (%) to total cost	0.2	0	0	5.1	10.3	6.8
2. Water Cess (%) to total cost	1.9	3.1	2.4	0.9	0.7	0.8
3. Total Cost (Rs.)	821	569	793	765	629.4	648
	(100)	(100)	(100)	(100)	(100)	(100)
4. Gross Returns (Rs.)	907	1027	1088	687	940	958
5. Net Returns (Rs.)	86.2	458	296	-78.0	311	310
6. BC ratio of Gr.	1.1	1.8	1.4	09	1.5	1.5
7. BC ratio of Nr.	0.1	0.8	0.4	-0.1	0.5	0.5
Large Farmers						
1. Electric Charges (%) to total cost	0	0.1	0	6.6	3.7	0
2. Water Cess (%) to total cost	2.8	2.9	2.5	0.2	1.0	0
3. Total Cost (Rs.)	645	635	754	709	818	421
	(100)	(100)	(100)	(100)	(100)	(100)
4. Gross Returns (Rs.)	1109	1128	1188	747.5	767	1378
5. Net Returns (Rs.)	464	493	435	39	-51	956
6. BC ratio of Gr.	1.7	1.8	1.6	1.1	0.9	3.3
7. BC ratio of Nr.	0.7	0.8	0.6	0.1	-0.1	2.3

Source: Field Survey.

(Reddy, 2003). Though user share of contribution of 15 per cent is imbibed in the PIM Act there is no evidence of any contribution from farmers, as revealed in the study. In fact, there are no efforts to collect this contribution.

CONCLUSIONS

The study is major finding is the context of rising pattern of per hectare/acre capital costs of major and medium irrigation schemes over the period, the present paper reveals that there were several systems of charging irrigation water in different parts of the country during the British Administration. The British tried to rationalize the high water rates by searching for principles of fair returns to investment by the government. At present in Andhra Pradesh there seem to be no rational economic principles to determine the water rates. The productivity and prices of rice had increased substantially in the selected districts of East Godavari, West Godavari, Krishna, Guntur, Kurnool and Nizamabad. The water rates ratio to gross value of output has been decreasing continuously since the inception of water rates in1865. The cost of recovery should be the main consideration in determining water rates. It also pointed out that the phase-wise rationalization of water rates to achieve full cost recovery is needed. In the context of recent reforms the main conclusions are;

1. The major benefit from the formal WUAs is improved irrigation facility, quantitatively as well as qualitatively. Though some benefits in terms of increased area under irrigation in canal systems and improved quality of irrigation is evident, the sustainability of these benefits is rather uncertain in the absence of efficient institutional structures.
2. Informal or socially embedded institutions are more effective than the formal WUAs. The reason is being that informal institutions reflect commitment and cohesion, as it is evolved from within the system. More importantly, it is flexible in adapting to the changing situations.
3. Self-sufficiency and resource strength is central to the sustainability of the institutions. The formal institutions are yet to generate their own funds, while informal ones have succeeded in generating surplus funds. Lack of proper devolution of powers to local bodies is hindering

the progress in this direction. Transfer of powers and responsibilities to the WUAs at the minor level should be done effectively though in a phased manner. Only under such circumstances innovations at the community level would become possible. Hitherto, the WUAs are entrusted with responsibilities without any rights. The main reason for the ineffectiveness of the formal WUAs is that lack of co-operation and interest among the farmers. This is mainly attributed to, apart from the above reasons, low economic incentive. It appears that institutions would be effective if the economic gains are substantial. Though this may be a typical case, economic gains from canal irrigation are low due to the reason that the prices of alternative sources such as groundwater or complementary inputs like electricity do not reflect their scarcity values. Reforms should initiate the process to convert all forms of water into an economic good through introduction of cost-based pricing as per volume and use (priority). This would help in realising the relative importance of each source of water. This dresses that need for an integrated approach towards water resources planning and management. Our study indicates that the farmers are not against the concept of WUA. On the other hand, their awareness about the WUA is rather not suficient. Therefore, efforts should be made to increase the awareness of the farmers about the benefits of WUA. Devolution of powers to the local level and making the WUA structure more flexible would go a long way in addressing the awareness problem.

4. The present arrangement of sharing 50 per cent of water cess to WUA is enough and hence they demand enhancement of this share further by 10 percent. It is opined that WUAs can collect the water charges with the assistance of revenue official. They also propose to enhance the financial viability of the WUAs through plantation and horticultural crops on common river beds, canal banks, etc. Further, social forestry can be taken up on the encroached lands. They also demand that the WUA should conduct the auction of fisheries.
5. Regarding water management, it is proposed that irrigation department should be decentralized in terms

of zone, district and project wise for efficient management of water resources.

6. The co-operation and co-ordination between president of WUA, EC and irrigation engineers should be strengthened. Water estimation and distribution shall be carried out by PCs, DCs and WUAs through general body according to the availability of water.
7. Laskars should be placed under the direct control of the WUAs. GB should have the authority to appoint the lascars in places where they are not there.
8. All works should be carried out by the WUA strictly adhering to the tender system. Quality control will be with a sub-committee of the WUA. Interestingly, WUA presidents put forth two peculiar demands. One is delinking of WUAs from the village PRI and two is WUAs should be represented in the legislative council (proposed to be revived soon in AP).

REFERENCES

ADB and IIMI (1986), Irrigation Service Fees, Asian Development Bank and International Irrigation Management Institute, Manila.

Ansari, N (1968), Economics of Irrigation Rates: A study in Punjab and Uttar Pradesh, Asia Publishing House, New Delhi.

Central Water Commission (1992), Water and Related Statistics, Statistical Division, New Delhi.

Deshpande, R.S. and V. Ratna Reddy (1990), "Social Dynamics and Farmers' Society: A case Study of Pani Panchayat," Indian Journal of Agricultural Economics", Vol. 45, No. 3, July-September.

Dhawan, B.D (1983), "Impact of Irrigation in Instability in Farm Output: A Case study of Tamilnadu State", EPW, Vol. 18. No. 39, September, 28.

Gadgil, D.R (1948), "Economic Effects of Irrigation", Report of a Survey of the Direct and Indirect benefits of the Godavari and Pravara Canals, Gokhale Institute of Politics and Economics, Publication No. 17, Pune.

Government of Andhra Pradesh (1959), "Report of the Land Revenue Reforms Committee, 1958-59", Part-I Vol. I and ii Land Revenue Department, Hyderabad.

————(1973), "Report of the Andhra Pradesh Irrigation Commission", Chapter Title- Water Rates, Hyderabad.

Government of India (1903), "Report of the First Indian Irrigation Commission (1901-03), Part-I and III, Provincial and General Preface, Calcutta.

————(1952), "First Five-Year Plan (1951-56)", Planning Commission, New Delhi.

————(1953-54), "Report of the Taxation Enquiry Committee", Planning Commission, New Delhi.

————(1957), "Report of Foodgrains Enquiry Committee", Planning Commission, New Delhi.

————(1963 & 1964), "Report of Committee to Suggest Ways and Means of Improving Financial Returns from Irrigation Projects, Ministry of Irrigation and Power, Part I & II, New Delhi, December.

————(1972), "Report of the Second Indian Irrigation Commission, Ministry of Irrigation and Power, Vol.1, New Delhi.

————(1993), "Draft of the Eight Five-Year Plan (1992-97)", Planning Commission, New Delhi.

Government of Madras (1858), "Report on the Direct and Indirect effects of the Godavari and Krishna Anicuts, published by H. Smith, at Fort Saint George Gazette Press, Madras.

————(1865), Order No. 986, Revenue, dated 11.05.1865, "A Collection of Papers Relating to Madras Irrigation Cess Act VII".

————(1869), Proceedings of Board of Revenue, No 7554 (Miss.) dated 11.10.1869, Paragraph 7.

————(1890), Order No. 272, Revenue, dated 9th September.

————(1894), G.O No. 562, Revenue dated 2nd August.

————(1923), Order No. 1089, Revenue, dated 19.07.1923 (Leach's Report).

————(1926), "Report of the Taxation Enquiry Committee (1924-25)", Vol. I, Chapter 1, Madras.

————(1931), Standing Orders of the Board of Revenue (Land Revenue Settlement and Miss.) Revised of several Years, Vol. 1, Chapter 1, Madras.

Government of Maharashtra (1962), "Report of the Maharashtra State Irrigation Commission,, Bombay.

Hemantha Kumar, U. (1992), "Irrigation Policy of the Andhra Pradesh State-1951-91: A case study of the Nagarjunasagar Command Area", Proceedings of Andhra Pradesh Economic Association, Tenth Annual Conference, Andhra University, Visakhapatnam.

Janakarajan, S. (1993), "Economics of Social Implications of Groundwater Irrigation: some Evidence from South India", IJAE, Vol. XLVIII, No. 1, January-March.

Khan, H.R. (1981), "Irrigation Water Pricing in Bangladesh", Proceedings of the Expert Group Meeting on Water Pricing, held at Bangkok, 13-19, Water Resources Series, 55, Washington,, D.C. USA, May.

Kimura, T. (1977), "Farm Water Management for Rice Cultivation", Part-IV, Country Reports, Chapter-5, Asian Productivity Organisation, Tokyo-Japan.

Maclean, C.D., (1987) & (1883), "Manual of the Administration of Madras Presidency", Vol. 1, Asia Education Service, New Delhi.

Nadkarni, M.V. et.al. (1979), Impact of Irrigation, Canal, Well and Tank Irrigation in Karnataka, Himalaya Publishing House, Bombay.

National Council for Applied Economic Research (NCAER) (1959), "Criteria for Fixation of Water Rates and Selection of Irrigation Projects", Asia Publishing House, New Delhi.

Nickum, J.E (1979), "The organization of Water Resources Development in the People's Republic of China", Agricultural Administration, 6th July paper to the Development Studies Association conference, University of Bath, Claluerton, Bath, U.K.

Prasad, Kamala and Rao, P.K. (1985), On Irrigation Water Pricing in India, (Mimeo).

Public Works Department (PWD), (1879), "Report of the Selection Committee on East India", London.

Raghavaiayangar, S.S. (1893), "Memorandum on the Progress of the Madras Presidency During the last Forty Years of British Administration", Madras.

Raju, V.T. and Rao, DVS (1991), "Study on the Structure and Performance of Water Markets in North Coastal Andhra Pradesh", IJAE, Vol. 42, No. 3, July-September.

Reddy, V. Ratna and Prudhvikar Reddy, P. (2002), "Water Institutions: Is Formalisation the Answer, A study of Water User Associations in Andhra Pradesh", Indian Journal of Agricultural Economics, Vol.57, No. 3.

Reddy, V. Ratna (2003), "Irrigation Development and Reforms", Economic and Political Weekly, Vol. 38, No. 12, 22-29, March.

Reddy, V. Ratna, Prudhvikar Reddy, P and Hemantha Kumar, U. (2004), Draft Report, on, "Formalising Irrigation Institutions: A Study of Water User Associations in Andhra Pradesh", Centre for Economics and Social Studies (CESS), Hyderabad.

Shah, T. and Raju, V.K. (1988), "Working of Groundwater markets in Andhra Pradesh and Gujarath: result of two village studies", EPW, Vol. 22, No. 13, March.

Small, L.E. et.al. (1989), Financing Irrigation Services: A literature Review on Selected Case Studies for Asia, IIMI, Colombo, Sri Lanka.

Uphoff, Norman, et al. (1991), Managing Irrigation: Analysing and Improving the Performance of Bureaucracies, Sage Publications, New Delhi.

Vaidhyanathan, A. (1987), "Irrigation and Agricultural Growth", IJAE, Vol. 42, No. 4, October-December.

19

PRICING OF CANAL IRRIGATION AND SERVICE DELIVERY AN EMPIRICAL STUDY IN FATEWADI CANAL COMMAND AREA IN GUJARAT

*Sonal Bhatt**

INTRODUCTION

Financing of the irrigation sector is becoming increasingly difficult in recent years due to resource crunch faced by the government at all levels and declining development assistance from international donor agencies. It is widely felt that the huge investment made in state run irrigation schemes is not generating sufficient returns. In the early 1970s, the National Irrigation Commission, appointed by the GoI argued that accepting social cost-benefit evaluation as the basis for investment decisions does not minimise the "importance of securing an adequate return from investment to irrigation projects and recommended that the financial return of a project should also be examined at the time of approval, and if

* Post Graduate Department of Economics, Sardar Patel University, Vallabh Vidyanagar, Gujarat

found inadequate to cover working expenses and indirect charges, water rates should be raised"(Vaidyanathan, A., 2001). According to a study by NIPFP, the recovery rate on irrigation in India was just 4.2% (GoI, 1997).Gulati (1989) found that the cost recovery rate for irrigation in Gujarat was 2.57%, which was much below the national average. Dholakia (2003) found that the cost recovery rate on irrigation water in Gujarat was as low as 6.3% between 1989 and1999. The returns on irrigation projects are so low that individually and collectively, they are not even able to meet the annual O&M costs. The Vaidyanathan Committee (GoI, 1992) observed that attempts to reduce the magnitude of subsidies in agriculture and augmenting investment must focus both on improving the efficiency of planning and management (thereby reducing costs) and raising water rates as well as more effective enforcement of the same. The price of canal water in most states is totally unrelated to the productivity or scarcity value of water, or the cost of delivering it. It accounts for just 8% of cropping expenses and is equal to barely 5% of the average incremental production of irrigated areas over rain fed areas (Raju and Gulati, 2004).

Water rates have not been revised since long in almost all the states, , and are falling in real terms (Gulati et al, 2003 and GoI, 1988). Bhatia (1989) found that very often, the cost of collection of water charges in Indian irrigation was higher than the total cess collected. The question of cost recovery in canal irrigation sector is also a politically sensitive issue. The stakeholders in this sector consist of states and central government, local self-governments, various groups of farmers like small and big, rich and poor, tail enders and up enders, NGOs, irrigation staff, bureaucracy, politicians and the people at large. Thus, the issue of financial viability is not an isolated one, but has many dimensions, which have to be looked at from the financial, operational, managerial, and socio-political points of view.

Many researchers have advocated augmenting investment resources in the water sector through private participation. These ventures are expected not only to reduce the immediate financial burden on the state exchequer, but also to achieve sustainable cost recovery and financial accountability in the long run (Raju and Gulati, 2002). This would necessitate pricing reforms in canal irrigation. However, in order to succeed, pricing reforms must be accompanied by institutional reforms because prices cannot be raised unless the performance of irrigation schemes in terms of user satisfaction is improved. New paradigms like Participatory Irrigation Management (PIM) have been suggested to improve the financial performance and service delivery to the users of canal irrigation (Saleth, R. Maria, 1999).

Benefit of canal irrigation accrues directly to the farmers and indirectly to the whole economy. The relationship between the consumer i.e. the user-farmer, and the state irrigation department i.e. the seller could be defined as purely commercial, i.e. where the entire cost of providing irrigation has to be borne by the farmers. Here, the concept of "consumers' surplus" could also be used whereby; the maximum price that the farmer is willing to pay for irrigation could be recovered from him. Conversely, it could also be defined as a purely non-commercial relationship, whereby, the cost is inconsequential, and irrigation cess is a sort of a tax, which ought to be charged according to the individual's 'ability to pay', taking into account, the welfare aspect. Ideally, a combination of these approaches could be used for pricing of canal water.

Earlier studies on canal irrigation have focussed mainly on the supply side and suggested that upward revision in canal water charges is essential in order to meet the higher cost of providing irrigation at the farm level (Sangal, 1991). Very few studies (Singh and Ghatak (1995) and Jairath (1998)) have discussed the demand side of the problem and the factors responsible for the farmers' unwillingness and lethargy to pay canal water cess. They advocate rationalisation of prices, and at the same time improvement in irrigation management to provide satisfactory service to the users. Jairath (1998) feels that high or low price has a reference to the quality of the irrigation service. Price of canal water does not derive its justification from cost considerations alone but also from the utility of irrigation to the farmer. The rational farmer may simply refuse to pay for the service that does not satisfy him. The farmer's capacity and inclination to pay for water follows from the provision of adequate and timely irrigation in terms of the credibility and controllability of water supplied. If this is assured, he might be willing to pay higher charges.

There is wide-ranging evidence on poor conditions of canals in India discussed extensively in the recent literature. Some questions that if water rates are considered low compared to production costs, what about production inefficiencies? It is considerce that subsidised water rates could be aptly characterised as third-rate irrigation service being sold at throwaway prices in the name of subsidy. Mitra (1996) opines that the users should not be burdened with higher cost of irrigation service emanating from inefficiency in state-run irrigation systems. Sidhu and Ramesh Chand (1984) found that the efficiency of irrigation water depends upon the allocation, timeliness and adequacy of its supply, which ultimately determines its productivity and profitability. They also found that there is a high correlation between the returns obtained from irrigation and the additional amount invested in improving irrigation facilities. Thus, the focus of this issue is the user farmer himself, who, if finds the irrigation service useful, satisfactory and worth paying for, would definitely show an inclination towards paying for it (Singh and Ghatak, 1995). It could be said that appropriate pricing is the necessary, but not the sufficient condition for improving the revenue performance. The sufficient condition is improving irrigation service, sincere and honest efforts to recover water charges and strong punitive action against the defaulters.

The dissatisfaction to the user farmers in terms of efficient service would reduce the willingness to pay canal water charges, reduce revenues for the irrigation department, reduce flow of funds for proper O&M of the physical system, deteriorate water delivery service to the farmers, and in turn, reduce further, the willingness to pay. This is a vicious circle. Contrarily, if the user were satisfied with the service, collection rate would improve; revenue would increase, resulting in better O&M and investment flow. Thus, by appropriate pricing reforms, the vicious circle could be broken and a positive cycle could be initiated. In irrigation schemes where Water Users' Associations (WUAs) are in practice in Gujarat, management, distribution and collection of charges at the farm level is said to be much better. Higher conformance to rules and willingness to pay higher irrigation charges has been observed (Singh Kartar, 1994 and Chaturvedi, 2003).

APPROACH

The willingness of any consumer to pay a higher price for a service or a good depends upon its marginal utility. Thus, farmers would be willing to pay canal irrigation charges provided the incremental value added by the use of water in terms of the income from the crop is more than the incremental cost of irrigation, i.e. canal water charges (Singh Kartar, 1978). Marginal utility of canal water depends on:

1. The quantity of water available in the canal
2. Timely availability of water for irrigation
3. Participation and dissipation of information to the users in decision making
4. The proportion of canal water charges in total agricultural cost
5. Cost of alternative sources of irrigation
6. Socio-political perceptions about consequences of non-payment

'Willingness to pay' for canal water has two dimensions. Firstly, a customer satisfaction dimension, which implies that willingness to pay would be higher if the service is perceived to be fair, efficient, consumer friendly, sensitive to consumer needs, transparent and useful. Second is the socio-political dimension, which refers to the ability of the socio-political and legal framework to reward those who pay by ensuring timely and sufficient water supply and punish the defaulters in a conclusive manner.

The cost of alternative sources of irrigation is an important determinant of willingness to pay, because any rational farmer would compare the opportunity cost of canal water vis-a-vis some other form of irrigation while deciding about its usage. The timeliness of water delivery, satisfactory service and socio-political perceptions of the farmers are qualitative aspects. The incremental value added in terms of farm output prices, proportion of irrigation cess in total input cost and expenditure on alternative sources of irrigation are quantifiable.

An empirically examinnation of the factors affecting farmers' willingness to pay higher canal water charges, a micro-level survey of hundred farmers in five head reach and five tail end villages of the Fatewadi Command Area was conducted in May 2004.

The Fatewadi project, one of the oldest irrigation systems in India; is built on the river Sabarmati near Fatewadi village in Daskroi taluka of Ahmedabad district and falls under the Ahmedabad Irrigation Circle. It covers a total of 87 villages across six talukas. The area irrigated by this project shows yearly variations according to fluctuations in water availability. It is a diversionary project, designed initially to take excess flow from Sabarmati, and subsequently, sewage discharge of Ahmedabad city. However, it carries Narmada waters as well, since year 2002. The command is topographically suited for lift irrigation rather than flow irrigation. Alternative sources of irrigation like tube-wells, wells and tanks are also used. Private ownership of irrigation sources is present and water markets and private water rights exist in varying degrees. Water Users Associations were not functioning in sample villages. Major crop grown is paddy in kharif and wheat in Rabi.

RESULTS AND DISCUSSION

The distribution of respondent farmers according to their land holding is presented in Table 1. Majority of the farmers in the sample were large, both at the head as well as tail regions. Marginal farmers were the least represented. The average size of land holding was higher in tail villages compared to head villages.

Table 1
Distribution of Respondent Farmers

	Marginal (Ma)	*Small (S)*	*Medium (Me)*	*Large (L)*	*Total*
Head	01	16	17	17	51
Tail	08	03	17	21	49
Total	09	19	34	38	100

Poor Accessibility to Irrigation

Table 2 presents the distribution of farmers according to the primary source of irrigation. In all, 62% of sample farmers used canal as the primary source of irrigation. This is a low percentage for a region classified as command area. It reveals the low utilisation of irrigation potential and difference between envisaged and actual command area.

Table 2
Distribution of Farmers According to Major Source of Irrigation

Reach	*Canal Lift*	*Canal Flow*	*Tube Well*	*Well*	*Pond*	*Total*
Head	49	2	–	–	–	51
	(96)	(4)				(100)
Tail	11	–	12	17	9	49
	(22)		(25)	(35)	(18)	(100)
Total	60	2	12	17	9	100
	(60)	(2)	(12)	(17)	(9)	(100)

(Figures in parenthesis indicate percentage).

In the head villages, 100% farmers used canal as the primary source of irrigation and a majority of 96% lifted water from the canal. The proportion of farmers using canal as primary source was found to be only 22% in tail end villages. This was mainly due to the unavailability or uncertainty of canal water supply. In the tail villages, 35% farmers relied on well as the primary source, followed by 24% of farmers using tube wells of a greater depth, while 18% used ponds. The tail end farmers definitely preferred to use canal irrigation compared to ground water, if it were available, because it was economical and superior in quality. But they were forced to supplement it with ground water, as canal water was insufficient or unavailable in time. Ground water use entailed, comparatively higher cost of withdrawal ground water, high capital cost of installation, unreliable supply of electricity etc. Even those tail end farmers having their own bore wells preferred to use canal water because it was more suitable for paddy compared to ground water which was saline. Out of 28 marginal and small farmers in the sample, about 20, i.e. 71% use canal as a primary source of irrigation (Table 3). Thus, the better physical and financial performance of this system is of utmost importance for the benefit of this group.

As can be seen from Table 4, all 51 farmers of the head villages use canal as a primary source of irrigation, whereas in tail villages, only 11 farmers out of 49 use canal as a primary source. This is either because canal water is insufficient or irregular. So, they are forced to invest in own ground water resources or purchase water from private water supplier. On the whole, a large majority of 38% of sample farmers relied on canal only as a secondary source of irrigation

Payment Behaviour

Almost 50% of the respondents reported that they paid canal water charges in the previous year, while the other 50% said that they did not pay. Out of the 50 who reported paid, 38 i.e. 74% of respondents in the head villages reported to have paid canal water charges. A majority of them (15) were large farmers. The respondents, who reported payment, revealed that they preferred to pay because they were sure about its availability. In the tail end villages, only 12 farmers out of 49, i.e. 25% reported payment of water charges in the previous year. They too were found to be expecting assured, sufficient and timely water supply (Table 5).

Table 3
Farmers Using Canal as Primary Source of Irrigation

Farm Holding	*Marginal*	*Small*	*Medium*	*Large*	*Total in Head or Tail*	*Grand Total*
Head	1	16	17	17	51 (100)	51
Tail	2	1	3	5	11 (22)	49
Total	3 (33)	17 (89)	21 (62)	22 (58)	62 (62)	
Total in category	09 (100)	19 (100)	34 (100)	38 (100)		100

(Figures in the bracket indicate percentages)

Table 4
Canal as Primary and Alternative Source of Irrigation

	Primary	*Alternative*	*Total*
Head	51	0	51
Tail	11	38	49
Total	62	38	100

Reasons for Not Paying Canal Water Charges

Reasons for not paying canal water charges in the previous year were sought from the 50% respondents, who reported non-payment. The qualitative responses were classified and tabulated in the form of broad categories as shown in the Table 6. In spite of the fact that no one complained about the insufficiency of water supply in the head regions, 13 farmers out of 50 (i.e. 26%) did not pay canal water charges for the previous year. According to 46% of them, out of which a majority were small and medium farmers, it was economically unaffordable. Similarly, out of 7 farmers in head villages who stated that they did not pay water charges due to various socio-political reasons, a majority of 54% (6 farmers) belonged to medium and large category. They gave spurious excuses for not paying such as, "no one came to collect the water charges," which, when checked up with other farmers of the same village, was found to be incorrect. This could be interpreted as the inability of the existing social, legal and political regulatory

Table 5

Distribution of Farmers According to Payment Behaviour

Farmers Reported			*Head*					*Tail*			*Grand*
	Marginal	*Small*	*Medium*	*Large*	*Total*	*Marginal*	*Small*	*Medium*	*Large*	*Total*	*Total*
Paid	0	13	10	15	38(74)	2	0	4	6	12(25)	50
Not Paid	1	3	7	2	13(26)	6	3	13	15	37(75)	50
Total	1	16	17	17	51(100)	8	3	17	21	49(100)	100

(Figures in the bracket indicate percentages)

mechanism to persuade or encourage the farmers to pay their dues through punitive actions against the defaulters, or incentives to those who paid regularly.

Thus, even in the absence of any sound economic motive, the farmers showed a tendency not to pay. Hence, socio-political reasons, and not economic ones, seem to be important for low water charges recovery in the head villages. On the other hand, in the tail villages, the unavailability of water per se, was cited as the main reason for non-payment by 98% of the farmers, while only 2% said that it was economically unaffordable to them.

Table 6
Reasons for Not Paying Irrigation Charges for The Previous Year

Reach	*Reasons*	*Farm Holding*				
		Marginal	*Small*	*Medium*	*Large*	*Total*
Head	Insufficient Availability	0	0	0	0	–
		(0)	(0)	(0)	(0)	
	Economically Unaffordable	1	2	3	0	6
		(100)	(67)	(43)	(0)	(46)
	Socio-political reasons	0	1	4	2	7
		(0)	(33)	(57)	(100)	(54)
	Total	1	3	7	2	13
		(100)	(100)	(100)	(100)	(100)
Tail	Insufficient Availability	6	2	13	15	36
		(100)	(67)	(100)	(100)	(98)
	Economically Unaffordable	0	1	0	0	1
		(0)	(33)	(0)	(0)	(2)
	Socio-political reasons	0(0)	0(0)	0(0)	0(0)	–
	Total	6	3	13	15	37
		(100)	(100)	(100)	(100)	(100)

(Figures in parenthesis indicate percentage of farmers)

When the reasons for non-paying tendency of farmers were discussed, the irrigation field staff in the division offices of Sanand and Bavla talukas opined that the general attitude of the farmers was against paying any kind of fees for the government facilities that they enjoyed. They happily paid water charges many times over to private water suppliers who provided them with assured water supply, but would be reluctant to pay canal water even if they were assured of the same quality of service. Again, it was very difficult to take punitive action against defaulters, in spite of existence of strict laws. Individual farmers could not be debarred from taking water from the canal even if they did not pay, because each section of the canal could not be physically guarded from misuse. If a section was debarred from drawing water due to non-payment of dues, the farmers would immediately involve the local politician, who would pressurize the higher irrigation authorities to reverse their stand. Thus, the quality of non-excludability of canal water coupled with political interference was the cause of difficulty in enforcement of punitive action.

Water Purchase

The sample villages were found to be water scarce, even after considering the supply of canal water from the Narmada. Some respondent farmers reported to have purchased water for supplementing their irrigation requirements from a private water supplier in the previous year (Table 7). The tendency to purchase water was highest in the case of marginal farmers in tail villages and small farmers in head villages. As land holding is higher, water requirement increases on a regular basis. Therefore, if canal water is found to be insufficient, a permanent provision for supplementing it is needed. During focus group discussions, it emerged that many large and medium farmers had made such provisions, while small and marginal farmers had to resort to purchasing water. This explains the relatively lesser proportion of farmers purchasing water in medium and large category. Other factors determining purchase of water are the ability to raise capital for construction of a tube well etc. and of course, the availability of groundwater on the site itself.

Table 7
Distribution of Farmers Reported Purchase of Water

Farm Holding	*Marginal*	*Small*	*Medium*	*Large*	*Total*	*Grand Total*
Head	1	9	6	9	25	51
	(11)	(47)	(18)	(24)	(47)	
Tail	7	2	9	10	28	49
	(77)	(11)	(26)	(26)	(53)	
Total who purchased	8(88)	11(58)	15(44)	19(50)	53(53)	
Who did not purchase	1	8	19	19	47	
Total of category	09	19	34	38	100	100
	(100)	(100)	(100)	(100)		

(Figures in parenthesis indicate percentage of farmers)

Improvements Desired In the Water Delivery System

Each and every farmer expressed dissatisfaction over the existing water delivery system, which emerges as one of the major reasons for their reluctance to pay canal water charges. Specific reasons for dissatisfaction were discussed. The parameters for dissatisfaction were denoted as in 'A' to 'D', from which the respondents were asked to choose either single or multiple parameters. Parameter "A" refers to timely delivery of water, "B" refers to sufficiency water, and "C" refers to information given to the farmers about delivery schedule. These three factors together contribute to the "utility" of irrigation water. Therefore, they are defined as "utility" factors. The answers of the respondents, who replied positively for one or more of the utility factors, have been clubbed together and treated as a singular category. The parameter "D" refers to the repairs of the canal network system, which is a factor indicating physical inadequacy of the system and has been tabulated separately. The responses are presented in Table 8.

Table 8
Farmers Responses about the Factors Responsible for Dissatisfaction

	Utility factors (A,B,C)	*System Design and Maintenance factors(D)*	*Both Factors*	*Total*
Head	11	6	34	51
Tail	13	3	33	49
Total	24	9	67	100

There were 67% of the respondents unhappy with the utility factors as well as the design and maintenance of the irrigation system. Further, it can be seen (Table 8) that 24% of respondents showed dissatisfaction only over utility factors, while 9% of respondents expressed dissatisfaction exclusively over the design and maintenance of the canal system.

Dissatisfaction on both the counts was expressed by almost equal number of respondents in head as well as tail villages. According to them canal water was not serving its purpose either because it was insufficient (due to source inadequacy or system inadequacy), or due to faulty distribution and operation, due to which tail enders were deprived of water. Water delivery was not timely, information about delivery schedule was not shared with them and commitments in this regard were not honoured. The faulty design and maintenance of the canal system also created hardships to the farmers. The head farmers suffered due to disrepair of canal because if the lining of the canal ways broken, water percolated down in the soil, causing water logging in that portion. The tail end farmers felt that if the canal in the upstream was repaired, it could hold more water, so that sufficient water could be made available to them. The farmers revealed that very often it so happened that, due to a breakage in the canal upstream, water overflowed into the fields nearby, due to which farmers in the head reaches complained to the authorities, who in turn stopped the canal water discharge. As a result, tail enders were deprived of their share of water. Thus, repairing of canal system could solve many of their problems. Many a times canal water was released much later than required and they have to pay both on the canal water charges as well as private water supply if the crops required water in the interval between canal waterings. Even if the water was received late or not received at all and the farmer did not take all waterings from the canal, he had to pay for it, and suffered economic loss.

In a focus group discussion with the field staff and senior officials of the Ahmedabad Irrigation Circle, it was revealed that the budget allocated for repairs and desilting of the canal network was insufficient. They were aware of the recommendations of the Gujarat Study Group appointed in 1988, which recommended the allocation of O&M expenditure per hectare, but said that they were not sure if these norms were being followed in their budget allocation. They said that water supply from Narmada was assured in the Fatewadi Command in kharif, but not assured in rabi season. It was originally envisaged only as a divergent scheme for flooded areas in the upper catchment and therefore, never had any surety of water supply except the sewage discharge from Ahmedabad city.

In the head villages, the replies of the respondents, as well as informal focus group discussions revealed that the most crucial factor for them was timely delivery of water. Even as they received sufficient quantity of water, its utility declined if it was not made available at the time when the crops required it. Therefore, responses from head region to 'timely delivery' as a desired parameter of service was analysed separately, in order to highlight the issue of lack of timely delivery. 38 out of 51 i.e. 75% of farmers in the head region were dissatisfied due to lack of timely availability of water (Table 9). The dissatisfaction was high across all categories of farmers. Numerous complaints were made like water not being made available to them at the time of watering for paddy due to which they were forced to buy water at higher rates from private tube well owners. The water delivery was often delayed and the schedule was not informed to them or not adhered to.

The major cause of dissatisfaction in the tail villages was the inadequacy of water supply. Responses referring to this are given in Table 10. A high proportion of tail end farmers i.e. 37 out of 49 (76%) were dissatisfied with the quantity of water supply. Farmers of one tail end village said that the entire irrigation in their village was done only with the help of bore wells and mixing rain water even though it was not suitable for paddy, since they rarely received water from the canal. Adequate canal water was their urgent need.

Table 9
Timely Availability at the Head Region as Specific Improvement Issue

Head	*Land size*	*A (timely availability)*
	Marginal	1 (100)
	Small	10 (63)
	Medium	14 (82)
	Large	13 (77)
	Total	38 (75)

(figures in parenthesis indicate percentage of farmers)

Table 10
Sufficient Water in Tail end as a Specific Improvement Issue

Tail	*Land size*	*B (insufficient water)*
	Marginal	7 (88)
	Small	3 (100)
	Medium	17 (100)
	Large	12 (57)
	Total	37 (76)

(figures in parenthesis indicate percentage of farmers)

Since the water was released from the Narmada Main Canal, delivery schedule was a bureaucratic decision. The staff on the ground level, which had no control whatsoever as to when and where the water would be worked released, had to face the brunt of this decision. The field staff felt that this decision was very often politically motivated, without considerations about actual and urgent requirement in a particular area.

Farm Input Cost

Based upon discussions with focus groups in the study villages, an estimate of farm input cost and output from two principle crops, viz, paddy in kharif and wheat in rabi, has been computed and presented in Table 11, which gives an idea about the proportion of irrigation cost in the total farm input cost and value of output.

Approximately two and a half to three hours of pumping is required to irrigate one bigha of land. Considering the price of diesel at say, Rs. 33 per litre (at the time of survey), the cost of one watering per bigha came to around Rs. 100/- .Approximately seven and five waterings per season were required for paddy and wheat respectively. Hence, cost of lifting was Rs. 700 for paddy and Rs. 500 for wheat per season per bigha, even though the canal water charge levied by the government was only Rs. 80 and Rs. 100 for paddy and wheat respectively.

The cost of canal irrigation, per se, was as low as 2.2% in case of paddy and 7.5% in case of wheat. However the cache is that, the cost of lifting water from the canal has also to be considered because the farmers spend on fuel and machinery for the same. If this element is considered, the cost of canal irrigation turns out to be 21.1% for paddy and 44.7% for wheat. Thus, canal irrigation by lift entails a huge cost to the farmers. Since it is not possible to default the price of fuel, but under the present regulatory framework, it is possible to escape paying canal water charges, the farmers may be tempted not to pay their irrigation dues.

The input cost per bigha was higher for paddy, than for wheat. While lifting water from the canal, if a 4 HP pump is used, it consumes approximately one litre of diesel per hour.

Cost of Purchasing Water

It is a custom in the area study that if irrigation is provided by a private tube well owner, he takes 1/3rd of the produce for it; hence, cost of irrigation if purchased from third party comes to Rs. 2500 per bigha for paddy and Rs. 1400 per bigha for wheat. The cost of

Table 11
Cost of Farm Inputs, Value of Production, and Irrigation [per bigha]
[one mound (md) = 20 kg]

	Cost element	Amount (Rs.) required per bigha For Paddy		Amount (Rs.) required per bigha For Paddy	
1.	Dry ploughing	3 times @ Rs, 125 per ploughing	375	Ploughing - Rs 125 per plough	375
2.	Wet ploughing	3 times @ Rs. 250 per ploughing	750	None	Nil
3.	Seed/ dharu	Lump sum	450	2.5 mounds per bigha @ Rs. 200 per mound	500
4.	Urea	80 kg per bigha when the price of a 50 kg bag was approximately Rs. 255	408	50 kg per bigha @ Rs.255 per 50 kg bag	255
5.	DAP	30 kg per bigha @ approximately Rs. 465 for a bag of 50 kg.	279	20 kg per bigha @ approximately Rs. 465 for a bag of 50 kg.	186
6.	Pesticides	Lump sum	100	–	Nil
7.	Harvesting	Lump sum(through labourers)	500	Lump sumBy harvester combine	200
8.	Other miscl. Labour (weeding etc.)	Lump sum	150	Lump sum	100
9.	Irrigation cess # *	Rs. 80 per bigha	Rs. 80	Rs. 100 per bigha	Rs.100
10.	Irrigation	Lifting cost for seven waterings	7x100 for lifting	lifting cost for fivewaterings for lifting	5x100
11.	Total input cost		3692		1341
12.	Output value	Approximately 50 md per bigha sold @ Rs. 150 per mound.	7500	Approximately 30 md per bigha sold @ Rs.140 per mound.	4200

	Cost element	*Amount (Rs.) required per bigha For Paddy*	*Amount (Rs.) required per bigha For Paddy*
13.	Canal irrigation as cess as % of input cost	2.2	7.5
14.	Total irrigation (incl. Lifting) as % of input cost	21.1	44.7
15.	Canal irrigation rate as % of value of output	1.1	2.4
16.	Tot. irrigation (incl. Lifting) as % of value of output	13.9	14.3

* The water rates levied by the irrigation department are on per hectare basis. They have been converted here on per bigha basis @ of 1 hectare =4.16 bighas.

The water rates shown here are those prevailing according to the GR in 2001. The rates were again revised in 2004 i.e. after the time period of the survey.

lifting canal water is estimated to be Rs. 700 for paddy and Rs. 500 for wheat. It is implied that the cost of purchasing water for irrigation is more than nearly three times in comparison with canal irrigation in case of both paddy and wheat. An impression was gathered in the focus group discussion with farmers that even as the irrigation rates levied by the department were quite nominal, the real factor that pinched the farmers was the cost of lifting water from the canal. As already discussed earlier, a total of 62% of the sample farmers lifted water from the canal and 53% of farmers purchased water, which is a costlier option than lifting canal water. Therefore, it can be safely deduced that the reluctance of farmers to pay water charges is associated not with his ability to pay, but his willingness to pay.

The respondents were asked to consider a hypothetical situation where changes in the canal water delivery system were made to their satisfaction. If so, would they be willing to pay canal water charges? Would they be willing to pay more or less than the current rates? The responses have been provided in Table 12.

Table 12
Willingness to Pay Higher or Lower Irrigation Charge

Head						
1.	Land size	Same	More	Less	Not Sure	Total
2.	Marginal	0 (0)	0 (0)	0 (0)	01	01
3.	Small	4 (25)	5 (31)	1 (6)	06	16
4.	Medium	8 (47)	1 (6)	1 (6)	07	17
5.	Large	6 (35)	3 (35)	1 (14)	07	17
	Total	18 (35)	09 (18)	03 (6)	21 (41)	51 (100)
Tail						
1.	Land size	Same	More	Less	Not Sure	Total
2.	Marginal	5 (63)	0 (0)	0 (0)	03	08
3.	Small	03 (6)	0 (0)	0 (0)	00	03
4.	Medium	11 (22)	0 (0)	02 (4)	04	17
5.	Large	15 (31)	03 (6)	01 (2)	02	21
	Total	34 (69)	03 (6)	03 (6)	09 (18)	49 (100)
	Grand Total	52	12	06	30	100

(Figures in parenthesis indicate percentage of farmers).

The above table shows, 35% of respondents in the head villages expressed a willingness to pay the current water rates. This implies that even at the current water rates, canal irrigation is found to be a cheaper by them. However, in the tail region, a larger proportion of 69% were ready to pay the current water charge. This may be due to severe insufficiency of canal water, as well as other difficulties in using ground water like the falling water table, high cost of extraction and salinity. Since the marginal utility of canal water in the head villages was lower, the willingness to pay the current price was lower. Conversely, the respondent farmers in the tail end villages who had faced acute scarcity of canal water in the past, perceived the marginal utility of canal water to be higher. Therefore, they showed a higher willingness to pay current water rates.

In the head regions, 18% of farmers were even ready to pay more for canal water than what they were paying presently. This may be because, having been exposed to both canal water and ground water usage, they found that canal water was cheaper and superior in quality, and therefore were willing to pay a higher price for it, provided the irrigation service was improved to their satisfaction. The same was not the case with tail end farmers, where only 6% of the sample farmers showed a willingness to pay more for canal water. This implies that even though the question was pertaining to a hypothetical situation of "betterment in water availability", the past experiences of the farmers still loom large over the minds of the respondents, who hesitate in saying that they are willing to pay more than the current water rates. Only a small minority of 6% of respondents wished to pay less than the present water rates. A majority of farmers felt that the upward revision of canal water rates was justified and were willing to pay more.

A sizeable proportion i.e. 41% in head region and 18% in tail region did not give a definite answer about their payment behaviour, regardless of positive changes in the water delivery system. It may be that in the head regions where water availability was sufficient and more certain, farmers did not give much thought about an alternative pricing scenario. On the other hand, the tail end farmers who attached a great value to canal water due to shortage in its supply were more forthcoming in answering this question, due to which the 'not sure' responses were comparatively fewer. Societal pressure at village level or institutional punitive actions could change this apathetic attitude. Some farmers said that the annual increase of irrigation charges by 15-25% was unjustified, because the farm output prices may not increase at such a constant pace every year. Some farmers said that the farmers who used lift irrigation were losers in spite of paying rates for lift irrigation (which is half that of flow irrigation). This is because they ended up paying more than flow irrigation due to ever-increasing price of diesel and kerosene, used as fuels in pumps for lifting water. They opined that the annual increase in irrigation charges should be linked with farm output prices of the previous year and also take cognizance of the yearly increase in prices of diesel and kerosene.

CONCLUSIONS

1. Canal water is a primary source of irrigation in the head villages whereas it is only a secondary source of irrigation for a majority of farmers at the tail end villages, even though they come under the command area.
2. Canal irrigation was found to be cheaper and superior in quality by the sample farmers.
3. The tendency of purchasing ground water from private tube well owners was high among sample farmers. In the head villages, 47% and in the tail villages, 53% of farmers reported to be purchasing ground water.
4. The reluctance of the farmers to pay canal water charges is associated not so much with their ability to pay, but their willingness to pay.
5. Non-economic factors are more prominent in determining the low willingness to pay amongst sample farmers. Insufficient water supply was found to be the main reason for non-payment in the tail end villages, while in the head villages, socio-political reasons and absence or ineffectiveness of the existing institutional framework was found to be the main reason for non-payment.
6. The proportion of canal water cess in total farm input cost was found to be as low as 2.2% in case of paddy and 7.5% in case of wheat.
7. The cost of purchasing water for irrigation is more than nearly three times the cost of canal irrigation, even after lifting, for both paddy and wheat.
8. On the whole, 52% of respondents expressed willingness to pay current water charges, while 12% of respondents agreed to pay more than that if irrigation service delivery were improved to their satisfaction..

9. Willingness to pay lesser than the existing water rates was found in only 6% of respondent farmers in both head and tail villages.

REFERENCES

Bhatia, (1989), "Financing Irrigation Services in India-A Case Study of Bihar and Haryana States", Quoted in Small et al.

Champati M., and S.C. Patnaik, (1984), "Fixing Water Rates for Major Irrigation Projects-A Case Study", Agricultural Situation of India, Vol. 39, pp-221-227.

Chaturvedi, (2003), "Financial Viability of Irrigation Co-operatives-An Exploratory Study in Gujarat", Development Support Centre, Ahmedabad.

Dholakia, (2003), "Some Issues Relating to Water Management and Cost Recovery in Gujarat", Proceedings of the Seminar on Networking of Financial Issues in the Water Sector", edited by Rohit Desai, Sardar Patel Institute of Economics and Social Research, Ahmedabad.

GoG, (2001), Narmada, Water Resources and Water Supply Department, Resolution No. WTR/1099/3453/63/P, Gandhinagar, Dated 30th January 2001.

GoG, (2004), Narmada, Water Resources, Water Supply and Kalpasar Department, Resolution No. WTR/1097/20/P, Sachivalaya, Gandhinagar, 18/05/2004.

GoI, (1988), "Report on Pricing of Irrigation Water in India" by Sub-group NO. 6 of Working Group No. 3, Vol. 1, National Commission on Integrated Water Resources Development Plan, December 1988, New Delhi, pp. 2-4, 2-5, 3-5, 3-6, 3-7, 3-15, 4-17, 4-19, 5-6.

GoI, (1992), "Report of the Committee of Pricing of Irrigation Water", Planning Commission, September 1992, New Delhi (also known as the Vaidyanathan Committee Report).

GoI, (1997), "Government Subsidies in India", Ministry of Finance Discussion Paper, based on a study by NIPFP and Presented by the Finance Minister in the Parliament in May 1997, National Institute of Public Finance and Policy, New Delhi.

Gulati, (1989), "Input Subsidies in Indian Agriculture– A State Wise Analysis", Economic and Political Weekly, June 24, p-A-57.

Gulati and Narayanan, (2001), "Subsidies Rise, Investments Fall", *Economic Times*, Ahmedabad, February 01.

Gulati, A. N. Svendsen, and N. R. Choudhury, (2003), "Major and Medium Irrigation Schemes: Towards better Financial Performance", *Economic and Political Weekly*.

Jairath, Jasween, (1998), "Irrigation Rates: Watery Arguments" Economic and Political Weekly, Vol. 33, April 18, pp-931-932.

Mitra, A.K., (1996), "Irrigation Sector Reforms-Issues and Approaches" *Economic and Political Weekly*, Vol. 31, March 30, 1996, pp-A-31 to A-37.

Raju K. V. and Ashok Gulati, (2004), "Pricing, Subsidies and Institutional Reforms in Indian Irrigation: Some Emerging Trends", Proceedings of the Silver Jubilee Symposium, IRMA, Anand, December 14th to 19th, 2004.

Saleth, R. Maria, (1999), "Irrigation Privatisation in India- Options, Issues and Experiences", *Economic and Political Weekly*, June 26, Vol. 34, pp-A-86-A-92.

Sangal, S.P., (1991), "Pricing of Irrigation Water in India", Economic and Political Weekly, Nov. 16, Vol. 26, pp. 2645-2651.

Singh Kartar, (1978), "Pricing Irrigation Water from Public Canals in India-A Proposal" Agricultural Situation of India, Vol. 52, January 1978, pp. 621-624.

Singh Katar, (1994), "Co-Management of Canal Irrigation-A Case Study in Gujarat "*Journal of Indian Water Resources Society*, Vol. 14, No. 1-4: January-October, pp. 26-33.

Singh, Katar and Ghatak, (1995), "Pricing of Canal Irrigation Water: An Application of the Contingent Valuation Method", Artha Vikas, Vallabh Vidyanagar, Vol. 31, July-December, pp. 2-7.

Vaidyanathan, A., (2001), 'Irrigation Subsidies', Paper Discussed at a Conference on Fiscal Policies to Accelerate Economic Growth Organised by the National Institute of Public Finance and Policy, May, 2001.

20

INSTITUTIONAL MECHANISM FOR TRADITIONAL AND MODERN PRACTICES IN THE MAINTENANCE AND MANAGEMENT OF IRRIGATION WATER

*V. Ragupathy**

INTRODUCTION

Tamil Nadu has a rich tradition of self governing institutions for the management of irrigation tanks. Particularly small rainfed irrigation tanks have been well managed for centuries by local communities. As per Tamil Nadu Panchayat Act minor rain fed tanks with less than 100 acres of irrigated area are vested with Panchayats.[1] But in practice Panchayats are not able to effectively generate peoples participation for the management of these tanks. But in some districts of Tamil Nadu, where tank irrigation is critical for the people informal local self governing institutions have been evolved.[2] These institutions are able to generate peoples participation without any outside support or intervention and

* Reader, Department of Political Science and Development Administration, Gandhigram Rural University, Gandhigram-624302.

manage tanks efficiently. It gives great hope for academics and policy makers interested in studying self governing local institutions for the management of minor irrigation tanks.

There are many studies on self governing community based institutions for common property resource management in many parts of the world. Scholars like Ostrom based on many empirical studies on self-governing natural resource management systems have found certain key determinants for the success of self governing organisations.[3] Restriction of membership of the stakeholders is one of such critical determinant. Members must be permitted to make and unmake operational rules and regulations for the management of their institutions. The violators of the rules must receive graduated sanctions. Exclusion of free riders from the organisation must be possible. The proportional equivalence between the benefits and costs or contributions must be maintained among the members of the group. Rights of the users to design and make their own local institutions are not to be challenged by external government authorities. If these conditions are met, it is possible for the stakeholders to evolve and manage their own institutions.

Olson helps us to understand the relationship between participation of stakeholders and factors like size of the group, the nature of benefits to the participants and the presence of leaders.[4] Risk and scarcity are other two important factors.[5] For survival needs people are forced to take collective action to solve common problems like water scarcity and crop failure. The need for assurance giving institution is insisted by scholars like Olson and Hardin.[6] Many self governing local organisations in Sivaganga district for tank management fulfil the most of the above mentioned conditions.

THE INVISIBLE WATER REPUBLICS AND LOCAL CITIZENSHIP IN SIVAGANGA DISTRICT

A study in Sivaganga district on the management of common property resources indicates about 65% of the villages have some form of local governance system for the management of irrigation tanks. More than 2000 small irrigation tanks are managed by local community organisations in Sivaganga district.[7] The cultural ecology of Sivaganga district plays a critical role in shaping the local governance system. David Moose, who studied the irrigation tanks of Sivaganga district regards the sandy type of soil, scanty rain fall and absence of alternative irrigation sources as the major contributors to the evolution of local institutions of irrigation tank management.[8] The areas of the Sivaganga district having light sandy soil are traditionally called as Mankala Nadu (Country with sandy soil) the soil does not have water or moisture retention quality. Moisture is very fastly evaporated. Rainfed farming is very difficult so irrigation is inevitable for agriculture. Hence construction and maintenance of irrigation tanks becomes a necessity. But the area has only light sandy soil and the risk of tank breach is very high.

During the rainy season the tanks must be constantly watched. Every year tank bunds must be strengthened. Because of these reasons there is a need for institutional arrangements for water management. Risk and scarcity plays a critical role in the evolution of little water republics in the district.

SIGNIFICANCE OF TANK IRRIGATION IN THE DISTRICT

Sivaganga district has land area of 4126.5 sq km. The rainfall is erratic and the average being 933 cm.[9] It receives most of it's rain from north east monsoon during August to November. The major part of the area of the district has sandy soil while a small portion has black soil. Because of the soil condition and shortage of rainfed agriculture is difficult and water has to be stored in irrigation tanks. Only 31% of the land area is under cultivation. Out of the land under cultivation 66.33% of the land under cultivation depends on irrigation tanks. Out of the total irrigated area 86% of the land area is under tank irrigation. Only 14% of the land area is under other forms of irrigations like well irrigation and river irrigation.10

In Tamil Nadu, the state average of tank irrigation is only 24%. The district has 4597 irrigation tanks. Out of this only 161 tanks are river fed and 4436 tanks are rain fed (96.4%). Out of this, 4025 tanks are under Panchayats and only 572 tanks are under public works department. It means most of the tanks are small in size (below 100 acres).

ROLE OF STATE AND PANCHAYATS

Out of the 4436 irrigation tanks 4025 tanks in the district are under Panchayats and only 572 tanks.11 As per rules rainfed tanks having below 100 acres of irrigated land is classified as Panchayat tanks.

The Panchayat Union and Village Panchayats are expected to manage the large number of tanks.

Section 133 of Tamil Nadu Panchayat Act gives the following responsibilities and powers to Panchayats.

(i) The management of turns of irrigation
(ii) Protection and maintenance of tanks
(iii) The regulation of distribution of water
(iv) Execute kudimaramat (maintenance by local people)
(v) Levy fee for kudimaramat for the maintenance of tanks
(vi) Management of fishery in the tanks

There are, however, innumerable conditions put on Panchayats for the management of tanks. The revenue department has several ways of controlling Panchayats and Panchayats found it impossible to do any of the above mentioned works. No Panchayat in the state was allowed to discharge any of the responsibilities given by the act.

Panchayats are expected to mobilise peoples participation for development activities. The 73rd Constitution Amendment act says Panchayats are self governing institutions. The eleventh schedule of constitution gives large number of functions to the Panchayats to manage natural resources. But in reality Panchayats are still dependent on state government for powers and financial resources. Panchayats are treated like agents to implement government sponsored programmes and schemes. The Inspector of Panchayats (District Collector) has the power to control the day to day administration of the Panchayats.

Because of these reasons Panchayats are not able to generate peoples participation for the management of irrigation tanks. But people depend on irrigation tanks for their livelihood in districts like Sivaganga. People have to make their own local institutional arrangements to ensure their food security. In this context in Sivaganga district most of the villages have evolved their own local institutions for the management of irrigation tanks.

THE KUDI OR LOCAL FAMILY MEMBERSHIP SYSTEM

In most of the villages of Sivaganga district, Ramnad district and Pudukkottai district local community organisations are functioning. The basic unit of the organisation is the kudi or family. Kudi, in the village means a family which is residing permanently and which is recognized as a member of the village community.[12]

The Kudi system operates in many villages with minor differences. Generally in a village there are two types of memberships practiced. Families having land are considered as farming members (Vivasaya Kudi). Families without land are called General Kudi or Podhu Kudi.

Generally Kudi is defined as family (married couple) living in the village at least for one year in a joint family or as a nuclear family. Many villages maintain Kudi register. New members are admitted in September. The farming Kudis are required to contribute cash, grain or labour for the management of irrigation tanks and for the general village fund. The general Kudis are required to contribute for the village general fund only.

General Kudis have all the privileges of all the common facilities and resources except a share in the tank water. Farming Kudis get a share in the water in the tank and contribute for the maintenance of tank.

VILLAGE ASSEMBLY

Most of the villages of Sivaganga district have village assemblies. The main tasks of the village assemblies are the maintenance of irrigation tank (Kudi Maramat), sharing of water, managing water scarcity, management of common funds, settlement of disputes, organizing festivals. Tank management is the primary task of the village assembly. Village assembly consists of the eldest male members of the Kudis. The village assembly is the policymaking and decision making body. It takes decision related to the management of irrigation tanks. Some of the major activities of the village assembly are given below.

Maintenance of Tank

- Annual maintenance work by the farming families.
- Regulation of desilting of tank bed
- Maintenance of field channels.
- Clearing the supply channel
- Preparation of tank bunds
- Strengthening of tank bunds
- Removal of encroachments
- Opening and closing of sluices
- Repair of sluices and surplus of weir.

Generating Resources

Fix to the rate of contributions from Kudis in the form of labour, cash, grain and kind.

Sale of Water

- Sale of fishing rights
- Sale of grazing rights
- Fines for violation of rules
- Receiving gifts

Management of Water Supply

- Deciding the cropping pattern based on the availability of water in the tank.
- Drafting the irrigation calendar
- Appointment of common irrigator
- Appointment of watch and ward staff
- Deciding method of irrigation duration and rotation
- Deciding sharing water in times of water scarcity.

Settlement of Disputes

- Settlement of disputes over sharing, duration, frequency violation of water supply.
- Controlling illegal use of water imposing fine and penalties for violation of rules.
- Using common irrigators and watch and ward staff to avoid disputes.
- Emergency functions during floods, protection and repair of tank bunds, supply channel, surplus weir during floods.
- Inter Village Management System

Many of the tanks of Sivaganga district are the system tanks. So there is a need for coordination of irrigation management among the cluster of villages. For this purpose above the villages, a federation of villages are still functioning in Sivaganga district in the name of Nadu or country. The coordination of the Nadu or Country is vested with the peasant representatives of the villages. This traditional institution plays a critical role in the coordination of water management at a regional level. It also takes activities related to common festivals, and settlement of disputes between villages.

PULIKANMAI-A CASE STUDY

Pulikanmai is a village located in Kalayarkovil Block of Sivaganga district. The village is located 2 km South West of Kalayarkovil town and 30 km away from the district headquarters.

Ecological Profile of the Village

The village has red sandy porous soil. The local people call this area as Mankala Nadu which in Tamil means sandy soil country. Even after heavy rains water is not retained in the land. There is very little ground water and mostly saline. The tank is the only source of irrigation. There is no scope for rain fed agriculture because of the scarce rainfall and the nature of the soil. The irrigation tank is considered the most valuable resource of the village. Even for the drinking water people depend on a well in the tank bed.

The irrigation tank of Pulikanmai has an irrigated area of 93 acres. The water bed of the tank is 96 acres. The tank has two slices or water regulators. The village has three major ponds and two small ponds. Twenty years back one pond was used for getting drinking water. The tank water is used for irrigation for about five months from September to January. In summer months tank bed is used for grazing. The entire Pulikanmai village is covered by a lush growth of trees. They believe thick cover of vegetation attracts rain in the area. A large number of peacocks and variety of birds are spotted in the village area. The supply channels and field channels are maintained well. There are three main supply channels to the Pulikanmai tank and there are numerous streams for the flow of rain water to the various water bodies of the village.

Socio Economic Setting of the Village

The village has a population of 850 out of whom 158 are SC people. The majority of the people are Udayar Christians. There are 121 Christian Udayar families, two maravar families, 32 families of Scheduled Caste, two families of carpenters, two families of barbers and three families of washer men. In total there are 121 Backward Class, 6 Most Backward Class and 32 Scheduled Caste families are in the village.[13] There is no big land lords in the Pulikanmai village. No single individual has more than six acres. In the tank irrigated area 82 families have land. The 93 acres of irrigated land is distributed among 82 families. Average size of land holding is about 1.25 acres. Small land holding is common in the village. The smallest land holding is 30 cents.

Among the 32 Scheduled Caste families, 12 families have land. Other 20 families do not have irrigated land. The landless families work in the nearby town and take up agricultural activities. They also get seasonal work in Pulikanmai itself. During harvest season the landless men get about 15 kg of paddy for their days work and women get about 10 kg of paddy. Each family can get work for about 12 days in harvest season. Each landless families could get about 300 kg of paddy as wages in harvesting season.

Kudi and Village Assembly System of Pulikanmai Village for Tank Management

Like many other villages of Sivaganga district in Pulikanmai village also a Local Community Organisation (LCO) is functioning and Kudi (family membership) and village assembly system are practiced. The Kudi is the family membership. Local Community Organisation consists of Kudi (family membership) village assembly and village executive committee. Families having land in the irrigated area of the tank is called as farming kudi. Family without land is called a general kudi. In the past SC families are not taken into kudi system. In recent years SC families with land in the irrigated area are admitted into the kudi system.

A kudi membership register is maintained. Every year in the month of July kudi register is updated. The kudi members form the village assembly. The village assembly has about 10 meetings in a year. The village assembly has an executive committee. Before 1977 traditional leaders from prominent families formed the executive committee. From 1977 a new system of selection of member for a period of one year by rotation was introduced. Village assembly selects five members in the month of September to the executive committee.

The executive committee is responsible for the day to day management of affairs related to the village and irrigation tanks. They collect he contributions, ensure participation in various activities related to tank maintenance, proper distribution of water management of water scarcity.

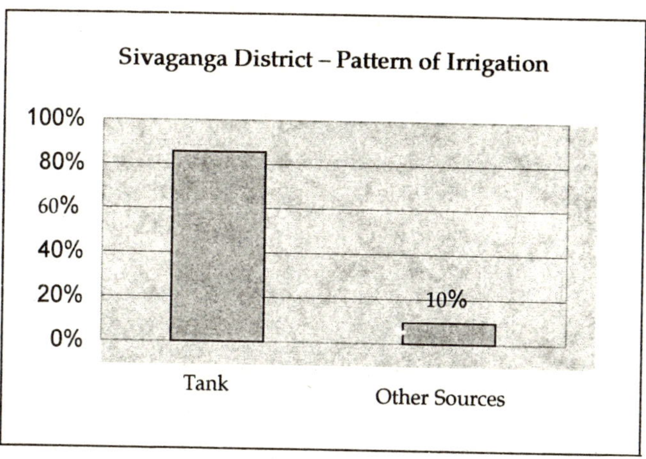

Participatory Management of Irrigation Tank

The kudi system facilitates participation of people in the management of irrigation tanks effectively. It is a self governing system without any external inducement or incentive. The system has been sustained for hundreds of year. The system, ensure livelihood security of the people of the village. Particularly during the years of reduced rainfall except for the system people could not have food security. The chances of crop failure is very much reduced. People are able to share water, change the cropping pattern and save the crop. Here major activities of the kudi system are discussed.

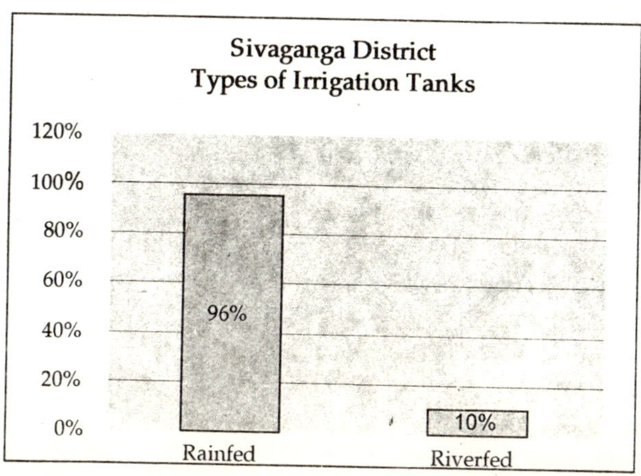

Kudimaramathu (Community Maintenance)

As in many other villages in Sivagangai and Ramnad Districts, the people in Pulikanmai village regularly undertake Kudimaramathu. Vivasaya Kudis in the village take the responsibility of doing kudimaramathu. Kudimaramathu in the village has the following functions:

1. Clearance of supply channels
2. Desilting the sluice areas
3. Clearance of field channels

Before the onset of the monsoon, a special meeting to discuss kudimaramat work is conducted. The meeting decides the date and the time of the work. Normally the clearing of the supply channels is done in the month of August. For the annual clearance work every kudi has to contribute labour. Every vivasaya kudi has to send one person with required implements like space, basket, pan and crowbar for the work. Both men and women are accepted. In case a kudi is not in a position to send a person it can send a labourer as alternative. On the same day the silt around the sluices of the supply channels are removed and it is ensured that the sluice is in good condition. In one single day the clearance work is done. Clearance of the field channels is also done on a fixed day before the opening of the sluices.

Disaster Management and Participation in Emergency Repair of Tank

During the rainy season and floods, the entire Pulikanmai village carefully watches the flood situation. The common irrigators, farmers and executive committee members take great care to prevent breaches. In case of overflow of water, specific methods are available as precautionary measures. In the year 1995, there was a breach in the tank. All the families were asked to come to work with empty jute bags. Each family was required to bring at least two such jute bags. Around two hundred men, women and children were engaged to repair the breach in the tank. Around three hundred jute bags filled with sand were used in the operation. Within four hours the entire operation was over. The cost of the operation could be around thousand rupees. It is impossible for any state agency or Panchayat to meet such an emergency situation mustering people's participation.

Distribution of Water

Water is a scarce commodity in Pulikanmai village. The water in the tank has to be economically used and coordination among the users is required to save water. In case there is no control over the use of water, chances for crop failure are high. In view of this the farmers of Pulikanmai have a practice of taking collective decisions regarding water in their village assembly. The village assembly decides the date for opening the sluices. It appoints tow common irrigators. Appointment of common irrigators helps the people to overcome difficulties in water distribution to the fields. The two irrigators can take care of the irrigation to all the fields served by the tank. Individual farmers are relieved of their responsibility of irrigating their farms. As the common irrigators are experienced people, they can use the water more efficiently and economically than other individuals.

Conflicts over use of water, particularly in times of water scarcity, are also reduced by the presence of common irrigators. Common irrigators are well aware of the rules, regulations and customs relating to the distribution of water. As the common irrigators have to collect their wages from the farmers, they ensure proper water distribution. The performance of the common irrigators is also monitored by the farmers and deficiencies, if any are reported promptly to the executive committee.

Scarcity Management

Water scarcity is a frequent problem in Pulikanmai. Water scarcity may result in damage to the standing crops. To mange water scarcity, the people of Pulikanmai have developed from experience a method of management. Food security and equity are the two basic principles governing the strategy for managing water scarcity. Water is considered a common property of the village and every kudi is assured minimum water supply to ensure minimum harvest. To manage acute water scarcity, special village assembly meetings are held. Depending on the availability of water, the area to be cultivated is determined. If the tanks if filled only to half it capacity, normally the assembly allows the farmers to cultivate only 50% of the landholding of the kudis. Kudis having less than 50 cents are permitted to irrigate their entire land area. If the water scarcity is acute, another system, viz, Kudi Eevu or share of the kudi is followed. In this system the concept of equity in water use is followed. Each kudi is permitted to irrigate only a small piece of land (about 50 cents) to ensure fulfilment of minimum food requirement for the year. The size of the land holding of the individual is not taken into consideration. In this system small and marginal farmers are benefited. But indirectly, the large farmers are also benefited, because unregulated use of water without any planning at a time of water scarcity may result in crop failure and economic loss. The theoretical problem of Hardin's tragedy of commons is skilfully avoided by the participatory approach of the kudi system of Pulikanmai.

Community Fishing

As per the rules the right to harvest fish in the Panchayat irrigation works must be given on lease to generate income for the Panchayat. In Pulikanmai village, the LCO has taken the responsibility of using the irrigation tank in the village to regulate fishing in the tank for the benefit of the entire community. Individual fishing or unauthorized fishing in the tank is strictly prohibited.

The executive committee members negotiate with the Panchayat and, as per the government rule, they pay the minimum fee prescribed and avoid public auction. On a date fixed by the village are assembly, the tanks is opened for community fishing. A member from each kudi is permitted to enter the tank for fish harvesting. Within three hours the harvest can be completed. The next day onwards people are allowed to fish in the tank without any restriction. For nearly a week, people can get fish for personal use. The system regulated the behaviour of the community for the entire year and permits them to harvest the benefit collectively by the year end. Every family in the village gets fish for nearly a week in a year.

Conservation of Trees

The people of Pulikanmai are aware of the importance of conservation of the watershed for the health of the irrigation tanks and the ecology of the village. Conservation of trees and the regulation of grazing are the twin policies followed by the LCO. The trees in the tank and in the poramboke lands are under the control of various government departments, the Panchayat and the Panchayat union. But, in Pulikanmai, regulations has been imposed by the LCO to conserve the trees in the village. Two guards have been appointed in the village to prevent any unauthorized cutting of trees. The entire village has taken the responsibility of protecting the trees. Any unauthorized cutting attracts penalty by the village assembly. Abundant thorny bushes are available in the poramboke lands of the village for fuel for the people of the village.

Regulation of Grazing

The LCO has imposed restrictions on grazing to conserve the top soil, to avoid silting of the tank and for the conservation of plants. Such restrictions are discussed in the village assembly and decisions are taken. At present, the following two types of restrictions are imposed on grazing:

1. Rearing or grazing of goat is strictly prohibited within the village land area
2. Grazing permits are issued by the LCO for cattle other then goats.

Goat rearing is a profitable income generating activity in rural areas. Pulikanmai village has abundant vegetation which is favourable for goat rearing. But the villagers have collectively imposed a ban on goat rearing, as goat rearing is considered a great threat to top soil, vegetation and conservation of trees.

Grazing by other animals is also restricted and stall feeding is practiced in the village. During summer, when getting fodder is difficult, grazing permits are issued to cattle other than goats. From this practice the village generates resource for the local community fund. In the year 1996-9, the village could collect around Rs.5000 from issuing permits for number of animals, types of animals and duration of grazing in a year. People strictly follow the rules relating goats. If a goat is purchased for a purpose of meat, it cannot be kept by a household for more than a week. Violation of this rule will attract penalty.

Collective Good

Wade argues, "In the South Indian context at least, villagers are likely to follow joint rules and arrangement sonly to achieve intensely felt needs that could not be met by individual response. These are likely to be concerned primarily with the defence of production (avoidance of crop or animal loss) secondarily with the enhancement of income, and finally (and a long stretch from the first two) with education, nutrition, health and civic consciousness.[14] The LCO, which is primarily concerned with the management of natural resources is also able to generate participation in the form of labour contribution, monetary contribution and adherence to prescribed rules for the development of the village in the following areas.

The villagers of Pulikanmai have decided to ban liquor (Sale or use) within the boundaries of the village. The mismanagement of the government run noon meal centre was collectively challenged by the people. The LCO tried to correct the corrupt organizers of the noon meal centre by persuasion. The food materials and eggs of the centre were not properly distributed to local children by the corrupt noon meal organizer who had the political patronage. The LCO repeatedly made several representations to the higher officials and finally got the noon meal organizer transferred from the village. Around Rs.700/- was spent from the village common fund for getting this transfer. The village did not have a ration shop. The villagers had to walk two kilometres to collect their ration goods from a nearby village. The village assembly decided to take efforts to bring a ration shop to the village and authorized the executive committee to take necessary steps and to spend money for the purpose from the village common fund. The executive committee worked hard for nearly two years to get sanction for the ration shop. A sum of Rs.1600/- was spent from the village common fund for this common purpose.

A local library and reading room was a felt need in the village. In a meeting of the village assembly it was decided to get a government sponsored local library in the village. For starting a local library, as per government rule, a minimum number of people have to take membership by paying the prescribed fee and the local community has to contribute a share of the expenditure on establishing a library.

The village assembly spent Rs.22,0000 on the supply of chairs, tables, books and payment of membership fee. Within fifteen days all the requirements for staring a rural library were fulfilled by the LCO. In the same way, the village has constructed a building in the local secondary school by its efforts. The kudis were asked to provide a person each to work in the construction project. The expenses on materials for the construction were met from the village common fund. Around Rs.15,000/- work of materials and

Rs.10000/- work of a labour were contributed towards the construction of the school building.

The village has the practice of collectively contributing for the funeral ceremony of anyone dying in the village. In case of a death, every kudi has to pay Rs. 10/- each and around Rs. 1200/- is immediately mobilized. The collection is given to the bereaved family. At least one member from each kudi has to attend the funeral ceremony. It is an arrangement of social security and display of communal solidarity. Such social action brings people together and personal conflicts and differences of opinion among families are reduced by such repeated processes of community action. It builds a base for harmony and cooperation among families for development work.

The LCO of Pulikanmai also operates a micro credit system. The major portion of the community fund is offered for credit to its members. In the year 1996-97, Rs. 25,000/- was distributed form the community fund among 50 kudis at an interest rate of 36%.13 As the interest rate was high, only people with real need took the loan and repayment was very prompt.[15] The income generated from the credit system was recycled again and again into the credit system. Normally the commercial money lender charges not less than 60%. The common fund in the village gives support to the families in the village during emergencies and saves them from the clutches of money lenders.

Assurance Provided by the LCO

As discussed earlier, 'assurance' is the key factor in facilitating collective action for the management of natural resources. Institutions are required to coordinate and predict the behaviour of individuals in the context of collective action.[16] If an institution can give sufficient assurance and if the participants are profited by their contribution, the scope for participation is high. In Pulikanmai village the farmers want to avoid crop failure due to water crop loss. Use of water without any restriction or coordination may result in crop loss. Such crop loss will ruin their economy. This is a key factor which influences the people of Pulikanmai to enter into an institutional framework. As per the law, the tank has to be managed and water has to be distributed by the Panchayats. But, in practice, not only in Pulikanmai, but also all over Tamil Nadu, the Panchayats do not undertake any such work.

In Pulikanmai, under threat of economic loss, people have developed a regulated behaviour and an institutional arrangement to coordinate behaviour. Their organization gives them assurance on efficient management of water. For managing scarcity of water and avoiding risk of crop failure, they depend on their organization. The organization, its rules and its conventions, are developed locally by the beneficiaries or stake holders over a period of time. The organization is fully autonomous. People are sure that their organization is not controlled by any other outside agency. The organization ensures that non contributors are not permitted to enjoy common benefits created out of the contribution of the rule abiding participations and no one can enjoy a free ride. The organisation is controlled by the stake holders and all major decision of the stake holders. The supply of water as per the rules to the fields of the participants is ensured by the organization. Appointment of common irrigators reduces the expenses of the individual farmers. It also prevents conflicts among farmers in water sharing. The kudi system links contribution to resource sharing. Without contributing to the maintenance of the irrigation tank, one can not get water from the tank. This assurance by the institution ensures a positive behaviour among the participants conductive to cooperation for collective good.

Another important assurance given by the institution is that of equal distribution of water among all kudis in case of scarcity. The interest of marginal farmers is also protected by this scheme. Every kudi is assured of minimum harvest of paddy even in a year of water scarcity. In the absence of such regulation, the entire crop in the tank area may fail in years of water scarcity.

The processes of decision making and implementation are very simple, transparent and quick enough to meet the requirements of the people. As there is not separate official machinery for this work, the cost of transactions is minimum and the scope for corruption is limited. As sanction for expenditure is given by the village assembly and the accounts of the organisation are discussed in the village assembly the scope for any misappropriation of funds is also nearly eliminated. The account books are well maintained. As the executive committee has only one year's tenure and membership is rotated, there is no scope for dominance by any individual in organisational matters.[17]

The system has been operating efficiently for several decades. The present structure of the organisation is a modernised form of the older organisation. In the new structure, traditional leaders have been replaced by an executive committee selected for one year. There is no electoral politics in the organisation and there is no intervention by any government authority. Because of these reasons the organisation is stable. This stability gives assurance to the individuals on the predictable behaviour of the organisation. The most important strength of the local organisation is the capacity to conserve and sustain natural resources for several generations. Because of the proper maintenance of the irrigation tank and other water bodies the people of the village are assured of getting the minimum harvest. The assurance given by LCO facilitates and sustains the participation of the people of Pulikanmai in the management of natural resources.

CONCLUSION

The tradition of participatory management of irrigation tanks are still found to exist in varying degrees in villages of Sivaganga district. Participation emerges in certain situations where in people are challenged by certain critical problems like drought, water scarcity, crop failure and floods. These factors forces people to take collective steps to solve their common problems. In such situations local stakeholders institutions are very much required. The local institutions must assure the stakeholders on it's capacity to protect their interests. Particularly institutions evolved by the stakeholders based on their needs are able to provide the required assurance to the people. Restriction of membership, equity, transparency autonomy and cost effectiveness are the key determinants of the sustainability of local institutions for tank management. It is very much encouraging to see that still thousands of locally evolved little organisations are able to protect and manage small irrigation tanks ignored by the state system. The activities of such little organisations reveals the potential for participatory institutions.

NOTES AND REFERENCES

1. Tamil Nadu Panchayats Act (1994), Section 133, Government of Tamil Nadu.
2. Ragupathy, V., Participatory Management of Natural Resources, Concept Publishers, New Delhi, 2003.

3. Ostrom, E., Governing the Commons: Evolution of Institutions for Collective Action, New York: Cambridge University Press, 1990.
4. Oslon, M., Logic of Collective Action: Public Goods and Theory of Groups, Cambridge: Harvard University Press, 1971.
5. Wade, R., Village Republic: Economic Conditions for Collective Action in South India, Orient Longman, 1987.
6. Hardin, G., The Tragedy of Commons, Science, 162, 1968.
7. Ragupathy, V., Local Community Organisations in Sivaganga District, Gandhigram, 1999.
8. Moose David, Local Institutions Ecology and History in South Indian Tank Irrigation, Centre for Development Studies, University of Wales, 1996.
9. District Revenue Profile, Collectors Office, Sivagangai 2004.
10. *Ibid.*
11. *Ibid.*
12. Group Discussion with Members of the Village Assembly LCO, Pulikanmai, May, 1998 all the details given on LCO.
13. LCO records Pulikanmai, 1997 all the Details regarding LCO of Pulikanmai are Gathered from LCO Records of Pulikanmai.
14. Wade, R., Village Republic: Economic Conditions for Collective Action in South India, Bombay: Orient Longman, 1987.
15. Pulikanmai LCO Accounts Book (1996-97), Pulikanmai: LCO, 1997.
16. Hardin, op.cit.
17. Discussion with LCO members, Pulikanmai, 1999.

21

TANK IRRIGATION IN INDIA: STATUS, TRENDS AND ISSUES

*A. Narayanamoorthy**

INTRODUCTION

Tanks are the most important traditional source of water used for irrigation and other livelihood purposes over centuries in India. Tanks are many in number and varied in size. Unlike the other surface sources of irrigation, it is a low cost source of irrigation and also predominantly managed by the farmers themselves. The command area of majority of the tanks is small and therefore, the problems associated with irrigation management are few. Besides being a low capital-intensive source of irrigation, tank irrigation is considered to be user-friendly with fewer environmental problems (see, Sengupta, 1985; Palanisami, 1990; Palanisami and Easter, 2000; Sakurai and Palanisami, 2001; Vaidyanathan, 2001). Though tank irrigation has been in practice in different parts of the country, it is mainly concentrated in the peninsular South Indian States of Andhra Pradesh, Karnataka and Tamil Nadu, where area under tank irrigation accounts for about 20 per cent of the net irrigated area even today (see, Dhawan 2000; Vaidyanathan

* Reader, Gokhale Institute of Politics and Economics, Pune-411 004, India. *Fax: +91-20-25652579; Tel: +91-20-25650287; E-mail: na_narayana@hotmail.com*

and Sivasubramanian, 2001; Raju, et al., 2004). The terrain condition (slope) and soil types that are prevailing in Southern India have favoured the concentration of the tanks.

However, in spite of having many advantages, the area under tank irrigation has consistently declined since independence. Surprisingly, the reduction in area under tank irrigation has been observed even in the years where rainfall is normal or above normal. While the area under tank irrigation declined from 4.56 million hectares (mha) in 1960-61 to 2.71 mha in 1999-2000, its share in the net irrigated area sharply declined from 18.49 per cent to 4.73 per cent during this period at the national level. This reduction is found to be more severe in States like Andhra Pradesh, Karnataka and Tamil Nadu, where tank irrigation is one of the important sources of irrigation even today (Narayanamoorthy, 2002; Raju, et al., 2002). The total area under tank irrigation of these three States together has declined from 2.43 mha in 1960-61 to 1.53 mha in 1999-2000, a reduction of over 37 per cent. What is astonishing is that the area under tank irrigation has declined despite the construction of thousands of new tanks since the first plan period (Vaidyanathan, 1994; 1999; 2001).[1] The continuous decline of tank irrigation has created tremendous hardship to the resource poor farmers mainly marginal and small farmers, who are mostly dependent on tank irrigation for crop cultivation (Rao and Chandrakanth, 1984; Bardhan, 2000; Deshpande, et al., 2004). As tanks have been playing an important role in increasing the recharging capacity of wells, its adverse impact on groundwater irrigation has also been realised in some parts of South India (Meinzen-Dick, 1984; Palanisami and Easter, 1991; Narayanamoorthy, 1993; Janakarajan, 1993; Palanisami and Balasubramanian, 1998; Shah, et al., 1999; Vaidyanathan, 2001).

A number of studies have been carried out mainly using field level data to find out the performance of tank irrigation in different regions of India. By and large, studies have coequally confirmed that performance of tank irrigation has been declining since the introduction of green revolution in India (Palanisami and Easter, 2000; Vaidyanathan, 2001). While analysing the trends and determinants of tank irrigation, the existing studies have shown that institutional, physical and technical factors are the main causes for the continuous decline of tank irrigation (see, Table 1). While some studies have shown that encroachment in supply channels and tank beds are the main reasons for the decline of tank irrigation, others have attributed to breakdown of village institutions, which have been managing the tank irrigation over centuries.[2] Rapid growth of groundwater irrigation in the tank command areas is also found to be one of the main reasons for the decline of tank irrigation (Palanisami and Ranganathan, 2004; Janakarajan, 1993). Importantly, watershed development works taken up in the upper region of the catchment area have also affected the performance of tanks by reducing the inflow of water in to tanks (Palanisami and Easter, 1991 and 2000).

Though the area under tank irrigation has been declining since independence, the reduction may not be the same across different time periods and the States. Similarly, factors determining tank irrigation are also expected to be varied across the States, as the agro-climatic factors including social factors that govern tank irrigation are not the same. Because of these variations, the trends and determinants of tank irrigation are also expected to be different across the States. What is the growth trend of tank irrigation across different time periods? What are the factors determining the area under tank irrigation? What is the extent of relationship between the growth of well irrigation and decline of tank irrigation? Does well irrigation affect tank irrigation uniformly across the States? Who are the losers and gainers of tank irrigation? etc., are the important questions that require empirical answers. Though there are studies focusing on tank irrigation using micro level data, not many studies are available covering all the macro level issues mentioned above and that too using time series data from 1950 to 2000 at the national level.[3] Therefore, in this study, an attempt has been made to study the trends and determinants of tank irrigation at the all-India level as well as across the States with the following objectives:

1. To study the growth pattern of tank irrigation at national and across States level using time series data from 1950 to 2000.
2. To study the nexus between rainfall and area under tank irrigation at a specific State, which has relatively larger area under tank irrigation.
3. To find out the losers and gainers of tank irrigation among different size of farmers.
4. To suggest policies to rejuvenate tank irrigation in India.

The study mainly uses secondary level data for its all analyses. It covers the period of fifty years from 1950-51 to 1999-2000.[4] The data have been mainly collected from the publication of Minor Irrigation Census, Indian Agricultural Statistics, Water and Related Statistics, Agricultural Census and Fertilisers Statistics as well as from the publications of Center for Monitoring Indian Economy. Besides these, data have also been collected from Season and Crop Reports of selected States wherever necessary. While the source-wise area under irrigation has been compiled from various issues of Indian Agricultural Statistics, farm size-wise area under irrigation has been compiled from the various issues of Agricultural Census. In order to analyse the issue of losers and gainers of tank irrigation, data from different time points have been used from the Agricultural Census, which is published by the Ministry of Agriculture with an interval of five years since 1970-71. For analysing the trends in area under tank irrigation, growth rate has been computed for different time points using log-linear function (lnYt = a + bt, where growth rate = antilog of b-1 x 100). Correlation analysis has been carried out to study the nexus between the rainfall (season-wise) and tank irrigated area.

The paper has been organised into seven main sections. Following the introductory section, the second section presents the review of various studies that are currently available on tank irrigation. Section three discusses the growth pattern of tank irrigation from 1950-51 to 1999-2000 at all-India level. State-wise growth pattern of tank irrigation covering the period from 1960-61 to 1999-2000 is discussed in section four. The nexus between rainfall and area under tank irrigation is analysed using the data of Tamil Nadu State in section five. An analysis focuses on the main beneficiaries of tank irrigation is presented in section six. The last section of the paper presents measures that can be considered to revitalise the tanks in India.

A GLIMPSE ON LITERATURE

Since the present study focuses mainly on the trends and development of tank irrigation using macro-level data, it is useful to review the literature available on this aspect to know the overall findings of the study. Therefore, before analysing the trends in area under tank irrigation in different regions in India, let us review the studies that have analysed the reasons for the poor and deteriorating performance of tank irrigation in India.[5] Tanks are broadly classified into two categories as system tanks and non-system tanks. Those tanks that receive water not only from their own catchment area but also from the reservoirs and dams through supply channels are called system tanks. Non-system tanks are rain-fed tanks that receive water only from its catchment area through rainfall. Non-system tanks are large in numbers in all the States when compared to system tanks.[6] Tanks were constructed by local kings, landlords and chieftains with the aim to provide water for irrigation and other purposes. These tanks are many centuries old and managed mainly by the community as most of the tanks are small in size.[7] The system of managing tanks through community participation is many centuries old. However, the performance of tanks has been deteriorating over the years owing to various reasons.

There are three kinds of studies available as far as the reasons for the decline of tank irrigation are concerned. While one group of studies report the technical and physical factors are the main reasons for the decline of tank irrigation, others have shown that institutional factors, which have direct bearing on the functioning of tank irrigation, are the main reasons for the reduction of tank irrigated area (see, Table 1). Another group of studies have shown that technical and physical as well as institutional factors are equally responsible for the reduction in area under tank irrigation.

Studies by Palanisami (1990), Palanisami and Easter (2000), MIDS (1986), Vasimalai (1998), Mohanakrishnan (1998), Thippiah (1998) and Shah et al., (1999) have indicated that insufficient alignment of channels, heavy siltation in tank beds and feeder channels, encroachment in supply channels and tank beds as well as large scale interruption in the catchment areas are the main reasons for the poor performance of tanks in different regions. Similarly, a study (Venkatram, 1980) carried out covering Andhra Pradesh, Tamil Nadu, Karnataka and Maharashtra found that lack of soil conservation and afforestation in the catchment area, increasing rate of siltation, unauthorised cultivation in the foreshore area and poor maintenance of tank bunds, sluices, drainage and field channels are the main causes for the degradation of tanks.

Table 1
Major Causes for the Decline of Tank Irrigation: Review of Selected Studies

Sl. No.	*Author(s)*	*Major Reasons Identified*
1.	Palanisami and Easter (1984)	Heavy siltation in tank bed and supply channels; encroachment in tank storage area; defective control structures; weak farmers organisation; absence of village leadership
2.	Palanisami (1990)	Heavy siltation in tank bed and feeder channels; encroachment on the Tank foreshore area; deforestation in the catchment area;
3.	Shankari (1996)	Encroachment in tank bed; village is fraction ridden; weeds and siltation in the supply channels; negligence of minor irrigation department; development of well irrigation.
4.	Barah, Reddy and Sudhakar (1991)	Less institutional support; development of well irrigation; severe financial stringency.
5.	Janakarajan (1993)	Changes in control over lands and land transfers; siltation and encroachment in tank bed and supply channels; sub-division and fragmentation of lands; rapid development of well irrigation.
6.	Reddy (1996)	Breakdown in Kudimaramath
7.	Sivasubramanian (1997)	Changes in land ownership; reduction of storage in the channels; encroachment in the supply channels; increasing private ownership of wells; deficient supply from the source for a long time.
8.	Thippaiah (1998)	Reduction in storage level due to siltation; the frequency of tank filling declined; encroachment in tank bed and foreshore area; poor physical and structural conditions of tanks; plantation in tank bed and catchment area causes reduction in storage.
9.	DHAN Foundation (1997)	Encroachments in supply channels and tank beds; problems in local village assemblies; segregated approach by the authorities; inadequate investment.
10.	Vasimalai (1998)	Breakdown of local institutions; encroachment in supply channels and tank bed; lack of farmers involvement; siltation; urbanisation; system failure; decline in tank-fed cultivation.

Contd...

Sl. No.	*Author(s)*	*Major Reasons Identified*
11.	Mohanakrishnan (1998)	Encroachment in catchment areas and surplus channels; inadequate funds allotment; loss of storage capacity; leadership problems.
12.	Pundarikanthan and Kallapiran (1998)	Heavy siltation in tank beds and supply channels; encroachment in tank foreshores, water spread and supply channels; lack of leadership at village level; poor linkages between communities and departments associated with tank irrigation; unscientific and indiscriminate development of wells in tank command areas.
13.	Selvaraj and Vasimalai (1999)	Conflicts among the villagers have reduced collective maintenance work; no rules are followed in sharing the tank water; tank water is exhausted in a shorter period.
14.	Shah *et. al.,* (1999)	Decades of siltation; encroachment in tank beds and supply channels; decline of water users participation.
15.	Vaidyanathan (2001)	Siltation; deterioration of inlet channels; encroachment of natural drainage channels; decline in community arrangements for tank management.

As regards physical and technical factors, many studies have pointed out that institutional factors are the main reasons for the poor performance of tanks (Janakarajan, 1993; Barah, et al., 1991; Reddy, 1996; Selvaraj and Vasimalai, 1999; Pundarikanthan and Kallapiran, 1998; S ivasubramanian, 1997; Mohanakrishnan, 1998). The reasons reported are breakdown in Kudimaramath,[8] conflicts among the villagers, unsatisfactory rules followed for sharing tank water, decline in water users participation, lack of village leadership, weak farmers' organisation, changes in land ownership and mode of cultivation, land transfer from high caste to middle and low caste farmers and sub-division and fragmentation of land.

Some of the studies have revealed that a rapid development of well or bore-well irrigation in the tank command area is one of the important reasons for the appalling condition of tank irrigation (Shankari, 1991; Barah et al., 1991; Janakarajan, 1993; Narayanamoorthy, 1993; Pundarikanthan and Kallapiran; 1998). Several studies report that these tanks have been traditionally managed by the rich and resourceful farmers for many years through community participation. But, owing to recent development in groundwater irrigation, community participation in tank related activities has drastically reduced (Janakarajan, 1993). Though some studies have analysed separately the role of physical and technical factors on tank irrigation, it is essential to include the institutional factors with any analysis on tank irrigation as technical and institutional factors do not operate in isolation, but they often interact (Janakarajan, 1993).

Since the Traditional Irrigation Institutions (TIIs) have played a crucial role in the management of tanks over centuries, many studies have attempted to find out the reasons for the fast decline in tank command areas. The various reasons have been identified for this. In those regions where tank is the only source of irrigation, the reasons reported by the studies are different from the areas where wells (dug wells) are being used to provide supplementary irrigation. Similarly, in those areas where groundwater irrigation is used predominantly using bore-wells for crop cultivation, the reasons for lesser participation of TIIs are totally different. Let us first understand the reasons for the decline of TIIs in the areas where tank is the only source of irrigation. One normally expects that TIIs might function well in those areas where alternative source of irrigation for cultivating crops other than tanks is absent or negligible. Despite the fact that collective participation of farmers is essential for better performance of tanks, farmers are not very much enthusiastic to participate in tanks related activities mainly because of conflicts among different group of farmers (see, CDF and NET, 1996; Shah, et al., 1999).[9] Though conflicts are commonly seen in the villages, it never hampered the involvement of irrigation institutions in tank related works in those days because the village institutions (panchayat, etc) were functioning very effectively. That in recent years, due to lack of dedicated leadership at village level many studies have pointed out community related problems (be it castes or temple or any other) could not be solved amicably in most cases (see, Shah et al., 1999). This has lead to friction among the society which ultimately reduced the farmers' participation in irrigation related works in many places.[10] Studies have also shown that the lack of leadership at village level is also one of the causes for the decline of tank irrigation (Mohanakrishnan,

1998; Vasimalai, 1998; Shah, et al., 1999). Importantly, the re-emergence of caste and politics nexus in the villages in recent times often makes TIIs less flexible in making the decisions pertaining to tank management.

The causes for the decline of tank irrigation are different in those areas, where wells are used for supplementary irrigation. Before 1960s, most of the wells in the tank command areas were dug-wells, especially in southern India. It is a well-known fact that the recharging capacity of dug-wells is less and also highly associated with the level of storage of water in tanks in each year. Generally, recharging capacity of dug-wells is higher when water storage is higher in tanks (see, Palanisami and Easter, 2000; Palanisami and Meinzen-Dick, 2001). Another peculiarity of dug-well is that unlike bore-wells, the pump-sets connected with dug-wells cannot be operated continuously because of its shallow nature. Hence, dug-wells are mostly used only for supplementary irrigation purpose. Because of these reasons, farmers could not completely ignore tanks.

Farmers have started largely neglecting the works associated with tank irrigation only after the massive development of bore-well/tube-well irrigation, which begun during the mid-sixties along with green revolution. While discussing about the various issues associated with tank irrigation, Vaidyanathan (1994) rightly surmises that ".....the people who have wells usually take lead in enforcing the collective regulations. But once they can appropriate water they need (by pumping), their interest in getting involved in communal maintenance and the enforcement of regulations flags" (p. 31).

Development of bore-well irrigation has affected the farmers' participation in tank related works in two ways. First, unlike dug-wells, supply of water from bore-wells is assured as it draws water from deep aquifer. This provides greater benefit to the farmers and hence, most of the resourceful farmers have started installing bore-wells wherever possible. This assured water supply from bore-wells discouraged the resourceful farmers in participating tank related works. However, these resourceful farmers could not completely ignore tank irrigation because the cost of water (per unit or per hectare) was higher at that time due to metered based tariff system (pro-rata) levied for electricity supplied for agricultural purposes. The problem became serious when State

Electricity Boards (SEBs) started introducing cheaper pricing policy (flat-rate or free electricity) for agriculture, beginning from the late eighties in different States.[11] As a result, the resourceful farmers have largely started reducing their involvement in various works associated with tank irrigation. Since irrigation institutions are normally managed and controlled by the resource owning large farmers, the small and marginal farmers could not manage and operate the irrigation institutions effectively because of their poor social and economic status. In the meantime, the small and marginal farmers are also benefited because of the cheaper pricing policy of electricity, as it has brought an extensive market for water, where water is sold for non-well-owning farmers by the well-owning farmers (Narayanamoorthy, 1993 &1996; Shah, 1993). Since electricity tariff levied for irrigation purpose was very low in many States, well owning farmers could sell water for low price that helped the resource poor farmers (marginal and small farmers) to cultivate crops without any serious problem. The availability of water with low price did not provide any incentive to the resource poor farmers to participate in the activities concerning maintenance/management of tanks. An extensive study conducted in Palar Anicut System (PAS), which supplies water for 317 tanks and irrigating command area of 80,000 acres in North Arcot and part of Chengalpattu districts in Tamil Nadu, has clearly shown that the degree of functioning of traditional irrigation institutions (TIIs) was poor in those areas where density of well is higher (Janakarajan, 1993).

On the whole, the literature survey presented above underlines four major reasons for the poor and deteriorating performance of tank irrigation in India. First, due to encroachments in catchment areas as well as in feeder channels, the water supply to tank has declined. Second, siltation in tank beds/basins has increased, which resulted in reduction in area irrigated by tanks due to poor maintenance. Third, the participation of farmers in managing the tanks has declined due to disintegration of traditional irrigation institutions that were managing the tanks over centuries. Fourth, rapid development of groundwater irrigation along with the supply of electricity for agriculture with cheaper tariff was also found to be responsible for reduction in the performance of tanks in recent years. Though the existing studies have brought out various reasons for the poor performance of tanks, not many studies seem to have studied the nexus between the level of rainfall and area under tank irrigation. The issue of who are the losers and gainers of tank irrigation is also not elaborately studied by the researchers so far. In addition, in recent time, no attempt is made to study the macro level (both at country and across States) trends and development in area under tank irrigation covering the time series data of five decades. Therefore, the present study makes a modest attempt at filling this gap using fifty years data.

GROWTH PATTERN OF TANK IRRIGATION IN INDIA

Though the area under tank irrigation has been declining over the years, the rate of reduction may not be the same across different time periods because of various reasons. In this section, let us study the trends in area under tank irrigation at the national level covering the period 1950-51 to 1999-2000. The primary objective of this analysis is to find out whether or not the growth trend in area under tank irrigation is same across different periods, i.e., pre and post-green revolution period as well as in different decades. Table 2 presents the area under tank irrigation along with other sources of irrigation for the country as a whole. It is clear that the area under tank irrigation has declined significantly over the last fifty years, while at the same time the area under other sources of irrigation has increased considerably. The share of tank irrigated area declined from 16.51 per cent in triennium ending (TE) 1952-53 to 5.18 per cent in TE 1999-2000, whereas the share of groundwater irrigation has increased from 30.17 per cent to 55.36 per cent during this period. Interestingly, tank is the only source where reduction has been noticed in both absolute (area) and relative (per cent to NIA) terms.[12] However, the area under tank irrigation has not consistently declined over the last 50 years in India. For instance, the area under tank irrigation has increased over 34 per cent (from 3.468 mha to 4. 651 mha) between TE 1952-53 and TE 1962-63, but started declining continuously thereafter (see, Figure 1). Among the three major sources of irrigation, tank is the only source, where the irrigated area has been declining continuously since early seventies. The reduction in area under tank irrigation is also coincided with the introduction of green revolution.[13]

Table 2
Source-wise Area under Irrigation from 1950-51 to 1999-2000, All-India

(Area in '000 ha)

Period	Net Irrigated Area (NIA)									
	Canals			Tanks	Groundwater					
	Government	Private	Total		Tube-well	Other-well	Total	Others	NIA	GIA
TE1952-53	7386	1227	8613	3468	-	6339	6339	2588	21008	23016
	(35.16)	(5.84)	(41.00)	(16.51)		(30.17)	(30.17)	(12.32)	(100.0)	
TE1962-63	9398	1170	10568	4651	431	6999	7430	2420	25070	28631
	(37.49)	(4.67)	(42.15)	(18.55)	(1.72)	(27.92)	(29.64)	(9.65)	(100.0)	
TE1972-73	12117	866	12983	3822	4866	7511	12377	2313	31494	38560
	(38.47)	(2.75)	(41.22)	(12.13)	(15.45)	(23.85)	(39.30)	(7.34)	(100.0)	
TE1982-83	15207	600	15808	3165	10212	8381	18593	2406	39971	51006
	(38.05)	(1.50)	(39.55)	(7.92)	(25.55)	(20.97)	(46.52)	(6.02)	(100.0)	
TE1992-93	16762	485	17247	2817	15081	10803	25884	3114	49395	65215
	(33.93)	(0.98)	(34.92)	(5.70)	(30.53)	(21.87)	(52.40)	(6.30)	(100.0)	
TE 1999-2000	17135	474	17609	2916	19904	12632	32536	3223	56285	75099
	(30.44)	(0.84)	(31.29)	(5.18)	(35.36)	(22.44)	(57.81)	(5.73)	(100.00)	

Note: NIA – Net Irrigated Area; GIA – Gross Irrigated Area; Figures in brackets are percentage to NIA.
Source: GOI, *Indian Agricultural Statistics* (various years); CMIE (2004); FAI (various years).

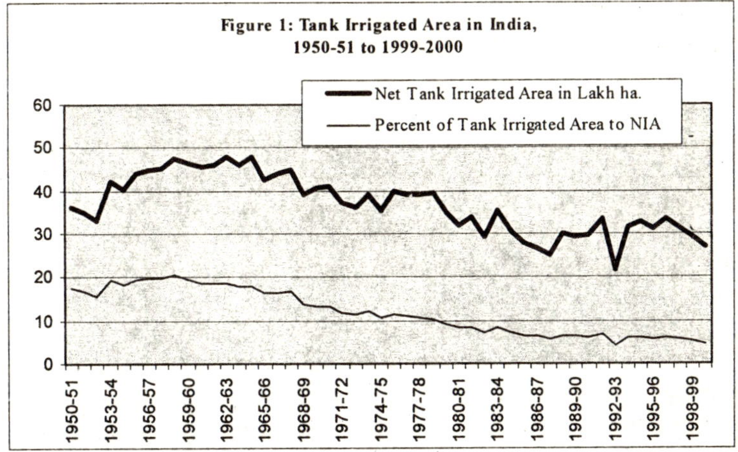

One of the important issues associated with tank irrigation is whether or not the rapid development in area under groundwater irrigation has made any impact on tank irrigation. Several field studies seem to suggest that the development of groundwater irrigation in tank command areas has reduced the performance of tank irrigation mainly due to decline in the participation of resourceful farmers in tank management activities. Macro-level data utilized in this study also seem to support the argument that there is a negative association between the area under tank irrigation and groundwater irrigation (see, Figure 2). However,

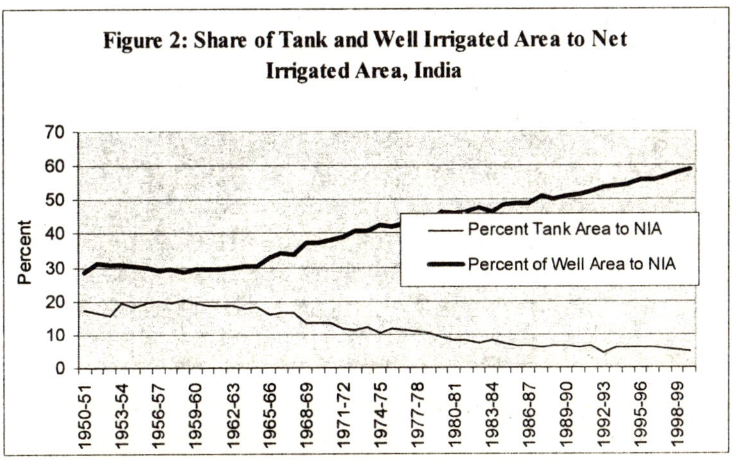

Rajeevan (1996), from his analytical study mainly based on the data from Tamil Nadu, indicates that "There is no firm evidence that availability of groundwater has had negative effects on maintenance of surface sources in particular group or group cohesion in general" (p. 21). It is difficult to pinpoint what kind of nexus exists between groundwater irrigation and tank irrigation using aggregate level data.[14] Thus, one needs to clearly re-examine the whole issue of whether or not groundwater irrigation has had any negative impact on tank irrigation using properly designed survey based study. An examination on this line would reveal the real cause for the decline in area under tank irrigation.

Another issue associated with tank irrigation is whether or not the growth rate in area under tank irrigation is the same across different time periods. Growth rate computed clearly indicates that the pattern of growth is not the same across different periods (Table 3). Growth in area under tank irrigation was not only positive but also statistically significant during the pre-green revolution period, i.e., 1950-51 to 1965-66. Among different time periods considered for the analysis, the growth performance in tank irrigated area was found to be very appreciable during the fifties (1950-51 to 1960-61), which was over 3.30 per cent per annum. The growth rate in area under tank irrigation was negative and also statistically significant during the post-green revolution periods, except the nineties (1990-91 to 1999-2000). This is possibly due to better performance of monsoon rain during the most part of nineties at the all-India level.[15] Interestingly, this positive growth was achieved despite significant growth in area under groundwater irrigation, especially from the source of tube-wells. This seems to contradict with the popular notion that tank performance is affected due to rapid development in groundwater irrigation.[16] Given the contradiction, the whole issue of negative impact of groundwater irrigation on growth of area under tank irrigation needs re-examination using field survey data. Be that as it may, the analysis clearly suggests that the area under tank irrigation has declined substantially in India, especially after the introduction of green revolution.

STATE-WISE TRENDS IN TANK IRRIGATION

Though the area under tank irrigation has declined considerably over years at the all-India level, the same trend many not be

Table 3
Growth Rate of Irrigated Area by Source, 1950-51 to 1996-97, All-India

(Per cent per annum)

Period	*Net Irrigated Area (NIA)*									
	Canals			*Tanks*	*Groundwater*					
	Govern-ment	*Private*	*Total*		*Tube-well*	*Other-well*	*Total*	*Others*	*NIA*	*GIA*
1950-51 to 1960-61	2.20[a]	0.56[ns]	1.98[a]	3.31[a]	nil	1.20[a]	1.29[a]	-1.29[ns]	1.65[a]	2.18[a]
1960-61 to 1970-71	2.47	-3.60[a]	1.93[a]	-1.59[b]	3.07[a]	0.83[b]	5.09[a]	-0.74[ns]	2.19	3.13[a]
1970-71 to 1980-81	2.01[a]	-0.36[a]	1.87[a]	-1.10[ns]	7.70[a]	1.20	4.16[c]	0.94[ns]	2.46	2.97
1980-81 to 1990-91	1.05[c]	-2.75[c]	0.91[c]	-1.52[c]	4.00[a]	1.97[a]	3.13[a]	2.27[a]	1.90[a]	2.30[a]
1990-91 to 1999-2000	0.33[ns]	-0.28[ns]	0.31[ns]	0.16[ns]	4.05[a]	2.25[a]	3.33[a]	0.39[ns]	1.91[a]	2.04[a]
1950-51 to 1965-66	2.37	-0.76[c]	1.99[a]	1.85[a]	nil	0.79[a]	1.87[a]	0.09[ns]	1.72[a]	2.11[a]
1965-66 to 1980-81	2.44[a]	-1.30[a]	2.17[a]	-1.40[a]	12.21[a]	0.85	4.81[a]	0.58[c]	2.65	3.09[a]
1980-81 to 1999-2000	0.65[a]	-0.61[ns]	0.60[a]	-0.20[ns]	3.96[a]	2.69[a]	3.43[a]	2.27[a]	2.13[a]	2.38[a]
1965-66 to 1999-2000	1.58[a]	-2.72[a]	1.37[a]	-1.29[a]	6.58[a]	1.74[a]	3.81[a]	1.33[a]	2.26[a]	2.59[a]
1950-51 to 1999-2000	1.95[a]	-2.64[a]	1.64[a]	-1.00[a]	9.17[*]	1.34[a]	3.81[a]	0.75[a]	2.18[a]	2.62[a]

Notes: a, b, c are significant at 1, 5 and 10 percent level respectively; ns – not significant; * - from 1960-61 to 1999-2000; NIA – Net Irrigated Area; GIA – Gross Irrigated Area.

Source: Computed from GOI, *Indian Agricultural Statistics* (various years); CMIE (2004); FAI (various years).

observed in different States due to variations in rainfall, development in other sources of irrigation, agro-climatic factors, etc. Which are the States gaining or losing in area under tank irrigation? Which are the worst affected States in terms of decline of tank irrigation? etc., are some of the questions which need to be answered using State-wise data. In order to study the Statewise position, a total of 12 major States where tank irrigation had considerable presence during the sixties have been considered.

Figure 3: Statewise Percentage Change in Tank Irrigated Area: TE 1999-2000 over TE 1962-63

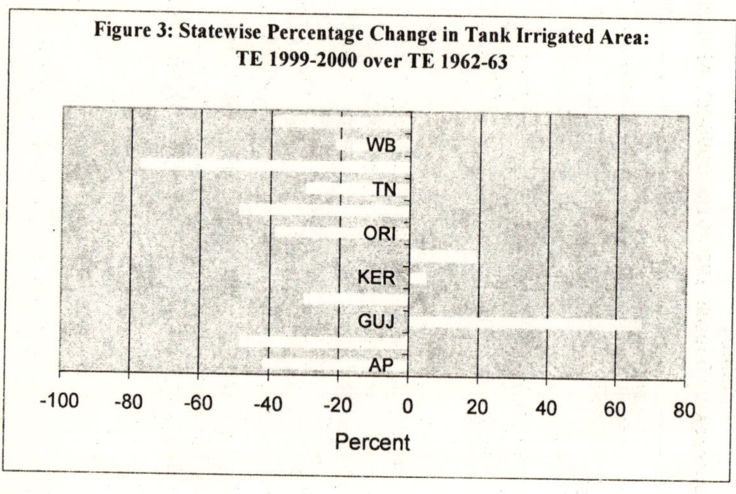

Table 4 presents Statewise trends in area under tank irrigation from TE 1962-63 to TE 1999-2000. As observed at the national level, the per centage of tank-irrigated area to NIA has sharply declined across all the States considered for the analysis. During the sixties, the area under tank irrigation accounted for nearly two-fifth of the total irrigated area in States like Andhra Pradesh, Karnataka, Orissa and Tamil Nadu, but it has declined to less than one fifth of the total irrigated area during TE 1999-2000.[17] Among the major tank irrigated States, drastic reduction is noticed in States like Orissa, Andhra Pradesh, Karnataka and Tamil Nadu (See, Figure 3).

Area under tank irrigation has also sharply declined in absolute term in almost all the States between TE 1962-63 and TE 1999-2000. Interestingly, the worst affected one are from south Indian States, where tank irrigation still plays a crucial role in

Table 4

Statewise Trends in Area under Tank Irrigation: TE 1962-63 to TE 1999-2000

Year	AP	Bihar	Gujarat	Karnataka	Kerala	MP	Maha-rashtra	Orissa	Rajasthan	Tamil Nadu	Uttar Pradesh	West Bengal	INDIA
Tank irrigated area ('000 ha)													
TE1962-3	1246	279	16	358	48	140	193	495	167	941	418	334	4651
TE 1972-3	911	141	38	367	74	143	205	179	185	923	352	303	3831
TE1982-3	932	97	41	306	58	152	276	243	113	615	173	312	3143
TE 1992-3	881	115	28	265	48	172	na	614	185	579	91	263	2927
TE1999-00	731	145	28	250	51	167	na	305	86	662	95	263	2825
Percent of tank area to NIA													
TE1962-3	41.00	13.96	2.35	39.49	14.75	15.09	17.65	50.67	9.22	37.77	8.23	23.50	18.55
TE 1972-3	29.48	6.21	2.74	30.07	16.87	8.74	15.38	20.75	8.44	34.13	4.91	20.32	12.16
TE1982-3	26.19	3.51	1.91	21.27	23.78	6.16	14.14	17.02	3.73	24.49	1.80	18.72	7.89
TE 1992-3	20.85	3.43	1.10	12.00	14.51	3.75	na	30.65	4.36	22.63	0.83	13.77	5.95
TE1999-00	17.03	4.00	0.89	10.13	13.71	3.05	na	14.59	1.55	22.21	0.76	13.76	5.01
Percent of tank area to NSA													
TE1962-3	10.77	3.38	0.17	3.48	2.47	0.86	1.07	8.42	1.23	15.08	2.42	6.12	3.45
TE 1972-3	8.02	1.71	0.39	3.62	3.39	0.78	1.21	3.07	1.22	14.70	2.04	5.32	2.75
TE1982-3	8.45	1.23	0.42	3.00	2.68	0.81	1.52	3.97	0.73	11.28	1.00	5.66	2.23
TE 1992-3	8.13	1.53	0.29	2.49	2.15	0.88	na	9.72	1.14	10.15	0.53	4.80	2.06
TE1999-00	6.97	1.95	0.28	2.43	2.24	1.11	na	5.02	0.53	11.90	0.57	4.82	1.99

Notes: NIA – Net Irrigated Area; na – data not available; NSA – Net Sown Area; AP – Andhra Pradesh; MP – Madhya Pradesh.

Sources: GOI (various issues), FAI (various issues); CMIE (2004).

protecting the livelihood of vast majority of marginal and small farmers including the agricultural labourers (Balasubramanian and Selvaraj, 2003).[18] Between TE 1962-63 and TE 1999-2000, the area under tank irrigation declined by about 41 per cent in Andhra Pradesh, 30 per cent in Karnataka, 29 per cent in Tamil Nadu and 38 per cent in Orissa. Reduction in area under tank irrigation has also been observed in States like Rajasthan, Uttar Pradesh and West Bengal. Importantly, Uttar Pradesh had over 4.18 lakh hectares of tank irrigated area during TE 1962-63, but the same declined to 95 thousand hectares during TE 1999-2000. Since Uttar Pradesh comes under northern region of the country, the reason for drastic reduction of area under tank irrigation might be different from the same observed in south Indian States.[19] Contrary to this, the area under tank irrigation has marginally increased in States like Gujarat, Kerala and Madhya Pradesh. Overall, two important points are emerging out from the analysis of State-wise area under tank irrigation. First, the area under tank irrigation has declined more drastically in those States where tank irrigated area accounts for a relatively larger share in the net irrigated area. Second, the area under tank irrigation has increased marginally in certain States where it accounts for very low share in the total irrigated area.

The trend in area under tank irrigation observed in some specific States provokes to rethink the issue on whether or not development in groundwater irrigation really affects the performance of tank irrigation. Though many studies carried out based on the data of south Indian States tend to support this issue, the trend observed from Orissa State seems to contradict with this. If groundwater irrigation has any impact on tank, the area under tank irrigation in Orissa State should not have declined by about 38 per cent between TE 1962-63 and TE 1999-2000, because the development of groundwater irrigation is very poor even today in Orissa including other eastern States due to poor institutional and infrastructural support.[20] This decline may have occurred due to change in rainfall pattern or disintegration of traditional irrigation institutions. Therefore, one needs to clearly examine why has the area under tank irrigation declined in certain States despite lower development of groundwater irrigation using more disaggregated or field level data.

Rainfall and Tank Irrigation Nexus

It is always believed that the total quantum of rainfall as well as its distribution has direct bearing on area under tank irrigation. In this section, an attempt is made to study the nexus between rainfall and tank irrigated area. Needless to mention that it is difficult to capture the nexus between the rainfall and area under tank irrigation using country level aggregate data because of varied nature of rainfall across different States. However, the nexus between the two can be studied relatively better manner by taking a single State where one can minimize the variation in the level of rainfall.[21] In view of the availability of time series data as well as large presence of tank irrigation, Tamil Nadu State has been purposively selected to study the nexus between rainfall and tank irrigation using forty years data, from 1960-61 to 1999-2000. Unlike other Indian States, Tamil Nadu State receives rainfall from both southwest monsoon (SWM) and northeast monsoon (NEM).[22] Therefore, the nexus between the two has been studied not only by taking into account the annual (total) rainfall but also the season-wise (SWM and NEM) rainfall.

As can be seen from Table 5, the area under tank irrigation in Tamil Nadu has declined from 9.34 lakh hectares in TE 1962-63 to 6.56 lakh hectares in TE 1999-2000, a reduction of nearly 30 per cent points. Interestingly, this reduction has taken place despite an increase in the annual rainfall in TE 1999-2000 over the period of TE 1962-63. The correlation results also show that there is a insignificant relationship between the level of rainfall and the area under tank irrigation (see, Table 6). Among the two major seasons, the area under tank irrigation has relatively stronger correlation (though statistically not significant) with northeast rainfall, which is understandable because the State normally receives higher rainfall during this season (see, Figure 4). Since the total rainfall of NEM and SWM is crucial for increasing the area under tank irrigation, it is expected that the total rainfall of these two seasons would have stronger relationship with area under tank irrigation in Tamil Nadu. However, contrary to our expectation, the correlation value between the two was insignificant, indicating that the rainfall of the two major seasons has not made any impact on tank irrigation in the State.

Table 5
Trends in Rainfall and Tank Irrigated Area in Tamil Nadu: TE 1962-3 to TE 1999-2000

Period	*SWM*	*NEM*	*Annual Rainfall*	*NEM+SWM*	*% of NEM+ SWM to annual Rainfall*	*Tank irrigated Area*
TE1962-63	348.3	428.3	961.7	776.6	80.6	933.9
TE1972-73	315.4	505.5	995.0	820.9	82.4	934.6
TE1982-83	273.1	379.3	772.1	652.4	84.2	687.4
TE1992-93	296.2	443.6	877.4	739.8	84.1	579.0
TE1999-00	275.3	628.1	1043.1	903.4	86.0	655.7

Notes: SWM – Southwest monsoon; NEM – Northeast monsoon.
Source: Calculated from Palanisami and Ranganathan (2004).

Table 6
Correlation between Area under Tank Irrigation and Rainfall: Tamil Nadu

Variables	*1960-61 to 1999-2000*	*1960-61 to 1979-80*	*1980-81 to 1999-2000*
Annual Rainfall and Tank Area	0.153	0.107	0.079
NEM Rainfall and Tank Area	0.189	0.111	0.276
NEM+SWM Rainfall and Tank Area	0.163	0.090	0.098
% of NEM+SWM Rainfall to Annual Rainfall and Tank Area	0.106	-0.084	0.090
SWM Rainfall and Tank Area	-0.030	-0.057	-0.328

Notes: SWM – Southwest monsoon; NEM – Northeast monsoon.
Source: Calculated from Palanisami and Ranganathan (2004).

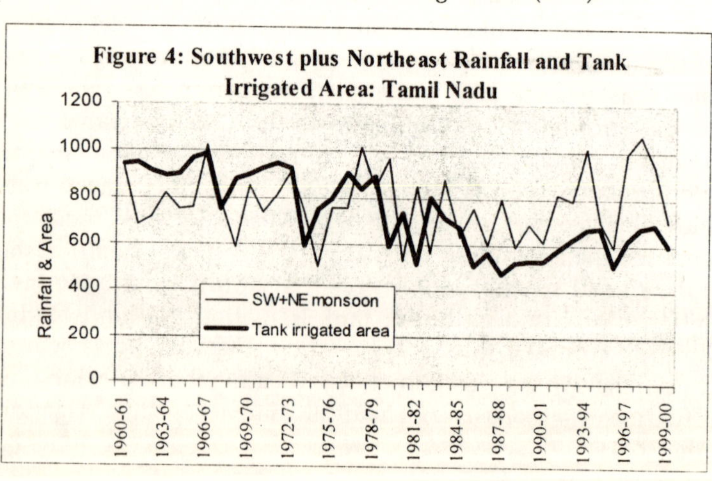

Similarly, since the total rainfall of NEM plus SWM is crucial for increasing tank irrigation, it is expected that the rainfall share of these two seasons to total annual rainfall will have relatively stronger correlation with tank-irrigated area. This is because of the fact that the rainfall received after the two major seasons (NEM and SWM) will have no impact on tank irrigation. Unexpectedly, the correlation value estimated between the two parameters turned out to be very weak (see, Figure 5). This suggests that even if the share of rainfall of the two main seasons to annual rainfall increases, it is not going to make any significant increase in area under tank irrigation. What appears from here is that any improvement in the level of rainfall is not going to make perceptible increase in area under tank irrigation, unless adequate arrangements are made to improve the flow of water supply in to the tanks as well as storage capacity.

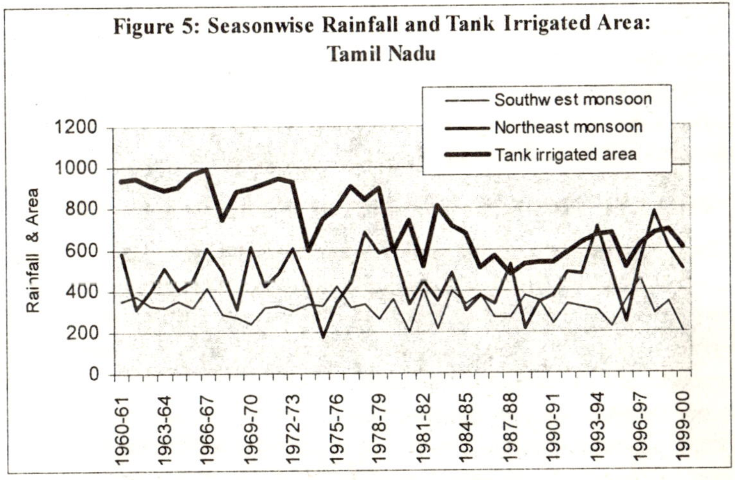

Figure 5: Seasonwise Rainfall and Tank Irrigated Area: Tamil Nadu

The weak relationship that exists between the rainfall and area under tank irrigation leads to a genuine question: Do other factors influence area under tank irrigation more than the level of rainfall? Through a recent detailed study on tank irrigation in Tamil Nadu, Palanisami and Easter (2000) have mentioned a number of possibilities for the low level of tank irrigation, which are presented below: "(1) The procedures used to calculate the water yield to tank storage are no longer applicable to the current situation and can lead to an over estimation of water availability. (2) There has been a change in the rainfall patterns over time and changes in catchment characteristics due to abuses in the upper part of the watershed. (3) Encroachment in the tank foreshore and along the feeder channels as well as siltation and tree planting in the tanks has reduced tank storage capacity. (4) In several cases, the construction of dams/anaicuts in upper watersheds has prevented the water supplies from reaching downstream tanks. (5) Low levels resource mobilization in the villages and low State budget allocations for O and M. (6) Community management of the tanks has declined due to conflicts between different groups in the villages. (7) Intensive groundwater development in tank command areas has caused tank management to decline. (8) The poor design of new tanks has resulted in a low level of performance" (pp. 55-56). Most of the reasons mentioned by Palanisami and Easter (2000) seem to support the findings that we arrived from the review of literature. What clearly emerges from the analysis is that increase in rainfall alone is not going to make any appreciable improvement in area under tank irrigation unless adequate measures are taken up to improve the flow of water supply and storage capacity of the tanks.

WHO ARE THE BENEFICIARIES OF TANK IRRIGATION?

Therefore the area under tank irrigation has declined across different regions in India. However, the reduction in area under tank irrigation might be different across farm-size categories. While the reduction in area under tank irrigation will have overall impact on agricultural growth, it is expected to have deleterious impact on the resourceless poor farmers namely marginal and small size of farmers. Who are the main beneficiaries of tank irrigation? Who will be the main losers due to decline in area tank irrigation? Is tank irrigation still important for marginal and small farmers? etc., are some of the important policy related issues, which are somehow not adequately addressed by the researchers in India. An attempt is made in this section to address these questions using the data from Agricultural Census of India, which is currently available for five time points namely 1970-71, 1976-77, 1980-81, 1985-86 and 1990-91.

Area under tank irrigation by farm size presented in Table 7 clearly shows that the resource poor farmers (less than 2 hectare farmers) still use major share of tank-irrigated area in India.

Table 7
Area under Tank Irrigation by Farm Size, 1970-71 to 1990-91, All-India

Size Class	*Tank irrigated area*						*Per cent of Tank Area to NIA*					
	1970-1971	*1976-1977*	*1980-1981*	*1985-1986*	*1990-1991*	*ACGR (%)*	*1970-1971*	*1976-1977*	*1980-1981*	*1985-1986*	*1990-1991*	*Reduction in % (1990-1 over 1970-1)*
1. Marginal (<1 ha)	737	742	941	805	940	1.22	16.78	13.24	13.69	9.99	9.94	-40.75
	(21.00)	(23.25)	(28.24)	(28.92)	(31.79)							
2. Small (1-2 ha)	668	627	742	636	682	0.10	14.09	11.56	11.21	8.31	7.51	-46.72
	(19.03)	(19.64)	(22.27)	(22.84)	(23.06)							
3. Semi-Medium (2-4 ha)	800	730	753	638	654	-1.00	12.11	10.23	8.64	6.59	5.96	-50.79
	(22.79)	(22.87)	(22.60)	(22.92)	(22.12)							
4. Medium (4 -10 ha)	828	732	635	509	503	-2.46	9.94	9.17	6.44	4.91	4.46	-55.15
	(23.59)	(22.93)	(19.09)	(18.28)	(17.01)							
5. Large (>10 ha)	477	361	260	196	178	-4.81	9.47	9.78	5.50	4.17	3.63	-61.68
	(13.59)	(11.31)	(7.80)	(7.04)	(6.02)							
All Total	3510	3192	3332	2784	2957	-0.85	12.06	10.70	9.05	6.88	6.47	-46.35
	(100)	(100)	(100)	(100)	(100)							

Notes: ACGR - Annual Compound Growth Rate in Area is 1990-91 over 1970-71; Figures in brackets are percentage to total area.
Source: GOI, *All India Report on Agricultural Census* (various years), Ministry of Agriculture, Government of India, New Delhi.

Marginal and small size group of farmers together accounted for about 40 per cent of tank-irrigated area in 1970-71, which further increased to nearly 55 per cent in 1990-91. On the other hand, the share of tank irrigated area used by large size group of farmers declined nearly by about six per cent points during this period, i.e., from 13.59 per cent to 6.02 per cent. It can be noticed from the table that tank irrigated area used by marginal and small farmers has not only increased over the last 20 years but accounted for nearly two third of tank irrigated area. Since the farmers belonging to marginal and small size group are mostly resourceless, they are unable to afford cost-intensive irrigation sources like groundwater as in the case of medium and large size group of farmers. Therefore, tank irrigation continues to play a crucial role among small and marginal size group of farmers even today. This is also by and large true across different States where tank irrigation has considerable presence even today (see, Table 8). In fact, farmers belonging to the category of below 2.00 hectares have gained in Andhra Pradesh and Karnataka States. In most of the tank-intensive States, farmers belonging to large group are only going out of tank irrigation between 1970s and 1990s. This means that the large farmers might be finding it difficult to stay in tank irrigation given the increasing risk due to poor tank performance and cultivation of high value commercial crops.

The significance of tank irrigation among different group of farmers can also be studied by relating with the net irrigated area of each size group of farmers. Unlike the absolute area under tank irrigation, the share of tank area to NIA among all size group of farmers has declined between 1970-71 and 1990-91. While the share of tank area to NIA declined from 16.78 per cent to 9.94 per cent among the marginal group of farmers between the two periods indicated above, the same has declined from 9.47 per cent to 3.63 per cent among the large size farmers during the same period. Data presented in Table 7 also shows that reduction in the share of tank irrigation during the last 20 years period is positively associated with the land size of the farmers. The overall reduction in the per centage of tank irrigation to NIA across different farm size groups is mainly due to significant improvement in the use of groundwater irrigation. Since the medium and large size group of farmers mostly own groundwater

Table 8
Area under Tank Irrigation by Farm Size in Selected States: 1970-71 and 1990-91

(Area in '000 ha)

State	Farm Size	*1970-1971* Tank	*1970-1971* NIA	*1990-1991* Tank	*1990-1991* NIA	*Percent of Tank Area to NIA* 1970-71	*Percent of Tank Area to NIA* 1990-91
Andhra Pradesh	< 1 ha	209	547	365	1208	38.21	30.22
	1-2 ha	182	526	279	1043	34.60	26.75
	2-4 ha	218	665	268	1090	32.78	24.59
	4-10 ha	259	799	213	911	32.42	23.38
	> 10 ha	217	533	70	295	40.71	23.73
	Total	1085	3070	1195	4547	35.34	26.28
Karnataka	< 1 ha	50	128	53	251	39.06	21.12
	1-2 ha	60	174	63	332	34.48	18.98
	2-4 ha	81	254	69	515	31.89	13.40
	4-10 ha	91	324	66	578	28.09	11.42
	> 10 ha	45	220	25	238	20.45	10.50
	Total	327	1100	276	1914	29.73	14.42
Maharashtra	< 1 ha	15	70	18	263	21.43	6.84
	1-2 ha	27	121	28	204	22.31	13.73
	2-4 ha	44	233	28	616	18.88	4.55
	4-10 ha	64	440	22	540	14.55	4.07
	> 10 ha	45	371	5	145	12.13	3.45
	Total	195	1235	101	1768	15.79	5.71
Orissa	< 1 ha	27	154	26	246	17.53	10.57
	1-2 ha	22	201	27	283	10.95	9.54
	2-4 ha	37	165	32	281	22.42	11.39
	4-10 ha	37	152	22	153	24.34	14.38
	> 10 ha	26	54	9	33	48.15	27.27
	Total	149	726	116	996	20.52	11.65
Tamil Nadu	< 1 ha	236	597	287	816	39.53	35.17
	1-2 ha	180	557	1	439	32.32	0.23
	2-4 ha	182	606	111	517	30.03	21.47
	4-10 ha	151	523	70	369	28.87	18.97
	> 10 ha	55	196	25	126	28.06	19.84
	Total	804	2479	494	2267	32.43	21.79
West Bengal	< 1 ha	52	172	66	856	30.23	7.71
	1-2 ha	59	219	0	685	26.94	0.00
	2-4 ha	69	265	38	538	26.04	7.06
	4-10 ha	43	165	11	174	26.06	6.32
	> 10 ha	2	27	0	3	7.41	0.00
	Total	225	848	115	2256	26.53	5.10

Note: NIA-Net Irrigated Area.

Source: GOI, *All India Report on Agricultural Census* (various years), Ministry of Agriculture, Government of India, New Delhi

structures, the reduction in the share of tank irrigated area would not have made any big burden for them. But the reduction would have made lot of hardships to marginal and small size farmers for whom other sources of irrigation are not available or affordable. On the whole, the farm size-wise analysis suggests that tank irrigation is still an important source of irrigation for marginal and small farmers.

CONCLUSIONS

Tank irrigation has many superior features than other sources of irrigation. In spite of this, the area under tank irrigation has been continuously declining particularly after the introduction of new agricultural technology at all-India level. The area under tank irrigation, however, has not declined uniformly across the major States. The worst affected States in terms of decline in area under tank irrigation are not only Andhra Pradesh, Karnataka and Tamil Nadu but also Orissa, Uttar Pradesh and West Bengal. The relationship between the level of rainfall and tank irrigated area is found to be insignificant in Tamil Nadu over the last forty years. Though the share of tank irrigated area to net irrigated area has declined across different size of farmers, the resource poor farmers (< 2.00 ha) utilize over 55 per cent of tank-irrigated area even today. It appears that the reasons for decline in area under tank irrigation might be different for different States. In view of the importance of tank irrigation in protecting the livelihood security of marginal and small farmers, many urgent measures need to be introduced to rejuvenate the age-old oasis. In so far tank irrigation (minor surface irrigation) has got least preference in terms of financial allotment.[23] Because of this, maintenance and other works could not be carried out on regular basis for a long time, which results in overall reduction in the storage capacity of many tanks. The State should provide more resource allotment to improve the overall condition of tank irrigation. Detailed information about the status of each tank is needed so as to take up maintenance work effectively. Therefore, all the tanks have to be surveyed and classified in terms of its sickness. Once the sickness condition of each tank is identified, financial allotment and treatment for tank rejuvenation can be done easily. The insignificant relationship observed between rainfall and area under tank irrigation suggests that other factors might play a dominant role in determining the area under tank irrigation. Several studies have also underlined that encroachments in tank beds, supply channels as well as in catchment areas are the main reasons for decline of area under tank irrigation. The level of encroachment has increased over the years due to socio-political reasons (Nadkarni, 2000). Therefore, concerted efforts need to be introduced to remove the encroachments. A joint effort by Village Panchayat and State agencies can help to solve all the issues associated with encroachments amicably. A large number of studies have shown that breakdown in traditional irrigation institutions due to conflicts among the villagers is one of the main reasons for the drastic decline in area under tank irrigation in most of the south Indian States. Despite the fact that user's participation is essential for improving the performance of tanks, State agencies have not made any serious effort to revitalize the irrigation institutions by providing support as in the case of canal command areas. Since it is difficult to improve the performance of tanks without users' participation, State agencies should make effort to revitalize the age-old irrigation institutions, which have been maintaining the system over centuries. Despite better rainfall in different years, the flow of water to tank has receded over the years. This is possibly because of various intervention programmes (soil moisture conservation programme, water harvesting programme, etc.,) taken up by different State agencies (forest department, public works department, irrigation department) at the foreshore and catchment areas of tank. While taking the intervention programmes by forest department and others in the upper catchment areas of tanks, it is seldom consulted with the concerned departments/village panchayats that look after tank irrigation. Some kind of coordination among various departments that are directly or indirectly linked with tank irrigation should be established so as to avoid any reduction in the flow of water supply to tanks. Given the fast growing demand for water from different sectors and decline of available water potential, it is essential to protect and rejuvenate all the tanks through all possible means so as to avoid water stress in the future.

NOTES

1. Considering the importance of minor irrigation in the overall development of the rural economy, many new tanks have been constructed since independence in India. However, the newly constructed tanks seem to be not performing well mainly due to faculty design and mismanagement (For details see, Palanisami and Easter, 2000; Vaidyanathan, 2001).
2. Generally, the risk associated with tank irrigation in terms of getting assured water supply for the entire crop season is very high, as the total tank storage is heavily dependent upon monsoon rainfall which is unpredictable. Since most of the tanks are not getting the designed number of fillings, crop failure is common phenomenon during below normal rainfall years. This in turn discourages the farmers not to involve in tank management activities.
3. Recently, using both time series and field level data, Vaidyanathan (2001) along with other scholars have systematically analysed the performance of tank irrigation specifically focusing on south Indian States. Similarly, Palanisami and Easter (2000) have also elaborately discussed the performance of tank irrigation in Tamil Nadu State using both time series as well as filed level data. However, both studies have not looked at the performance of tank irrigation at the all-India level using time series data.
4. Though the study covers the period of 1950-51 to 1999-2000 at the national level, the State level analysis is carried out covering the period only from 1960-61 to 1999-2000. This is done specifically to avoid the overestimate (double counting) of area under tank irrigation among different States as many States were reorganised during fifties.
5. The review has been done taking the studies that have focused the tanks located in the rural regions used mainly for irrigation purpose. It is obvious that the roles of tanks located in urban and rural regions are totally different. While the tanks located in rural regions mainly used for irrigation purposes, the same is not true in the case of tanks located in urban areas.
6. For instance, in Tamil Nadu State, system tanks account for only 9 per cent of the total tanks (39200), whereas the non-system tanks account for 91 per cent. The same trend is observed in Andhra Pradesh and Karnataka as well (see, Vaidyanathan, 2001).
7. An in-depth overview about the tank irrigation in south India can be seen from the edited volume of Vaidyanathan (2001).
8. Kudimaramath system is a traditional village level irrigation institution, which looks after the repair and maintenance of tanks,

distribution of water and other associated activities. The system of Kudimaramath is predominantly practiced in south Indian States like Tamil Nadu and Andhra Pradesh. Though the system of Kudimaramath has weakened over the years due to various problems, it still works relatively well in certain places where alternative sources of irrigation have not yet well developed. To know the historical background of Kudimaramath and its evolution, readers are advised to refer Reddy (1996).

9. Conflicts among the villagers sharing the tank water have been aggravating in recent years. It occurs not only because of problem in sharing tank water but mostly due to caste and other petty problems.

10. Elaborative discussions on this can be seen from Palanisami and Easter (2000); CDF and NET (1998).

11. Unlike other sectors of the economy in India, electricity is presently supplied for agricultural purpose mostly on flat-rate (FR) tariff system (horse power based annual fee). The cheaper tariff system was introduced by many States during the eighties mainly to attract the votes of the farming community, which is numerically large in India. Currently, most of the States supply electricity to agricultural sector by following FR tariff system, while some States (Punjab and Tamil Nadu) provide free electricity too. It is argued that the present system of electricity supply encourages the over-exploitation of groundwater, as the marginal cost of water is near zero under this tariff system. Some arguments on this can be seen from Shah (1993); Saleth (1996); Narayanamoorthy (1997).

12. It is to be noted here that the tanks in India are managed by different agencies depending upon the size of the command area of the tanks. Generally, the minor level tanks are managed by the village level institution (Panchayat Union), whereas the large tanks are managed by the State agency (Public Works Department or Irrigation Department). Studies carried out in Tamil Nadu State using field level data have highlighted some variations in the performance of tanks managed by these two agencies (see, Palanisami and Easter, 2000). It is worthwhile to carry out studies in other States as well to find out whether the performance of tanks managed by different agencies is same or otherwise.

13. Even though the time series data suggests a substantial reduction in tank irrigated area after the introduction of green revolution, it is not always prudent to relate directly tank irrigation with the period of green revolution, as suggested by the anonymous referee of the paper. It is necessary to study as to what has happened to other determining factors (such as investment on tank irrigation, rainfall,

groundwater, cropping pattern, etc.,) of tank irrigation during this period to pinpoint the exact reasons. Importantly, as pointed out by Vaidyanathan (2001), meager distribution of plan outlay to minor irrigation during different plan periods may have also impacted the area under tank irrigation.

14. Though some studies have shown that the development of borewell irrigation has been affecting the tank performance, one may not be able to see this phenomenon commonly across different regions in India. Wells and tanks have different linkages at different regions. Wells have been developed as a supplementary source when the tanks become a source of uncertainty, as most tanks generally supply water only for shorter duration (2-3 months). Only in those regions where tanks are close to urban influence, tanks have been used as full source of domestic water supply as tank irrigation could not be practiced much because the command area was not used for cultivation activities. This has resulted in poor management and in some cases abandonment of tanks. However, wells are still complementing the role of tanks in most regions, except in regions with urban influence.

15. It is not prudent to compare the rate of rainfall with area under tank irrigation at aggregate level due to wide variation in both rainfall and tank irrigation across different regions in India. In order to understand whether area under tank irrigation declines due to variation in rainfall or because of any other factors, ideally one should compare the good rainfall year with area under tank irrigation of that same year at a relatively micro-level (may be at block level). Due to data constraint, we could not relate the block level rainfall with area under tank irrigation. However, since all-India average rainfall was good throughout the nineties, it might have made some impact on tank irrigation. The data presented below show the country level rainfall received during the nineties:

Year	1990	1991	1992	1993	1994	1995	1996	1997	1998	1999	2000
% to Normal Rainfall	119	91	93	100	110	100	103	102	106	96	92

16. The issue of negative impact of groundwater irrigation on tank irrigation also contradicts with economic logic. It is well known fact that the cost of tank water is very low as compared to groundwater at any given point of time. Given the huge cost advantage of tank irrigation, farmers cannot afford to neglect it as argued by many scholars.

17. While analysing area under tank irrigation across different States in India using time series data, one should also look at the definition of tank irrigation used by different States; whether it includes the

area supplemented by wells or not. If so, what discount factor one should use to get the real tank irrigated area is one of the researchable issues that need to be studied. In any case, if one deducts the area supplemented by wells from the total tank irrigated area, the area irrigated by tank might go down further. Unfortunately, we could not check the definition of each State due to data constraint.

18. The livelihood issue of tank irrigation is elaborately discussed in various articles published by CDF and NET (1996).

19. Due to urban agglomeration in different parts of India, tanks situated near the cities are encroached for housing and other non-agricultural purposes. This could be one of the reasons for the drastic reduction of area under tank irrigation in Uttar Pradesh State. Therefore, one must segregate the tanks into two categories as the tanks located near urban centres and rural centres to find out where exactly reduction has taken place in area under tank irrigation. This can be done selectively by taking a few districts where area under tank irrigation declined drastically.

20. As of today, Orissa State has utilised only 6.33 per cent of its ultimate groundwater potential, which is 4.20 million hectares (CWC, 2002). This is one of the lowest among the States in India. Moreover, as of 1999-2000, the share of Orissa State's groundwater irrigated area to its net irrigated area is also very low (40 per cent) as compared to the national average of 59 per cent (CMIE, 2004).

21. Ideally speaking, one may not be able to clearly study the relationship between rainfall and area under irrigation without analysing how the intensity of rainfall has changed over the years. For example, atleast 25 mm of rainfall should fall within one hour to have an effective run-off from rainfall in a moderate catchment. Unfortunately, due to data constraint, we could not study the intensity of rainfall while studying the nexus between rainfall and tank irrigation. The author is thankful to the anonymous referee of the paper for highlighting this point.

22. Tami Nadu State receives about 32 per cent of rainfall through south-west monsoon and about 52 per cent from north-east monsoon. This is totally different from the average rainfall pattern of the country, which receives nearly 74 per cent of rainfall only through south-west monsoon.

23. The distribution of outlay for minor surface irrigation from each five-year plan is not available separately. However, Vaidyanathan (2001) rightly surmises that "Rehabilitation of existing tanks and other small local diversion and lift irrigation facilities which were badly in need of repair as well as the construction of new works in this category obviously had low priority under the plans" (p. 2).

REFERENCES

CWC (1998 & 2002), Water and Related Statistics, Central Water Commission (CWC), Ministry of Water Resources, Government of India, New Delhi.

CMIE (2004), Agriculture, Centre for Monitoring Indian Economy (CMIE), Mumbai, India.

Bardhan, Pranab (2000), "Irrigation and Cooperation: An Empirical Analysis of 48 Irrigation Communities in South India", Economic Development and Cultural Change, Vol. 48, No. 4, July, pp. 847-865.

Barah, B.C. (ed.) (1996), Traditional Water Harvesting Systems: An Ecological Economic Survey, New Age International Publishers, New Delhi.

Barah, B.C.; D.N. Reddy and T. Sudhakar (1996), "Decline of Traditional Water Harvesting Systems", in Barah, B.C. (ed.), pp. 149-162.

Balasubramanian, R and K. N. Selvaraj (2003), Poverty, Private Property and Common Pool Resource Management: The Case of Irrigation Tanks in South India, Working Paper No. 2-03, South Asian Network for Development and Environmental Economics (SANDEE), Kathmandu, Nepal.

CDF and NET (1998), Proceedings of the Seminar on Conservation & Development of Tank Irrigation for Livelihood Promotion, A Joint Publication of the Conservation & Development Forum (CDF) and the Network of Institutions Engaged in Tank Irrigation Management (NET), University of Florida, Florida, USA.

Deshpande, R.S.; L. Venkatachalam and A. Narayanamoorthy (2004), "Policies for Access, Equity and Equality: Physical Resources and Human Resources", in Sundaram, K.V.; M. Moni and M.M. Jha (eds.), Natural Resources Management and Livelihood Security: Survival Strategies and Sustainable Policies, Concept Publishing Company, New Delhi, pp. 401-447.

DHAN Foundation (1997), Renovation of Rainfed Irrigation Tanks with Farmers Involvement in Andhra Pradesh, proceedings of the Consultative Workshop, DHAN Foundation, Madurai, India.

Dhawan, B. D. (2000), Studies in Traditional and Modern Irrigated Agriculture, Commonwealth Publishers, New Delhi.

FAI (various years), Fertiliser Statistics, The Fertiliser Association of India (FAI), New Delhi.

Government of India (GOI) (various years), Indian Agricultural Statistics, Directorate of Economics and Statistics, Ministry of Agriculture, New Delhi.

Janakarajan, S (1993), "In Search of Tanks: Some Hidden Facts", Economic and Political Weekly, Vol. 28, No. 26, June 26, pp. A53-A60.

Janakarajan, S. (1993), "Economic and Social Implication of Groundwater Irrigation: Some Evidence from South India", Indian Journal of Agricultural Economics, Vol. 48, No. 1, January-March, pp. 65-75.

Meinzen-Dick, R. S. (1984), Local Management of Tank Irrigation in South India: Organisation and Operation, No. 3, Cornell Studies in Irrigation, Cornell University, Ithaca, USA.

MIDS (1986), Tank Irrigation in Tamil Nadu: Some Macro and Micro Perspective, Madras Institute of Development Studies (MIDS), Madras.

Mohanakrishnan, A. (1998), "Policy Issues in Tank System Management", in CDF and NET (ed.), pp. 11-16.

Nadkarni, M.V. (2000), "Poverty, Environment, Development: A Many-patterned Nexus", Economic and Political Weekly, Vol.35, No. 14, April 1, pp. 1184-1190.

Narayanamoorthy, A. (1993), "Competition for Borewell Water and its Impact on Tank: Some Observations", proceedings of the 15th Congress on Irrigation and Drainage, The Hague, The Netherlands, International Commission on Irrigation and Drainage, New Delhi, pp. 557-566.

Narayanamoorthy, A. (1996), "Electric Pumpset and Groundwater Management: Macro and Micro Evidence of India", Water Resources Journal, No. 189, June, pp. 23-33.

Narayanamoorthy, A. (1997), "Impact of Electricity Tariff Policies on the Use of Electricity and Groundwater: Arguments and Facts", Artha Vijnana, Vol. 39, No. 3, September, pp. 323-340.

Narayanamoorthy, A. (2002), "Indian Irrigation: Five Decades of Development", Water Resources Journal, No. 212, June, pp. 1-29.

Palanisami, K (1990), "Tank Irrigation in South India: What Next?", ODI/ IIMI Irrigation Management Network Paper 90/2e, London and Colombo.

Palanisami, K. and R. Balasubramanian (1998), "Common Property and Private Prosperity: Tanks Vs. Private Wells in Tamil Nadu", Indian Journal of Agricultural Economics, Vol. 53, No. 4, October-December, pp. 600-613.

Palanisami, K. and K. W. Easter (1991), "Hydro-Economic Interaction Between Tank Storage and Groundwater Recharge, Indian Journal of Agricultural Research, Vol. 46, No. 2, pp. 174-179.

Palanisami, K. and K. W. Easter (2000), Tank Irrigation in the 21st Century: What Next?, Discovery Publishing House, New Delhi.

Palanisami, K. and R. Meinzen-Dick (2001), "Tank Performance and Multiple Uses in Tamil Nadu, South India", Irrigation and Drainage Systems, Vol. 15, pp. 173-195.

Palanisami, K. and C.R. Ranganathan (2004), Value of Groundwater in Tank (surface) Irrigation Systems, Water Technology Centre, Tamil Nadu Agricultural University, Coimbatore, India.

Pundarikanthan, N. V and S. N. Kallapiran (1996), "Challenges in Tank Irrigation", in CDF and NET (ed.), pp. 69-72.

Rao, V. M. and M.G. Chandrakanth (1984), "Resources at the Margin: Tank Irrigation Programme in Karnataka", Economic and Political Weekly, Vol. 19, No. 26, June 30, pp. A54-A58.

Rajeevan, K. (1996), "Irrigation Organisation and Technical Change", in CDF and NET (ed.), pp. 17-24.

Raju, K.V.; A. Narayanamoorthy; G. Gopakumar and H. K. Amarnath (2004), Water Resources: State of the Indian Farmer, Academic Foundation, New Delhi.

Raju, K.V; G.K. Karanth; M.J. Bhende; D. Rajasekar and K.G. Gayathridevi (2003), Rejuvenating Tanks: A Socio-Ecological Approach, Books for Change, Bangalore, India.

Reddy, M. A. (1996), "Kudimaramat" in Barah, B.C. (ed.), pp. 101-119.

Sakurai, T. and K. Palanisami (2001), "Tank Irrigation Management as a Local Common Property: The Case of Tamil Nadu, India", Agricultural Economics, Vol. 25, pp. 273-283.

Saleth, R. M. (1996), Water Institutions in India: Economics, Law, and Policy, Commonwealth Publishers, New Delhi.

Selvaraj, S. Martin and M.P. Vasimalai (1999), "Sharing the Tank Water: Types and Issues in Vallakulam Cascade", Tank Cascade (A Development Journal of Irrigation Tanks in South India), Vol. 1, No. 1, June, pp. 21-30.

Selvarajan, S; A. Ravishankar and P. A. L. Prasanna (2001), Irrigation Development and Equity Impacts in India, Policy Paper: 14, National Centre for Agricultural Economics and Policy Research, New Delhi.

Sengupta, Nirmal (1985), "Irrigation: Traditional vs Modern", Economic and Political Weekly, Vol. 20, Nos. 46-47 (special number), November, pp. 1919-1964.

Shah, Tushaar (1993), Groundwater Markets and Irrigation Management: Political Economy and Practical Policy, Oxford University Press, Delhi.

Shah, Tushaar; R. Seenivasan; C.R. Shanmugam and M.P. Vasimalai (1999), Sustaining Tamil Nadu's Tanks: Field Notes on PRADAN's Work in Madurai and Ramnad, DHAN Foundation, Madurai, Tamil Nadu, India.

Shankari, Uma (1996), "Tanks: Major Problems in Minor Irrigation", in Barah, B.C. (ed.), pp. 131-148.

Sivasubramaniyan, K. (1997), "Irrigation Institutions under Two Major System Tanks in Tamil Nadu", Review of Development and Change, Vol. 2, No. 3, July-December, pp. 355-374.

Thippaiah, P. (1998), Study of Causes for the Shrinkage of Tank Irrigated Area in Karnataka, ADRT Unit, Institute for Social and Economic Change, Bangalore, India.

Vaidyanathan, A. (1994), "A Question of Harvesting Water", Down to Earth, March 15, pp. 30-33.

Vaidyanathan, A. (1999), Water Resources Management: Institutions and Irrigation Development in India, Oxford University Press, New Delhi.

Vaidyanathan, A. (ed.) (2001), Tanks of South India, Centre for Science and Environment, New Delhi.

Vaidyanathan, A and K. Sivasubramanian (2001), "Tank Irrigation in South India: An Overview", in Vaidyanathan, A. (ed.), pp. 5-30.

Vasimalai, M. P. (1998), "A Perspective of the Tank Irrigation Programme in Tamil Nadu, India", in CDF and NET (ed.), pp. 1-10.

Venkatram, B. R. (1980), Administrative Feasibility of Tank Irrigation Authority, Economics Programme, ICRISAT, Hyderabad, India.

22

PARTICIPATORY MANAGEMENT OF LIFT IRRIGATION IN HARYANA

*C.S. Singhal**

INTRODUCTION

The National Water Policy emphasizes on the progressive participation of farmers in the management of irrigation. As a part of the implementation of the water Resource Consolidation Project (WRCP) started in 1994, the Government of Haryana introduced PIM from February 1995. A gradual turnover of the irrigation system at the watercourse level to the Water User Associations for operation and maintenance is the policy of the state Government. The Participatory Irrigation Management attempts to organise farmers and mobilise their skills and resources to encourage them to take an active role in operation and maintenance of watercourses. It has been decided to give priority to rehabilitation and modernisation of watercourses where farmers are willing to form WUAs and take over maintenance responsibilities. For effective functioning of WUAs, the state Government has approved the draft Bye-Laws and MoU in October, 1998.

In Haryana, there are about 6000 watercourses. As per the estimate (HIRMI, 2000) there are 2799 WUAs have been registered

* Associate Professor, NIRD, Hyderabad.

under the Societies Registration Act, 1860 and of them 991 WUAs were formed under the guidance of Command Area Development Agency (CADA) and 1808 WUAs formed by Haryana State Minor Irrigation and Tube wells Corporation (HSMITC) and Haryana Irrigation Department. Till 2000 only 376 WUAs were handed over to farmers.

So far no comprehensive study has been conducted on the Participatory Irrigation Management of lift irrigation area in Haryana. Hence the present study was conducted to study the nature and extent of participation of stakeholders and planning of rehabilitation of irrigation works, operation and maintenance of the watercourses and suggest remedial measures to enlist farmers and officials participation for effective functioning of WUAs under Lift Irrigation in Haryana.

METHODOLOGY

The study was confined to five Water User Associations (WUAs) spread over three villages i.e. Chang, Sai, and Nimriwali in Bhiwani District of Haryana. The purposive random sampling method was used to select WUAs. The sample of the study consists of three functional and two nonfunctional WUAs. These WUAs were formed by HSMITC. The criteria for selection of functional WUAs were (1) Registration (2) Share money deposited by more than 50 per cent share holders (3) completion of lining works of watercourses & its handed over to the Association. The non-functional WUAs were those who had not fulfilled any one of these criteria.

The data were collected through individual interviews and focus group discussion with members of WUAs, executives of the Association, few influential people of the village and members of Panchayat. The small and marginal farmers, scheduled and backward castes were also covered in the sample. About 15-20 people from each WUA area were interviewed individually and in-group after establishment of rapport. About 20 officials in Haryana Irrigation Department, Haryana Irrigation Research and Management Institute were also interviewed to get desired information.

IRRIGATION SYSTEM UNDER WUAs

The lift irrigation improved the quality of life of the people of this area. The formation of WUAs was an effort to enlist farmers participation to improve the condition of water courses, control the water logging and optimum utilization of water, theft and ensure the irrigation water for tail enders.

The stakeholders revealed that as a result of lift irrigation and lining of water courses more water was available for drinking purposes. It benefited in irrigation. As a result of increase in availability of water for irrigation, the cropping pattern has changed as cotton and wheat are also being produced now in this area.

In village Neemriwali, two WUAs were studied namely 4500 L (Dharrn Sinchai Samiti) and 10,500 (Bhaichara Sinchai Samiti). The shareholders of Neemriwali Village informed that coverage of irrigation under these two WUAs did not increase. The irrigation water was reported as available only in one WUA (4500L). Even in this WUA also out of 100 ha command area, in 40 ha no irrigation water was available. However drinking water supply position improved in this area.

In another WUA (10,500) area, water did not came as the construction of Satluz Yamuna Cannal is not constructed by Punjab Government. The another reason of not reaching water was the high level of tail and theft of water.

In Rewari Khera hamlet (tail) of chang village water did not reach the field. The MITC has abruptly left the work in between causing concern to the villagers. In 20 ha. lining is not completed. Only rehabilitation work of old watercourse was undertaken in this village.

In one of WUA (154500L) in Chang village the condition of water course was not good. The water courses were broken at several places causing concern to the farmers. However, being near the head of minor, there was no shortage of water. The productivity of agriculture produce in this WUA's area has improved considerably and the cropping pattern has changed.

In another WUA (148170L) in Sai Village, HSMITC did not take-up any rehabilitation work. So shareholders did not get any benefit of irrigation facilities.

The stakeholders were not aware about the By-laws and MoU. As the HSMITC did not complete the work or not taken up the rehabilitation work as a result most of the shareholders losing their interest in the activities of the WUAs. Another reason for losing interest were non involvement in the decision related to the designing and construction activities and non release of, HSMITC's contribution to WUAs.

The meetings of the WUAs were not regular. In Neemriwali village, MITC left the construction work incomplete. In Rewari Khera Hemlet the leveling of watercourse was not corrected in-spite of reminders by shareholders, de-motivated them. The water flows smoothly only in small part of water course, while in large portion it was not flowing smoothly.

Initially, shareholders were highly motivated and majority of them contributed their share. In Sai village where shareholders were not asked to contribute their share as shown in the Table 1 given below.

Table 1
Financial Contribution by Shareholders under WUAs

Village/WUAs	*Contribution by shareholders (Rs.)*	*Government Contribution*
Neemariwali		
4500 L	1200	NIL
10500	NIL	NIL
Change		
154500 L	8000	NIL
Lohani		
1750 R	33000	NIL
3800 L	30000	NIL
3200	NIL	
Sai		
137981 L	NIL	NIL

Source: Field Study, August, 2003.

This amount was kept in the bank with joint names. In Rewari Khera (tail) Hemlet Shareholders were critical about the fact that officials did not considered their view in designing and even charged shareholder @ Rs. 100/ha for broken water course. There was misunderstanding among officials and shareholders in Rewari Khera. The shareholders of this village were even critical about the leakage of funds in linning. The shareholders cited a case where in about 100 feet linning were shown on papers were not actually undertaken in the field and on repeated request to concerned officials on this matter no action was taken, informed the shareholders. The shareholders were not satisfied with the norms of designing and quality of material. They even tried once to return the sub standard material being used during construction activities. The shareholders were critical about use of quality material in construction activities.

The shareholders were cleaning the water course themselves in 154500 L WUA in chang village. There were not many complain about the Warabandi but theft causing serious problem in availability of irrigation water at tail in the villages under study. Some of the issues emerged out of the study of WUAs in Lift Irrigation are as follows:

POLITICAL WILL

The political leaders at various level can influence and enlist the participation of farmers in irrigation management. At present the political leaders (Members of Legislative Assembly and Parliament) are telling farmers that Government will give full assistance in construction of linning of water courses. The farmers need not pay any thing that many demotivate farmers to contribute even the minimum required contribution they are supposed to make. There should be massive training programme of political leaders and the exposure visit to successful cases of PIM, so that they can appreciate the utility of farmers ' involvement in irrigation management.

DEPENDENCY SYNDROME AMONG FARMERS

The most of the farmers were not interested to contribute anything towards strengthening of WUAs and lining water courses. They expect Government to do every thing. In Bhiwani district in about 40 ha. Government provided sprinkler sets in high land where there was no source of irrigation. Government runs these sprinklers for more than 2 years. Afterwards it was expected to take-over the operation and maintenance of these sprinkler system by shareholders. But none of the farmers came forward on the request of irrigation officials.

When asked for farmers why they could not take up this matter for their own benefits, they expressed their apprehension about the security of sprinkler system, contribution-from fellow farmers for maintenance cost and in such case it was threatening for them to formally takeover the sprinkler system, hence did not come forward to take up the system.

Later, when no farmer turned up to take up the responsibility to run the sprinkler system, the irrigation department has also withdrawn the sprinkler. Ultimately all farmers of the assured irrigation area has became rainfed due to withdrawal of sprinklers by government and lack of initiatives of farmers in operationalisation of irrigation system. Now some farmers are repenting. The individual sprinkler system is successful in the area but group one is not successful.

As a result of lack of understanding and confidence among farmers and officials of irrigation department, there was wastage of sprinkler sets. The farmers disillusioned, developed dependency syndrome and suffer due to wrong promises by politicians.

AMENDMENT OF ACT

At present the PIM Act and Bye-laws are approved by a core-group constituted by Government of Haryana. There is administrative order to form PIM at watercourse level. This is not empowering the farmers to participate in decision making about the construction, operation and maintenance of water courses. The PIM Act should be passed through State Legislative Assembly with well-defined job responsibilities and power of different stakeholders above water course level. The proper financial devolution of power should be given to farmers of WUAs.

CLOSE DOWN OF HSMITC

The Haryana State Minor Irrigation Tube-well Corporation (HSMITC) was created in 1971 for design, construction of watercourses and formation and strengthen of WUAs. The Government of Haryana has abruptly closed the MITC without prior handing over of its liabilities and responsibilities to an other department. This was the major setback in undertaking the watercourse lining and setup and strengthening of WUAs in lift irrigation area of the state. Since the nodal agency i.e. HSMITC has been closed down, no one is found responsible to take care of the proper functioning of the WUAs created by HSMITC.

The Government is now planning to hand over the unfinished work of MITC to CADA, which will take some time to take over the uncompleted construction work and formation of WUAs in the water course level.

BUREAUCRATIC SET UP

The officials were reluctant in apportion their power of construction, operation and maintenance. Hence, they hardly take interest in strengthening of WUAs. As a compulsion or essential requirement they form the WUAs but were found quite ignorant about the Act and By-laws and their specific responsibilities in PIM. There is need to orient concerned officials through intensive periodical training and exposure visit of successful cases within and/or outside state.

There should be clear cut instructions by competent authority about the job of each functionaries involved in implementation of PIM. At the circle level one Officer should be made nodal officer to coordinate PIM activities. Officer should be give feed back also to higher level.

WATER AVAILABILITY IN WATER COURSES

It is found that due to various reasons very often irrigation water is not made available to the farmers thereby farmers are loosing their interest in the WUAs. In Neemariwali, Rewari Khera village (tail village) water is not reaching to the farmers.

In Neemariwali village water should come from Satluz Yamuna link canal (SYLC). But due to non completion of this link canal (SYLC) farmers did not get irrigation water and suffering. The farmers requested to construct the SYLC on priority basis so that they can get assured irrigation. At present due to non-availability of water, farmers are loosing interest in WUAs and agriculture activities have lost proposition as stated by farmers in Neemariwali village.

At time there is political interference in release of water causing unequal distribution of water among the shareholders need to be checked by development of proper norms and by taking stand by concerned officials.

WATER THEFT

Water theft has been recurring phenomenon in Neemariwali and Chang village at small scales. The concerned officials were at time afraid to visit the minor on the complained of the shareholders, as there was political interference. It is difficult for Govenment officials alone to get hold of the theft unless all concerned cooperate. So, the massive awareness should be created, simultaneously WUAs can play a role in containing the water theft if given proper powers.

DESIGN OF WATER COURSES

The design of the water courses as per the prescribed norms was done by engineers alone. This was causing dissatisfaction among the large number of shareholders (farmers). The shareholders should have to be taken into confidence while designing the watercourse-lining etc. so that they own the project on its completion.

Involvement of Shareholders in construction and rehabilitation activities

The construction/rehabilitation work were undertaken by private contractor under the supervision of engineers of HSMITC. The shareholders involvement in supervision and monitoring was ignored by private contractors and officials as reported in village Rewari Khera. This created feeling of hatered and suspicion among the shareholders and officials of HSMITC.

PROVISION OF EXTENSION SERVICES

For optimum utilisation of irrigation water there is need for change in cropping pattern. The agriculture department needs to help farmers in this regard. However, in the field such coordination need to be developed for mass awareness through extension agencies.

ORIENTATION OF OFFICIALS

It was heartening to know that most of the district irrigation officials were not even aware of activities under participatory irrigation management/WUAs etc. Hence there is need for massive awareness/training for officials working at district/block level.

In this regard Haryana Irrigation Research Management Institute (HIRMI) at Kurukshetra need to organise training programmes for all concerned. The core faculty at the HIRMI should be appointed having background of behavioural/ participatory aspect. At present there is no core faculty at HIRMI. Hence, its training and research aspect is very weak, need to be looked into on priority basis.

PRIs AND NON-GOVERNMENTAL ORGANIZATION INVOLVEMENT

PRIs and NGOs could play an important role in mobilisation and in awareness generation and streghtening of WUAs. In the study area no NGO were involved. The role of PRIs were also negligible. Hence PRIs and NGOs should also be involved in strengthening the PIM.

SUGGESTIONS FOR FURTHER IMPROVEMENT

1. The officials, representatives of PRIs and NGOs need to be oriented by organization of training periodically by HIRMI. HIRMI should be strengthen by appointment of core faculties in the areas of behavioural/participatory aspects.
2. Due to close down of HSMITC, the construction and rehabilitation of water courses has been stopped. Government should assign this task to irrigation department/CADA or any other agency to complete the unfinished work in time bound basis.
3. Condition of most of the minors need to be improved so that shareholder can get sufficient irrigation water. Government on priority basis should also take up the rehabilitation of old watercourses.
4. In none of the WUAs majority of farmers contributed their share for operation and maintenance while Government has not contributed their share making the shareholders de-motivated and weak financial aspect. Hence Government should release their contribution to WUAs without further delay.
5. There was hardly any effort to involve shareholders in designing and construction of water courses/rehabilitation activities. Hence necessary changes should be made in the PIM by-laws so that it becomes mandatory to get approval of design and financial expenditure from WUAs.
6. The extension services should be provided to the shareholders in a coordinated way by all concerned departments like agriculture, horticulture, soil conservation, etc.
7. Non-governmental organizations and PRIs should be involved in mobilisation and creation of mass awareness about PIM.

REFERENCES

Singhal, C.S. (2004), Participatory Irrigation Management-A Study of WUAs in Haryana, NIRD, Hyderabad, Research Report.

———-(2004), Farmer's Participation in Lift Irrigation Management in Haryana, International Conference on "Citizenship and Governance Participation, Learning and Social Transformation Re-examining Rhetoric, Action and Impact, Bhu, Varanasi, Priya, New Delhi.

Tanwar, B.S. (2000), Lift Irigation in Haryana, Status Paper, Hirmi, Kurukshetra.

23

MARKET PRICING OF GROUNDWATER IN AGRICULTURE IN KARNATAKA

*P.S. Srikanthamurthy & Others**

GROUNDWATER MARKETS–A RECENT DEVELOPMENT IN KARNATAKA

About 90 per cent of domestic water needs are met by groundwater in Karnataka. In recent years it has been seen that groundwater markets are slowly developing as niche markets endemic in a few pockets. Groundwater markets for non-agriculture purposes, mainly for domestic use, are developing in the semi-urban fringes meeting the needs of urban and semi-urban areas. The groundwater market for agriculture purpose is highly localized and need based. The groundwater sold for agriculture purpose is in larger volume, but the seller realizes much lower value for groundwater than that sold for non-agriculture purpose.

Groundwater markets for agriculture, generate benefits to buyers such as higher and risk-free water-income flows thus increasing cropping intensity, changing crop pattern in favor of

* P.S. Srikanthamurthy, S.C. Deepak, M.G. Chandrakanth and N. Nagaraj Department of Agricultural Economics, University of Agricultural Sciences, GKVK Bangalore, *E-mail: mgchandrakanth@yahoo.com*

high value-low water intensive crops and the use of modern farming technologies. Water markets offer an opportunity to small and marginal farmers who own an irrigation well to increase water utilization for which their own land would not permit and thereby to spread its overhead cost over a larger command area. The cumulative benefits of water markets are greater than the said effects, the benefits mole and more widely distributed if water markets are efficient. Existing literature on groundwater market is partial as it largely covers the situation in Gujarat. In addition, most of the work is on the working and structure of groundwater markets. Some studies have examined the impact of different forms of groundwater market on crop pattern, productivity, equity and employment. A few studied the reliability and efficiency of groundwater markets. But, none have quantified the impact of water management under different forms of groundwater markets.

GROUNDWATER MARKETS AND EFFICIENCY

This study explores the efficiency and equity in groundwater markets by considering sample groundwater market farmers in hard rock areas of peninsular India (Karnataka) who are selling groundwater for agriculture purpose, those selling for non-agriculture purpose (mainly for silk filatures/domestic purpose) and comparing their efficiency of water use with farmers who are using their groundwater on their own farm (not involved in selling groundwater) and farmers who are buying groundwater for agriculture purpose. The Nash Equilbrium describing the bargaining power buyers and sellers of groundwater is also analyzed.

HYPOTHESES

Ambiguous property rights and price variation due to end use pricing characterize natural resource markets. Groundwater markets are no exception to this. This study aims at to reflect the differential groundwater use efficiencies among farmers with the hypothesis that farmers who use groundwater on their farms and also sell groundwater for non-agriculture purpose are relatively more efficient than (i) farmers who are using groundwater for their agriculture and also sell groundwater for agriculture purpose, (ii) farmers who are buying groundwater for agriculture purpose, (iii) farmers using groundwater for their agriculture and not involved in any selling of groundwater for any purpose. The hypothesis underlying the study is that groundwater markets bring efficiency in groundwater use in groundwater scarcity areas, since the scarcity rent for groundwater renders farmers to be relatively more efficient than those who are not involved in groundwater market.

SAMPLING

Field impressions pertaining to GWM (Groundwater markets) activity in EDZ (Eastern Dry zone) of Karnataka, suggested that the GWM activities are intensive here due to high initial investments for bore wells, high risk of well failure and increasing demand for groundwater for agriculture and non agriculture purposes. Sidlaghatta taluk where GWM are discernable as compared to other Taluks of Kolar district is selected. The taluk is classified as 'dark' area (Department of Mines and Geology, Government of Karnataka, 2001) as more than 85 per cent of the recoverable recharge is being utilized. The Sidlaghatta kasaba hobli was selected for this study as it had concentration of water markets reflected in preliminary survey.

Based on intensity and functioning of water markets eight villages were selected randomly from the sample hobli. A sample of 120 respondents were selected for detailed economic investigation, comprising of following categories (Table 1).

Farmers Not Selling Water (FNSW): This sample of 30 farmers are those who own irrigation wells, use their water for irrigation and are not involved in either selling or buying groundwater for irrigation.

Farmers Selling Water For Agriculture (FSWFA): This sample of 30 farmers own irrigation wells, use part of this water for irrigating their crops and also sell part of groundwater for irrigating crops of neighboring farmer buyer. In turn they receive rent for water let out in terms of crop share of 1/3 to ¼ of the value of gross returns realized by the farmer buyer using the purchased water.

Farmers Buying Water For Agriculture (FBWFA): This sample of 30 farmers buy water for agriculture from their neighbor. Thus, FSWFA and FBWFA are a pair of sample farmers of which the first sells and the second buys water for irrigation. Thus, FBWFA may not own any wells, may own irrigation well/s, which may have failed and/or may not be sufficiently yielding groundwater for irrigation.

Farmers Selling Water For Non Agriculture Purposes (FSWFNA): This sample of 15 farmers own irrigation wells use a part of this water for irrigating their crops and sell a part of this water for non-agriculture purposes like use in processing of silk yarn by filatures, for small hotels or for domestic use.

Buyers of Water For Non Agriculture Purpose (BWFNA): This is a sample of 15 buyers buying water from FSWFNA for purposes like processing silk yarn (filature), hotels and households for domestic use.

Table 1
Distribution of Sample Respondents in Groundwater Markets in Sidlaghatta Taluk, Karnataka (2004)

Villages	*FNSW* *No. of sample farmers*	*FSWFA* *No. of sample farmers*	*FBWFA* *No. of sample farmers*
1. Anur	02	02	02
2. Anur Hunasehalli	05	02	02
3. Bachalli	06	05	04
4. Bommanalli	03	03	04
5. Devaramallur	03	03	03
6. Pindipapanalli	02	04	04
7. Sonnenalli	05	03	03
8. Veerapura	04	08	08
Total	30	30	30

YIELD OF IRRIGATION WELL

Bore well was the predominant groundwater extraction structure in study area. All of the 93 functioning wells in the sample, were bore wells, and there were no functioning dug wells. Yield of bore well was estimated by recording the number of seconds to fill a bucket (of known volume) of water. This was extrapolated to obtain the groundwater yield in gallons per hour. The yield of bore well in liters is then converted to gallons per hour by using conversion of 4.5 liters =1 gallon.

Volume of groundwater sold for agriculture by FSWFA (= volume bought by FBWFA) in acre inches is obtained as (yield of well of farmer selling water in Gallons per hour (GPH) X (frequency of irrigation per month of the buyer) X (duration of the buyer's crop in months)/22611.

Volume of water sold for non-agriculture purpose by FSWFNA is obtained as (no. of tankers filled per day) X (tanker capacity in liters) X (no. of days of water sold in a month)X (no. of months where water was sold in a year) X (4.54 liters per gallon)/22611 gallons per acre-inches.

For FSWFA total groundwater extracted includes water used on their own farm plus the water sold for agriculture to buyers.

COST OF PRODUCTION

Government of Karnataka imposed a flat charge of Rs. 300 per HP per year up to 10 HP pump set since April 1997. Current charges are Rs. 525 per HP per year up to 10 HP. Beyond 10 HP pump sets farmers are required to install electrical meters. There were no explicit costs of irrigation as there were no operational costs of payments towards electricity utilization earlier. Accordingly, as no variable cost is incurred by farmers in lifting groundwater other than repair and maintenance of the irrigation pump set and well there would theoretically no charges towards the cost of irrigation. However due to frequent failure of irrigation wells, farmers are forced to invest on additional wells due to externality and to cumulative interference.

Thus the annual cost of irrigation = Amortized cost of irrigation well+ Amortized cost of conveyance + Amortized cost of pump set and electrical installation + annual cost of repairs and maintenance. The amortization is over the average life of irrigation wells in each category of sample. The labor cost of irrigation was merged with the cost of other cultural operations. The annual cost of irrigation pertains to each irrigation well on the farm and this is added across all the wells on the farm. This total cost of irrigation is then appropriated over individual crops according to the volume of water used in each crop. Cost of irrigation per acre- inch = [Total amortized cost of irrigation]/[total acre-inches of water used in that year]. The cost of irrigation is the cost per acre-inch of water multiplied with the number of acre-inches of water applied to the crop.

The proxy for negative externality (per acre inch) is taken as amortized cost per acre-inch of water used by FNSW minus amortized cost per acre-inch of water used by intended category of farmers. The amortized cost per well is the total amortized cost of all wells/Total number of wells. Amortized cost per functioning well is the total amortized cost of all wells/Total number of functioning wells. The proportion of well failure is equal to the Number of failed wells divided by Total number of wells. The net return per acre-inch of groundwater used is equal to Total net returns/Total groundwater used (acre-inches). Cost of cultivation is the summation of cost of human labor, bullock labor, machine hours, seeds, fertilizers, manure, plant protection, bagging, transporting and packing, the apportioned annual cost of irrigation for each crop on the farm. Proportion of well failure, gross irrigated area (GIA), groundwater used per acre, Net returns per acre, amortized cost per functioning well and all wells, irrigation cost per acre of GIA were calculated to analyze economics of irrigation. Negative externality (per well) = Amortized cost per functioning wells minus amortized cost per well.

ECONOMIC PROFILE OF FARMERS INVOLVED IN GROUNDWATER MARKET

Among farmers selling groundwater for non-agriculture purposes, 70 per cent were small farmers. Among buyers using groundwater for agriculture, 70 per cent belonged to marginal and small farmers (Table 2). Thus, groundwater sale for agriculture is by both large farmers (60 per cent) and marginal and small farmers (40 per cent). Considering buyers of groundwater for agriculture, marginal and small farmers are the predominant. Average size of holding of farmers in groundwater market is below 5 acres, being 2.17 acres for farmers selling water for non agriculture purpose, 3.45 acres for farmers buying groundwater for agriculture purpose and 4.71 acres for farmers selling water for agriculture purpose.

FBWFA had smaller total holding size compared to FSWFA. Moreover, FBWFA had 83 per cent of total land holding as dry land. The FSWFNA had smaller total holding size, but 74 per cent of that area was under irrigation. Small land holders with higher proportion of dry land could not afford to invest on well drilling due to their modest resource base, forcing dependency on water markets. Groundwater sellers sold water to all categories of farmers promoting equity in access to groundwater not only for small farmers but also the large farmers. Similar results were reported by Prasad (1991) Raju and Rao (1991). FSWFDP had a smaller holding size of 2.17 acres and still they invested in irrigation wells, as they were located in the city outskirts and had other sources of income, better knowledge and well informed.

Table 2
Socio-Economic Features of Groundwater Seller and Groundwater Buyer Farmers in Sidlaghatta Taluk, Karnataka (2003-04)

Sl. No.	*Particulars*	*FNSW (n=30)*	*FSWFA (n=30)*	*FBWFA (n=30)*	*FSWFNA (n=15)*
1.	Age (yrs)	45.43	42.2	43.93	38.53
2.	Family size (number)	7.47	6.6	5.5	4.63
3.	Work force available per family for agriculture (number)	4.07	4.0	4.03	3.93
4.	Schooling				
	(a) Primary (1-7) (Number)	7 (23.34)	4 (13.34)	10 (33.34)	3 (20.00)
	(b) Secondary (7-10) (Number)	13 (43.34)	21 (70.00)	9 (30.00)	4 (26.67)
	(c) College (Number)	4 (13.34)	2 (6.70)	2 (6.67)	8 (53.34)
	(d) Illiterates (Number)	6 (20.00)	3 (10.00)	9 (30.00)	0 (00.00)
	Total	30 (100)	30 (100)	30 (100)	15 (100)
5.	Total land holding size (acres)	6.12 (100)	4.71 (100)	3.45 (100)	2.17 (100)
6.	Dry land area (acres)	3.00 (49.00)	2.00 (42.46)	2.87 (83.18)	0.42 (19.35)
7.	Area irrigated by irrigation well (acres)	2.23 (36.43)	2.03 (43.09)	0.37 (10.72)	1.60 (73.73)
8.	Orchards (mango/eucalyptus)*	0.89 (14.54)	0.68 (14.43)	0.21 (6.08)	0.15 (6.91)
9.	Marginal farmers (below 2.5 acres)	2 (6.67)	4 (13.34)	8 (26.67)	3 (20)
10.	Small farmers (2. 5 to 5 acres)	8 (26.67)	8 (26.67)	13 (43.34)	9 (60)
11.	Large farmers (above 5 acres)	20 (66.66)	18 (60)	09 (30)	03 (20)
12.	Total	30 (100)	30 (100)	30 (100)	15 (100)

Note: Figures in parentheses show per centage to total.

FNSW: Farmers not selling water, FSWA: farmers selling water for agriculture purpose, FBWFA: farmers buying water for agriculture, FSWFNA: farmers selling water for non-agricultural purposes.

*Considering that 12,10,3 and 2 farmers in FNSW, FSWFA, FBWFA and FSWFNA had mango/eucalyptus orchard, the area is 2.22, 2.04,2.10 and 1.13 acres per farm respectively.

FEATURES OF IRRIGATION WELLS

In the case of FNSW, FSWFA, FSWFDP all the farmers owned functioning bore wells but in the case of FBWFA only 5 sample farmers (17 per cent) owned (Table 3). The sample of 30 FBWFA had eight irrigation wells of which five wells were functioning.

Table 3
Particulars of Irrigation wells of Sample Farmers in Groundwater Market in Sidlaghatta Taluk, Karnataka (2003-04)

Sl. No.	*Particulars*	*FNSW (n=30)*	*FSWFA (n=30)*	*FBWFA (n=30)*	*FSWFNA (n=30)*
1.	Number of farmers owning well	30	30	05	15
2.	Total Number of Functioning wells	38 (82.60)	35 (83.40)	05 (62.50)	15 (83.40)
3.	Total Number of Non functioning wells (includes dug wells and borewells)	08 (17.39)	07 (16.67)	03 (37.50)	03 (16.67)
4.	Total number of bore wells	46 (100)	42 (100)	08 (100)	18 (100)
5.	Average age of wells as on 2003 (years)	10.14	9.23	8.78	8.86
6.	Gross irrigated area using water from own wells (acres)	191.5	186	24.5	73
7.	Gross irrigated area per functioning well	5.03	5.31	4.60*	4.66
8.	Average number of all wells per farm	1.54 (3 wells for 2 farms)	1.40 (4 wells for 3 farms)	0.26 (1well for 4 farms)	1.20 (5 wells for 4 farms)
9.	Average number of functional wells per farm	1.27 (5 wells for 4 farms)	1.17 (7 wells for 6 farms)	0.17 (1 well for 6 farms)	1.0 (1well for 1 farm)
10.	Average depth of the well (feet)	375.55	420.30	381.24	362.67
11.	Average yield of wells (gallons/hr)	2500	2900	1550	2750
12.	Number of farms without wells	00	00	25	00
13.	Amortized cost Per irrigation well	14058	15547	13483	15427
14.	Amortized cost Per functioning irrigation well	15725	17417	17544	16885

Note: Figures in parentheses show per centage to total

FNSW: Farmers not selling water, FSWA: farmers selling water for agriculture purpose, FBWFA: farmers buying water for agriculture, FSWFNA: farmers selling water for non-agricultural purposes

* Includes area irrigated through purchased water and from own wells

Thus, FBWFA suffered the highest failure rate of wells and eventually resorted to buying groundwater from neighbouring farmer. Even though gross irrigated area per well is around five acres and average age of well was 9 to 10 years, for all categories of farmers, wide variation existed in the total area irrigated by all wells under each category as the number of functioning wells varied from 8 wells for 30 FBWFA to 46 wells for 30 FNSW. The negative externality per irrigation well was the highest for FBWFA (Rs. 4061) while for other categories it ranged from Rs. 1500 to Rs. 1900. There were 25 FBWFA whose main source of irrigation was the purchased water. Considering this predicament, the plight of FBWFA has been reduced by FSWFA resulting in equity through groundwater market.

CROPPING PATTERN

Mulberry a hardy low water perennial is the major crop preferred by farmers in groundwater market. The FBWFA had 87 per cent of the gross irrigated area under mulberry cultivation followed by FSWFA (70 per cent), FSWFDP (64 per cent) and FNSW (56 per cent) (Table 4). The area under other crops such as tomato, chillies, carrot, beetroot and potato, which require relatively more water, was higher for FNSW over other categories of farmers. While, there was no conditionality of FSWFA on FBWFA regarding crops to be irrigated with purchased water, both buyers and sellers were relatively efficient in using water compared to FNSW. The area irrigated under perennials was higher for FSWFA compared to FBWFA and thus, water sellers have reliable and sustainable sources of income.

For the three groups of farmers participating in groundwater market (FSWFA, FNSW, FSWFNA), income from farming ranges between 20 and 30 per cent of the annual income, while for FBWFA, the crop farming contributes 15 per cent of their annual income (Table 5). While, FSWFA realized 17 per cent of their annual income from sale of groundwater, FSWFNA realized 41 per cent. FSWFA not only sold higher volume of 47 acre inches per farm for agriculture, but also realized lower returns (of Rs. 29,069 per farm), when compared with FSWFNA who sold half the volume sold by FSWFA (25 acre inches) but realized an income (Rs. 85000) twice that of FSWFA. This illustrates that end use price of the resource and not so much the volume of water sold,

Table 4
Cropping Pattern of Different Categories of Sample Farms in Groundwater Market in Sidlaghatta Taluk, Karnataka (2003-04)

(Area in acres)

Crops (Season Wise)	*FNSW (n=30)*		*FSWFA (n=30)*		*IX. FBWFA (n =30)*		*FSWFNA (n=15)*	
	Net cropped Area	*per cent to total*	*Net cropped Area*	*per cent to total*	*Net cropped Area*	*per cent to total*	*Net cropped Area*	*per cent to total*
			Kharif					
Tomato	14.5	7.57	08.5	4.56	03.0	2.31	04.5	6.16
Chillies	03.0	1.56	07.5	4.03	00.0	00.0	00.0	00.0
Carrot	02.0	1.04	03.5	1.88	00.0	00.0	03.0	4.10
Beet root	10.0	5.22	0.00	00.0	02.0	1.54	00.0	0.00
Maize	5.50	2.87	00.0	00.0	00.0	00.0	00.0	0.00
Ragi	00.0	00.0	00.0	00.0	03.0	2.31	00.0	0.00
Potato	03.0	1.56	00.0	00.0	02.0	1.54	02.5	3.42
			Rabi					
Tomato	06.0	3.13	06.0	3.22	02.0	1.54	02.0	2.73
Ragi	00.0	00.0	02.0	1.07	00.0	00.0	00.0	0.00
Potato	14.0	7.31	03.0	1.61	02.0	1.54	03.5	4.79
Carrot	02.0	1.04	05.5	2.95	1.5	1.18	02.5	3.42
Beet root	06.0	3.13	00.0	00.0	02.0	1.54	00.0	0.00
Maize	02.0	1.04	00.0	00.0	00.0	00.0	00.0	0.00

Contd....

Crops (Season Wise)	FNSW (n=30) Net cropped Area	per cent to total	FSWFA (n=30) Net cropped Area	per cent to total	IX. FBWFA (n =30) Net cropped Area	per cent to total	FSWFNA (n=15) Net cropped Area	per cent to total
			Summer					
Tomato	12.0	6.26	14.5	7.79	01.5	1.18	04.5	6.16
Carrot	02.0	1.04	00.0	00.0	01.5	1.18	02.0	2.73
Ragi	02.0	1.04	00.0	00.0	02.0	1.54	01.5	2.05
Chillies	05.0	2.61	03.0	1.61	02.0	1.54	00.0	0.00
			Annuals					
Mulberry*	35.5	55.61	43.5	70.16	37.5	86.87	15.5	63.69
GIA	191.5	100	186	100	129.5	100	73.0	100
			Perennials					
Mango*	20.0	75.47	14.5	70.73	02	30.76	00.0	00.0
Eucalyptus*	6.50	24.52	06	29.26	4.5	69.23	03.0	100
Total perennia	26.5	100	20.5	100	6.5	100	03.0	100
			Rain fed					
Ragi	58.0	62.36	35	59.32	49	56.98	04.0	61.53
Maize	35.0	37.63	24	40.67	37	43.07	02.5	38.46
Rain fed total	93.0	100	59	100	86	100	6.5	100

Note: FNSW: farmers not selling water, FSWA: farmers selling water for agriculture purpose, FBWFA: farmers buying water for agriculture, FSWFNA: farmers selling water for non-agricultural purposes.

* For Mulberry, 1 acre of crop is considered as 3 acres, for mango and eucalyptus, 1 acre of crop is considered as 2 acres while computing cropping intensity

Sources of Income for Farmers in Groundwater Market

is the key for shaping marginal productivity of groundwater. Groundwater for silk filatures and domestic use fetched much higher price compared to agriculture. The, the sale of groundwater added considerably to the total net returns of the FSWFA. This sale of groundwater which added around 17 per cent of the gross returns is equivalent to the contribution of dairy income. Considering the net returns from different sources of income of FBWFA, FNSW and FSWFA, farmers derived major part of the returns from sericulture, which contributed 30 to 40 per cent of the total net returns. Livestock rearing is main source of income for FBWFA, which adds 46 per cent to the total net returns.

Table 5
Sources of Farm Income in Different Categories of Sample Farms in Groundwater Market in Sidlaghatta Taluk, Karnataka (2003-04)

(Net returns in Rs. Per farm per year)

Source of income per farm	*FNSW (n=30)*	*FSWFA (n=30)*	*FBWFA (n=30)*	*FSWFNA (n=15)*
1. From Farming (irrigated agriculture)	47149 (30.98)	41584 (24.58)	10683 (14.90)	43783 (20.96)
2. From Dairy	26884 (17.67)	29485 (17.42)	32965 (46.01)	9154 (4.38)
3. From Sericulture	61426 (40.37)	58691 (34.69)	20431 (28.51)	45500 (21.78)
4. From Sale of ground water	–	29069 (17.18) from selling 1396 acre inches for irrigation	–	85000 (40.69) from selling 368.46 acre inches for non agriculture purpose
5. From other sources (business, job)	16666 (10.95)	10334 (6.10)	7563 (10.56)	25428 (12.17)
Total annual income per farm	152125 (100)	169163 (100)	71642 (100)	208865 (100)

Note: Figures in parentheses show per centage to total.

FNSW: farmers not selling water, FSWA: farmers selling water for agriculture purpose, FBWFA: farmers buying water for agriculture, FSWFNA: farmers selling water for non-agricultural purposes.

Income from dry land, perennial crops like mango were not considered here as the major focus of the study was dynamics of water markets.

ECONOMICS OF IRRIGATION

The net return per acre of GIA is the highest being Rs. 18345 for FSWFNA followed by Rs. 17009 for FNSW, Rs. 16173 for FSWFA and Rs. 7208 for FBWFA (Table 6). FBWFA are relatively more efficient in using water for irrigation than FSWFA, as they used

Table 6
Economics of Irrigation in Different Categories of Sample Farms in Groundwater Market in Sidlaghatta Taluk, Karnataka (2003-04)

Sl. No.	*Particulars*	*FNSW (n=30)*	*FSWFA (n=30)*	*FBWFA (n=30)*	*FSWFNA (n=15)*
1.	Number of sample farmers	30	30	30	15
2.	Cropping intensity (per cent)	174.8	201.7	163	244.2
3.	Irrigation intensity (per cent)	260	286	272	287
4.	Gross irrigated area (acres)	191.5	186	129.5	73
5.	Gross irrigated area per farm (acres)	6.38	6.20	3.65	5.54
6.	GW extracted and used for farming from own wells (acre inches)	4949	3946	592.25	1100
7.	GW Marketing (acre inches)	–	1395.75 sold	1395.75 pur-chased	368.46 sold
8.	GW Marketing per acre of GIA (acre inches) (=7/4)	–	7.50 sold	10.77 pur-chased	5.04 sold
9.	Total GW extracted/used including purchased/sold groundwater (acre inches) (=6+7)	4949	5341.75	592.25 + 1395.75 =1988	1100 + 368.46 =1468.46
10.	Water extracted per well (acre inches)	107.58	127.16	74.00	81.55
11.	GW used on per farm basis (acre inches)	164.96	131.53	66.26	36.66
12.	GW used per acre of GIA (acre inches) (=9/4)	25.91	21.21	15.41	15.64
13.	Amortized cost of irrigation per acre inch of groundwater from owned irrigation well (Rs.)	131	122	182	189
14.	Amortized cost per well (Rs.)	14058	15547	13483	15427
15.	Total net returns of sample farms (Rs.)	3257250	3008250	933420	1339245
16.	Net returns per farm (Rs.)	108575	100275	31114	89283
17.	Net returns per acre inch of Groundwater used (Rs./acre inch) (=15/9)	658	762	470	1217
18.	Net returns per acre of GIA (Rs.)	17009	16173	7208	18345

Note: FNSW: farmers not selling water, FSWA: farmers selling water for agriculture purpose, FBWFA:
Farmers buying water for agriculture, FSWFNA: farmers selling water for non-agricultural purposes.

15.41acre inches when compared to 21.21 acre inches used by FSWFA per acre of gross irrigated area. Both FSWFNA and FBWFA were the most efficient in using groundwater for irrigation, as they were judicious in using scarce water in irrigating crops at their critical stages of growth giving protective irrigation using around 15 acre inches per acre. This was due to their awareness of the marginal productivity of groundwater due to their exposure to groundwater market. As mentioned earlier, end use pricing prevails in price differentiation for natural resources and groundwater is no exception to this. In corroboration, the net return per acre-inch of groundwater used for irrigation was the highest for FSWFNA, followed by FNSW, FSWFA and FBWFA in that order.

INCOME FROM SALE OF GROUNDWATER

The FSWFA realized Rs. 762 per acre-inch of groundwater from their own farm, and realized Rs. 624 for every acre-inch of groundwater sold (Table 7). This shows that FSWFA, by selling their groundwater, realized almost the same amount of return (Rs. 624) as they could realize by using groundwater on their farm (Rs. 762). Thus, FSWFA gets an additional Rs. 624 per acre inch purely by selling groundwater and this additional Rs. 138 per acre inch is not large enough to cover the additional risks and uncertainty in farming. The notional price realized by FSWFA and FSWFNA per acre-inch of groundwater is Rs. 624 and Rs. 6910 respectively.

The cost of purchased water for agriculture was about 50 per cent of the net returns (Table 8). It is in order to appreciate the contribution of purchased groundwater for FBWFA, who are irrigating 4.29 acres using purchased water for every acre of irrigated area from their irrigation well. On an average, a FBWFA realizes a net return of Rs 7208 per acre of gross irrigated area. The irrigation cost is the highest for FBWFA being two to four times higher than the cost incurred by other categories of farmers. When this cost is compared with the risk of initial well failure and the relatively higher investment and the interest payments on borrowed investment, FBWFA felt it most appropriate to buy water for irrigation. Considering the sale of groundwater by FSWFA and FSWFNA, both the categories sold about 25 per cent of total groundwater extracted by them but realized differential net return due to the price differential obtained, a strong function of the end use. The FSWFA and FSWFNA sold 46.52 and 24.56 acre-inches of groundwater per farm, and 7.5 and 5 acre inches per acre of their gross irrigated area.

Table 7

Net returns from agriculture and from sale of groundwater in groundwater market in Sidlaghatta taluk, Karnataka (2003-04)

Sl. No.	*Particulars*	*FSWFA (n=30)*	*FBWFA (n=30)*	*FSWFNA (n=15)*
1.	Total net returns on sample farms (Rs)	3008250	933420	1339245
2.	Net returns per farm (Rs)	100275	31114	89283
3.	Total net returns including returns from sale of water among sample farms (Rs)	3880320	–	2614245
4.	Net returns per farm including returns from sale of groundwater (Rs)	129344	–	174283
5.	Net returns (per farm) from selling groundwater per year (Rs)	29069	-	85000
6.	Total GW extracted including sold/bought water on all sample farms (acre inches)	5341.75	1988	1468.46
7.	Total GW extracted and used for farming from own irrigation wells on all sample farms (acre inches)	3946	592.25	1100
8.	Total GW purchased/sold (acre inches) on all sample farms	1395.75	1395.75	368.46
9.	Net returns per acre inch of water from farming (Rs/acre inch)	563	470	912
10.	Net returns per acre inch of groundwater from farming and sale of groundwater (Rs/acre inch)	726	-	1780
11.	The addition to the net returns per acre inch of groundwater extracted (Rs)	163	-	868
12.	Net returns per acre inch from using own water in farming (Rs/acre inch)	762	-	1217
13.	Notional price realized per acre inch of groundwater sold (Rs/acre inch)	624*	-	6910**

Note: FNSW: farmers not selling water, FSWA: farmers selling water for agriculture purpose, FBWFA: farmers buying water for agriculture, FSWFNA: farmers selling water for non-agricultural purposes.

*Table (4.15) offers greater details.

** Table (4.16) offers greater details.

Table 8
Economy of Groundwater Buyers for
Irrigation in Sidlaghatta Taluk, Karnataka (2003-04) (n=30)

Particulars	*Quantity*
1. Number of farmers buying groundwater for irrigation	30
2. Gross irrigated area of all sample buyers (acres)	129.50
(A) Gross area irrigated with purchased ground water (acres)	105 (81)
(B) Gross area irrigated with ground water from own irrigation wells (acres) (only 5 farmers among FBWFA owned irrigation wells)	24.50 (19)
(C) Area irrigated with purchased groundwater per acre of area irrigated using own groundwater	4.29
3. Average gross irrigated area irrigated from purchased water (acres)	3.5
4. Net returns per acre inch of Groundwater used (Rs./acre inch)	470
5. Payment made for groundwater purchased per farm per year (Rs)	34745
6. Net returns per farm per year from agriculture after paying for purchased groundwater (Rs)	31114
7. Net returns per acre of GIA per year from agriculture after paying for purchased groundwater (Rs)	7208

Note: Figures in parenthesis indicates per cent.

GROUNDWATER SOLD FOR DIFFERENT PURPOSES

Considering the two groups of farmers FSWFA and FSWFNA selling groundwater, both groups of farmers sold around 25 per cent of the water extracted, but realized different prices depending on the end use to which groundwater was sold (Table 9). The estimated price realized by FSWFA is Rs. 624 per acre inch as they sold for irrigation while FSWFNA realized ten times this price, since they sold groundwater for non agriculture purposes (silk filatures/domestic use). The FSWFNA extracted around 45 per cent lower groundwater extracted by FSWFA, but realized 192 per cent higher net return than FSWFA. Thus, end use and end use price play a key role in determining the volume of groundwater extracted, sold and the revenue realized in water market. FSWFNA selling mainly for silk filatures and for domestic purposes, as well as FSWFA sold around 25 per cent of their groundwater extracted.

INSTITUTIONAL ARRANGEMENTS IN GROUNDWATER MARKET

The only mode of groundwater sale and method of payment for agriculture in the groundwater niche market was through crop share contract. The groundwater price ranged from 1/3rd to 1/4th of the gross value of produce cultivated using the purchased groundwater. About ninety per cent of the buyers opted for one-year contract and ten per cent opted for seasonal contract. Easy monitoring of groundwater supplies, payment, recovery that lead to little or no disputes, were the major reasons for preferring crop share method of payment for groundwater. In addition, this method of share crop contractual payment for groundwater is in vogue in other parts of Kolar district where groundwater niche markets have been in existence since the last ten years as opined by the farmers. About 17 per cent of the sellers also indicated that crop share method offered them remunerative net return to groundwater. However, they did not indicate other alternative modes of groundwater pricing.

Table 9
Details of Groundwater Sold for Different Purposes in Groundwater Market in Sidlaghatta Taluk, Karnataka (2003-04)

Sl. No.	*Particulars*	*FSWFA (n=30)*	*FSWFNA (n=15)*
1.	Total water extracted by all sample farms (acre inches)	5341.75	1468.46
2.	Groundwater extracted per farm (acre inches)	178.05	97.89
3.	Groundwater extracted per acre of gross irrigated area (acre inches)	29	20
4.	Total volume of groundwater sold (acre inches)	1395.75	368.46
5.	Volume of groundwater sold per farm (acre inches)	46.52	24.56
6.	Volume of groundwater sold per acre of gross irrigated area (acre inches)	7.50	5.04
7.	Proportion of groundwater sold (per cent) out of groundwater extracted	26.12	25.09
8.	Net return (per farm) from selling groundwater (Rs.)	29069	85000
9.	Estimated price realized per acre inch of groundwater (Rs per acre inch)	624	6910
10.	Price realized per acre inch of groundwater sold for non agriculture use per rupee of price per acre inch of groundwater sold for agriculture purpose	–	11.07

Note: FSWA: farmers selling water for agriculture purpose, FSWFNA: farmers selling water for non-agricultural purposes.

ESTIMATED PRICE OF GROUNDWATER IN NICE WATER MARKETS ACCORDING TO END USE

Considering the estimated price of groundwater in agriculture (Table 10), in the crop share contract basis, the groundwater price ranged from Rs. 567 to Rs. 656 per acre inch. At the average amortized cost of Rs. 122 per acre inch of groundwater, the estimated price to amortized cost ratio ranged from 4.65 to 5.37. The price of groundwater for non agriculture purposes, ranged from Rs. 4 to Rs. 6.9 per 100 litres of water for purposes like household use, hotels and silk filatures (Table 11).

NASH EQUILIBRIUM MODEL OF GROUNDWATER PRICING

The absence of competitive markets in transaction of groundwater between owners of private irrigation systems and non- owners is widely observed. Topography constrains the deliverable area of water. Capacity of water pump, length of conveyance facilities such as pipelines also limit the deliverable area. Thus, the number of potential buyers that the owners of irrigation systems can expect is physically limited, and the number of potential sellers that the non-owners can expect as well. Usually the owner farmer and non-owner farmer in close location are a pair of seller and buyer for water transactions. The price is determined under bilateral bargaining between them. Thus, the price is not necessarily a competitive one. There is also a concern about water seller's exploitative pricing behaviours towards buyers as the owner is treated as water lord.

Assumptions of Nash equilibrium

- The two players are highly rational
- Both have equal bargaining skills
- Each has full knowledge of the tastes and preferences of the other.
- Each individual wishes to maximize the utility to himself of the ultimate bargain

According to Nash Equilibrium, bargaining power among the buyers and sellers of ground water is the key determinant in ground water markets. Accordingly if the bargaining power of the seller is measured in terms of two variables namely GIA of sellers and total water extracted by sellers, bargaining power of buyer is measured by the GIA of buyers. However as crop share is 1/3rd to 1/4th across different crops, the dependent variable of water price tends to be uniform. Thus, for performing regression analysis, variation in groundwater price as induced by price-cost ratio as equal to price per acre inch of ground water charged by the sellers

Table 10
Estimated Price of Groundwater in Different Crops in Groundwater Market, Sidlaghatta taluk, Karnataka (2003-04)

Crop	*Yield of the crop (tonnes/acre)*	*Groundwater price as crop share (or proportion of crop yield)*	*Quantity of crop share (tons/acre)*	*Ground water bought and used per acre (acre inches)*	*Estimated value of groundwater used per acre as equal to Market value of the crop share quantity (Rs)*	*Estimated price of groundwater sold for agriculture in Rs per acre inch*	*Amortized cost per acre inch of ground water*	*Estimated price - amortized cost ratio*
Mulberry	8.56	1/3	2.85	5.4	3420	3420/5.4= 633	122	5.19
Tomato	23.30	1/4	5.82	19	12460	12460/19=656	122	5.37
Potato	6.21	1/4	1.55	12	6800	6800/12=567	122	4.65
Carrot	6.06	1/4	1.51	11	6795	6795/11=617	122	5.06
Beetroot	6.82	1/4	1.71	9.5	6014	6014/9.5=633	122	5.19

Table 11
Price of Groundwater Purchased for Non-agricultural Purposes from Farmers in Groundwater Market in Urban Fringe of Sidlaghatta town Karnataka (2003-04)

Groundwater buyers	*Number of groundwater buyers in the sample (n=15)*	*Groundwater purchased in litres per day*	*Groundwater bill(Rs/ day)*	*Price per litre of groundwater (Rs)*	*Price per 100 litres of groundwater (Rs)*	*Justification regarding the groundwater price*
1. Silk filatures	08 (53.66)	1400	95	0.069	6.90	Power tiller driven tankers are the main mode of transporting water with tanker capacity ranging from 1400-1500 litres per tanker
2. Hotels	04 (26.34)	1100	70	0.064	6.40	Bullock drawn, cart mounted tankers are the main mode of transport, the tanker capacity ranging from 250 to 375 litres.
3. House holds	03 (20)	400	20	0.040	4.00	Bullock drawn, cart mounted tankers are the main mode of transport with tanker capacity ranging from 250 to 375 litres

Note: Price of the tiller mounted tankers ranges from 90 to 100 Rs per lo

divided by the amortized cost per acre inch of groundwater of the seller, is considered as the dependent variable. This price-cost ratio is taken as a surrogate for market price. Nash bargaining model is thus constructed as:

Ratio of groundwater price to its amortized cost = f (GIA of sellers, total water extracted by sellers, GIA of buyers). Empirically Nash bargaining model was assumed to follow the quadratic price-cost ratio (Y) function: Y = a +b1x1 + b2x2 +b3x3 + c1x12 + c2x22 + c3x23, where, x1 = Gross irrigated area of seller; x2 = Total water extracted by seller

x3 = Gross irrigated area of buyer;

Since price-cost ratio is a dependent variable, elasticity of price-cost ratio with respect to each of the explanatory variables is considered as $(b_i + 2c_i\bar{x}_i)\ \bar{x}_i/\ \bar{y}_i$ (see Table 12).

The results of the above Nash Equilibrium model indicated that for one per cent increase in GIA of the seller, the price-cost ratio increased by 0:098 per cent. For one per cent increase in total water extracted by sellers, price-cost ratio increased by 0.85 per cent and for one per cent increase in GIA of buyers, the ratio increased by 0.3 per cent. Thus, total water extracted by sellers and GIA of buyers are the key explanatory variables in price determination in ground water market.

The Total water extracted by sellers give greater bargaining power for the seller in price determination than the GIA of buyers and GIA of the seller. This also shows that the ground water seller by judiciously distributing the extracted water between his farm and neighboring buyer farm is able to reap larger benefits to the tune of Rs. 29069 per farm.

SUMMARY AND IMPLICATIONS

Considering the volume of groundwater used per acre of gross irrigated area, it was considerably lower for farmers involved in water market viz. 21 acre inches for FSWFA, 15 acre inches for FBWFA and 16 acre inches for FSWFNA when compared to 26 acre inches for FNSW. At the same time, the net returns per acre-inch of groundwater used for irrigation was higher for farmers involved in groundwater market, viz.,. Rs. 762 for FSWFA, Rs.908 for FBWFA and Rs. 1217 for FSWFNA when compared to Rs. 658 for FNSW.

Table 12
Nash Equilibrium Model of Groundwater Niche Market
Dependent Variable: Ratio of Groundwater Price to its Amortized cost (n=30)

Items	*Inputs* *Intercept*	*GIA of seller* (X_1)	*Total water extracted by seller* (X_2)	*GIA of buyer* (X_3)	X_1^2	X_2^2	X_3^2	X^2
Estimated Co-efficient	-1.0938	2.6861	-0.1334	4.1523	-0.208	0.0004	-0.4928	
	(1.7440)		(0.0884)	(2.1150)	(0.1064)	(0.0002)	(0.245)	0.7328
t value		1.5402	-1.5079	1.963	-1.960	2.313	-2.009	

Note: Figures in parentheses are standard errors.

These outcomes indicate that the water markets have helped the farmers in using the groundwater more efficiently. In addition (25) farmer water buyers not owning any irrigation bore wells could buy water for irrigation realized higher net returns, than those owning irrigation wells and 70 per cent of these buyer farmers were marginal and small farmers, is an equity implication.

Groundwater markets have the potential to promote further water use efficiency among sellers and buyers by installing drip/ sprinkler systems, since the seller is aware of higher marginal productivity of water and the buyer is aware that he has to pay for his water through his nose.

There has been increased extraction of groundwater in this region. This has repercussions on sustainability of groundwater use. Hence, to promote sustainability of groundwater use and increasing the accessibility of water to farmers, groundwater literacy should be promoted including: (i) role of desiltation and tank rehabilitation process in augmenting groundwater recharge in hard rock areas, (ii) type of well to be constructed, the depth and volume of availability of groundwater, in relation to the existing and projected interference problems in aquifers where groundwater mining is apparent, (iii) quality, type and HP of the pump to be used, (iv) cropping pattern to be followed in relation to water requirement of crops and the volume of availability of groundwater every year.

By selling water for agriculture and non-agriculture purpose, farmer realized higher marginal productivity of water when compared with farmers who did not sell water. Even with this realization of higher marginal productivity of water, farmers who sold water for agriculture or non-agriculture purposes realized 17 per cent and 41 per cent of their total net return from sale of water respectively. This is an additional income not at the sacrifice of their original returns from agriculture. Thus, groundwater market in groundwater scarce area has not only helped farmers to achieve economic water use efficiency but also complemented additional net return per unit of water extracted. Therefore groundwater markets in rural areas can be facilitated by considering groundwater markets as legal with well defined property rights to water.

Farmers who are able to sell their groundwater irrespective of the purpose, are relatively more efficient than farmers who are just using groundwater for irrigating their own crops. It is crucial to note however that, farmers selling groundwater have not abandoned their agriculture, and have continued to farm in addition to groundwater sale. Thus, farmers participating in groundwater market are more efficient in water use in addition to conserving groundwater than those who are not participating in groundwater market. Thus, groundwater markets are promoting efficiency in groundwater use and in addition are a conservation strategy for the scarce groundwater in hard rock areas.

REFERENCES

Deepak, S.C., (2004), An Economic Analysis of Groundwater Markets in Eastren Dry Zone of Karnataka, Un published MSc(Agri Econ) Thesis, Dept of Agri Economics, UAS, GKVK, Bangalore 560065.

Kajisa Kei and Sakurai Takeshi, (1999), Price Determination under Bilateral Bargaining with Multiple Modes of Contract: a study of Groundwater Market in India. (from google.com)

Saleth, R.M., (1994), Water Institutions in India -Economics, Law and Policy. Commonwealth Publishers, New Delhi.

Shah, T., (1991), Water Markets and Irrigation Development in India. Indian Journal of Agricultural Economics, 46(3): 335-348.

24

SUSTAINABLE MANAGEMENT OF GROUNDWATER USE FOR IRRIGATION: A VILLAGE STUDY

*Sarah Ahmed**

INTRODUCTION

Groundwater has been used as a major source of water by all civilizations for their sustenance. In India, water utilization projections for 2000 put the groundwater usage at about 50% of the total water usage. The National Water Policy (1987) states that water, a basic human need, is a prime natural resource and precious national asset. As such water should be used efficiently by making optimum use of available water through regulation and water-saving technologies like controlled irrigation, instead of continuous floods and innovative participatory approach towards water-management. The policy calls for controls on the exploitation of groundwater through regulation and an integrated and coordinated development and use of surface- and ground water.

Groundwater crisis however has cropped up more as a result of unthoughtful human actions rather than purely due to natural factors. For example, during the past two decades, the water level

* Lecturer in Economics, Department of Economics, Faculty of Arts The Maharaja Sayajirao University of Baroda.

in several parts of the country has been falling rapidly due to a high rate of extraction. The number of wells drilled for irrigation of both food and cash crops has increased rapidly and indiscriminately. India's rapidly rising population and changing lifestyle has also increased the domestic need for water. The water requirement for the industry also shows an overall increase. Intense competition among the users-agriculture, industry, and domestic sectors -has driven the groundwater table to recede rapidly in most parts of India. In addition, the quality of groundwater is also getting badly affected because of saltwater ingress and seepage of polluted surface water. Discharge of untreated wastewater through bores and pollution from unscientific disposal of solid wastes also contaminates groundwater, thereby reducing the quality of fresh water resources.

An uncontrolled use of the bore well technology has led to the extraction of groundwater at such a high rate that often recharge is not sufficient. The causes of low water availability in many regions are also directly linked to the thinning of forest cover and soil degradation further intensifying the problem of evaporation and percolation, which greatly influences the level of water table underground.

Many parts of India, including Gujarat have experienced a fall in water table due to excessive withdrawal to meet the growing demand for irrigation and domestic use and a corresponding slow rate of recharging. The annual rainfall received by India is distributed unevenly across the different states. In Gujarat the rainfall is around 1243mm as per the statistics given by Central Ground Water Board, Ministry of Water Resources GOI [Gujarat State profile]. The available groundwater resources for irrigation is 17.32 billion cubic meter per year (Table 1).

Table 1
Gujarat State

1. Rainfall	1243mm.
2. Total Replenishible Ground water Resource	20.38 BCM/Yr
3. Provision for Domestic, Industrial & Other uses	3.06 BCM/Yr
4. Available Ground Water Resources for irrigation	17.32 BCM/Yr
5. Net Draft	9.55 BCM/Yr
6. In storage fresh water resources	104 BCM

Source: Central Ground Water Board, GOI.

As the natural terrain in the East and Northeast is hilly, the runoff rate of rainwater is high preventing little percolation and therefore with slow recharging, the groundwater possibilities are low; however in the alluvium basin especially of Cambay, the groundwater yield is high, ranging between 75-150 m3/hr.

Farming in Gujarat, especially in central and north Gujarat including Kutch and Saurashtra, mainly depend upon groundwater as a source of irrigation. According to the 1991 District Census Handbook for Vadodara district the gross irrigated area in the district is 156800 hectares of which 131800 hectares is irrigated by wells alone. This shows that well irrigation is the major source followed by tank irrigation.

However over exploitation of groundwater has led to depletion of groundwater tables at an alarming level. This has also resulted increase in the cost of irrigation. In coastal Gujarat depletion of water table has resulted in salt-water ingress in the groundwater aquifers, making the resource unfit for either drinking purpose or for farming. The problem associated with water range from depletion of ground water tables to water logging and rising water tables of polluted and saline water. Therefore over-use or overexploitation of groundwater resource has a double negative impact on the rural economy viz. (1) on the health of rural population due to degradation in quality of drinking water and (2) on the crop production in the form of increase in cost of irrigation as an input and more seriously on non-usability due to excessive salt-content or other mineral content.

In view of these problems it is imperative to use and manage the groundwater resource in a sustainable manner so as to prevent a catastrophe in the rural economy in terms of health and livelihood risks of the rural population.

The National Water Policy actively promotes watershed development as the basis of soil and water conservation. It encourages participatory irrigation management through water-users' association at village level. Even the Ninth Five Year Plan (1997-2002) lays emphasis on recognizing water as a scarce resource to be utilized prudently. It emphasizes the need for promoting the conjunctive use of surface water and groundwater, community participation in the management of water, progressive reduction of conveyance losses, bridging the gap between irrigation potential created and utilized, restoration of old irrigation tanks, and rainwater harvesting. There are several approaches to harness groundwater resource, for example, recharge with imported surface water, recharge with rainwater, vegetative treatment of the Catchments and Domestic Rainwater harvesting. (Tushaar Shah et al, Monograph 2000). In Gujarat, over the years a number of institutional measures in different forms have been adopted and implemented over the years like Participatory Irrigation Management, Irrigation Management Transfer, Water Users Association, Pani Panchayats and the like. However these efforts have resulted in high payoffs but with high costs[Tushaar Shah, 2005].Even legal provisions in different forms which would involve low costs but also result in low payoffs, therefore we hardly see any such provisions being practically implemented.

However, fundamental to the problem of water scarcity is to know how to monitor or regularize the use of water for irrigation, which is the largest user of freshwater. A study in this direction was carried out in two villages of Padra Taluka of Vadodara district. This taluka has 82 villages. It is predominantly an agriculture based region, however a couple of villages are have industries also. In the absence of surface water for irrigation this region depends solely on groundwater resources for irrigation. According to the Statistical information from the Taluka Panchayat Office, Padra, there are a total of 1395 irrigation wells catering to Gross Cropped Area of approx.45690 hectares. The distribution of wells and tube wells is given in Table 2.

Table 2
Distribution of Wells and Tube wells in Padra Taluka

Type of well	*Working condition*	*Closed condition*	*Total*	*Percentage of closed wells/tube wells*
1. Wells	339	303	642	47.19
2. Tube wells	694	59	753	7.83
Total	1033	362	1395	25.94

Source: Taluka Panchayat Office, Padra Dec.2001.

The above table shows that 47% of the total wells and about 8% of the total tube wells are inoperative. This is mainly due to receding level of water table underground. This indicates over-exploitation of the available groundwater resource. Traditionally the cropping pattern consisted of paddy, vegetables and tobacco but of late, the farmers find growing paddy, which is a water demanding crop, an expensive proposition and therefore limit themselves to only tobacco while a few also grow vegetables and cotton, depending on the individual farmers' irrigation resources.

The primary data collected shows that the major crops grown in Ghayaj village are Tobacco and Cotton and in Ranu village the major crops grown are Wheat, Vegetables and Cotton. Almost all the farmers grow hybrid variety of cotton and tobacco. Tobacco has a maturity period of more than 180 days and optimally requires around 5-6 irrigation or around 250-300 mm. Cotton has a maturity period of 6-8 months and requires more frequent irrigation with an interval of 15-20 days. The optimum requirement of water for Cotton is around 500-600 mm of water or around 10 times irrigation with a depth of 50-60 mm per hectare.

A comparison between farms in Ghayaj and Ranu is possible for Cotton crop, which is the common crop in both these villages. It is noted that farmers in Ghayaj village in the farm size 5-9.99 & 10-14.99 use more water in both tube well owned farms as well as non tube well owned farms. Farmers not owning tubewells in the 0-4.99 size class in Ghayaj village however, use much less water as compared to the same category in Ranu village.

To know the volume of water used by the farmers, an analysis was made regarding the delta and the average number of waterings in each size class for both the villages. The results are given in the following tables.

In Ghayaj village, all the respondents use more water than optimally required for tobacco in both the categories i.e. tube well as well as non-tube well owned farms. In the category of tube well owned farms, it is observed that farmers use double the required amount of water for tobacco in size-class 0-4.99,5-9.99 & 15-19.99, whereas in size-class10-14.99 and 20 and more, farmers use almost about four times the required amount of water for tobacco.

In the category of non tube well owned farms, farmers in the size-class of 5-9.99 and 15-19.99 use twice the volume of water required, while farmers in the category of 10-14.99 use three times the required amount of water. A common trend that is observed on both tube well owned and non tube well owned farms is that the farmers in the category of 10-14.99 size-class use water much in excess than the farmers on the other size-class who are also using water that is more than the required amount.

Table 3
Volume of Water Used Crop-Wise, Size-Class Wise and by Ownership of Tube-Wells (Village Ghayaj)

		Tubewell		*Non-tubewell*	
Farm size	*Crop*	*Vol. of water used*	*Vol. of water optimally required*	*Vol. of water used*	*Vol. of water optimally required*
	hect. mts.	*hect. mts*	*hect.mts*	*hect. mts*	
A. 0-4.99	Tobacco	0.69	0.345	13.84	5.72
Cotton	nil	nil	2.05	3.12	
B. 5-9.99	Tobacco	2.7	1.12	12.40	6.08
Cotton	0.08	0.137	nil	nil	
C. 10.14.99	Tobacco	20.29	5.97	7.83	2.61
Cotton	1.21	1.21	nil	nil	
D. 15-19.99	Tobacco	6.42	3.48	2.67	1.11
Cotton	3.93	4.92	0.166	0.25	
E. 20>	Tobacco	17.54	4.71	nil	nil
Cotton	11.73	9.31	nil	nil	
Total (all size)	Tobacco	47.64	15.62	36.74	15.52
Cotton	16.95	15.57	2.21	3.37	

In the case of cotton crop, we find that farmers are very particular in the use of water. All farmers show usage of water that is less than optimally required, except size-class 20 and more, on tube well owned farms using water marginally higher than that required optimally.

In the non-tube well category, only farmers in the marginal farm size class viz.0-4.99 acres and medium farmers in size class 15-19.99 grow cotton. The usage of water is much less than optimally required. Farmers in this village were earlier also growing paddy and vegetables but with the depletion of water table and according to the farmers perception of lack of availability of adequate water, they had to shift to cultivation of tobacco and cotton. However, the analysis reveal that if water is used judiciously for tobacco crop, substantial amount of water can be released for growing other crops, or alternatively huge amount of water can be preserved for future use and maintain sustainability for the future.

In Ranu village, the farmers are growing three crops i.e. wheat, vegetables and cotton. The water table in this village is at a depth of 300 feet. In this village, in the category of tubewell owners, farmers growing wheat are using more water than optimally required, however the deviation from the optimal requirement is not very large. Farmers are using only marginally more than required. For cotton crop, it is observed that farmers across all size-class are growing less than optimally required. However, in the case of vegetables the usage of water is far in excess of that optimally required. Farmers in almost all the size-class are using water more than 3-4 times than that optimally required.

Table 4
Volume of Water Used Crop-Wise, Size-Class Wise and by Ownership of Tube-Wells: Village Ranu

Farm size	*Crop*	*Tubewell*		*Non-tubewell*	
		Vol. of water used hect. mts.	*Vol. of water optimally required hect. mts*	*Vol. of water used hect.mts*	*Vol. of water optimally required hect. mts*
0-4.99	Wheat	0.581	0.249	1.36	0.58
	Vegetables	11.484	4.30	14.39	6.07
	Cotton	1.22	1.52	4.21	3.15
5-9.99	Wheat	1.11	0.666	1.17	0.51
	Vegetables	16.29	4.89	8.02	3.00
	Cotton	1.80	1.80	0.33	0.415
10.14.99	Wheat	1.94	1.16	nil	nil
	Vegetables	17.64	7.22	1.39	0.625
	Cotton	1.67	2.78	1.946	0.834
15-19.99	Wheat	nil	nil	0.55	0.33
	Vegetables	nil	nil	3.25	1.62
	Cotton	nil	nil	1.11	1.39
20>	Wheat	nil	nil	nil	nil
	Vegetables	14.65	4.39	nil	nil
	Cotton	nil	nil	nil	nil
All sizes	Wheat	3.63	2.07	3.08	1.42
	Vegetables	60.06	20.8	27.05	11.31
	Cotton	4.69	6.1	7.6	5.78

In the category of non tubewell owned farms also, the farmers are using more water for vegetables, however the deviation from optimum level is just about twice the amount required. In the case of wheat, across the farm size analysis shows that farmers are using marginally more than that optimally required. However it is surprising to note that the non-tubewell owned farms are using more water than that optimally required in all the size-class except in small size farms of 5-9.99 size-class where the usage of water is less than that optimally required.

The above analysis shows that the use of water is not done judiciously. There is a lot of wastage of water which can be controlled and preserved for future use. The per acre productivity is quite low for tobacco and also for other crops. The use of water is in excess as compared to the optimum levels required, despite the fact that the supply of water has become quite expensive for the farmers because they have to pump up water from a depth of about 300 feet. Many times this burden is shifted on to the non-tube well owners who purchase water from them on hourly basis. This tendency undermines equity aspects among the farmers, as the non-tube well owners who are relatively poor have to bear the burden of increased cost due to water scarcity. The looming fear among the respondents is that the water from the tube well might become inaccessible, judging by the depth to which they have to dig to draw a fair amount of water. Almost all the farmers have reported that just 5-6 years back the depth of tube well required was about 150-200 feet, but at present they have to dig up to a depth of 250-300 feet. This very well indicates that the groundwater resource is fast depleting in this region. Adequate steps need to be taken not only to recharge the aquifers but also to enlighten the farmers in proper usage of water; otherwise agriculture may well become unsustainable. At present there are practically no measures to recharge the wells or gauge the volume of water flooded in the fields. Installation of water meters and change in the system of irrigation from flooding to sprinkler or some such suitable method is required to be done. It is quite evident that the village has almost crossed the level of sustainability and effective measures need to be taken to replenish the wells which are fast drying up. Many of the respondents are also troubled over the fact that after digging new wells at greater depth they find the water salty which is bad for the crop as well as for the fields in the long run. This is specially noticed in Ghayaj village. The respondents feel that the much promised commissioning of the Narmada Sardar Sarovar canals is their last hope of reviving farm production at a sustainable level.

On being asked about any formal or informal arrangement regarding sharing of water, the farmers responded in the negative except that the tube well owners sell water on hourly basis. So the only arrangement prevalent is advance intimation by the non-tube well owners for supply of water to the farms. The tube well owner accordingly decides the timings for the distribution of water to the individual subscribers. This is a form of informal market that exists which in partially regulates the use of water, though not optimally.

The farmers are aware that the groundwater aquifer should be recharged for sustained supply of water in future. On being asked about their opinion on rainwater harvesting, the response was positive with about 50% of the respondents, however with the reservation that making provisions to collect rainwater and supplying it to the field can be expensive for the farmers. Another option which finds favor with a majority of the farmers is the concept of khet talavdi whereby the farmers can have individual talavdis on their own farms to collect the surface run-off; however many of the farmers were apprehensive about its viability. A section of the farmers were not at all receptive to the idea of managing the supply of water through either of the concepts mentioned above. In their opinion, water supply through Narmada canal system is the only solution to the problem.

CONCLUSIONS

1. The groundwater resource in the sampled villages is depleting at an alarming rate.
2. Effective measures need to be taken to recharge the existing but non-functional wells.
3. The system of irrigation should be changed from flooding to drip or sprinkler or some such suitable method according to the crops grown.
4. The farmers can be encouraged to adopt the concept of having khet talavdis on their farms. Also Institutional arrangements for water recharge need to be developed through mobilization of villagers for self-management of the institutions in order to reduce the risks of drying of tube wells.
5. In order to conserve the moisture within the soil and prevent rapid evaporation, thereby reducing the demand for water, the farmers can be encouraged to adopt the practice of green

manuring and mulching. Also the farmers should be informed, through demonstrations about the optimum use of water for different crops.

6. Ghayaj village is in close proximity to the town of Padra, so the groundwater might be accessed more for domestic and industrial use in this town, with the result that the underground aquifers get depleted faster than it is replenished. Therefore attention should also be paid to water-supply and water-use management in this town.

Note: Acknowledgements are due to IWMI-TATA-SPIESR Collaborative research Program 2002 for financial support to the empirical research project entitled: Sustainable Groundwater Use: a village-level study of Padra Taluka, from which part of the data is taken and forms the basis of this paper.

REFERENCES

Tushaar Shah, David Molden, R. Sakthivadivel and David Seckler (2000),: The Global Groundwater Situation: Overview of Opportunities and Challenges, International Water Management Institute.

Tushaar Shah (2005), The New Institutional Economics of India's Water Policy, International workshop on 'African Water Laws: Plural Legislative Frameworks for Rural Water

Management in Africa', 26-28 January (2005), Johannesburg, South Africa

GOI, Ministry of Water Resources, Central Ground Water Board: data from following *website: cgwb.gov.in/gw_profiles/st_Gujarat.htm.*

Ahmed Sarah (2002), Sustainable Groundwater Use: a Village-level Study of Padra Taluka, Research Project, (Unpublished).

25

IRRIGATION AND EQUITY– A REGIONAL ANALYSIS

*Ch. Radhika Rani**

INTRODUCTION

The state of Andhra Pradesh was selected for the study which has a heritage of irrigated agriculture dating back to several centuries. It has three major rivers Krishna, Godavari, Pennar whose dependable flows are 2746 TMC of water of which 1846 TMC has been tapped for irrigation, creating an irrigation potential of 4.8 million hectares under major and medium projects. The major irrigation projects that were taken up across the rivers like Nagarjuna Sagar, Srisailam, Prakasham, KC. Canal can be dated back to 60's and 70's and not much investment has been taken up afterwards in terms of Public investment in irrigation except for some minor irrigation projects. The Net Irrigated Area (NIA) under some of the major and medium projects was also shrinking and there is shift in the area under tube wells in all the three regions resulting in increase in the private investment. Therefore, the data is collected for all the three regions from 1955 to 2005 regarding the area under different sources of irrigation to understand region wise shift in irrigation. The data was also collected size class wise

* Asst. Professor (CAS & DM), NIRD, Rajendranagar, Hyderabad-500 030.

for two periods 1970-71 and 2000-01 to find out whether there is an increase or decrease of area under irrigation between these two time periods. For comparison of equity impacts of irrigation distribution farm sizes are also standardized into five size groups. They are 0 to 1ha, 1 to 2 ha, 2 to 4 ha, 4 to 10 ha, and more than 10 ha as marginal, small, semi medium, medium and large farmers respectively.

SOURCE WISE IRRIGATION

It is seen from Table 1 that the area under tank irrigation in Andhra region as a proportion of GIA has declined from 25% to 13% over the past five decades. Though the area under tank irrigation in Rayalaseema and Telangana regions was more during the decade 1955-65 than the Andhra region, over a period of five decades the area under tank irrigation in these two regions has come down rapidly than the Andhra region. During the decade 1955-56 the area under canal irrigation in Andhra region was 50 per cent of GIA. Whereas, in the other two regions it was 17 and 18 per cent of GIA, during the same period. There was a decline in the area under canal irrigation as a proportion of GIA, in the last decades among all the three regions. However, in Telangana and Rayalaseema regions the area under tank irrigation which was 53 per cent and 33 per cent respectively of Gross irrigated Area was replaced by the area under tube wells and other wells by around 50% of GIA by 2005. Whereas, in the Andhra region the area under, canals and tanks which was 75% of GIA during 1955 was partly replaced by area under tube wells (around 12 per cent) during 2005. This shows that Tank irrigation which was the major source of irrigation of Telangana and Rayalaseema region during 1950s has come down drastically by 2005 which was replaced to a maximum extent by tube well irrigation. Whereas, in Andhra region also canal irrigation has not made much dent after the first decade i.e. 1955 to 65. Infact it has come down by 2005 from 50 to 42 per cent of GIA which was replaced to some extent by tube well irrigation. The net irrigated areas (NIA) as a per cent of GIA has declined gradually in the Andhra region over the past five decades from 82 per cent to 78 per cent. However, the per cent of area irrigated more than once as a proportion of GIA has increased from 19 per cent to 23 per cent over the past five decades in Andhra region.

Table 1
Proportion of Area Irrigated Through Different Sources to the Gross Irrigated Area

Year	*Tanks*	*Project canals*	*Tube wells*	*Other wells*	*Other sources*	*Net Area Irrigated*	*Area irrigated more than once*
			Andhra Region				
1955-65	0.25	0.50	0.005	0.04	0.02	0.82	0.19
1965-75	0.21	0.49	0.02	0.03	0.02	0.78	0.19
1975-85	0.18	0.51	0.05	0.03	0.01	0.80	0.19
1985-95	0.15	0.48	0.55	0.04	0.03	0.79	1.02
1995-05	0.13	0.42	0.12	0.04	0.04	0.78	0.23
			Rayalaseema Region				
1955-65	0.33	0.17	0.00	0.21	0.04	0.77	0.22
1965-75	1.06	0.96	0.00	1.33	0.17	3.48	1.09
1975-85	0.26	0.27	0.00	0.39	0.05	0.97	0.30
1985-95	0.12	0.22	0.08	0.36	0.02	0.81	0.18
1995-05	0.08	0.15	0.32	0.22	0.03	0.80	0.19
			Telangana Region				
1955-65	0.53	0.18	0.00	0.14	0.03	0.76	0.14
1965-75	0.32	0.20	0.00	0.16	0.02	0.76	0.20
1975-85	0.27	0.18	0.00	0.22	0.02	0.73	0.26
1985-95	0.19	0.18	0.04	0.32	0.03	0.78	0.21
1995-05	0.10	0.11	0.22	0.28	0.02	0.75	0.24
			Andhra Pradesh				
1955-65	0.34	0.36	0.00	0.09	0.31	0.83	0.16
1965-75	0.26	0.36	0.01	0.11	0.02	0.78	0.21
1975-85	0.23	0.41	0.03	0.14	0.02	0.85	0.24
1985-95	0.16	0.35	0.06	0.17	0.03	0.79	0.20
1995-05	0.11	0.27	0.18	0.18	0.03	0.76	0.23

Similar trend was observed in Telangana region where there was a decrease in NIA and increase in the area irrigated more than once over the past five decades. Whereas, in Rayalaseema region there was a continuous decline in the both NIA and area irrigated more than once as a proportion of GIA in the last five decades except for a slight fluctuation in between. This shows that the region Rayalaseema has effected more in terms of area under irrigation when compared to the other two regions in the last five decades. For the whole of state the NIA as a proportion of GIA has come down during the last five decades from 83 to 76 per cent. There was a decline in area under all sources of irrigation except tube wells.

REGION-WISE GROWTH IN DIFFERENT SOURCES OF IRRIGATION

Table 2 given below shows that the CGR of area under tanks in Rayalaseema was highest when compared to the other two regions with 61.36 per cent, followed by Andhra and Telangana regions with 14.36 per cent and 2.36 per cent respectively during 1955-65. However, there was a gradual decline in the growth rate of all the three regions in the last five decades. Insignificant increase in the growth rate in the area under tanks has observed in the regions Andhra and Rayalaseema during 1985 to 1995. The CGR of area under canal irrigation in Andhra region during 1955-65 was 18 per cent. While, it was 13 per cent in Rayalaseema and 5 per cent in Telangaṅa regions during the same period. There was some significant growth in the area under canal irrigation in AP in the later decades due to the increase in the availability of water resources through some major and medium irrigation projects on Godavari and Krishna rivers. Particularly, there was some significant increase in the area in Telangana region till 1975-85 with 7.69 per cent. A substantial investment in medium and minor irrigation besides other private sources made an impact in this region during this decade. In Rayalaseema though increase in the growth under canal irrigation was observed till 1985-95 due to modernization of KC canal, the growth was not much significant. Though there was increase in the area under canal irrigation in Andhra region during 1985-1995 it was not significant. The area under canal irrigation has declined to 4.9 per cent, 11.9 per cent and 9.9 per cent in the last decade in all the three regions respectively. This is mostly due to inability of tail end irrigation, failure of drainage system because of improper maintenance of canals and there was not much investment on irrigation during this period. The present irrigation efficiency of the state is only 35-40 per cent. This compares poorly with both the internationally accepted standards of 57 per cent as well as the best practice efficiency of nearly 70 per cent achieved by the superior irrigation systems of Israel. Another issue is the gap between the irrigation potential created (the planned command area) and utilized (the actual irrigated area). Currently this gap is between 20 to 25 per cent Though the state government has proposed to take up Palavaram project on the river Godavari which is proposed cover the entire Godavari districts and also provides drinking water supply to vizag district there was much controversies surrounding it regarding the extent of area it will submerge, impact on environment and the cost of rehabilitation. However, the allocated dependable supply of Krishna water is almost fully harnessed through the construction of several reservoirs and barrages across it. The only river that can still be exploited in the state is Godavari. Out of dependable flows of the Godavari system 720 TMC has been utilized so far. The balance 775 TMC has still to be tapped. To realize the potential of water from Godavari from the proposed Ichampally and Polavarm project 530 TMC can be tapped. And part of this flow can be diverted to Krishna waters which will benefit the Telangana region also.

Table 2
Region-wise and Decadal-wise Growth Rates of Area under Different Sources of Irrigation

Year	*Tanks*	*Project Canals*	*Tube wells*	*Net Area Irrigated*	*Gross Area Irrigated*
		Coastal Andhra			
1955-65	14.31 NS	18.01 NS	92.05 NS	13.83 NS	16.68 NS
1965-75	-4.90 NS	4.62*	35.43 NS	6.03*	2.10*
1975-85	-0.10*	1.62**	9.46 NS	1.41 NS	2.47 NS
1985-95	33.41 NS	33.14 NS	48.71 NS	35.82 NS	36.43 NS
1995-05	-13.03*	-4.90*	9.41 NS	-4.00*	-5.40*
		Rayalaseema			
1955-65	61.36*	13.84*	-9.79 NS	9.07**	7.71*
1965-75	-0.71 NS	2.44 NS	81.34*	3.88*	13.07 NS
1975-85	-38.81 NS	-20.11 NS	34.52*	-24.74 NS	-23.25 NS
1985-95	31.29 NS	30.21 NS	122.06 NS	40.50 NS	42.34 NS
1995-05	-29.91**	–11.67*	16.74 NS	-3.97 NS	-4.17 NS
		Telangana			
1955-65	2.36 NS	-5.11 NS	-84.28 NS	4.54*	4.46*
1965-75	-3.06 NS	21.45 NS	60.78 NS	9.91 NS	14.00 NS
1975-85	-5.00 NS	7.69**	20.43*	0.17 NS	-00.98 NS
1985-95	-6.79 NS	-2.75 NS	7.48 NS	7.68 NS	9.17 NS
1995-05	-8.98 NS	-9.90 NS	22.25**	1.02 NS	0.41 NS

The CGR of area under tube wells was positive and very high in coastal Andhra during 1955-65; with 92 per cent. However, it was negative in the other two regions during this decade with 9.7 per cent and 84 per cent respectively. Though the CGR of area under tube wells was positive and very high during all the periods, it was very high with 48 per cent and 122 per cent respectively in Andhra and Rayalaseema regions during 1985-1995. During the recent decade i.e. 1995-05, the CGR of area under tube wells was more in Telangana region followed by Rayalaseema and Andhra regions respectively.

An interesting observation is the CGR of GIA in Andhra region was positive and high from 1955 to 1995 and it was negative with 5 per cent in the last decade. Similar is the case of Rayalaseema region. In Telangana though the CGR of GIA was positive, it was very low with 0.41 per cent and has declined to a large extent from the period 1985-95.

INTER HOUSEHOLD EQUITY

Generally the tank irrigated command is dominated by small and marginal farmers. There was a steady deterioration in the area under tank irrigation more in Andhra and Rayalaseema regions than Telangana regions. It is seen from the Table 3 that the area under irrigation has decreased for marginal and small farmers of Andhra and Rayalaseema regions during 1970-71 to 2000-01. The area under irrigation for medium and large farmers has increased in these two regions between these two time periods. The situation is reverse in Telangana region where the increase in tube well was more than the other two regions and there was an increase in the area under irrigation for small and marginal farmers and decrease in the area for large farmers between these two time periods. This shows that as the area under canal irrigation as proportion of GIA has increased large farmers are benefiting more with increase in the area under irrigation and as the area under tube wells irrigation as a proportion of GIA has increased, small and medium farmers are benefiting more with an increase in the area under irrigation.

MAIN ISSUES

- The area under tank irrigation as a proportion of GIA has declined rapidly in Rayalaseema and Telangana regions than Andhra region.

Table 3
Region- wise and Class- wise Change in the Area under Irrigation

	Andhra		*Rayalaseema*		*Telangana*		*Total*	
	1970-71	*2000-01*	*1970-71*	*2000-01*	*1970-71*	*2000-01*	*1970-71*	*2000-01*
0-1 ha	0.387747	0.382758	0.351626	0.40768	0.330369	0.438921	0.368674	0.400518
1-2 ha	0.317207	0.277944	0.253351	0.489282	0.211906	0.487526	0.282537	0.355194
2-4 ha	2.466618	2.064423	2.000988	2.000988	1.38172	1.518373	1.959385	1.727558
4-10 ha	5.008397	3.721665	3.816563	3.816563	3.366571	2.509753	4.47559	2.898886
<10 ha	11.76352	12.08296	7.75203	5.054256	5.967491	6.770379	9.547951	7.454183
Total	1.01129	0.770163	0.76603	0.955893	0.748288	0.904573	0.915943	0.837987

* Data is given in Annexure-I.

- The area under canal irrigation has declined in all the three regions but more rapidly in Rayalaseema followed by Telangana and Andhra regions, during the last decade.
- The NIA and GIA has declined in all the three regions but more rapidly in Andhra region followed by Rayalaseema and Telangana regions during the last decade.
- The area under tube wells has increased to a large extent in all the three regions during the last decade and it was more in Telangana followed by Rayalaseema and Andhra regions.
- In terms of area under irrigation the Rayalaseema region has effected more in the last five decades followed by Telangana and Andhra regions.

RECOMMENDATIONS

- The gap between the irrigation potential created and utilized can be bridged through the standardization of canals, drains and field channels so that the entire command area can be irrigated. This will be achieved by ensuring between recovery of operating costs and increasing O&M allocations.
- The participation of the farmers in the maintenance or operation of irrigation schemes started recently with the enactment of irrigation Act of AP 1997. This should be extended to the bore wells also by enhancing investment in digging community bore wells in all the three regions.
- Since the state government is now coming with a water policy involving all the major medium and minor irrigation projects and other sources of irrigation into it, it is the time to come out with a concrete rehabilitation package which should be uniform in all the regions as far as possible and also a policy on afforestation.
- Rehabilitation package should be on land on land basis which eliminates corruption to some extent.
- Priority should be given to Rayalaseema region followed by Telangana and Andhra regions.
- Since the canal irrigation which is a public investment is benefiting more of large farmers and tube well irrigation which is a private investment is benefiting more of small and medium farmers, the other forms of input and output subsidies in agriculture should be more oriented towards small and marginal farmers which ensures some equity.

REFERENCES

Government of India, (1997), Agricultural Statistics at a Glance. Directorate of Economics and Statistics, Department of Agriculture and Cooperation, Ministry of Agriculture, New Delhi.

Jasveen Jairath (1999), "Participatory Irrigation Management Experiment in Andhra Pradesh". EPW 31(40) 2834-2837.

Sampath, R.K. (1992), A Farm-Sizewise Analysis of Irrigation Distribution in India.The *Journal of Development Studies*, Vol. 29 (1): 121-147.

Sampath, R.K. (1990), On Some Aspects of Irrigation Distribution in India. Land Economics, Vol.66 (4): 448-463.

Selva Rajan st. al (2001), Equity Impacts of Irrigation Development in India.

World Bank Technical Paper No. 449 "Transferring Irrigation Management to Farmers in Andhra Pradesh, India, 1999.

Annexure I

Region	0-1 ha		1-2 ha		2-4ha		4-10ha		<10 ha		Total	
	No	Area	No	Area	No	Area	No	Area	No	Area	No	Area
(Holding size) 2000-01												
Andhra	2240545	857586	530059	622746	254740	525891	82390	306628	7124	86079	3114858	2398948
Rayalseema	328692	134001	160432	160823	110095	167858	48560	123054	8165	41268	655944	627012
Telangana	974873	427892	484720	475276	303440	460735	128678	322950	19567	132476	2011278	1819348
Total	3544110	1419479	1175211	1258845	668275	1154484	259628	752632	34856	259823	5782080	4845308
(Holding size) 1970-71												
Andhra	784968	304369	191070	248997	105101	259244	47757	239186	8356	98296	1137252	1150092
Real Seema	174296	61287	38966	44158	20235	40490	7692	29357	1355	10504	242544	185796
Telangana	313280	103498	60318	66386	93744	129528	18373	61854	4491	26800	429888	321680
Total	1272544	469154	290354	359541	219080	429262	73822	330397	14202	135600	1809684	1657568

PART III

DEVELOPMENT AND MANAGEMENT OF DRINKING WATER

ACCESS TO DRINKING WATER IN INDIA STATE AND MARKET INTERVENTIONS

*Samanta Sahu** *& Rajashree Padhi***

INTRODUCTION

Access to drinking water in India has remained a challenge even after five decades of planning and development and despite massive outlays in this sector. Some of the main challenges the country is facing for the provisioning of drinking water are people in the rural areas who do not have assured sources of drinking water, deprivation among the urban population, water pollution and contamination, problem of fluorosis, and presence of high level salinity in water. In addition to this, in the present context it is widely seen that while the State is withdrawing away from its welfare services, privatization of such services including water has become the new *'mantra'* for the policy makers.

* Research Scholar, Department of Political Science, University of Hyderabad, Hyderabad & Research Associate, Administrative Staff College of India, Bella Vista, Hyderabad.

** Research Scholar, Department of Political Science, University of Hyderabad, Hyderabad & Research Fellow, National Institute of Rural Development, Rajendranagar, Hyderabad.

On the other hand, the process of water service delivery in India is one top-down approach from the center to the states, where the third tire of the government remains ineffective. All the programmes and the policies are decided at the central level and passed on to the States for implementation. The problem is even worse in the urban areas where decentralization is non-existent[1]. Furthermore, in the entire process of drinking water provisioning, the role of NGOs, civil society organizations and ultimately the community participation is poorly defined and at times non-existent. The beneficiaries of water services, the consumers and members of community are overlooked while plans and projects are being finalized. Even though there is administrative decentralization, sufficient fiscal and political decentralization is lacking.

It is at this backdrop, the present paper is a modest attempt to understand the debates on various issues determining people's access to drinking water, debates on centraliasation vs. decentralization, and privatization vs. community participation. Further, it tries to identify, allocating huge resources and implementing numerous programmes without community participation and consulting the communities can secure people with access to safe drinking water. Therefore, the central focus of the paper would be, decentralization with community participation could be a better option than centralized State planning and privatization that can bring desired results and provide people with assured sources of access to drinking water in India.

CURRENT SCENARIO

Most studies have shown how iniquitous access to water has been prevalent in India. For example, the National Commission on Urbanization too observed how water supply system was unequal, unjust, and highly biased in favour of the rich. It stated how the wealthiest and often the most politically powerful member of the society had access to drinking water. Due to commodification of water, people having purchasing power (socially affluent people, politically powerful and higher castes) have primary access to water. The poor who constitute the majority lacking purchasing capacity gets little share and in the process water becomes a scarce resource for them. Thus, the 'social construction of power structure' makes water 'artificially scarce' for the poor.

Statistics shows that in rural areas around 50% of households were served by a tube well/hand pump, 26% by a well, and 19% by a tap. Only about 31% of rural households reported having their source of water within their premises, the rest had to go out to fetch their drinking water. (10th five year plan 2002-07: 601). On the other hand, in the urban areas, around 70% of households reported to have been served by tap, and 21% by tube well/hand pump. Around 66% urban households were having their principal source within their premises, while 32% had it within 0.2 km. (Ibid: 634).

Consequently, drinking water has become a 'politically contentious issue'[2]. Water has become a major problem for the common man and the political leaders across their political affiliations try to attract attention by raising the issue of drinking water. The issue of drinking water evokes strong response in the electoral political discourse of the country where election promises have drinking water as the top priority (Water Aid India: 2005).

NATIONAL POLICY FOR DRINKING WATER

Since the dawn of Independence the declaration of Five-Year Plans, the Indian State has taken many initiatives and has launched many programmes and schemes for providing safe drinking water to its people (The programmes are briefly discussed later). Simultaneously, there has been an enormous increase in the plan outlay of the country[3]. There are also several policies drafted at the national level to facilitate the State provisioning of drinking water. Besides several central programmes for drinking water, the central government introduced the 'Technology Mission'[4] in 1986, drafted the National Water Policy 1987, introduced Sector Reforms for water in 2002, and redrafted the National Water Policy 2002. All these major policy documents on water gave top priority to drinking water availability to all.

INSTITUTIONAL MECHANISM FOR DRINKING WATER

Water resource management is a State subject under the Indian Constitution (entry 17, State list). After the 73rd & 74th Constitution Amendment act 1992-93, the provision of drinking water to the people has been delegated as the primary responsibility of the local government for both rural and urban areas for better service delivery. There are numerous institutions dealing with drinking water supply both at the central and State level. Broadly they can be categorized under two heads: (a) Macro level, and (b) Micro level.

(a) Macro Level Institutions: The institutions coming under this category are the Ministry of Health, Ministry of Urban Development (nodal agency for water supply in urban areas), Ministry of Rural Development[5] (nodal agency for water supply in rural areas), Central Water Commission (deals with surface water), Central Ground Water Board (deals with ground water), Central Pollution Control Board (deals with water quality) and many other dealing with rainfall and precipitation, irrigation, hydel power. Apart from this, the other institutional arrangements for drinking water are the Five Year Plans and the budgets under macro level.

(b) Micro Level Institutions: The micro level institutions that deal with drinking water are the local-self government both in the urban and rural areas, i.e. municipalities and the gram panchayats and finally the community itself.

Apart from the above, every State in India has water related departments. Lack of effective and appropriate planning, delay in investment decisions and above all malfunctioning of state run departments has led to a massive water crisis in post-independent India. (Ghosh, 2003: 44).

DRINKING WATER AND THE FIVE-YEAR PLANS

Although at the beginning of Five-Year Plans in India (during early 1950s) it was realized that providing safe drinking water to the people was vital for the development of the country still, the plan outlay was not significant. During the Fifth Five-Year Plan, it was realized that no improvement in the standard of the people could be brought about without providing safe drinking water. From this plan the Minimum Needs Programme (MNP) was introduced and provision of safe drinking water was included in it. During the sixth plan a massive programme was introduced for providing drinking water facilities in the rural areas under the MNP. More than 80 per cent of the 2.31-lakh villages were identified as 'problem villages' (where the sources of water was more than 1.6 km or at the depth of 15 miters or contaminated water) out of which 1.92 lakh were provided with water supply. During 8th plan it was implemented in a more decentralized manner with the involvement of people and local institutions and the role of government was restricted to planning, monitoring and partial financial support.

Significantly, in 9th plan, the role of the government shifted from that of a service provider to a facilitator in policy framework, institutional capacity building and financial reforms. In the light of the Constitution (74th Amendment) Act, while State agencies may continue to plan and implement capital works, the responsibility of distribution may be progressively decentralised to local bodies, and where feasible, to the private sector, within the policy guidelines of the State Governments to strengthen the delivery and management of drinking water. It was also decided that, plan funds would be used not only for direct intervention but also as a leverage instrument. This strategy has to be increasingly adopted to activate and sustain the funds flow from outside the government sector.[6] The Tenth Plan shifted the role of government from direct service delivery to that of planning, policy formulation, monitoring and evaluation, and partial financial support.

Apart from the Five-Year Plans, the government at the national level has from time to time launched various programmes to provide the people with safe drinking water. Some of the programmes can be summarized here;

WATER SUPPLY PROGRAMMES

National Water Supply and Sanitation Programme (NWSSP, 1954)

The National Water Supply and Sanitation Programme (NWSSP) was formulated by the central government during 1954 with an idea to deal with the provision of drinking water supply to groups of villages. It gave priority to areas of greater water scarcity and salinity and those in which water borne diseases are endemic. The expenditure during 1st and 2nd plan under this programme was Rs 33 crores approximately and 11,000 villages were provided drinking water supply through pipes. (Third Five Year Plan, 1961-66, p. 345).

Village Water Supply Programme (VWSP)

Village Water Supply Programme was primarily intended to deal with the rural drinking water supply at the village level. It was to be undertaken at the block level through panchayat samities and village panchayats. Funds for this programme were routed through the organization at the block level. The programme covered primarily the following areas:

(a) Backward areas,
(b) Areas not covered by Community Development Programme,
(c) Pre-extension blocks,
(d) Blocks, which have completed their 1st and 2nd stage in the Community Development Programme. (Ibid, p-346).

Urban Water Supply Programme (UWSP)

Urban Water Supply Programme like the VWSP was implemented in order to provide drinking water supply to the urban areas. It was being executed by municipalities and corporations with loans provided by the central and State governments. A sum of Rs. 89 crores has been provided for Urban Water Supply and drainage during 3rd plan. (Ibid, p. 347).

Minimum Needs or MNP

The programme was incorporated in the 5th plan and among other basic necessities rural drinking water supply was included in it. Under this programme, certain areas were identified as 'problem areas' or 'problem villages'[7] based on the following three guidelines:

(a) Those which do not have assured sources of drinking water supply within a reasonable distance i.e. 1.6 k.m.
(b) Those where the sources of drinking water supply are endemic to water borne diseases like cholera and guinea worms.
(c) Those, which suffers from excess of salinity, iron and fluorides.

Accelerated Rural Water Supply Programme (ARWSP)

The Accelerated Rural Water Supply Programme was introduced as a central scheme to accelerate the programmes of rural drinking water supply in the State sector in 1972-73 as a part of overall programme of special social welfare scheme. Under this scheme, 100% assistance was given to the States/UTs for extending the drinking water supply to areas or villages where water problem was more acute[8]. Preference here was given to villages inhibited predominantly by tribal, SCs/STs and other backward classes (OBCs). (Fifth Five-Year Plan, Vol. II, P. 263).

Rajiv Gandhi National Drinking Water Mission (RGNDWM)

During the mid-term appraisal of the 7th plan (1985-90), with a view of tackling the problem of drinking water supply in rural areas a 'Technology Mission' on drinking water in villages and related management was launched by the Government of India in August 1986. It was launched in order to provide scientific and cost-effective content to the centrally sponsored Accelerated Rural Water Supply Programme. The programme had the following objectives:

(a) To cover all residual problem villages by 1990.

(b) To supply potable water at 40 liter per capita per day (lpcd), and 70 lpcd including cattle demand in desert areas of Rajasthan.

(c) To evolve cost effective technology mix to achieve these objectives within the constraint of plan allocation, and

(d) To take conservation measures for sustained supply of drinking water. (Seventh Five Year Plan, 1985-90, Vol. I, p. 55).

The 'Technology Mission' was later renamed as Rajiv Gandhi National Drinking Water Mission (RGNDWM). (Eighth Five Year Plan, Vol. II, p. 375).

Accelerated Urban Water Supply Programme (AUWSP)

During eighth plan (1992-97), in 1993-94 a new programme was launched for urban drinking water supply. This programme was targeted to small towns with population less than 20,000 in 1991 census, for drinking water supply as they had been excluded or not given due importance earlier. (Ninth Five-Year Plan, Vol. II, P. 265). Funding for this programme was envisaged by central and State governments in the ratio of 50:50.

Swajaldhara (2002)

Reforms in rural drinking water supply were adopted during 1999 and also a few pilot projects launched. They were intended to implement during Ninth Five Year Plan and, with the experience gained there on, the reform initiatives were to be scaled up during the Tenth Plan. They were intended to adopt a demand-driven approach and institutionalize community participation for the sustainability of drinking water supply and sources in rural areas. With this background, on December 2002, Swajaldhara was launched and sector reforms initiatives scaled through out the country. The Sector Reform initiatives primarily suggested the institutional set up to give prominence to the Panchayati Raj institutions at the district and the gram panchayat level. As a result Swajaldhara has two streams: (i) Swajaldhara-I, which has gram panchayat as the lowest unit for implementing reforms initiatives; and (ii) Swajaldhara-II, which has the district as the unit for implementation. Under Swajaldhara panchayats/communities are given powers to plan, implement, operate, maintain and manage all water supply and sanitation schemes and the ownership of drinking water assets are to be with appropriate levels of panchayats.

Jawaharlal Nehru National Urban Renewal Mission (JNNURM)

JNNURM is the latest programme under the auspicious of the Government of India for the 63 cities around the country, which will be funded for a holistic development of these cities. Of the different components, water supply in these cities has been given the most priority area of investment.

Despite several drinking water schemes as discussed earlier there are several number of habitations either partially or not covered and insufficient drinking water facility. There is inequitable distribution and consumption of water among various settlements and towns. The worst victims are the people of below poverty line both in rural and urban areas. Thus, there is a mismatch between State policies, plan outlays and the people's access to potable drinking water. Water accessibility to all has still remained an unfinished goal of the government. Mismanagement of water resources in India by the government machinery has ended up pleading for private investment and withdrawal of State from its primary responsibilities as service provider.

Market Approach for Drinking Water

Over a period of time with a centralized planning and development it was realized that it did not yield the desired results (See Annexure I). Therefore, to overcome this situation there has been policy shift from the omnipotent role of the State to that of a facilitator[9]. The turning point was during 1991-92 when the Indian State opened up its economy for becoming a free market as a part of Structural Adjustment Programme and adopted de-regulation, liberalization, and privatization. Since 1991, large-scale changes have been initiated in India's political-economic scenario with the liberalization, privatization and globalization process. While the process began in 1991 itself in sectors like power, but in the water sector it was little delayed. Following the structural adjustment programmes in India since 1991, the Eighth Five Year Plan (1992-97) made a significant departure from the past while giving a significant role towards privatization in water sector and emphasized the public-private management.

It was declared that 'water has to be managed as a commodity as any other resources' (Eighth five year plan: 379), which was a major shift from earlier plans. Further, private sector efforts for construction and maintenance of drinking water projects were encouraged since the Eighth Plan. In the Tenth Plan (2002-2007) it was further stressed that, water supply to consumers would normally be demand based rather than supply based which the consumers are willing to maintain, operate and finance. This indicates, that those having purchasing capacity will have easy access to drinking water and it's a dream for the people having no purchasing capacity. The features of privatization are also reflected in the National Water Policy (2002), which encouraged private sector participation in planning, development and management of water projects.

The management, operations and financing the water supply, which is traditionally seen as a primary responsibility of the government, is termed as the 'old view'. And consequently, a 'new view' has emerged, on policies for water resource supply on an efficient, equitable and sustainable basis, where water is primarily treated as having an economic value (Briscoe & Garn: 2002).

The privatization of water sector in India has taken place in two ways. First, the outright privatization of water systems by management contracts. The second that is more insidious and have a far-reaching impact is through the water sector reforms. These policies pushed by the donor agencies with an underlying objective of converting potable drinking water from fundamental right to a mere commodity which can not be claimed but can be bought or sold in the market.

The market forces have taken hold and there is a big private sector with many multinationals and transnational corporations operating in the country in water sector. They are operating both in terms of water supply, distribution and maintenance, and bottled water business. It is estimated that, the bottled water business in India is worth Rs. 12,000 crores (Ghosh, 2003: 43). In the process the privatization of water had adverse effects on the poor. The growth rate of the market intervention is predicated upon the rolling back of the State from its main responsibility i.e. to provide safe drinking water to the people. (*Down To Earth,* 15th February 2003)

There is also a debate on the scarcity of water. But, the problem is whether it is scarcity that is denying people drinking water or it is the State policies that have deprived the people. When the State does not have enough water to provide its people for drinking, how has it leased out water bodies like rivers, to the multinational companies to be exploited to their own ends.

A consensus has slowly been build that there is water scarcity and in the interest of water use and efficiency, water needs to be both priced and traded (Lee: 1999, Prasad & Ramachandaiah: 1999). The rationale behind this is, trading in water rights promotes water use efficiency as markets allocates water to the highest valued use (Dinar & Subramanian: 2002). "Nor it is surprising that even today we find it difficult to treat water as a commodity, as an economic good, even when we recognize it as a scarce resource" (Lee: 1999). Thus, the scarcity of water is transformed by the multinational companies and donor agencies into a market opportunity for water corporations (Shiva: 2002).

When State is abdicating away from its welfare role like provisioning of safe drinking water to the citizens, is privatization or marketisation the only answer? In such scenario, when market is controlling the resources, what happens to the equity and social justice aspect (drinking water is a Fundamental Right under Article-21 of the Constitution, Supreme Court: 2000) and the sustainable use of water resources?

A market-based approach to the management of water resources has serious socio-economic implications. Such an approach towards an inevitable necessity of life has put a question mark on the equity and sustainability of it. Moreover there are important issues of equity, social justice and sustainability, which are not the concerns of the market forces (Iyer: 2003).

The depletion and deterioration of water resources has taken place with the rise in the power of transnational corporations driving the local communities and indigenous people over their share of water (Gleick: 2002a). This approach to water delivery, it is further argued has led to the inequality in access to safe water which has increased rather than decreased (Roowen, 2002). Thus, access to water is denied to many because of the policies pursued by the State itself (Prasad & Ramachandaiah: 1999).

THE WAY FORWARD: DECENTRALIZATION AND COMMUNITY PARTICIPATION

Thus, the question arises, is it end or is there any other alternative approach to look into the water problem in order to ensure access to safe drinking water to the people? Further the major issue now is the rights over water or the ownership of the water. Who owns the water resource?

In the ancient days, due to the uniqueness of water to life, water was considered as a 'social assets' (Shiva: 2002, Petrella: 2001, John: 1997). Further, going a step forward, Vandana Shiva (2002) argued that water is a natural right to any human beings, which can be used but not owned. The laws in India recognize only the use rights and not ownership or proprietary rights over flowing water (Iyer: 2003, Milliman: 2002). It is argued that, the original natural rights over water belong to the people and not to the government or the State (Singh: 1992). But in the case of ground water, the ownership of land simultaneously carries the ownership of water (Singh: 1992).

An analysis of two colonial laws in India clearly states, how water a primary possession of the indigenous communities moved in the hands of the State and subsequently became a private property. The Indian Easement Act (1882), states that, "a land owner has the right to appropriate water which is below his or her land and no action will be taken against him even if he intercepts, abstracts or directs water which remains under the land of another" (Mosse, 2003: 245). This Act virtually made the ownership of water, a part of the ownership of land. The community ownership of water was transferred to the private individuals who were landowners. On the other hand, a similar Act, The Forest Act (1894), gave the State the right to acquire forestland and along with it, the underground water resources (Ibid). These two Acts during the colonial period, gradually shifted the 'rights of the communities' to own and access water into the hands of the State to the individual and subsequently private ownership.

Therefore, the primary need in the present context is decentralization and people's participation. On the other hand, in a centralized system of planning and plan execution, everything is planned at the central level without involving the local people. The decisions are made without considering the local context; the real problem faced at the local level and thus it becomes a limited success or sometimes a failure. So, a participatory approach and decentralized decision-making to water management and distribution would make it sustainable for the future[10]. However, decentralization by itself has no meaning, if the members of the community are not able to access information they need and independence in functioning and decision-making. The new coalition of energized and organized communities created will find innovative solutions to the problems at the local level.

The community level participation in water supply and management is often ignored by the international development community (Gleick, 2002b). But, a consensus has strongly been built that public participation is a necessary input for both efficiency and equity because private ownership and operation often neglect social goals (Howe & Dixon, 2002).

The 'International Conference on Fresh Water'[11] asserted that, 'decentralization is the key to sustainable water management'. It states, "local authorities, if delegated the power and means and if supported to build their capacities, can provide for increased responsiveness and transparency in water management, and increase the participation from different groups in the community"(Ghosh, 2003: 38). The countries attending the conference reached a consensus for an integrated water resource management to bring all water users to the information sharing decision-making tables. (Ibid: 39).

It is now amply clear that, the State provisioning of drinking water to the citizens have not yielded the desired results; on the other hand, private provisioning also leads to inequitable access to drinking water. Market solutions to the problem of access to drinking water may not be a viable options in all situations and conditions.

In a nutshell it can be argued that, the State institutions can be strengthened through people's participation for better water service delivery. A stronger State and its decentralized institutions strengthened through people's participation from planning to implementation to the decision-making level would make water accessibility easier and sustainable. This kind of arrangement would empower the people to decide for themselves and what kind of solutions they need for the problem they face at the local context, making it sustainable in the long run.

NOTES

1. Water Aid India, Drinking Water and Sanitation Status in India: Coverage, Financing and Emerging Concerns, Water Aid India, New Delhi, 2005.
2. There are numerous such instances around the country, to sight an instance the recently concluded Assembly Elections in Kerala where the Mail Pool Theme was on Privatization of Water. It can be cited here that one of the significant movement against MNC like Coca-Cola in Plachimada, Kerala has been Continuing for the last several years. The movement Highlights the unrest between State and Community and community ownership over water resource, The Hindu, April 13th 2006, Hyderabad.
3. The outlay in the first five-year plan was Rs. 33.59 billion, which rose to Rs. 7800.00 billion in the ninth five-year plan in the urban areas. It is much more in the rural areas.
4. The Motive of Technology Mission was to provide scientific and cost effective content to the Centrally Sponsored Accelerated Rural Water Supply Programmes, Seventh five year plan, 1985-90, p. 55.
5. The Department of Drinking Water Supply iş the Nodal Department Responsible for Processing the Release of funds under the Pradhan Mantri Gramodaya Yojana for Potable Drinking Water to the Rural Areas in the States yet not Covered or Partially Covered.
6. The Percentage Share Gradually Increased from the 1.28% to 1.38% between First to Eighth Plan and then Dramatically Improved to 2.17% of the total Public Sector Outlay in the Ninth Five Year Plan, 9th Five Year Plan, Planning Commission, Government of India.
7. Out of the total of about 5.76 lakh Villages in the Country, about 1.2 lakh were Identified as Problem Villages. (Fourth Five Year Plan), 1974-79, Vol. I, p-89.

8. The Number of Habitations Covered under it are 26, 803 against a Target of 45,527 and the Population covered is 10.5 million as against a target of 21.6 million till the end of January 2002. Economic Survey 2001-02.
9. During the Period Preceding Indepence till early 1990's, the State in India played the Central role in Planning and Execution in all Sectors. That Changed with the Introduction of Structural Adjustment Programme during early 1990's.
10. Different Community Based Organizations (CBO) and their success in mobilizing people's participation, like the Tarun Bhagat Sangh in Rajasthan, Shradha in Himachal Pradesh, Service Centre in West Bengal, are at the helm of people's participation in water resource management for sustainable use to the future. Anil Agarwal and Sunita Narain, (Ed) Dying Wisdom: Rise, Fall and Potential of India's Traditional Water Harvesting Systems, CSE, New Delhi, 1997.
11. The International Conference on Fresh Water was held in Bonn, Germany, from 3-7 December 2001. It was attended by 118 Countries, 47 International Organizations Including UN and EU and 73 Organizations from the Civil Society.

REFERENCES

———Five-year Plans, Planning Commission, Government of India, New Delhi.

———National Commission on Urbanization, New Delhi.

———National Sample Survey Organization, New Delhi, 1998.

———Rural Water Supply Hand Pumps Projects, The World Bank, Washington D.C., 1986.

———The Private Sector in Water Competition and Regulation, Finance, Private Sector and Infrastructure Network, The World Bank, Washington D.C., 1999.

———Water Aid India, Drinking Water and Sanitation Status in India: Coverage, Financing and Emerging Concerns, Water Aid India, New Delhi, 2005.

Agarwal, Anil and Sunita Narain, (Ed) *Dying Wisdom: Rise, Fall and Potential of India's Traditional Water Harvesting Systems,* CSE, New Delhi, 1997.

Ghosh, Buddhadeb and Prabir De, *India Infrastructure Database, Vol. 1,* Bookwell, New Delhi, 2005.

Ghosh, Gouri Shankar, Kamal Mazumdar, Amarender Kumar Dubey & Jagdish Chander, *Water Supply in Rural India (Policy and Programme),* Ashish Publishing House, New Delhi, 1995.

Iyer, Ramaswamy R., *Water: Perspective, Issues and Concerns,* Sage Publications, New Delhi, 2003

John, Jacob, *Economics of Public Water Supply System: (with Special Reference to Kerala),* Rawat Publications, Jaipur, 1997.

Krishna Iyar, V.K. Water Patriotisms-Plachimada Terrorism, *Mainstream,* Vol. XLII, No. 14, March 27, 2004, p. 7-9.

Krishnan, Rekha, *Swajaldhara*: More Empty Promises, *Economic and Political Weekly,* Vol. 38, No. 10, March 2003, p. 937-939.

Lee, Terence Richard, *Water Management in the 21st Century: The Allocation Imperative,* Edward Elgar Publishing, UK, 1999.

Mosse, David, *The Rule of Water: Statecraft, Ecology, and Collective Action in South India,* Oxford University Publications, New Delhi, 2003.

Mujeri, Mustafa K., and A.V.S. Reddy, (ed), *Rural Water Supply in Asia,* CIRDAP, Dhaka, 1997.

Prasad, Sheela and C. Ramachandaiah, New Economic Policy: Implication for Water, *Economic and Political Weekly,* Vol. 34, No. 21, 1999, p. 1251-56.

Ramachandraiah, C., *Right to Drinking Water in India,* CESS, Hyderabad, Working Paper No. 56, May 2004.

Saleth, R. Maria, (Ed), *The Management of Water Resources–(3): Water Resources and Economic Development,* Edward Elgar Publishing, UK, 2002.

Shiva, Vandana, *Water Wars: Privatisation, Pollution and Profit,* India Research Press, New Delhi, 2002.

Singh, Chattrapati, *Water Law in India,* The Indian Law Institute, Bombay, 1992.

———, *Water Rights and Principles of Water Resource Mnagement,* Tripathi (for the Indian Law Institute), Bombay, 1991.

Vaidyanathan, A. and H.M. Ondshoom (ed.), *Managing Water Scarcity: Experiences and Prospects,* Manohar Publications, New Delhi, 2004.

Table 1
State-wise Household with Drinking Water within Premises –2001 census

Sl. No.	State / Union Territory	*Households with Source of Drinking Water within Premises* Proportion of Households
	India	39.00
1.	Punjab	85.50
2.	Lakshadweep	83.50
3.	Chandigarh	77.20
4.	Delhi	74.90
5.	Kerala	71.60
6.	Goa	61.70
7.	Pondicherry	60.50
8.	Maharashtra	53.40
9.	Andaman & Nicobar Islands	47.10
10.	Gujarat	46.50
11.	Uttar Pradesh	46.00
12.	Sikkim	45.70
13.	Uttaranchal	44.80
14.	Haryana	44.50
15.	Daman & Diu	42.40
16.	Bihar	39.60
17.	Assam	37.90
18.	Rajasthan	32.90
19.	Himachal Pradesh	32.60
20.	Arunachal Pradesh	32.40
21.	West Bengal	32.10
22.	Karnataka	31.70
23.	Jammu & Kashmir	31.60
24.	Andhra Pradesh	31.30
25.	Tamil Nadu	27.10
26.	Madhya Pradesh	24.60
27.	Nagaland	22.90
28.	Tripura	21.80
29.	Meghalaya	20.10
30.	Jharkand	20.00
31.	Mizoram	19.60
32.	Chhatisgarh	19.00
33.	Orissa	19.00
34.	Dadra & Nagar Haveli	15.80
35.	Manipur	11.90

Source: India Infrastructure Database, 2005 (Vol. 1).

Table 2
State-wise Household with Tap as Drinking Water – 2001 Census

	Households with Tap as Source of Drinking Water	
Sl. No.	*State/Union Territory*	*Proportion of Households*
	India	36.70
1	Chandigarh	91.90
2	Pondicherry	89.30
3	Himachal Pradesh	84.10
4	Andaman & Nicobar Islands	76.20
5	Delhi	75.30
6	Daman & Diu	72.90
7	Sikkim	70.30
8	Goa	69.00
9	Arunachal Pradesh	67.80
10	Uttaranchal	65.90
11	Maharashtra	64.00
12	Tamil Nadu	62.50
13	Gujarat	62.30
14	Karnataka	58.90
15	Jammu & Kashmir	52.50
16	Haryana	48.10
17	Andhra Pradesh	48.10
18	Nagaland	42.00
19	Rajasthan	35.30
20	Meghalaya	34.50
21	Punjab	33.60
22	Mizoram	31.90
23	Manipur	29.30
24	Dadra & Nagar Haveli	28.20
25	Madhya Pradesh	25.30
26	Tripura	24.60
27	Uttar Pradesh	23.70
28	West Bengal	21.40
29	Kerala	20.40
30	Chhatisgarh	15.50
31	Jharkand	12.60
32	Assam	9.20
33	Orissa	8.70
34	Bihar	3.70
35	Lakshadweep	3.10

Source: India Infrastructure Database, 2005 (Vol. 1).

27

DECENTRALISATION OF DRINKING WATER AND ROLE OF PANCHAYATS

*Y. Bhaskar Rao**

INTRODUCTION

The Government has accorded the highest priority for rural drinking water to ensure universal access as a part of policy framework. Private sector investment or community support were absent towards this end. India has declared 2003 as the 'Year of Freshwater' in keeping with the United Nations Resolution declaring 2003 as the 'International Freshwater year'.

In 1986, the Government of India created the National Drinking Water Mission and subsequently, renamed it as the Rajiv Gandhi National Drinking Water Mission in 1991, with a mandate to provide finance, policy and technological support and develop a multi pronged approach relating to the rural water supply.

Sources of water in the country include, rainwater, snowfall, the river systems and ground water. Rainfall in India is dependent largely on the southwest and northeast monsoons, on shallow

* Asst. Professor (CPR), NIRD, Hyderabad

cyclonic depressions and disturbances and on violent local storms. In India, the per capita average annual freshwater availability has been reduced from 5,177 cubic meters in 1951 to about 1,869 cubic meters in 2001 and is estimated to further come down to 1,341 cubic meters in 2025 and 1,140 cubic meters in 2050. This is a cause of concern for all of us. About 85% of the drinking water systems are based on ground water sources, and utilizes only 5% of total groundwater extraction. The ground water depletion has also aggravated deterioration of its quality due to excess fluoride, arsenic, iron and brackishness, in many parts of the country.

The rapid and accelerated use of ground water to meet demands from various sectors has led to an alarming decline in ground water level in many areas and consequent stress on ground water resources. The tubewell and pump technology has been responsible for raising agricultural production and meeting the domestic and industrial water needs. Over-exploitation of ground water and non-completion of surface water schemes have resulted in marked lowering of ground water levels. The situation is grave in seven per cent of the country's area where the stage of ground water depletion exceeds 85 per cent of the annual replenishible recharge. Besides, it has reached the critical stage in 11 per cent of the area where over 65 per of the ground water is developed. The major impact may be seen in Rajasthan, Gujarat, Andhra Pradesh, Tamil Nadu, Karnataka, Madhya Pradesh, Orissa, Haryana etc. In general the demand for fresh water from different sectors has been growing against a finite available quantity, resulting in scarcity situation spread over 1203863 sq. km. in 1205 blocks of 223 desert and drought prone districts. Further, a study shows that per capita availability of water, which was around 5000 cubic meter per year at the time of independence, has drastically fallen to 2000 cubic meter per year at present. Apart from the increasing mismatch in supply–demand scenario caused mainly due to rapidly growing population and ever increasing desire for higher levels of water supply, gross mismanagement of water resources, coupled with negligence of traditional systems and also the practice of collection and use of rainwater by individuals/communities for their domestic use, further aggravated the situation in the form of falling water levels and its quality and prevalence of water related diseases in these areas.

FINANCIAL FEASIBILITY

Financial viability is essential for sustainable service delivery in the Sector. An estimate shows that providing safe drinking water to all rural habitations in the country with minimum satisfactory level of service would require an investment of about Rs.450 billion. Further, at the present price, about Rs.2000 Crore per year would be required to keep the existing assets functional. The permissible O & M expenditure at this stage is only up to 15% of the State and Center funds put together. Hence, the resource gap between available funds and estimated total expenditure would work out to the tune of Rs.1300 Crore. In view of increasing constraints, it would be a difficult task for the Government alone to mobilize such huge investment single handedly following "supply driven" approach. Therefore, a productive and pro-active involvement of user community and other stakeholders in resource mobilization and community participation in the implementation process as well as O & M of the systems has become imperative. To translate the above strategy into practice, it requires massive efforts to transform the prevailing mindset of the major partners–the users, community, PRIs Government Agencies, NGOs etc.

POLICY INTERVENTION

Institutional arrangements for implementation and delivery of services are critical from its sustainability point of view particularly with community based approach. The institutional arrangement may be adopted depending upon local situation from Gram Panchayat System/VWSC, NGO with technical support needed from Government at the local level. The Institution's capability needs to be built-up properly for the success of the system. The local setup of institution should be up-linked with the district and State level department/institution responsible for monitoring the programme on behalf of the State Government. The district/ state level departments should shoulder responsibility to disseminate the information of latest developments and technologies elsewhere.

After the 73rd & 74th Amendments to the Constitution of India, the subject of rural water supply was placed under the Panchayati Raj Institutions (PRIs–rural local bodies) making them responsible for it in their respective areas. Government of India is supporting the process of empowerment and capacity building of the PRIs to enable them to discharge their responsibilities. In every village, Gram Sabha, the general constitutional body of the village having all adult villagers as members, has to be fully involved in reviewing the implementation and management of drinking water systems and sources. However, there is need of adequate legislation to provide the basis for effective institutional arrangement for efficient and equitable water allocation and administration without overlaps/gaps to avoid confusion and jurisdictional uncertainties to facilitate consistent enforcement. It should also cover legislators to be well informed about water related issues because water resource development and management is a highly specialized area, particular effort is required to present materials to decision makers in a form that is easily used.

There is also need for an appropriate level of framework for inter-agency coordination particularly as competition for water increases and the fragmentation of responsibilities do exist supported with inadequate data collection, dissemination arrangement and week coordination mechanisms on water related issues. World Bank, in its Review Report, (1998), has made the following findings of much interest on India's Water Resources Management Sector:

(i) Water is becoming an increasingly scarce resource in India, its finite and fragile water resources are depleting, yet it continues to be used inefficiently on a daily basis in all sectors, while various sectoral demands are growing rapidly.

(ii) The current approach emphasizes development of water resources and construction of new infrastructure under a top-down, supply-oriented and fragmentary framework.

(iii) The present institutional arrangement in India, including central, state and local institutions, and both formal and informal structures, do not enable comprehensive water allocation, planning and management.

(iv) Existing organizations, furthermore, lack capacity in key management areas as well as effective mechanisms for implementation.

(v) Appropriate economic incentives for efficient water use and conservation are lacking on various levels, thereby impacting negatively on water provision and usage in these sub-sectors.

(vi) The absence of appropriate direct water pricing, and lack of adequate application of other economic and financial incentives at the sub-sector level has also served as an obstacle to the smooth transfer of water between sectors and states.

(vii) Supporting technological and informational systems, to enable effective planning and management of water are also weak.

(viii) The current situation in water service delivery in India is, in general, characterized by a vicious circle of inadequate financial allocations to the sector (particularly for O & M) and inefficient and bloated service institutions, which have led to poor quality and unreliable services, user dissatisfaction with their services, and an unwillingness of users to pay for those services. The inadequate resources generated by the sector due to low prices and user unwillingness to pay for services further undermine sector financial resources contributing to a perpetuation of the circle. The end result is a sector that has become unsustainable.

(ix) The vicious cycle of the drinking water sectors (both rural and urban) varies due to the varied and desegregated institutional structures that make up the RWSS and UWSS sub-sectors within the states in India.

Further, GWSS Assessment Report–2000 identified similar sector constraints viz. (i) financial difficulties, (ii) institutional problems, (iii) inadequate human resources, (iv) lack of sector coordination, (v) lack of political commitment, (vi) insufficient community involvement, (vii) inadequate O & M, (viii) poor water quality, (ix) insufficient information & communication and lack of hygiene education are equally applicable to India.

PROGRAMME SUSTAINABILITY

Rural water supply is a State subject. States have been taking up projects and schemes for the provision of safe drinking water from their own resources. However, recognizing the importance of providing safe drinking water in rural habitations, Government of India has been providing financial assistance to State

Governments under the Centrally Sponsored Scheme *"Accelerated Rural Water Supply Programme"* (ARWSP) will be working with the objectives like.

- To ensure coverage of all rural habitations with access to safe drinking water;
- To ensure sustainability of drinking water systems and sources;
- To tackle the problem of water quality in affected habitations;
- To institutionalize the reforms initiative in the rural drinking water supply sector.

Coverage Norms

- 40 lpcd of drinking water for human beings;
- 30 lpcd of additional water for cattle in areas under the DDP;
- One hand pump or stand post for every 250 persons; and
- Availability of water source within 1.6 km. In plains and 100 meters elevation in hilly areas.

Funding Pattern

- Under ARWSP, funds are provided to States for making provision of safe drinking water in rural habitations. 15% of these funds can be spent on operation and maintenance (O &M) of the existing drinking water system and sources. State Governments should match funds released by this Department on 1:1 basis.
- Upto 20 per cent of the funds can be utilized by the State Governments (a) to take up projects under the Sub-Mission programme to tackle water quality problems like fluorosis, arsenic, brackishness, et. And (b) to ensure source sustainability by conserving water, recharging aquifers, etc. Out of this, 15% should be utilized for projects for quality and 5% for projects for sustainability of sources. Funding is done in the ratio of 75:25 between Central and State Governments.
- 20 per cent of the annual outlay of ARWSP has been earmarked for implementation of Swajaldhara and Sector Reform Projects. The funding pattern for these projects is 90% from Government of India and 10% by way of community contribution.
- About 5 per cent of the annual allocation is also earmarked to States covered by the Desert Development Programme (DDP) under ARWSP. This is funded as 100% Central grant to the States.

- Further, 5% of the ARWSP allocation is earmarked to meet the contingencies arising due to natural calamities and emergent situation. This is funded at 100% grant from the Central Government.

IMPLEMENTING AGENCIES

State Governments decide the implementing agencies for the programme. The agencies may be the Public Health and Engineering Department (PHED), Rural Development Department or the Panchayati Raj Department. Implementation is also taken up by the Government Boards/Nigams/Agencies in some States, for example, the Gujarat Water Supply and Sewerage Board (PHED), Jal Nigam in Uttar Pradesh and Water and Drainage Board in Tamil Nadu.

VILLAGE WATER AND SANITATION COMMITTEE (VWSC)

As per the 73rd Amendment to the Constitution of India, the subject of rural water supply vests with the Panchayati Raj Institutions (PRIs). The Panchayats and VWSC are to play a major role in providing safe drinking water and managing the systems and sources in their respective areas. They can be involved in the implementation of schemes, particularly in selecting the location of hand pumps, stand posts and spot sources; in Operation and Maintenance (O & M), etc. Government of India has laid emphasis on empowering and capacity building of the PRIs and VWSC to enable them to discharge their responsibilities in drinking water supply. The Department of Drinking Water Supply has given maximum financial flexibility in the guidelines for implementation of Rural Water Supply Programmes in respect of the North Eastern States as 10% of the total Central outlay for the programme is earmarked for the NE States. To ensure that the unutilized funds released to the North Eastern States do not laps, a Non-lapsable Central Pool of Resources has been created. Any unutilized funds of Government of India share are credited to in this Pool under which the State Governments can take up various projects.

BHARAT NIRMAN

Drinking Water Supply is one of the six components of Bharat Nirman, which has been conceived as a plan to be implemented in four years, from 2005-06 to 2008-09 for building rural infrastructure. The proposed coverage in respect of rural drinking water supply under Bharat Nirman is:

- Coverage of 55067 Left Over Habitations of Comprehensive Action Plan (CAP) of 1999.
- Coverage of 2,16968 quality affected habitations
- Coverage of Slipped Back Habitations
- For transparency and accountability, a computerized list of habitations will be prepared, which would be web-enabled.

COVERAGE OF LEFT OVER HABITATIONS OF COMPREHENSIVE ACTION PLAN (CAP) 1999

The Habitation Survey conducted in 1991 was updated in 1999 and a CAP 99 for coverage of these left over habitations was formulate. There are still 55067 uncovered habitations (4588 not covered and 50479 partially covered) of CAP 99 as on 1.4.2005. Under Bharat Nirman it is proposed to cover all uncovered habitations of CAP 99 by 2008-09.

COVERAGE OF QUALITY AFFECTED HABITATIONS

The results of the quality survey of 2000 reveal that there are 2,16,968 quality affected habitations. It is proposed to cover all (2,16,968) quality-affected habitations during the 4-year period of Bharat Nirman. While tackling water quality problem, priority will be given to arsenic, fluoride and salinity affected habitations. 90% quality affected habitations in these categories will be covered through surface water alternate sources and 10% through treatment by low cost technologies. In respect of iron, however, 30% quality affected habitations will be covered by surface water sources and 70% by treatment. In regard to nitrate-affected habitations, 100% habitations will be covered through surface water alternate sources.

COVERAGE OF SLIPPED BACK HABITATIONS

Fresh habitation survey was conducted in 2003. The survey prima facie, reveal that there are incidences of slippage from fully covered to partially covered/not covered due to various factors like –

- Sources going dry or lowering of the ground water table
- Sources becoming quality affected
- Sources out living their lives

- Systems working below rated capacity due to poor operation and maintenance
- Increase in population resulting in lower per capita availability
- Emergence of new habitations, etc.

The 10th Plan Working Group assessed slippage of 2.8 lakh habitations.

SWAJALDHARA

The Government of India has been emphasizing the need for taking up community based rural water supply programmes and with this end in view a beginning was made in 1999 by sanctioning Sector Reform pilot Projects on an experimental basis. With the experience gained, the reform initiatives in the rural drinking wager supply sector have now been opened up through out the country by launching the Swajaldhara programme in 2002. The key Principles of the programme are:

- Adoption of demand responsive, adaptable approach along with community participation based on empowerment of villages to ensure their full participation in the project through a decision making role in the choice of the drinking water scheme, planning, design, implementation, control of finances and management arrangements;
- Full ownership of drinking water assets with appropriate level of panchayats;
- Panchayats/communities to have the power to plan, implement, operate, maintain and manage all water supply and sanitation schemes;
- Partial capital cost sharing either in cash or kind including labour or both, 100% responsibility of operation and maintenance by the users;
- An integrated service delivery mechanism; and
- Taking up conservation measures through rain water harvesting and ground water recharge system for sustained drinking water supply.

SECTOR REFORM

It has been realized that to strengthen the socio-economic conditions of rural India, mere administrative decentralisation or increased investment is not enough. The power of people's participation has been recognized and brought to the fore. Despite good investments, and improvement in the rural water supply and increased outlay by the Government, particularly in the last one decade, general satisfaction is rather limited at the community level. Earlier emphasis on hand-pumps fitted to tube-wells and bore-wells had resulted in an impressive increase in the total rural water supply coverage. However, the availability of potable drinking water in rural areas, especially during the summer months, is still not satisfactory. Though about 1 lakh habitations are covered every year, the number of problem habitations has not declined proportionately. Hence, Government of India realizes that sustainability of sources and systems is key to people's satisfaction. Systems are falling idle and into disrepair. This is due to the perception of the rural people that water is a social right to be provided by the Government, free of cost.

The Government tried to drive home the principles that water is an economic and social good and should be treated as such. It should be managed at the lowest appropriate level with users involved in the planning and implementation of projects. With this aim in view, Government of India has brought about policy changes by introducing reforms in the rural drinking water supply sector. ARWSP was improved in April 1999 to include proposals to mobilize community participation in rural water supply programmes, and 20 per cent of the annual outlay has been earmarked for providing funds for such projects.

This shift envisages a demand-responsive approach, community participation and decentralization of powers for implementing and operating drinking water supply schemes. To ensure people's participation, the Central Government adopted the following approach:

- Adoption of a demand–responsive and adaptable approach based on empowerment of villagers based on empowerment of villagers to ensure their full participation in the project through a decision-making role in the choice of scheme design, control of finances and management arrangements;
- Partial capital cost sharing either in cash or kind or both and 100 per cent responsibility of O & M by the users.

In this new approach, the government plays the role of a facilitator. Efforts are being made to create awareness through Information Education and Communication (IEC) amongst the people about the need for their effective participation in this programme. The community should be willing to be involved in the implementation of the water supply schemes, for which they should have a feeling of ownership of the assets created. The various activities, that can be taken up at Panchayat level are:

(i) To assess the existing condition of rural drinking water supply and sanitation;

(ii) To estimate the requirement of safe drinking water and sanitation facilities;

(iii) To identify the appropriate agency for a particular programme;

(iv) To assist the implementing agencies in identification of (a) individual beneficiaries (b) community groups which form weaker sections, (c) location of the source/facility;

(v) To enforce the stipulated norms under various schemes;

(vi) To ensure the maintenance and use of the facilities;

(vii) To prevent misuse and deliberate attempt to damage the facilities;

(viii) To report appropriate authorities about the operational problems;

(ix) To repair and maintenance of existing sources/facilities;

(x) To obtain basic input materials like hand pumps, construction materials like cement rings, hume pipes, pre-fabricated sanitary structures, etc.;

(xi) To draw technical and administrative help towards implementation of schemes successfully.

DEVOLUTION OF FUNCTIONS, FUNCTIONARIES AND FUNDS

As per the 73rd Amendment, the devolution of powers, funds and functionaries for water supply at the three levels of the PRIs are presented below:

CONCLUSION

The real problem is enforcement and implementation by Panchayati Raj in the village level, identification of location and installation of water tap points in the village, more specifically in SC/ST and other community habitations. It needs basically understanding between the community, officials and elected representatives of panchayats in providing safe drinking water as a whole of the village. The gram Sabha should be strengthened

Scheme	Role and Functions	Powers	Staff and Org. support	Funds	Remarks
Zilla Panchayat					
Water Supply and Sanitation Piped water supply (of specified sizes) Bore/Dug well, spot sources-well/ hand pump Sanitation programme	1. Supervision of the implementation of programmes/ schemes under overall guidance of the PS 2. Selection of spot/ site and beneficiaries, monitoring the release of assistance under development programmes, administering the programme. 3. Demonstration, implementation and dissemination of information. 4. To send reports/ returns as prescribed review of progress, coordination with other departments.	1. Power to check attendance of all GP level officials, for ward the tour diaries/ works details to the PS 2. Allocation and real-location of work within the GP 3. To recommend transfer to P.S. 4. Can call for details from the line department, receive the complaints and inform the panchayat Samiti. E.O. to be Controlling Officer of all departmental officer in the block for tour diaries, attendance, annual reports etc.	GP level senior-most line functionary to attend the GP meetings other officials too will have to attend when asked for and give details. Department Officer to be the member/ invitee of the Committee on these programmes	1. Funds from ZP/ PS 2. Untied funds from DPC 3. Local resource Grants from Govt. for earmarked funds	Liaise with PS/ ZP if any such office/ agency/ institution is located in an area which caters to the requirements of more than one GP/PS. In consultation and coordination with Z.P GP.

Contd.....

Scheme	*Role and Functions*	*Powers*	*Staff and Org. support*	*Funds*	*Remarks*
Panchayat Samiti					
Water Supply and Sanitation	To plan, supervise and implement all programmes in the block area under ZP's guidance and control.	1. Identifying location and feasibility through involvement of Gram Sabha 2. To construct wells, tanks, taps allotted by Zilla Parishad and Panchayat Samiti 3. Treatment of water and proper distribution of water for households. 4. Maintenance of drinking water schemes, collection of water charges. 5. Reporting progress of schemes to Zilla Parishad and Panchayat Samiti	Gram Panchayat Secretary and Senior functionary allotted by Zilla Parishad and Panchayat Samiti		In consultation with Zilla Parishad or Panchayat Samiti and own resources
Gram Panchayat					
Drinking water and Sanitation	1. To prepare plans and projects for drinking water 2. To coordinate with Panchayat Samiti and Zilla Parishad to get funds and grants 3. Periodic monitoring and supervision 4. To monitor and providing quality safe drinking water 5. Maintenance of assets and local infrastructure				

giving more power to plan formulation and identification of the location as well as points. In fact gram Sabha should have been an ultimate and sovereign authority at grass root level but in reality which has become a formal organization with just deliberative function. It needs certain controlling and supervisory powers of course we know that it is easy so that to do it in practice. However, it should be positive towards implementing in nature. Coordination between officials and elected representatives of PRIs in delivery of services particularly providing of safe drinking water to the rural people is the need of hour. Both officials and non-officials at the grass root level should always cooperate and coordinate each other in finalizing plans and implementation for the economic development and social justice. In the question of identification of digging points and installation of taps related to limited geographical area, horizontal harmony between technical functions tend to be stronger than in the state services. The local departments need to understand each others activities day-to-day with nature of flexibility responses. The elected representatives in the PRIs are bound to see the technology part which has been implementing by the technocrats at the village level in providing safe drinking quality water and try to avoid

REFERENCES

WHO-UNICEF-WSSCC (2000), Global Water Supply and Sanitation Assessment–2000, Report.

World Development Indicators, 1998.

"India–Water Resources Management Sector Reforms", World Bank, 1998.

Barot, J.M. "Sustainability of Rural Water Supply", 21st WEDC Conference Kampala, Uganda, (1995).

"Towards Conserving the Elixir of Life" India Toady– August 04, 2003.

"Guidelines for the Implementation of Rural Water Supply Programme", Ministry of Rural Development (GOI), New Delhi (2000).

"Guidelines for Implementation of Schemes and Projects on Sustainability Under ARWSP and PMGY – RWS", Ministry of Rural Development, (GOI), New Delhi, (2002).

"Guidelines on Swajaladhara", Ministry of Rural Development (GOI), New Delhi (June, 2003).

Occasional Paper 5, Powers and Functions of PRIs, A Framework, Ministry of Rural Development, Government of India, 1994.

Occasional Paper 6, Renewing Local Self-Government in Rural India, Ministry of Rural Development, Government of India, 1994.

Panchayati Raj in India: Status Report Task Force on Panchayati Raj, Rajiv Gandhi Foundation, New Delhi, March 2000.

Dinesh Chand, Towards Sustainable Rural Water Supply, Kurukshetra, Vol. 51, No. 12, October 2003, p. 3-8.

Durga Prasad, P., Drinking Water: Management Strategies and Future Agenda, Kurukshetra, Vol. 51, No. 12, October, 2003, p. 10-15.

Yogesh Kumar, Community Initiatives in Water in a Decentralized Governance System of Madhya Pradesh, Presented in a National Seminar on Water, Organized by CESS, Hyderabad, 2003 p. 448-460.

28

MANAGEMENT OF DRINKING WATER SUPPLY SYSTEM BY RURAL LOCAL SELF GOVERNMENTS

A.C. Jena & E.V. Parkash Rao***

INTRODUCTION

The provision of safe and adequate drinking water is one of the basic factors in determining quality of life of people and in the process their capabilities. It is well known that occurrence of a high proportion of diseases and death in rural India has been traced to unsafe drinking water. Therefore, supply of portable drinking water forms part of the strategy for improvement of health and nutrition status. So access to safe water at an affordable price should be viewed as a basic entitlement of the people. To protect this entitlement of the people, the central and state governments have been making laudable effort during the past six decades in this regard. However, the problem still persists. As population increases and their economic activity expand in scale and diversity, the demand for fresh water also continues to rise.

* Formerly Professor, National Institute of Rural Development, Hyderabad, & Currently working as a Consultant to the Institute; and

** Assistant Professor, National Institute of Rural Development, Hyderabad.

Rural drinking water supply has always received high priority in India's Five Year Plan, the provision of drinking water in the rural areas has been especially highlighted in national programmes. A substantial investment to the tune of about Rs.55,000 crore has been made in the rural water supply system by the central and state governments since First Five Year Plan and significant achievement has been made in terms of providing access to potable drinking water (Annual Report 2005-06, MORD, GOI).

POLICY AND APPROACHES

A National Water Supply and Sanitation Programme was introduced in the social sector in the country in 1954. After that during Fourth Five Year Plan, Government of India provided assistance to the states to establish special investigation divisions in Fourth Five Year Plan to carryout identification of problem villages. Further taking into account the magnitude of the problem and to accelerate the pace of coverage of the problem villages the Government of India introduced the Accelerated Rural Water Supply Programme (ARWSP) in 1972-73 to assist states and UTs with 100% grants-in-aid to implement drinking water supply schemes in such villages. The entire programme was given a mission approach when the Technology Mission on Drinking Water Supply Management called the National Drinking Water Mission (NDWM) was introduced in the social sector in 1986. NDWM was renamed as the Rajiv Gandhi National Drinking Water Mission (RGNDWM) in 1991 and further in 1999 Drinking Water Department was created to handle the programme.

The main objectives of the Mission were: (a)To ensure coverage of rural habitations with safe drinking water supply; (b) To ensure sustainability of the systems and sources; (c)To pursue quality of water by institutionalizing water quality monitoring and surveillance as well as support the states for tackling the quality problem.

A Committee was appointed to study the RGNDWM programme in 1993. The Committee submitted its report in 1994, observed that: quantitative coverage of habitation for provision of water supply was satisfactory. However, adequate attention was not paid to sustained quality management and involvement of community. Several deficiencies were pointed out in technical, organizational, administrative and operational aspects of implementation of mini and sub-mission programmes. The salient observations of the Committee are:

- The mission mode revealed several practical shortcomings and failures at the implementation level. Right from the sanctioning stage of the scheme important software aspects like training, awareness building, community participation were not focused and acted upon appropriately.
- The mini-mission and sub-missions have not attempted to develop appropriate operation and maintenance models to suit varied hydro-geological, agro-climatic, techno-economic conditions and societal perceptions in the rural areas.
- The operation and maintenance of rural water supply systems continues to be weak. Community participation has not been given the importance it deserves and the involvement of NGOs has only been marginal.
- The sub-mission have not succeeded in tackling quality problems in a systematic manner except in the case of guinea worm eradication.
- The hand pump attached deflouridation and iron removal plants ended up as a failure due to inappropriate technology unsuited to community perceptions and community involvement.
- The de-salination plants prove costly, and failed due to poor planning and implementation of technology inappropriate to the rural setting. High cost of operation and maintenance were the other factors that led to the failure.
- The sub-mission on conservation of water and recharging groundwater aquifers has not been given any prominence.

The Committee suggested additional professional expertise in the fields of IEC, HRD, MIS and micro-level planning are required to manage the drinking water scheme. In order to provide safe drinking water Government of India has been providing financial assistance to state governments under centrally sponsored scheme Accelerated Rural Water Supply Programme (ARWSP). The objective of the existing drinking water supply programme is to cover all rural habitations with access to safe drinking water, to ensure its sustainability, to tackle the problem of water quality in habitations and to institutionalize the reform initiatives. The norms to provide 40 lpcd drinking water to all human beings, one hand pump for every 250 population and availability of water source within 1.6 Km in plains and 100 meters elevation in hilly areas. Under ARWSP, 15 per cent of the fund can be utilized for O&M activities, 20 per cent of the fund is allocated for quality problems like flurosis, arsenic, iron and brackishness etc. Out of this, 5 per cent is earmarked for sustainability of the sources. Further, 20 per cent of the fund is earmarked for Swajaldhara scheme, 5 per cent for natural calamities and another 5 percent earmarked for DDP programmes.

Implementing Agencies: The State Government decides the implementing agencies for the programme. The agencies may be Public Health Engineering Department (PHED) or Panchayati Raj Department. Implementation is also taken up by the government Boards/Corporations/Agencies in few states. For example Gujarat Water Supply and Sewerage Board is the implementing agency in Gujarat. UP Jal Nigam is the agency for Uttar Pradesh and Tamil Nadu Water and Drainage Board for Tamil Nadu.

Financial Progress: The Government of India and the state governments together have so far invested about Rs. 55,000 crores on rural drinking water supply schemes. The central outlay for Rural Water Supply Sector for 2005-06 was Rs. 4050 crores, indicates the importance given to this sector.

According to 2001 census, 94 percent of rural habitations are covered with portable drinking water supply. The existing coverage of rural habitations with drinking water facilities as on 23-1-2006 is presented in Table 1.

PROBLEMS IN COVERAGE

A number of habitations, which are covered earlier have slipped back NC/PC due to numerous reasons : (a) sources going dry or lowering of the ground water table (b) source becoming quality affected; (c) drinking water systems becoming old (d) drinking water systems working below the capacity due to poor operation and maintenance system (e) increasing population resulting in lower per capita availability (f) increase in population; and (g) emergence of new habitations.

Table 1
Coverage of Habitations with Drinking Water Facilities in the Country (as on 23-1-2006)

Type of Coverage	*No. of Habitations*
1. Fully covered (FC)	1373827
2. Partially covered (PC)	44521
3. Not covered (NC)	3935
4. Uninhabited/urbanized	381
Total	1422664

Source: Annual Report, 2005-06, Ministry of Rural Development, Govt. of India.

SWAJALADHARA (2002)

Apart from the above scheme, a scheme called Swajaldhara was also introduced in December, 2002. It was realized that mere administrative decentralization and increased investment is not enough for making success of implementation of drinking water supply. Despite increased investment and improvement in water supply general satisfaction is limited at community level. Earlier emphasis on hand pumps fitted to the tube wells and bore wells had resulted in an impressive increase in coverage of water supply. However, availability of drinking water in summer months is not satisfactory. Though every year around one lakh habitations have covered the number of problem habitation has not come down. Hence, Govt. of India realized the sustainability of the sources and systems is key to the peoples' satisfaction by involving people and their institution is of paramount importance. There is a perception of the rural people that water is a social right to be provided by the government free of cost. The Govt. of India tried to drive upon the principles that water is an economic and social good. It should be managed at the appropriate level locally with users' involved in planning and implementation of projects. With this aim in view Govt. of India introduced policy changes by introducing reforms in drinking water supply system. ARWSP was improved in April 1999 to include proposals to mobilize community participation in Rural Water Supply Programmes and 20% of the outlay has been earmarked for that. The shift envisages a demand responsive approach, community participation and decentralization of programmes for implementation and operating the drinking water schemes. With the experience (initiated in 1999) gained, reform initiative in rural drinking water supply was introduced as Swajaladhara Programme on 25-12-2002. It was envisaged that by the end of 10th Five Year Plan this programme will cover the whole country. The key principles of the programme are:

(a) Adoption of demand responsive acceptable approach along with community participation based on the empowerment of villages to ensure their full participation in project through decision making role in the choice of the drinking water scheme, planning, design, implementation, control of finances and management arrangements.

(b) Full ownership of drinking water assets with appropriate level of panchayats

(c) Panchayats and community to have the powers to plan, implement, operate, maintain and manage all water supply and sanitation schemes

(d) Partial cost sharing either in cash or kind including labour or both, 100% responsibility of O&M by the users

(e) An integrated service delivery system/mechanism

(f) Taking of conservation measures through rainwater harvesting and ground water recharge systems, watershed for sustained drinking water supply.

Keeping in view of decentralization of powers have been delegated to District Water and Sanitation Committee (DWSC) as envisaged in guidelines June 2003. As per guidelines Swajaladhara will have two dharas–first dhara (Swajala Dhara I) will be for a Gram Panchayat (GP) or a group of GPs or an intermediate Panchayat (block level) and second Swajal Dhara (Swajala Dhara II) will have a district as projection.

Funds under Swajaladhara are now allocated to states/UTs. State/UT make district wise allocation and furnish details to the department of Drinking Water Supply, Govt. of India, on receipt of such information, the funds are released directly to DP/DWSM by the Department of Drinking Water Supply. The total fund made available for Swajaladhara programme is Rs.80429 lakh, out of which Rs.23257 lakh has been spent (about 29 percent) during 2002-03 to 2005-06 since last four years. The number of schemes taken up is 29565 but the number of project completed is only 5075. The contribution from the public is less than 10 percent. This indicates the tardy implementation of the programme in the country as a whole. It is also observed that the proper community initiative contribution is not forthcoming in implementation of the programme as envisaged. Therefore, there is need to strengthen the community based approach as well as PRIs to take the responsibility of implementation of Swajaldhara programme. It is further observed that only in the place of acute shortage of drinking water in difficult areas the community initiative is taking place. Hence, still many people feel water is a right to be provided by the government free of cost.

Another scheme called Bharat Nirman has been conceived to develop rural infrastructure within four year period (2005-06 to 2008-09). Drinking water supply is one of the components of Bharat Nirman. It envisages the coverage of all uncovered habitations, stepped back habitations, habitations facing quality water problem and partially covered villages. Under Bharat Nirman the coverage of Drinking Water is: (a) coverage of 55067 left over habitations of Comprehensive Action Plan (CAP) of 1999; (b) Coverage of 2,16,968 quality affected habitations; (c) Coverage of slipped back habitations; (d) for transparency and accountability a computerized list of habitations will be prepared, which would be web based. The Drinking Water Programme under Bharat Nirman is to be implemented through existing Accelerated Rural Water Supply Programme (ARWSP) of the Ministry of Rural development.

PERFORMANCE OF THE DRINKING WATER SUPPLY– SOME CASE STUDIES

KBK-Orissa Experience

A study of KBK districts in Orissa finds that due to installation of hand pumps and piped water supply, the availability of safe drinking water is 40 LPCD in sample villages of KBK districts. The burden of carrying water by women for long distance has reduced from 21/2 Km to 200 mts. The villagers no longer lose wages for carrying drinking water from far off places. The girl child school attendance has increased. There has been reduction in diarrhea cases in the sample villages. It was also found every caste/tribe has got access to common water supply systems. Now 95 percent of the villagers are depending upon drinking water sources created out of state and central government schemes.

However in many villages the operation and maintenance of drinking water supply was poor. Lot of water is being wasted in the villages due to lack of proper maintenance. (a) There was no evidence of social mobilization among the commitments to maintain the drinking water system and there was lack of awareness among the commoners; (b) The villagers were not involved in planning, implementation monitoring of drinking water supply. It was all the while top down approach; (c) There was no formation of village water and sanitation committee in the village; (d) In some villages over exploitation of ground water is there for agricultural purposes. This may lead to shortage of ground water; (e) Therefore, NGOs, CBOs, PRIs should create awareness on water consumption for crops; (f) Many villages are affected with water quality problem specially excess iron and fluoride above permissible limits.

In KBK districts Panchayats were not involved in maintaining the water supply systems. A recent circular during February 2006 was issued to handover the drinking water supply system for O&M to Panchayats the preparatory work on transfer of power has yet to be taken up. In addition Panchayats have not been trained to take up the task to maintenance of water supply schemes. There was no coordination between the RWSS PHE Dept, and the Self Employed Mechanics (SEMs) and Panchayats. Erratic power supply affects the water supply system. Illegal connections taken by some people without any payment is noticed in study villages.

A Panchayat has 20-25 hand pumps in its area but are not maintained properly due to : (a) adequate spare parts were not available for O&M resulting in continuous breakdown of water supply system; (b) un-hygienic conditions found in terms of stagnant water around the hand pump; (c) the maintenance expenditure for a pipe system is 40,000 and for a hand pump Rs.800-1000 per annum. At present the expenditure on O & M is borne by RWSS (PHED) of Rural Development, Govt. of Orissa, is not adequate and therefore causes problem in proper maintenance of the facility.

OBSERVATIONS

Inspite of huge investment on drinking water facilities, poor M & O and lack of local participation, many schemes are not functioning properly. The reason being non-involvement of the local governments or Gram Panchayats which are the primary stake holders in drinking water projects. This is the reason for many villages fully covered under water supply schemes are slipping back to the position of partially covered and even uncovered villages.

Integrated Rural Water Supply and Sanitation Project in Karnataka

A study in Karnataka (Partners in Development – A Case Study of Integrated Rural Water Supply and Sanitation Project in Karnataka) finds partnership arrangements among government, user groups/ NGOs/PRIs. The study shows that there is a success of multipurpose project (assisted by World Bank) for supplying safe drinking water in Mysore district of Karnataka (the project covers 110 selected villages). The project supplies an average 55 LPCD of water in the project villages which is above national average. The completed water schemes have been handed over to village water supply committee (VWSC) for operation and maintenance. This was possible with the help of people/NGOs and Panchayat involvement in the project. The VWSCs are fully responsible for O&M in drinking water supply system. The O&M expenses are met through water charges collected from the water users. The monthly water charges are fixed by the VWSC keeping in view of varying O&M expenses. The VWSCs are collecting water charges at the rate of Rs. 50 per month per individual household connection. Those who use public tap pay a nominal charge of Rs.10 per month. Evidently, VWSCs, stand post committees, local youth and water users are intensively involved as partners in managing drinking water systems. The project has been an integrated project of water supply, sanitation and health communication.

OBSERVATIONS

Community participation was one of the inbuilt system in the project. It was made mandatory that local water users' should contribute 30% of the cost towards drainage construction. Women participation was greater in the project villages. Street level committees were formed to oversee the water supply systems. The IEC components were strong in project villages due to involvement of Panchayat extension staff, NGOs and VWSCs.

Rainwater harvesting was encouraged among community by NGOs/PRIs and government functionaries for maintaining constant groundwater recharge at source for sustainable water supply. Local youth were trained for O&M activities and was helpful in managing the drinking water system.

The above evidences from the field shows that involvement of local people and local institutions at all stages of the development of drinking water facilities and sanitation in the villages and creating a sense of ownership in the management of drinking water supply among the users, is the key for the success of the schemes on a sustainable basis.

INVOLVEMENT OF PRIs IN MANAGEMENT OF DRINKING WATER SUPPLY–A SUGGESTIVE FRAMEWORK

From two case studies discussed above, it is clear that to make drinking water schemes successful and sustainable, it is necessary to involve Gram Panchayats at all stages of the project and also should own the responsibility for operation and maintenance, supply and distribution with equity. For this purpose a suggestive frame work is presented here to promote involvement of local bodies in the management of drinking water on a sustainable manner.

ROLE OF GRAM PANCHAYATS

In some states, the gram Panchayats have already been assigned the responsibility for the management of the drinking water facilities. Within the Gram panchayat, a Village Water and Sanitation Committee (VWSC) is constituted to manage the functioning of the scheme.

In order to make VWSC function effectively, they should be empowered to perform the following functions: (a) to participate fully in planning and implementation of the project; (b) to mobilise community participation and decision making in all phases of scheme through Gram Sabha; (c) to mobilise community contributions towards capital costs both in cash and kind (land, labour or materials); (d) collection of user charges for O&M of the system; (e) to ensure supply and distribution of water with equity and proper access, and (j) promoting awareness for integration of drinking water, sanitation and hygiene in the village.

ROLE OF DISTRICT PANCHAYATS

Under the programme Swajaldhara, the district Panchayats have already been provided certain functions as follows:

(a) The District Panchayat will cover Swajaladhara II for trigger projects in the district. The District Water and Sanitation Mission (DWSM)may be constituted for the purpose;

(b) The District Water and Sanitation Committee (DWSC) will be under the DWSM;

(c) The drawing and disbursing officer DDO of the District panchayat/DWSM shall also act as the DDO for all funds received under Swajaladhara;

(d) The District Panchayat/DWSM shall meet not less than four times a year. It will consider and take note of all schemes under Swajaladhara which have been technically scrutinized and approved by the DWSC. It will review implementation, progress and total sanitation campaign in the district;

(e) There will be a Water Sanitation Committee (DWSC) which is headed by CEO of District Panchayat in performing the functions of DWSC. DWSC will have district level officers such as the Executive Engineers of Drinking Water, and District Panchayat, District Educational Officer, District Health Officer Project Director, DRDA, District Panchayat Officer, District Social Welfare Officer, District Public Relation and Information Officer; and,

(f) More members from reputed NGOs will be co-opted into the Committee as members with prior approval of the SWSM. The Executive Engineer of Drinking Water of District Panchayat Officer shall be chosen by the DWSM and concurred to by the Executive Committee of SWSM shall be Member Secretary of the DWSC.

The functions of the District Water and Sanitation Committee (DWSC) will be as follows: (a) Formulation, management and monitoring of the project; (b) Scrutinizing and approval of the schemes submitted by Block Panchayat/Gram Panchayat; (c) Selection of agencies and NGOs and enter into agreements for social mobilization, capacity building, communication, project management and supervision; (d) Sensitizing the public representative officials and general public about the Swajaladhara principles; (e) Engaging institutions for imparting training for capacity building of all stakeholders, undertaking communication campaign; (f) Interaction with SWSM, State Govt. and Govt. of India; (g) The DWSC will utilize the office, equipment, vehicles/ telephone etc of Zilla Parishad/line departments of the official activities. Recurring cost can be met from project fund; (h) A core group consisting of different professionals may be positioned to assist DWSC in implementing project activities. The expenditure on core group can be met from project fund.

DELEGATION OF POWERS

Keeping in view the concept of decentralization of power, Govt. of India has delegated powers to States. All projects and schemes proposed under ARWSP are approved by the State Level Scheme Sanction Committee.

As per the 73rd Amendment to the Constitution of India, rural water supply is a subject of the PRIs. The panchayat and VWSC due to play a major role in providing safe drinking water and managing the systems and sources in their respective areas. They can be involved in implementation of schemes, particularly in selecting the location of hand pumps, stand posts and spot sources, in operation and maintenance (O&M). Govt. of India has laid emphasis in capacity building for PRIs and VWSC to enable them to discharge their responsibilities in drinking water supply.

The institutional setup suggested under Swajaldhara scheme should be extended to all states and UTs of India. There should be full integration of DWSM/DWSC with Zilla Parishad, Block Panchayat and Gram Panchayats of India and their function should be clearly defined. There should be comprehensive training programmes for all elected representatives at three levels and DWSC and VWSC members in all aspects of managing drinking water supply systems.

SUSTAINABILITY ASPECTS

This is an important aspect for the success of the water supply schemes on a long term basis. Central Ground Water Board (CGWB) and National Geophysical Research Institute (NGRI) have been engaged in the programme since the inception of the mission. Further, state governments have been advised that upto 5 percent of the funds released under ARWSP should be used for the submission on sustainability.

The reasons for taking up sustainability in drinking water sector: (1) Fast depletion of groundwater level leading to quality problems like arsenic and flurosis; (2) Sources go dry due to deforestation leading to reduced recharge of aquifers; (3) Poor maintenance of existing water supply systems; (4) lack of participation of people in the operations and maintenance of the systems; and (5) Neglect of traditional water management practices and systems.

In order to overcome these problems, Govt. of India aims to concentrate on (a) control on over extraction of groundwater; (b) funds for repairs and rehabilitation; (c) emphasis on community participation; (d) promotion of water as socio-economic good (e) stronger links with watershed development programmes and water conservation; (f) encouragement of rain harvesting structures.

The Ministry of Urban Development is emphasizing to promote rain harvesting structures in urban areas. Ministry of Water Resources has been requested to promote Water harvesting measures. All MPs have been requested to encourage water harvesting schemes from MP local area fund schemes. Further, household community and institution level rainwater harvesting structures, bankable schemes are being worked out for funding by NABARD. A manual on recharge of groundwater has been finalized for circulation to States and all concerned. All these efforts indicate that there is a need for convergence of funds of water conservation schemes, rain harvesting structures with drinking water schemes at the village level.

TWELFTH FINANCE COMMISSION

The maintenance of drinking water system is a major issue in the rural areas. Though crores of rupees have been invested on drinking water system and assets created there is lack of fund for O&M expenses for the drinking water supply system. That is why 12th Finance Commission has recommended that Panchayat need to be encouraged to take over and maintain all such schemes. Some of the existing schemes may require special repairs to make them full functional. The PRIs may take over the assets and utilize these grants for repair/rejuvenation and maintenance to make them fully operational. Even after this the PRIs may not be able to bear the entire cost of O&M of water supply for an initial period of five years. They should at least recover 50 percent of the recurring costs in the form of water charges. They have also recommended to ensure the environmental sanitation in the village. Against this background the Commission recommended that the grants-in-aid allocated by them for the PRIs in each state, priority should be given to incur expenditure on O&M cost of water supply and sanitation. This will facilitate Panchayats to take over the schemes and operate them. It is to be noted that the TFC (Twelfth Finance Commission) Report 2005, recommended a sum of Rs.20,000 crores to Panchayats to meet the O&M cost of the water and sanitation schemes.

REFERENCES

Annual Report (2005-06), Ministry of Rural Development, Govt. of India.

Basic Rural Infrastructure and Services for Improved Quality of Life, Vol. 1, NIRD: 1999.

P. Durga Prasad, P. Sivaram, "Good Governance and HRD, Case Studies of User Managed Decentralised Safe Drinking Water and Sanitation Project in India (mimeo)– Partners in Development– A Case study of Integrated Rural Water Supply and Environmental Sanitation Project in Karnataka".

Quick Evaluation of the Projects and the Programmes implemented under RLTAP in KBK Districts of Orissa, NIRD:2006 (Mimeo).

Rajiv Gandhi National Drinking Water Mission (1994), Report of the Expert Committee on Rural Sanitation Programme, Ministry of Rural Areas & Employment, Govt. of India, New Delhi.

Swajaldhara–Guidelines, Dept, of Drinking Water Supply, Ministry of Rural Development, June 13, 2003.

29

DRINKING WATER ACCESSIBILITY IN RURAL AREAS THE CASE OF WEST BENGAL

*Dilip Kumar Ghosh**

INTRODUCTION

Tenth Five Year Plan (2002-07), in conformity with this essentiality, affirmed that "in line with the National Agenda for Governance, safe drinking water is to be provided in accordance with the stipulated norms on a sustainable basis to all habitations by March 2004". The stipulated norm is to ensure supply of 40 liters of safe drinking water per capita per day within a walking distance of 1.6 kilometers or elevation difference of 100 meters in hilly areas. In course of mapping the progress in the area of rural drinking water supply, the mid term appraisal of Tenth Five Year Plan makes the following observation:

"In 2002, 15798 of the country's 14.22 lakh habitations still remained to be covered and 1,33,305 partially covered habitations needed to be upgraded to the fully covered category... A large number of habitations have slipped from being fully covered to partially covered status and from partially covered to not covered status".

* Senior Research Officer, (SIPRD), Kalyani, Nadia-741 235.

Most of uncovered habitations are located in remote and difficult areas generally inhabited by the marginalized sections of population. According to the Mid term Appraisal of Tenth Plan, per capita availability of water was about 5200 cubic meter in 1951; then it had fallen to 2200 cubic meter in 1999 and further to 1820 cubic meter in 2001 reflecting the effect of rising population. If the present trend continues, there is every likelihood that India will fall below the water stress level in the coming years. In terms of the prevailing norms, a per capita availability of 1700 cubic meter is necessary to remain free of water stress. If per capita water availability falls below 1000 cubic meter, then it would lead to water scarcity.

Keeping in view the importance of drinking water, the present the paper students only the rural segment of West Bengal because in this State around 72 per cent population live in the rural areas. It is imperative that the whole rural population should have the access to safe drinking water. West Bengal is selected for the study mainly for two reasons. First one is that the Left Parties under the leadership of the Communist Party India (Marxists) has been governing the State since 1977. The second one is the fact that since 1978, the Panchayati Raj institutions in the State have been given the responsibility to implement different schemes of rural water supply. In West Bengal, the panchayats are vested with the responsibilities to maintain and install sources of safe drinking water. Uninterrupted functioning of the panchayats for a period of around twenty-eight years has resulted in institution building in the rural areas. This also generates high expectation from the panchayats.

In this backdrop, the present study is undertaken with the following objectives:

(i) to examine the accessibility of the rural households in safe sources of drinking water;

(ii) to study the institutional arrangements in West Bengal for ensuring safe drinking water to the people living in the rural areas; and

(iii) to portray the effectiveness of the service delivery mechanism of providing drinking water in the rural areas.

The paper consists of three sections. Section I presents the all India situation of coverage of safe drinking water in 15 major states, namely, Andhra Pradesh, Assam, Bihar, Gujarat, Haryana,

Karnataka, Kerala, Madhya Pradesh, Maharashtra, Orissa, Punjab, Rajasthan, Tamil Nadu, Uttar Pradesh and West Bengal. The prevailing situation in West Bengal districts regarding accessibility of the rural households in the safe sources of drinking water is discussed in Section II. Section III explores the institutional arrangements made by the State Government for delivering safe drinking water to the people.

SECTION I

The WHO-UNICEF Sponsored Assessment Report of Water Supply and Sanitation in India (2002) reported that "between 69 to 74 per cent of India's rural population take their drinking water from protected sources, leaving an unserved population of 26 to 31 per cent". The coverage of population with protected source is also not uniform. On the basis of Census 2001 data, Table 1 is constructed to indicate the statewise coverage of rural households with safe drinking water. The distribution of households has been made on the basis of location of the source of drinking water. According to Census definition the type of drinking water source is treated within the premises if the water source used is located within the house or within the premises of the census house where the household lives. Likewise, a source is considered to be located near the premises if a household in the rural areas has to cover a distance of 500 meters or less for fetching water from the source. In case of location of the source considered to be away from the premises, the households have to cover a distance of more than 500 meters for collecting water.

From Table 1 shows that the coverage of safe drinking water is the best in Punjab as most of the rural households have the sources within their houses. This is the most desirable situation. Of the 15 major states, 9 are below the all-India average of 22.98 per cent. Taking water sources within the boundary of the houses requires some favourable economic status and Punjab has that advantage. For enhancing the access to safe drinking water sources, most of the states manage to install water sources within 500 meters of the premises of the households. These sources are community sources where the ownership lies with the State. In this respect eight states (excluding Punjab) are below the all-India coverage figure of 39.35 per cent. This implies that in these states people have to traverse longer distance for collecting water in comparison with the states having better access. Inter-state difference in access to this basic need is of great concern. To ensure physical access to the source it is necessary that the distance to the source of water is not too much. Location of the water source away from the premises means minimum walking of a distance of more than 500 meters. All India average of households having the water sources away

Table 1
Accessibility to Safe Drinking Water in Rural Areas: All India (2001)

State	Source within premises (in %)	Rank 1	Source near premises (in %)	Rank 2	Source away from premises (in %)	Total coverage (in %)
Andhra Pradesh	17.58	9	45.32	4	13.95	76.85
Assam	21.84	7	24.04	13	10.91	56.80
Bihar	27.56	6	38.90	7	7.95	74.42
Gujarat	28.79	5	38.39	9	9.69	76.87
Haryana	30.19	4	36.49	10	14.45	81.13
Karnataka	13.37	11	48.00	3	19.15	80.52
Kerala	9.52	13	4.58	15	2.78	16.88
Madhya Pradesh	7.11	14	42.41	5	13.35	62.87
Maharashtra	32.07	3	29.04	12	7.30	68.41
Orissa	6.05	15	38.60	8	18.23	62.88
Punjab	81.66	1	12.16	14	3.08	96.91
Rajasthan	16.89	10	31.18	11	12.37	60.44
Tamil Nadu	10.67	12	66.07	1	8.55	85.29
Uttar Pradesh	36.56	2	40.88	6	7.79	85.23
West Bengal	20.67	8	49.23	2	17.10	87.00
All India	22.98	–	39.35	–	10.90	73.23

Source: Tables on Houses, Household Amenities and Assets, Census of India, 2001.

Note: (i) Rank 1 indicates rank of state on the basis of percentage of households having safe drinking water sources within their premises. Rank 2 is the rank of a state on the basis of percentage of households having drinking water sources near the premises.
(ii) Total coverage is calculated by adding three locations of safe sources of drinking water.

from the premises is 10.90 per cent– in absolute term, the number of households is quite sizeable. Covering this distance on daily basis with water–containers is surely an arduous task. As women are the principal water-carriers in their households, it is unpleasant too. The distance norm of 1.6 kilometer in plain areas is also on the higher side; it requires downward revision to alleviate drudgery of women. National Perspective Plan for Women (1988-2000) recommended that "Ideally, as in the case of urban services, the Government should be in a position to supply potable water to all households through pipelines and taps right at the door-steps". If this becomes possible, it will not only broaden the access to safe drinking water, but also reduce the sufferings of women in drawing and fetching water for their families. In general in government programmes, priorities and norms are not adequately sensitive to the issues of women.

The primary responsibility of providing drinking water facilities rests with the State Governments. The Union Government provides financial assistance under the Central Scheme of Accelerated Rural Water Supply Programme (ARWSP) for supplementing the efforts of the States. Annual Report (2005-06) of the Ministry of Rural Development, Government of India mentioned that "substantial investment to the tune of about Rs. 55,000 Crore has been made in the rural water supply sector by the Central and State Governments since the First Five Year Plan and as a result significant achievement has been made in terms of providing access to potable drinking water". ARWSP was introduced in 1972-73. In 1986, the Government of India reoriented the programme with the mission approach and created National Drinking Water Mission. This Mission was renamed as Rajiv Gandhi National Drinking Water Mission in 1991. Later on, the Department of Drinking Water Supply was set up within the Ministry of Rural Development in the year 1999. The issue of huge unspent balances under ARWSP is a major concern in programme implementation. The Twentieth Report of the Standing Committee on Rural Development of 14th Lok Sabha expressed its concern by saying that "as per the Performance Budget of the Department for the financial year 2006-07, unspent balance of funds under ARWSP as on 31st December, 2005 was Rs. 2113.30 Crore". Of this amount, Andhra Pradesh, Bihar, Gujarat, Karnataka, Maharashtra, Rajasthan, Uttar Pradesh and West Bengal together account for Rs. 1386.72 Crore i.e. around 66 per cent. Incidentally, in all these states, accessibility of the rural households to the sources of safe drinking water is not of very high order. The coexistence of the problem and low expenditure can only accentuate the sufferings of the inhabitants having natural right to safe drinking water. Regarding unspent balance of fund, the Standing Committee on Rural Development of 14th Lok Sabha strongly recommended to the Department of Drinking Water Supply of the Government of India "to take all the desired actions to ensure that every paisa earmarked for the drinking water sector is meaningfully utilized". The Department's argument for such huge unspent balance is non-receipt of monthly progress report from the states. The Standing Committee negates this contention in the Twentieth Report. It is relevant here to quote the argument:

> "The Committee are not inclined to accept the casual reply of the Department stating non-receipt of monthly progress reports as the reason for unspent balances in this era of technological advancement. The Committee have repeatedly been expressing their concern over the under-spending with various State Governments. In spite of that, there seems to be little improvement in this regard".

This is the state of affairs with regard to implementation of ARWSP, the major programme for ensuring drinking water to the rural inhabitants. For understanding the status of drinking water supply, the Government of India conducted a survey of rural habitations in 1991. The result was validated in 1994 and on that basis a Comprehensive Annual Plan, 1999 was prepared by updating the habitation survey results validated in 1994. Again, a fresh habitation survey has been conducted in 2003 and the results of the same are being validated. In Table 2, the coverage of habitations during two surveys is compared. In the second survey, number of habitations increased due to emergence of new habitations in different states. A habitation is defined as a locality within a village where a cluster of families reside. Forming a habitation, minimum of 100 people are required. Both the surveys classified the habitations into three categories viz. fully covered, partially covered and not covered. Habitations which have a safe drinking water source point (either private or public) within 1.6 kilometers in plain or an elevation difference of 100 meters in hill areas but the capacity of the system ranges between 10 lpcd to 40 lpcd, are categorized as partially covered. Those habitations having less than 10 lpcd are categorized as not covered.

Table 2
Coverage of Rural Habitations with Safe Drinking Water

Category	*1991 Survey*	*2003 Survey*	*Change over previous survey (in%)*
Fully covered	1116103 (78.45)	869997 (57.72)	-22.05
Partially covered	268496 (18.87)	389409 (25.83)	+45.03
Not covered	38065 (2.68)	247943 (16.45)	+551.37
Total	1422664 (100.00)	1507349 (100.00)	+5.95

Source: Department of Drinking Water Supply, Government of India.
Note: (i) Figure within brackets are in percentage.
(ii) Only habitations as per ARWSP norms are considered.

From Table 2 it can be noticed that throughout the country many habitations slipped back either to partially covered or not covered category from fully covered, or to not covered habitations from partially covered habitations. A number of factors are responsible for such phenomena. Prominent among these are emergence of new habitations, increase in population resulting in reduction of per capita availability of drinking water, poor operation and maintenance works leading to functioning of the system below rated capacity etc. In view of this situation, the National Government included drinking water in the Bharat Nirman Programme launched in the year 2005. The goal of this programme is to build up a strong rural India by strengthening its infrastructure. The proposed strategy for coverage of habitations in the Bharat Nirman Programme is given in box 1. The duration of Bharat Nirman period is four years i.e. from 2005-06 to 2008-09.

As the Bharat Nirman Programme is a time bound programme, it is expected that it can deliver the benefits to the rural people through provisioning of safe drinking water. With presentation of all India scenario, let us now focus the situation in West Bengal in the next section.

Box 1
Strategy Prescribed in the Bharat Nirman

(i) Coverage of 55067 uncovered habitations of comprehensive Action Plan 1999.
(ii) Coverage of water quality affected habitations.
(iii) Coverage of slipped back habitations based on 2003 survey with priority to tackling problems of arsenic, fluoride and salinity.

SECTION II

The Economic Review 2005-06 of the Government of West Bengal reflects the vision of the State Government in the sector of rural water supply in following words:

> "The State's vision is to achieve full coverage of all scarcity habitations by March 2007 and all quality affected habitations by March 2012. Based on this vision, the State's policy for rural drinking water with a clear statement on required reforms would be formulated. The reform measures envisaged would increasingly involve the beneficiaries through the panchayats in operating and maintaining the water sources. Capital cost of the drinking water projects will continue to be largely state financed although the local community would have to contribute to the cost to an extent".

With this vision in view, the Public Health Engineering Department of the State Government took initiatives to bring as many habitations as possible under the fully covered category. As on April, 2004, it became possible to cover 66799 habitations fully out of 96020 habitations (Habitation Survey 2003); in percentage, 69.57 per cent only. According to the norms of coverage, a habitation is said to be covered fully if following three provisions are satisfied. The provisions are:

(i) there is one spot source for 250 persons or part thereof;

(ii) in case of piped water supply there is one public stand post for 250 persons or part thereof and assured water supply of 40 litres per capita per day; and

(iii) the drinking water source exists within 1.6 kilometer of the habitation in the plain or 100 meters elevation in hill areas.

The Economic Review 2005-06 of the State Government reported that piped water supply coverage in rural areas has increased from 30.30 per cent of the rural population on April 1, 2004 to 31.05 per cent on April 1, 2005. That is an increase of 0.75 per centage points within one year. Considering the total rural population of 57734690 in the State (Census 2001), the increase in estimated number of population benefited is around 4.33 lakhs. This is no doubt a mean achievement. With this introduction about the existing scenario in West Bengal, so far as rural water supply is concerned, the district disaggregated analysis is made hereafter mainly with Census 2001 data. Table 3 presents the coverage of households with safe sources of drinking water i.e. tap, hand pump and tube well. In case of hand pump, water is taken out manually by operating the hand pump. Tube well means where sub-soil water is taken out through electricity or diesel-pump.

Table 3
Coverage of Households on the Basis of Safe Sources of Drinking Water (Rural, 2001)

District	*Tap*	*Hand pump*	*Tube well*	*Total coverage*	*Rank*
1. Burdwan	10.57	69.67	13.05	93.29	9
2. Birbhum	6.58	71.24	9.67	87.49	10
3. Bankura	4.08	54.00	12.52	70.60	14
4. Midnapur	7.64	63.77	13.40	84.81	13
5. Howrah	12.94	72.30	13.90	99.14	2
6. Hooghly	7.17	83.85	7.91	98.93	4
7. North 24 Parganas	7.64	77.27	14.03	98.94	3
8. South 24 Parganas	6.07	79.28	13.91	99.26	1
9. Nadia	4.14	87.16	7.54	98.84	5
10. Murshidabad	3.47	85.02	9.80	98.29	6
11. Uttar Dinajpur	1.88	88.16	3.55	93.59	8
12. Dakshin Dinajpur	2.31	88.93	4.61	95.85	7
13. Malda	4.31	71.41	9.47	85.19	11
14. Jalpaiguri	17.03	23.61	5.49	46.13	15
15. Darjeeling	27.30	5.95	4.22	37.47	17
16. Cooch Behar	3.57	73.64	7.91	85.12	12
17. Purulia	2.87	21.98	16.05	40.90	16
West Bengal	7.02	69.24	10.73	86.99	–

Source: Tables on Houses, Household Amenities and Assets, Census of India, 2001, West Bengal.

Note: (i) Coverage figures are in percentage.
(ii) Ranking of districts is made on the basis of total coverage value.

From Table 3 it can be noticed that the coverage of tap i.e. piped water supply is the least in rural West Bengal. In urban West Bengal this source covers up maximum households. Rural piped water systems are either single village based or multi-village based scheme. In this system, water is pumped from the groundwater source into an overhead holding tank from which a small network of pipes for carrying water to neighbourhood is laid down. In this system, the source points are mostly erected near the premises of the households. Very few rural households have tap within their premises. Contrary to this scenario, in urban areas majority of the households have their water-source points within their premises. As mentioned earlier, according to the Government of India norm, household is said to be served by piped water system if there is a source point of water (tap) at no more than 1.6 kilometer from the house and this source has the capacity to deliver 40 litres per capita per day. While the third condition is that this source should be shared by no more than 250 persons. From the health point of view, piped water is considered to be more safe than other sources as it goes through at least secondary and usually tertiary treatment. For clarity of concept of treatment, it is pertinent here to quote the definition given by Gleick (2000):

> "Secondary treatment separates solids from liquids and includes process to remove microbiological contaminants. Tertiary treatment removes additional solids and nutrients and inactivates pathogenic organisms. Chemical coagulation, filtration and disinfections are all tertiary process".

The coverage of piped water network in rural areas, however, remains low mainly because of its expensive nature in comparison to the other two sources i.e. hand pump and tube wells. In rural areas, hand pumps are the most popular source for safe drinking water. Water is drawn manually through operation of pump by suction. Generally through hand pump water can be lifted from a depth of 30 feet. Groundwater depletion and aquifer pollution may be the threats to get sustainable extraction of water for drinking purpose and domestic usage. Major contributing factor for depletion of ground water resources is the unregulated withdrawal of water for irrigation purposes. For this reason, particularly during summer, hand pumps become derelict or water discharge becomes low. Next to hand pump, open wells are the most prominent source of drinking water as they are inexpensive to construct and does not require skill and improved equipments for lifting water. These wells are appropriate in the areas where the water table is high and gets recharged regularly. But open wells are susceptible to biological contamination. Hence this source is not recognized as safe source for drinking water. Sealed bore-wells fitted with hand pumps, however, are the safe source as the possibility of contamination is controlled. In West Bengal, open will is a predominant source of drinking water in districts like Bankura, Purulia and Jalpaiguri. In Table 4, proportions of households in the rural areas using water from unsafe sources are depicted. Even in 2001, around 1.60 per cent of the rural households consume contaminated water of tanks, ponds, rivers etc.

Table 4
Proportions of Rural Households Using Unsafe Sources of Drinking Water, 2001

District	*Well*	*Tank/ Pond/ Lake*	*River, Canal*	*Springs*	*Any other source*	*Households using surface water*
1. Burdwan	6.14	0.12	0.10	0.17	0.18	0.57
2. Birbhum	11.71	0.09	0.42	0.20	0.09	0.80
3. Bankura	27.50	0.20	1.10	0.44	0.16	1.90
4. Midnapur	14.14	0.40	0.24	0.23	0.18	1.05
5. Howrah	0.48	0.15	0.02	0.16	0.05	0.38
6. Hooghly	0.61	0.07	0.02	0.19	0.18	0.46
7. North 24 Parganas	0.08	0.35	0.04	0.20	0.39	0.98
8. South 24 Parganas	0.03	0.40	0.01	0.20	0.10	0.71
9. Nadia	0.40	0.04	0.06	0.18	0.47	0.75
10. Murshidabad	1.18	0.08	0.06	0.17	0.22	0.53
11. Uttar Dinajpur	6.04	0.03	0.06	0.13	0.15	0.37
12. Dakshin Dinajpur	3.72	0.08	0.02	0.22	0.11	0.43
13. Malda	14.28	0.11	0.06	0.17	0.17	0.51
14. Jalpaiguri	51.89	0.20	0.56	0.49	0.73	1.98
15. Darjeeling	34.27	0.10	0.92	25.34	1.90	28.26
16. Cooch Behar	13.52	0.06	0.27	0.08	0.95	1.36
17. Purulia	52.19	0.90	4.47	1.24	0.30	6.91
West Bengal	11.41	0.22	0.36	0.73	0.29	1.60

Source: Tables on Houses, Household Amenities and Assets, Census of India, 2001, West Bengal.

Note: (i) Proportion of households using surface water for drinking purposes is calculated by adding the proportions of households using water sources other than wells in table 4.

(ii) All figures are in percentage.

From Table 4 it can be concluded that barring Darjeeling district, in Purulia many people use surface water (obviously unfiltered) for drinking purposes. As a major part of Darjeeling district is in mountainous region, there are many natural springs in these areas. Local people use these as sources of drinking water. In wells and surface water, possibility of contamination is high. When water is contaminated with virus, bacteria, parasites or chemicals, chances of water–borne diseases increase. The common water borne disease are cholera, typhoid fever, bacillary dysentery, amoebiosis, giardiasis, etc. In view of India Assessment Study (2002) on water supply and sanitation, "morbidity and mortality due to water-borne diseases have not declined commensurate with increase in availability of potable water supply, largely owing to the fact that quality of water is not maintained at consumer point and that safe water may become contaminated during storage due to poor handling practices and poor personal hygiene". Expansion of safe sources of drinking water as well as awareness building among the local inhabitations can give relief from this uncomfortable situation.

According to Census 2001 data on household amenities, in West Bengal 234371 households spreading over all the districts in the state have to collect their drinking water from the sources like river, canals, tanks, ponds etc. Considering the mean size of a household as 5, total population using unsafe surface water for drinking will be around 11.71 lakh. More than three-fourths of these households are located in seven districts. In table 4(a) the district wise details are provided for having an idea on the seriousness of the problem.

From Table 4(a) it can be noticed that of total households using tank/pond/lakes etc. as sources of drinking water, around 31 per cent live in North and South 24 Parganas districts where Sundarbans delta is located. As salinity in water in these two districts is of high order, people use these sources for getting sweet water. Again, in the western plateau of the state comprising districts like Purulia, Bankura and part of Midnapur, 66.70 per cent of the households use rivers and canals as sources of drinking water. In these three districts there are canal networks of Kangsabati Project and Damodar Valley Corporation. Mainly the poor people and those living in the fringe areas collect water from these sources. In these areas, the Public Health Engineering Department of the State Government has no scheme for treatment of the surface water to be used for drinking and domestic usage.

If the household coverage in the rural and urban areas is compared according to the sources of drinking water, the rural-urban divide is glaring. Urban areas have better coverage with safe sources than the rural areas. The primary factor making the difference is the infrastructural facilities in urban areas. The municipalities in urban segment are traditionally entrusted with

Table 4(a)
Concentration of Households Collecting Drinking Water from Surface Sources

Surface water sources	*Jalpaiguri*	*Burdwan*	*Bankura*	*Purulia*	*Midnapur*	*North 24 Parganas*	*South 24 Parganas*	*West Bengal Total*
Tank/Pond/Lake	1238	2185	1094	3714	6723	5236	5365	34236
	(3.61)	(6.38)	(3.19)	(10.85)	(19.64)	(15.29)	(15.67)	
River/Canal	3226	2014	6126	18242	3847	431	121	42302
	(7.62)	(4.76)	(14.48)	(43.12)	(9.09)	(1.02)	(1.02)	
Total	4464	4199	7220	21956	10570	5667	5486	76538
	(5.83)	(5.49)	(9.43)	(28.69)	(13.81)	(7.40)	(7.17)	

Source: Census 2001.
Note: Figures within brackets are in percentage.

the delivery of safe drinking water to its citizen. In rural areas, the Public Health Engineering Department of the State Government has this responsibility. With the passing of the Constitution (73rd Amendment), 1992 and inclusion of the Eleventh Schedule under Article 243G of the Constitution, the responsibility of rural water supply has been given to the panchayats in appropriate tier. Discrimination in respect of infrastructure building relating to supply of drinking water still persists. In Table 5, the source wise coverage is shown to focus the differences; for example, the coverage of tap water in urban areas is eight times higher than that in rural areas.

Table 5
Distribution of Drinking Water Sources on the Basis of Coverage of Households, 2001

Source → Areas ↓	*Tap*	*Hand pump*	*Tube well*	*Well*	*Tank, Pond, Lake*	*River, Canal*	*Sp-rings*	*Any other*
Rural	7.02	69.24	10.73	11.40	0.21	0.36	0.73	0.30
Urban	56.66	22.92	12.71	6.50	0.20	0.05	0.46	0.50

Source: Tables on Houses, Household Amenities and Assets, Census 2001, West Bengal.
Note: All figures are in percentage.

Table 5 shows that in case of safe sources of drinking water, urban areas are better than the rural areas. In urban areas, the coverage of households with safe sources is 92.29 per cent, while in case of the rural areas, the corresponding figure is around 87 per cent. Regarding unsafe sources of drinking water, the rural households are more exposed than their urban counterparts. Particularly in case of the slum-dwellers who live in the fringe areas of newly built up township, the possibility of using unsafe sources is high. Expansion of infrastructure can help to reduce the proportion of unsafe use of water. World Health Organization (WHO) upholds the right to water for every individual and explains the approach in following manner:

> "A rights-based approach has implications for a range of actors concerned directly or indirectly with water issues. Governments, as primary duty bearers, must take concrete steps to respect, protect and fulfill the right to water and other water related rights

and to ensure that anyone operating within their jurisdiction–individuals, communities, civil society and the private sector–do the same".

On this yardstick, the service-providers in the rural areas are not doing justice to the local residents. The practice is quite in contravention with the WHO proclamation that 'water that is consumed must be safe'. On the basis of location of the sources of drinking water the rural areas are in disadvantageous state. Table 5(a) narrates the situation. It can be seen from the table that there is biasness of the State against the rural areas in the matter of delivering piped water at the zero distance of the houses of its inhabitants. While a majority of the urban households have the facility of getting tap water within their houses, in rural areas this is very rare. Commissioning a piped water supply project is a costly affairs in consideration of the other sources like installation of hand pump or a tube well or a well; even the recurring costs in case of piped water supply schemes are comparatively higher than the other sources. This in fact poses as the prohibitory factor in rapid expansion of the piped water network in villages. Along with this cost factor, inadequate infrastructural development in the rural areas hinders the expansion of rural piped water schemes; for example, the availability of reliable power supply, good road connectivity, technical personnel etc. may be cited. Predominance of hand pumps in the rural areas is due to low technical requirement for operation and maintenance and easy availability of spare parts. In case of any piped water supply scheme, ensuring regular supply of chemicals and spare parts is an essential inputs for maintaining the system. This also can raise problems in expansion of rural piped water schemes. Whatever be the impediments in the path of expanding the accessibility of piped water supply, the investment by the State has to be made in the interest of providing clean and safe drinking water to the rural masses. If we go into the equity aspects of drinking water to see whether all village people have equal access to drinking water within their premises or near the premises, it can be seen that this also is not in a desirable situation. Marginalised sections of the society like scheduled castes and scheduled tribes have less access than those who do not belong to these communities. This aspect is presented in Table 6. For the purpose of table 6 only the proportions of households having drinking water sources within premises and near premises are considered (data taken from census 2001). The logic of considering only these two locations of the spot sources is that of three locations as defined by the census, in these two cases the households have to travel least.

Table 5(a)
Distribution of Households on the Basis of Accessibility to Drinking Water Sources and their Location, 2001

Area/Location	*Tap*	*Handpump*	*Tube well*	*Well*
Rural				
Within premises	1.33	17.25	2.08	2.71
Near premises	4.27	39.37	5.59	6.27
Away from premises	1.41	12.61	3.06	2.43
Urban				
Within premises	33.97	10.46	4.94	3.90
Near premises	17.78	9.40	5.24	1.77
Away from premises	4.90	3.05	2.53	0.82

Source: Tables on Houses, Household Amenities and Assets, Census 2001, West Bengal.
Note: All figures are in percentage.

Of the total rural households, only 23.40 per cent have the source of drinking water (any source of eight type sources identified by the census 2001). In case of urban areas the corresponding percentage is 53.43 per cent. If exclusively, the scheduled castes households are considered, the corresponding percentages are 22.52 per cent in rural areas and 38.68 per cent in urban areas. Similarly for the households belonging to the scheduled tribe communities, the percentages are 9.70 per cent and 38.50 per cent respectively for the rural and urban areas. Tribals normally live in the fringe areas where very little of amenities are available. For example, the common scenario is no proper road communication (absence of all weather roads), no primary school, no health center, no potable water source, no toilets etc. According to census 2001, only 67.75 per cent of scheduled tribe households in the rural West Bengal have safe sources of drinking water; in case of scheduled castes it is 86.07 per cent while in case of non SC/ST households this is 89.87 per cent. If the collection of water from surface sources like rivers, ponds, tanks etc. is considered, 3.51 per cent of ST households use these for drinking purposes while the corresponding figures for SC and non-SC/ST households are 1.17 per cent and 1.56 per cent respectively. Exposure of ST peoples to contaminated water sources is very high. This calls for urgent change.

From Table 6 it can be seen that very few scheduled tribe households have the facility of drinking water within the boundary of their houses. Two disparity indices are calculated to show the differences in availability of drinking water to the scheduled castes (SC) and scheduled tribes (ST) on one side and non SC/ST households on the other side. The value of disparity index closer to 100 shows more equality than the value away from 100. When this value is equal to 100, it shows perfect equality. The disparity index of SC to non SC/ST assumes value higher than 100 in cases of Uttar Dinajpur and Darjeeling districts where more SC households have drinking water facility within their houses than non SC/ST households. The situation in respect of ST households is depressing in all the districts. Even in the districts where the tribals have high share in total households, their situation with regard to availability of drinking water within premises is no better. These are the examples of deviation in equality aspect. The cases of Burdwan, Birbhum, Bankura, Howrah, Dakshin Dinajpur, Malda and Purulia can be cited. In this districts, values of disparity index of ST to non SC/ST are less than even 30 where optimum value should be 100 to ensure equality among all people irrespective of their caste and community status. When near premises criterion is considered, it can be seen in table 6 that the coverage of ST households is the highest among three groups. This signifies the dependence of the less privileged on community sources. From table 6 it can be noticed that on an average 66.54 per cent of ST households have to travel between 0-0.5 km depending on the number of family members and the extent of domestic use of water. In West Bengal there are as many as eight districts where much more higher proportion of ST households than the State average have to travel the distance of up to 0.5 km to get a source of drinking water. The socially disadvantaged groups are even more likely to travel extended distances if the location of water source is beyond 500 meters; this is the case of location of source away from the premises. In West Bengal 23.76 per cent of ST households and 20.60 per cent of SC households have to walk a distance of more than 0.5 km daily to collect water. The corresponding proportion for non SC/ST households is 19.94

Table 6
Distribution of Households on the Basis of Location of Sources of Drinking Water: Rural, 2001

Distri·	*Within premises*					*Near premises*				
	Non SC/ST	*SC*	*ST*	*DI_1*	*DI_2*	*Non SC/ST*	*SC*	*ST*	*Share of SC house holds*	*Share of ST house holds*
Burdwan	33.11	10.51	5.07	31.74	15.31	50.03	67.91	73.88	32.09	8.27
Birbhum	17.86	4.61	2.11	25.81	11.81	64.40	76.19	81.17	30.55	7.44
Bankura	18.44	5.45	2.08	29.55	11.28	58.86	70.23	71.12	32.96	11.38
Midnapur	9.04	4.25	4.87	47.01	53.87	63.74	66.40	67.39	16.68	9.56
Howrah	6.26	2.78	6.32	44.41	100.96	71.83	72.33	67.62	22.28	0.40
Hooghly	24.56	5.36	3.50	21.82	14.25	40.21	76.59	78.24	29.27	6.31
North 24 Parganas	38.12	35.48	12.89	93.07	33.81	43.43	44.87	65.43	30.24	4.19
South 24 Parganas	4.78	3.59	3.54	75.10	74.06	63.56	63.18	59.03	36.41	1.43
Nadia	59.30	56.88	31.45	95.92	53.03	31.59	33.10	55.26	30.50	2.84
Murshidabad	29.02	18.85	10.41	64.95	35.87	54.59	61.97	69.33	12.37	1.48
Uttar Dinajpur	38.21	44.92	27.97	117.56	73.20	51.63	47.38	59.41	30.42	6.08
Dakshin Dinajpur	28.84	17.49	6.46	60.64	22.40	57.01	67.16	74.73	30.65	18.62
Malda	28.92	16.15	5.38	55.84	18.60	55.47	61.67	65.75	17.96	8.15

Contd....

District	*Within premises*					*Near premises*				
	Non SC/ST	*SC*	*ST*	DI_1	DI_2	*Non SC/ST*	*SC*	*ST*	*Share of SC house holds*	*Share of ST house holds*
Jalpaiguri	46.50	45.65	22.96	98.17	49.38	38.92	39.97	58.49	40.75	21.48
Darjeeling	25.39	32.60	14.43	128.40	56.83	37.85	42.18	52.66	18.74	15.47
Cooch Behar	71.36	65.59	60.86	91.91	85.28	21.06	25.27	27.14	54.16	0.58
Purulia	9.39	4.59	2.56	48.88	27.26	59.70	64.76	65.11	17.59	21.17
West Bengal	25.57	22.52	9.70	88.07	37.93	54.48	56.88	66.54	27.48	7.36

Source: Tables on Houses, Household Amenities and Assets, West Bengal, 2001.

Note: (i) All figures are in percentage.
(ii) DI1 -> Disparity Index (Non SC/ST, SC).
(iii) DI2 -> Disparity Index (Non SC/ST, ST).
(iv) Disparity Index (Non SC/ST, SC) = $\frac{\text{Proportion of SC household coverage}}{\text{Proportion of Non SC/ST household coverage}} \times 100$

Likewise Disparity Index (Non SC/ ST, ST) is calculated.

per cent. The maximum distance one may have to travel for getting water is 1.6 km as Government of India norm is to provide access to safe water source within 1.6 km. Regarding distance norm, the observation of a World Bank study (2006) may be referred here. To quote:

> "Under the 10th Five Year Plan, the Government of India reaffirmed its norms where rural consumers were considered to be served by infrastructure if they were obtainable within 1.6 km distance or an elevation of 100 metres in hilly areas. In contrast, the World Health Organisation classifies rural consumers as having access to improved drinking water supply if the distance traveled to the source is no longer than 1 km. While the United Nations is testing this norm through National Health Surveys in select countries, India still uses wider criteria for access to improved water sources. The disparity in definition of covered populations can produce inflated coverage statistics for India."

If the distance of sources of drinking water can be brought down to 1 km from the houses of the inhabitants, it would have beneficial impact on women particularly, as they are the main carriers of water for their family members. In the reduced distance norm women have to travel less with heavy load on their heads. This can save their time and reduce their several hazards including the health hazards.

Next to equity aspect, let us focus on water quality which has emerged as a major issue. The Standing Committee on Rural Development of 14th Lok Sabha opined that " it is not sufficient to simply provide for drinking water in rural areas, but the Government should simultaneously focus on quality of the water as it has major linkages with health and well-being of the people." In West Bengal, arsenic contaminated habitations have been identified in eight districts. The districts are Burdwan, Hooghly, Nadia, Howrah, Murshidabad, Malda, North and South 24 Parganas. According to the State Government's estimate, ' a total population of 2.86 Crore including urban population of 1.20 Crore are affected by arsenic contamination'. In rural areas 79 blocks in 8 districts are affected. For combating the water quality problems, the government of India in the Department of Drinking Water Supply decided in February 2006' to retain up to 20 per cent of ARWSP funds at the Center to provide focused funding to those States which have reported drinking water quality problems for treatment of water contamination'. (20th Report of Lok Sabha Standing Committee, 2006).

According to the norms, if ground water contains the presence of arsenic beyond permissible limit of 0.05 mg/litre, then it will be arsenic affected. Identification of arsenic affected areas is not an one time exercise; regular monitoring as well as examination of water quality is necessary for checking the extent of damage. The State Government in Public Health Engineering Department has set up district level laboratory in arsenic affected districts and a Central Laboratory in Kolkota. The State Government is endeavouring for an elaborate arrangement to mitigate the problems of arsenic affected areas. Tenth Five Year Plan of the Government of West Bengal indicated that "the State Government has taken up a massive water testing programme and capacity building at community level for management of safe water source in collaboration with UNICEF involving the suffering mass". In arsenic affected blocks, apart from continuous water quality monitoring, the State Government has undertaken a number of river based water supply schemes in the affected districts through best use of surface water. For providing safe drinking water to its citizens, the State Government has the primary responsibility. After the effect of 73rd Constitutional Amendment, this responsibility also came to the panchayats in their respective jurisdiction. The next section discusses institutional arrangements and community involvement in the sphere of drinking water.

SECTION III

In West Bengal, the panchayats at three tiers are guided by the West Bengal Panchayat Act, 1973 (henceforth, the WBP Act). In this Act, there are provisions for formation of different standing committees both in the district tier (Zilla Parishad) and in intermediate tier (panchayat Samiti). The Zilla Parishad and panchayat Samiti act through these committees to discharge their duties and responsibilities enshrined in the WBP Act (section 109). At present there are ten standing committees in each of these two tiers. Of these committees, one is Public Health and Environment Standing Committee. This Standing Committee is looking after rural water supply along with public health and environment. It needs mentioning that all 29 subjects of the Eleventh Schedule of the Constitution of India are distributed over these ten standing committees. According to the provisions of the WBP Act, the head of a standing committee is responsible for the financial and executive administration in respect of the schemes and programmes under the purview and control of the standing committee within the budgetary provisions of the respective panchayat tier.

Well before the 73rd Amendment of the Constitution, in the WBP Act it has been laid down that the duties of a gram panchayat shall be to provide for supply of drinking water and the cleansing and disinfecting the sources of supply and storage of water (subsection 2 of section 19). In subsection 1 of section 19 of the WBP Act it is also provided that a gram panchayat has to prepare an annual plan for each year for development of human resources, infrastructure and civic amenities of the area under its jurisdiction. Provision of safe drinking water falls within provisioning of civic amenities. Through an amendment in the WBP Act in the year 2003, the State Government made provision to constitute five sub committees (Upa Samities) in each and every gram panchayat of the State (section 32A). Of these sub committees, Education and Public health Sub Committee is responsible for drinking water supply in the villages under the jurisdiction of a gram panchayat. According to the provisions of the WBP Act, a gram panchayat shall delegate its power, functions and duties to the respective Upa Samities while upa Samities shall function under the overall guidance and budgetary provisions specified by the gram panchayat. The Department of Panchayats and Rural Development, Govt. of West Bengal, through an administrative order clearly demarcates the activities of each tier of panchayat in respect of ensuring supply of safe drinking water to the rural people. This functional demarcation has been made on the basis of the principle of subsidiary i.e. functions which could be performed best by the lowest tier, should be entrusted only to that tier and not to a higher tier. In Table 7, demarcation of functions of three tier panchayat bodies in drinking water supply is presented.

From the activities spelt out in Table 7, it can be seen that the identification of beneficiaries and the location where the scheme is to be installed is the responsibility of the gram panchayat, the tier nearest to the people. The gram panchayat's concern is mainly with the hand pumps sunk by it or vested in it by other departments/agencies. In gram panchayat there is only one

Table 7
Activities of Different Panchayat Tiers

Activities	*Zilla Parishad*	*Panchayat Samiti*	*Gram Panchayat*
1. Identification of beneficiaries and selection of location	To take up large scheme of piped water supply and to select place and to identify beneficiaries on the basis of recommendation of the panchayat samiti(s)	To select place for piped water schemes (may be one or more) and to identify beneficiaries in consultation with the concerned gram panchayat(s)	To receive proposals from the gram sansads for new sinking and re-sinking of tube wells (hand pumps) and to select places for these purposes.
2. Preparation of schemes and estimates	Public Health Engineering officials at district and sub division level will do	–	Nirman sahayak will prepare schemes and estimates.
3. Execution of schemes	Large rig bore schemes to be taken up	Sinking of rig-bore tube well to be taken up. To hand over the scheme after completion of installation of tube well. In some cases handing over the tube well to the gram panchayat	Repairing and maintenance of all tube wells (hand pumps) in the gram panchayat jurisdiction. If necessary, to allot small amount of fund to gram unnayan samiti (gus) for execution of work.
4. Fund allotment	Alloting fund in favour of panchayat samiti	Allotting funds to gram panchayats pumps), small	For petty repairing and maintenance of tube wells (hand
amount of fund may be given to gram samiti			unnayan
5. Submission of utilization certificate of funds received	Submission of utilization certificate to appropriate department	Submission of utilization certificate to Zilla Parishad	Submission of utilization certificate to Panchayat Samiti.

Source: Department of Panchayats and Rural Development, Govt. of West Bengal.

technical person known as Nirman Sahayak (junior engineer level). There are no employees of PHED posted at gram panchayat to render support to the panchayat for maintenance of the tube wells.

On the other side Zilla Parishads have a good set up of engineering personnel. For this reason the district tier can take up major drinking water supply schemes in the district. At the intermediate tier (block level), there is one technical person in the rank of sub-assistant engineer (diploma holder engineer) and 2-3 fitters of PHED for looking after all tube wells/hand pumps installed by the panchayat Samiti in different gram panchayat areas. In addition if necessary they have to support the activities of the gram panchayats in maintaining drinking water sources.

For decentralized governance and participatory development, the government of West Bengal endows the gram panchayats with the responsibilities to identify the beneficiaries for most of the programmes implemented in the rural areas. The gram panchayats in turn have to honour the suggestion/recommendations made by a gram sansad, a village constituency based forum of people. The specific provision of the WBP Act (subsection 3 of section 19) in this respect is quoted below:

> "A gram panchayat shall not omit or refuse to act upon any recommendation of a gram sansad relating to prioritization of any list of beneficiaries, or schemes or programme so far as it relates to the area of the gram sansad unless it decides in a meeting for reasons to be recorded in writing that such recommendation or recommendations are not acceptable or implementable under the provisions of the Act or any rule, order, or direction there under; such decision of the gram panchayats shall be placed in the next meeting of the gram sansad"

It needs to be added that, all electorates of village constituency are the members of a gram sansad. According to the provisions of the WBP Act, in a year two meetings of a gram sansad are prescribed where people can deliberate upon the activities and make suggestion for new schemes for the benefit of the local area. The intention of the WBP Act is that " a gram sansad shall guide and advise the gram panchayat in regard to the schemes for economic development and social justice undertaken or proposed to be undertaken in its area…" For furtherance of decentralization process and active participation of people in implementation, maintenance and equitable distribution of the benefits of different programmes and schemes, the Government of West Bengal through an amendment in the WBP Act in 2003 created Gram Unnayan Samiti (GUS) in each gram sansad. In GUS provisions have been made for wide representation of people from different sections. The purpose is to ensure holistic representation of local people. Gram Unnayan Samiti's involvement in planning and implementation is just coming up.

With the introduction of Swajaldhara programme in December 2002, it also became necessary to broad base the participation of people in rural water supply systems. A notable feature of Swajaldhara is involvement of the panchayats and/or village water and sanitation committees in planning, execution, operation and maintenance of rural drinking water supply schemes. In Swajaldhara, community contribution is the must. This creates some confusion at the village level because the villagers are accustomed to a supply driven approach where everything is done either by the Government or by the panchayats and people in turn get water free of charge. To turn the ground conditions gram sansad and their gram unnayan samitis are being utilized for adequate sensitization of the communities regarding the demand driven and participatory approach of Swajaldhara. As per the principles of funding pattern in the new programme, the Government of India's share in capital cost is restricted to 90 per cent while remaining 10 per cent will come as community contribution. The incorporation of community contribution has been made for inculcating the feeling of ownership among the villagers. The community contribution may take the form of cash/kind/labour/ land or combination of these. According to a report of the Department of Drinking Water Supply, Govt. of India as on 15th February 2006, in West Bengal it became possible to mobilize Rs. 51.88 lakh as community contribution for 85 schemes. Swajaldhara guidelines prescribe that operation, maintenance and management costs of the water supply schemes have to be borne fully by the concerned panchayats or VWSC or user group as may be the case. The government will not take any responsibility towards these expenditures.

For capturing the grassroots reality, a quick survey has been organized to assess the functions of the village panchayats and intermediate panchayats in providing drinking water to the villagers. Unbundling of the activities indicates that no user groups are formed in twenty-five surveyed gram panchayats. For this reason every gram panchayat has to spend around 30 per cent of their total allocation for rural water supply on repairing of hand pumps/tube wells. Piped water supply schemes are in the jurisdiction of PHED and maintenance works are done by the department. No household connection is there, stand-posts are only erected. Regarding periodicity of supply it varies from area to area. In some areas supply is for only two hours in a day when the people has to collect their required water for the day. In all the surveyed gram panchayats there is no system of periodic chlorination of tube wells and other drinking water sources for disinfection. There is no system of periodic testing of water quality also. In one arsenic affected block (Karimpur II in Nadia district) there is a water testing laboratory set up at block headquarter in collaboration with a local non-government organization. In West Bengal, water-testing laboratories have been set up at district and sub district levels in arsenic affected districts only. The arsenic testing laboratory at Karimpur II block caters to the need of four blocks, namely, Karimpur I, Karimpur II, Tehatta I and Tehatta II.

In respect of imposition of user charges, the West Bengal Panchayat Act makes a provision in section 47 that a gram panchayat may levy a water rate where arrangement for the supply of water for drinking purpose is made by the gram panchayat. Feedback from the field on this issue is not worth noting. Lack of awareness among the heads of the village panchayats, heads of intermediate tier panchayats and the general members in different areas of management of rural water supply is of great concern. In water sector their main thrust is on irrigation water; that too at no cost. Their logic is simple– for sustaining the livelihood of the rural people first priority should be to promote agricultural activities through assured irrigation. Uncontrolled exploitation of ground water creates problem for the hand pumps/tube wells installed for drinking purpose (as their depth is mostly shallow). This happens particularly in summer season and during inadequate rainfall when groundwater aquifers are not sufficiently recharged. Change in attitude of the panchayat members towards the need for provisioning of drinking water instead of thrusting on more use of water for irrigation can improve the situation. It is expected that formation of user groups and community involvement in drinking water can put pressure on the panchayats to alter their priorities. Another advantage of formation of user groups will be mobilization of own resources for the gram panchayats as the water rate can be levied. When operation and maintenance of hand pumps/tube well undertaken now by the gram panchayats, can be handed over to the user groups, local youths can get at least marginal employment through rendering their services. Initially, the panchayats may make arrangements for the training of some youths at the government initiatives. The discussion on institutional arrangement may be concluded with an observation of the World Bank Study (2006). In our field level exploration similar experiences are gathered. To quote from the study:

> " Currently, there are unclear and overlapping responsibilities for policy making, financing, regulation, ownership of assets and provision of services, resulting in a low level of accountability and deteriorating rural wss (water supply and sanitation) services. It is important to unbundle and restructure the State institutions and agencies in line with the shifted role of the State as a facilitator and the devolution of funds, functions and functionaries to the PRIs and user communities."

Incomplete devolution, that is, devolution of functions without appropriate devolution of funds and functionaries may hamper the services the Government wants to deliver. Swajaldhara programme attempts to reduce these gaps through transfer of ownership of the assets to the community.

CONCLUSIONS

With the increasing thrust from the Government of India to hand over the management of drinking water sources to the community and the panchayats, it becomes necessary to take up capacity building measures for the panchayat members and the members of VWSC. Capacity building here does not mean training only; it includes support services of technical persons of the PHED, exposure visits to better performed areas, training of rural youths regarding operation and maintenance of the pumps etc. In other words, it includes all steps required to build up a conducive environment to own and manage a system.

In spite of the persuasion from the State Government, own resource mobilization by the panchayats does not assume encouraging trend. The panchayats need to be motivated extensively for imposing user charges for using a stand-post of piped water or hand pump or tube well installed by any tier of the panchayat set up. Any shift in the present approach of the panchayat leaders towards this issue calls for political willingness to impose user charges. In urban areas, the households using piped water are paying user charges to the municipalities. The vicious cycle of low quality service, low revenue and low willingness to pay user charges will continue in rural areas, so long the State Government continues to promote water as a social right. Lack of funds in the hands of the panchayats sometimes restrains them from undertaking repairing works on urgent basis. This only adds to the miseries of the people using the water source. Suggestion is that the State Government may think to allot around thirty to fifty thousand rupees as revolving funds to the gram panchayats for taking up repairing works quickly. The quantum of funds may be determined on the basis of population in a gram panchayat. The system of financing may be like this : as soon as the available fund is exhausted, the gram panchayat will submit utilization certificate to the panchayat samiti to get further fund of equal amount.

Representation of at least one third women in the VWSC as members is a prerequisite on the basis of Swajaldhara guidelines. For need based planning at the village level, fulfillment of this requirement seems to be essential as women are the principal carriers of water and they know where the shoe pinches. This is not practiced, however. Some form of enforcement from above, may be the district level or state level can help to increase women representation in these committees, Self help groups formed with women at the village levels may be entrusted with the maintenance of hand pumps/tube wells. Skill training to women may be imparted through Swarna Jayanti Gram Swarozgar Yojana Training Fund. This fund is available with District Rural Development Agency at the district and panchayat samiti at the block. Involvement of SHGs in Swajaldhara programme may yield better prospects for formation of user groups. Whatever be the models and modalities, it needs to be emphasized that access to safe water is a human right and it must be respected and fulfilled by the State.

REFERENCES

FIAN International Secretariat, Germany-Update on the Enjoyment of the Right to Food and Water in India, 2004.

Gleick, Peter–The World's Water (2000-2001), Published by Island Press, Washington in 2000.

Government of West Bengal-Economic Review, 2005-06.

Government of West Bengal, Development and Planning Department, Tenth Five Year Plan, Volume I, 2002.

Lok Sabha Secretariat, New Delhi -20th Report of the Standing Committee on Rural Development, 2006.

Ministry of Rural Development, Government of India - Annual Report, 2004-05.

Planning Commission, Government of India–

(i) National Perspective Plan for Women 1988–2000.

(ii) Tenth Five Year Plan, Volume II.

(iii) Mid term Appraisal of Tenth Five Year Plan

(iv) India Assessment (2002), Water Supply and Sanitation, A WHO-UNICEF sponsored study, 2002

United Nations Millennium Project–Report of the Task Force on Water and Sanitation, 2005, Published by Earth Scan, London, UK, 2005.

World Bank–India: Water Supply and Sanitation, Bridging the Gap between Infrastructure and Service, 2006.

World Health Organisation–The Right to Water, 2003.

30

GOVERNMENT INITIATIVES IN RURAL AND URBAN WATER SUPPLY A STUDY IN THREE DISTRICTS OF KBK REGION IN ORISSA

P. SivaRam[*], *R.R. Hermon*[**], *A.V. Yadappanavar*[***] & *T. Vijaya Kumar*[****]

INTRODUCTION

The KBK region of Orissa State is one of the most backward regions in terms of socio-economic development. The undivided districts of Koraput, Bolangir and Kalahandi popularly called as KBK districts. These three districts have been divided into eight districts, namely, Koraput, Malkangiri, Nawrangpur, Rayagada, Bolangir,

* Associate Professor, Center for HRD, NIRD, Rajendra Nagar, Hyderabad-500 030 *E-mail: siva@nird.gov.in*

** Assistant Professor, Geomatic Cell, NIRD, Rajendra Nagar, Hyderabad, *E-mail: rrh@nird.gov.in*

*** Associate Professor, Center for Gender Studies, NIRD, Rajendra Nagar, Hyderabad -50030.

**** Assistant Professor, Center for HRD, NIRD, Rajendra Nagar, Hyderabad -500 030.

Sonepur, Kalahandi and Nuapada in 1992. In the year 1988, to develop this area, the Government of Orissa had formulated and implemented a special programme on 'Area Development Approach for Poverty Termination' (ADAPT) in 15 blocks of two districts, namely, Koraput (8 blocks) and Kalahandi (7 blocks). This was implemented in order to provide employment round the year to the rural poor. In due course, the Government felt that short term strategies were not appropriate to address the multi-faceted backwardness of the region. Therefore, a Long Term Action Plan (LTAP) for the three undivided districts of KBK districts was prepared in 1993 in consultation with the Government of India. The LTAP was conceptualized for a period of seven years from 1995-96 to 2001-02 with two objectives in view: (a) drought and distress proofing, and (b) poverty alleviation and development saturation. LTAP envisaged an outlay of Rs. 4557.03 crores. However, LTAP did not take off for want of availability of funds. In 1998, a Revised Long Term Action Plan (RLTAP) was submitted to the Government of India. The revised project was envisaged for a period of nine years from 1998-99 to 2006-07 with an outlay of Rs. 6251.06 crores.

The RLTAP was prepared in a sub-plan mode to address the peculiar socio-economic problems of this chronically poor region, which is also geographically contiguous. This project envisages an integrated approach for speeding up the socio-economic development of this region by synergizing effectively the various developmental activities and schemes under implementation both in Central as well as State Governments. The critical gaps in the development efforts as well as resources are sought to be bridged through Additional Central Assistance (ACA)/Special Central Assistance (SCA) as a special dispensation.

The focus of RLTAP is not only on economic development but also on social development. The relationship between drinking water and food, health and nutrition is well known fact to improve the quality of life of the people. The social development programmes under RLTAP are–Urban Drinking Water Supply, Rural Drinking Water Supply, Mobile Health Unit and Emergency Feeding programme. These programmes have been implemented in all the eight districts of KBK region.

OBJECTIVES

- To study and analyse Rural and Urban Drinking supply schemes under RLTAP
- To study the impact of the Rural and Urban Drinking supply schemes
- To study the implementation and other constraints and to suggest remedial measures

STUDY AREA

Three districts, namely, Koraput, Navarangpur and Malkangiri are randomly selected for the study. The social development programmes viz. Urban Drinking Water Supply, Mobile Health Units and Emergency Feeding were selected for the study. The sample blocks and villages are selected in consultation with the concerned officials of departments-Public Health Engineering Department (PHED), Rural Water Supply (RWSS), Health and Family welfare and Social Welfare. From each block two or three villages are selected for in-depth study. The sample Districts, Blocks and villages are presented in Table 1.

Table 1
Distribution of Sample Districts, Blocks and Villages

Name of the Sample District	*Name of the Sample Blocks*	*Name of the Sample Village*
1. Koraput	1. Laxmipur	1. Teksahi
		2. Laxmipur
	2. Jeypore	3. Kenduguda
		4. Ghat Bagra
2. Navarangpur	3. Papadhandi	5. Dalaiguda
		6. Minza
	4. Ummar Kot	7. Badabarandi
		8. Jamarunda
3. Malkangiri	5. Katimela	9. M.V.55**
		10. Jorapalle
	6. K. Gumma	11. M. V. 36
		12. K.Gumma

Source: Field study: ** Malkangiri Village (MV).

METHODOLOGY

Participatory rural appraisal tools such as Group Discussions (GDs), focus group discussions (FGDs), semi-structured interviews and transects are used to gather qualitative information on the RLTAP social development programmes from providers' point of views and villagers'/users' point of views. Also, quantitative tool like structured interview schedule is used to gather information about nature and extent of RLTAP initiatives. The secondary data is collected from the concerned departments. The field work was conducted in the month of May, 2006.

BRIEF PROFILE OF THE KBK REGION

The KBK districts account for 19.72 per cent of population to the State population. Nearly, 90 per cent of the people of this region are still living in the villages. The region has low population density (152 persons/sq. km.), while compared to 236 persons of the State indicates difficult living conditions and an undeveloped economy. The region is dominated by the tribal communities including four primitive tribal groups (PTG), namely, Bondas, Dadai, Langia Sauras and Dangaria Kandhas. The literacy rate (36.58 per cent) are also far below the State (63.61) literacy average. The female literacy is only 24.72 per cent whereas the state female literacy is 50.97 per cent. As per BPL survey (1997), about 72 per cent families below poverty line live in this region. Other socio-economic indicators including population composition and density, net area irrigated, hospital beds, and connectivity of villages to growth centres and service centers are also far from satisfactory. According to the report of "the Committee on the Constitution of Separate Development Board in Orissa", 96 per cent of Community Development (CD) Blocks in these districts are either 'very backward' or 'backward'. To be specific, 49 CD Blocks of KBK districts are regarded as "very backward" and 28 CD Blocks are considered as "backward".

Government Initiatives in Drinking water Supply

Urban Drinking Water Supply Programme

The Public Health Engineering Department (PHED) is a nodal agency for planning, implementation and monitoring of urban water supply under RLTAP. In Koraput, Jeypore and Navarangpur towns were covered under urban water supply programme for increasing drinking water storage capacity. These schemes are to cover uncovered areas and also to meet the demand of increasing population.

Programme Strategy

- Augmentation of existing storage capacity of water.
- To increase the duration of drinking water supply timing.
- To increase the area of coverage by laying distribution pipelines in uncovered areas.

Programme Implementation

The planning is done by PHED at the district level. The same is sent to Head office of PHED at Bhubaneswar for approval. After approval of the scheme, it is implemented in the district through PHED concerned officials. FGDs with the PHED concerned officials revealed that some times, the approvals are delayed. Because of this, the implementation is affected.

In the sample districts, the urban water supply O&M is being done by PHED. The entire cost of O&M is borne by the PHED. The collection of water user charges of household piped water supply connections is also about 60 to 80 per cent in sample districts except in Jeypore town of Koraput town.

Impact

The FGDs with the users in the urban areas/towns revealed the following benefits accrued due to them:

- The availability of water increased from 30 LPCD to 60 LPCD through household connections and public stand posts.
- The Public Stand Posts have increased from 1015 to 1181 across the sample towns. Majority of the lanes/streets are provided the public stand posts except newly developed housing complexes and lanes.
- Group Discussions with District level Doctors and Health officials in the project area as well as data from the hospitals revealed that there is a reduction in water borne diseases. The people have confirmed the fact during group discussions.
- The FGDs with women revealed that they are satisfied with new augmentation water supply as they are getting adequate water supply even in summer seasons. They mentioned that time taken to fetch drinking water has drastically reduced.

- Household connections have increased from 26,734 to 29,432. Middle and rich income groups have taken the household connections. These people have deposited Rs. 3000/- for each household piped water connection and they are paying Rs. 50 to Rs. 70/- (each household connection) as water charges per month. It is recorded that water charges collection is better in Koraput, while compared to other urban areas/towns. The officials mentioned that water tampering is common phenomenon in Jeypore town.
- The people mentioned in group discussions that school drop outs are reduced among girl children. And these girl children are regularly attending the schools. This is a positive development indicator.

Field Observations based on FGDs and Transects of Three Study Districts

- People revealed that the schemes were planned and executed without the user's involvement/people's participation.
- New and uncovered areas of towns required pipelines of different diameter for giving adequate connections which was not given due attention.
- FGDs revealed that tampering of distribution systems particularly in Jeypore is resulting in wastage and contamination of water, leading to costly repairs.
- The tail enders are not receiving adequate water supply.
- Erratic electricity supply is a major cause in disrupting drinking water supply in the towns.
- The water pressure is always low for household residing at higher altitude.

Rural Drinking Water Supply Programme

In sample districts, the villagers used to collect drinking water from tanks, springs, and rivulets. Waterborne diseases such as diarrhea, malaria, gastroenteritis etc, were prevalent in the areas, which lead to loss of wages. This frequently and badly affected their livelihood patterns. In addition, by and large, the women folk were engaged in fetching drinking water from far away (2 to 3 Km. from homes) though the male family members usually go for earning their livelihoods. To cope up with this situation, Department of Rural Development in charge of Rural Water

Supply, Government of Orissa had implemented the Accelerated Rural Water supply schemes (ARWSP) schemes, and Minimum Needs Programme (MNP) of Ministry of Rural Development, Government of India in the villages of sample districts. These schemes covered majority of revenue villages. However, the inaccessible and remote villages and habitations still remained to be covered under these schemes. Therefore, Government of Orissa had introduced rural water supply schemes (RWSS) to cover inaccessible, remote and uncovered villages under RLTAP by providing hand pumps, tube wells, construction of sanitary wells, spring based water supply system and piped water supply schemes. The objectives of RWSS are as follows.

Programme Strategy

- The norms for installation of hand pumps / bore wells in KBK districts were relaxed. Against a bore well/hand-pump installed for 250 population or at a distance of 1.6 Km, as per Government of India norm, in KBK districts, was relaxed to with 150 population or even less than that.
- To ensure provision of safe drinking water in all rural habitations especially inaccessible and remote villages and habitations.
- Awareness creation through IEC, especially, in personal hygiene and water handling practices.
- Completed schemes are to be handed over to the Gram Panchayats for O&M.
- To improve health status of people by reducing the water-borne diseases in rural areas of KBK districts.

Programme Implementation

At the district level, the executive engineer, Rural Water Supply (RWSS) prepare the plans with Engineers and officials of blocks in the district. While preparing the plans, the priority is given to inaccessible and remote villages/habitations. The prepared plans are then sent to the District Level Committee (DLC) of RLTAP for scrutiny and approval. Subsequently, the approved plans are sent to the Chief Engineer, RWSS at the state Head quarters for sanction of the rural water supply schemes and the same will be informed to the concerned Executive Engineers at the district level. Based on the approvals, the Executive Engineers initiate the process of implementation. The officials mentioned that they have been getting financial approvals in time. Therefore, the schemes are being implemented in time in the rural areas of sample districts.

Field Observations based on FGDs and Transects of Three Study Districts

- Social Mobilization is not used for bringing villagers together and creating awareness towards ensuring participation in drinking water supply schemes of RLTAP. In the practical sense, social mobilisation is the pre-cursor to the community participation, which is lacking in the sample villages of study districts. Therefore, Social mobilization process should be initiated by Department responsible for implementation of RWSS in partnership with NGOs, CBOs and PRIs for ensuring community participation, which in turn bring the ownership feeling among villagers towards drinking water supply systems.
- In all the drinking water supply schemes of RLTAP, there is no people's/users' participation because the drinking water supply schemes do not have in-built system of involving villagers in planning, implementation and monitoring. Therefore, the sustainability of water schemes is doubtful.
- The people of sample villages are primarily depending on agriculture for their livelihood. Paddy and sugarcane being the main crops (in the sample villages of Koraput and Navarangpur districts), which are high water consumption crops causing over exploitation of ground water, leading to shortage of drinking water. There are still uncovered remote and inaccessible villages in study districts. Hence, there is an urgent need to provide at least one source of drinking water in uncovered villages/habitations.
- Though RWSS maintains the water schemes in the villages, it does not have adequate funds for maintenance of water supply systems.
- In February 2006, water supply schemes of RLTAP were handed over to Gram Panchayats for O&M. The villagers indicated that PRIs are attending to only small repairs but they are unable to carry out major repairs of water supply schemes as they do not have adequate funds for O&M. In addition,

they are not oriented and trained to maintain water supply schemes. In some of the sample villages, the deep tube well schemes with stand posts through piped water supply were running with electricity. The power supply is erratic, and some times the power supply would not be there two to three days. During the time, the villagers depend on hand pumps or tanks and springs causing water borne diseases.

- Wherever piped water supply is there, only the main village is getting adequate water and the hamlets located at a distance are not getting adequate water.
- Self Employed Mechanics (SEMs) of RWSS are p[aid Rs.300-400 per month, which is low. They attend only minor repairs and cover 20 to 25 hand pumps in a Gram Panchayat.
- SEMs also do not have adequate spare parts for O&M causing continuous breakdown of water supply systems.
- SEMs skills need to be upgraded so as to enable them to undertake major repairs in remote and hilly areas. These people should be provided with adequate tool kits.
- It is noticed that there is a lack of coordination between RWSS and Panchayats and SEMs at village level.
- Village Water and Sanitation Committees (VWSCs) are under formation stage in the study villages.
- It is observed that there are unhygienic conditions in terms of stagnant water around tube-well/hand pump resulting mosquitoes breeding in and causing Malaria.
- It is also observed that taps are broken and some times illegal house connections were also taken in some sample villages, especially, in the context of piped water supply.

The maintenance of a tube well is Rs. 1000 per annum and for a piped water supply per 1000 population costs around Rs. 40,000 per annum. At present, the said amount is being borne by the RWSS.

Impact

- The primary health center data indicated that the water borne diseases have come down drastically in the sample villages because of installation of hand pumps and piped water supply under RLTAP. The villagers are regularly earning wages. The overall impact is that the quality of life has been improving in the sample villages of KBK districts.

- The availability of Safe Drinking water is 40 LPCD in the sample villages of KBK districts, which is stated by the villagers in FGDs.
- The burden of carrying water by women for long distances is reduced from the distance 2.5 km to 200 meters. The saved time by the women folk is being utilized for income generating works in SHGs, which supplements to their income.
- School attendance of Girl children has increased to the extent of 15 to 20 per cent.
- The PHCs, records shows a reduction of 50 to 60 per cent of diarrhea cases in the sample villages.
- The field evidences revealed that every caste and tribe is having access to fetch the drinking water from hand pumps, piped water supply and tanks. As such, there are no separate water systems for upper and lower castes in the sample villages.
- At present, 95 per cent of villagers are depending on water supply sources for drinking water created under RLTAP and other state and centrally sponsored drinking water schemes.

CONCLUSIONS

The Government initiatives through drinking water supply programmes have started giving positive results in the sample villages. Before RLTAP, the villagers used to walk down three to four km for fetching a pot of water in the sample villages. At present, no doubt the safe drinking water is available within village vicinity. As a result the water-borne diseases have come down in the sample villages. This is a positive sign development indicator. It is clearly visible that drinking water infrastructure development has been occurring in the sample villages. However, the sustainability of the schemes will depend on the people's participation in O & M of the system and also ability to generate resources through user charges to meet the expenses on the O & M.

REFERENCES

Downloads of RLTAP of KBK districts.

Records of PHED for Urban Water Supply Programme of Koraput, Jeypore and Navarangpur.

Records of Department of Rural Development in-charge of Rural Water supply of Koraput, Malkangiri and Navarangpur.

Records of Department of ICDS of Koraput, Malkangiri and Navarangpur.

Records of Department of Social welfare of Koraput, Malkangiri and Navarangpur.

Records of Department of Health and Family Welfare of Koraput, Malkangiri and Navarangpur.

Table 1.1
District wise status of Swajaldhara Programme, from 2002-2005

(Rs. in Lakh)

Sl. No.	*District*	*Funds received*	*No. of Schemes take up*	*No. of schemes completed*	*No. of Schemes in progress*	*Expenditure*	*No. of installments Received*	*Percentage of fund utilization*
1.	Anugul	26.41	53	7	46	19.50	4	73.83
2.	Bargarh	28.28	10	0	10	11.313	4	39.39
3.	Bhadrak	91.31	86	0	86	25.775	4	28.22
4.	Bolangir	95.34	79	8	71	84.40	5	89.57
5.	Boudh	35.495	33	0	33	20.877	3	58.81
6.	Cuttack	135.30	108	0	108	66.86	4	49.41
7.	Deogarh	12.30	4	0	4	7.75	3	63.00
8.	Dhenkanal	45.90	10	0	10	36.12	5	78.69
9.	Gajapati	24.73	12	0	12	12.71	3	51.39
10.	Jagatsinghpur	1.89	2	0	02	0.00	1	0.00
11.	Jajpur	84.69	22	0	22	0.00	3	0.00
12.	Jhasuguda	10.345	14	10	04	7.47	3	72.20
13.	Kalahandi	16.69	6	0	06	5.46	4	32.71
14.	Kandhamal	6.67	15	3	12	2.41	4	36.13
15.	Kendrapara	56.29	8	2	06	29.52	4	52.44
16.	Keonjhar	20.295	19	1	18	5.51	4	27.14
17.	Khurda	38.66	16	0	16	5.85	4	15.13
18.	Koraput	19.56	32	0	32	11.88	3	60.73
19.	Malkangiri	150.54	183	59	124	108.69	4	72.20
20.	Mayurbhanja	86.03	20	0	20	23.45	2	27.45
21.	Nawarangpur	47.04	8	0	8	19.54	3	41.53
22.	Nayagarh	35.33	13	0	13	0.00	3	0.00
23.	Nuapara	8.10	11	0	11	5.84	1	72.09
24	Puri	14.09	11	0	11	9.45	3	67.06
25.	Rayagada	50.34	17	1	16	36.25	2	72.01
26.	Sambalpur	48.615	31	18	13	36.63	3	75.34
27.	Sonepur	27.85	13	0	13	15.90	2	56.98
28.	Balasore	52.68	20	0	20	0.00	1	0.00
29.	Ganjam	89.87	17	0	17	57.00	1	63.42
30	Sundargarh	11.58	4	0	07	10.15	1	87.65
	Total	1372.22	877	109	768	676.305		49.28

Source: Annual Report, 2004-05, Rural Development Department Government of Orissa.

Table 1.2
Identification and Sensitization of Stakeholders in Swajaldhara Programme

Districts	Need Identification					Methodology					Interest				
	1	*2*	*3*	*4*	*Total*	*i*	*ii*	*iii*	*iv*	*Total*	*A*	*b*	*c*	*d*	*Total*
1. Bolangir	34	64	2	–	100	25	60	–	15	100	85	72	15	11	100
	(34)	(64)	(2)		(100)	(25)	(60)		(15)	(100)	(85)	(72)	(15)	(11)	
2. Mayurbhanj	22	78	–	–	100	32	60	–	8	100	75	80	12	16	100
	(22)	(78)			(100)	(32)	(60)		(8)	(100)	(75)	(80)	(12)	(16)	
3. Khurda	35	65	–	–	100	31	62	–	7	100	85	88	21	25	100
	(35)	(65)			(100)	(31)	(62)		(7)	(100)	(85)	(88)	(21)	(25)	
Total	91	207	2	–	300	88	182	–	30	300	245	240	48	52	300
	(30.3)	(69)	(0.66)		(100)	(29)	(60.6)		(10)	(100)	(81.6)	(80)	(16)	(17.3)	

N.B. :
(i)

Need identification	*Methodology*	*Interest*
1. By the villagers	(i) Discussion	(a) Availability
2. By RWSS Deptt.	(ii) Meeting	(b) Time Savings
3. By Gram Panchayat	(iii) Training	(c) Health & Sanitation
4. By NGO/ Others	(iv) Other Methods	(d) Others

(ii) Figures in the parenthesis indicate the percentage.
(iii) Since one single respondent has given multiple responses, figures in the parenthesis do not match with the total.

Table 1.3
Participation of Primary Stakeholders in Swajaldhara

Districts	*Planning & Designing*					*Implementation*						*Decision Making*					
	1	*2*	*3*	*4*	*Total*	*i*	*ii*	*iii*	*iv*	*v*	*Total*	*a*	*b*	*c*	*D*	*e*	*Total*
Bolangir	69	65	21	28	100	27	21	18	3	31	100	6	18	27	12	37	100
	(69)	(65)	(21)	(28)		(27)	(21)	(18)	(31)	(31)	(100)	(6)	(18)	(27)	12	37	
Mayurbhanj	65	64	23	32	100	24	18	15	6	37	100	7	9	21	15	38	100
	(65)	(64)	(23)	(32)		(24)	(18)	(15)	(6)	(37)	(100)	(7)	(19)	(21)	15	38	(100)
Khurda	64	63	28	39	100	33	24	13	9	21	100	12	21	33	18	16	100
	(64)	(63)	(28)	(39)		(33)	(24)	(13)	(9)	(21)	(100)	12	21	33	18	16	(100)
Total	198	192	72	99	300	84	63	46	18	89	300	25	58	81	45	91	300
	(66)	(64)	(24)	(33)		(28)	(21)	(15)	(6)	(30)	(100)	(8)	(19)	(27)	15	30	

N.B.

(A) PLANNING & DESIGNING	*IMPLEMENTATION*
1. Choice of the Scheme	(i) Construction activity
2. Location mapping	(ii) Procuring of raw materials.
3. Estimate preparation	(iii) Day to Day supervision
4. Others	(iv) Others
	(v) In involvement

(A) DECISION MAKING

(a) Control of finance
(b) Voluntary contribution
(c) Management and coordination
(d) Others
(e) No involvement

(B) Figures in the parenthesis indicate percentages of the total. Since a single respondent has given multiple responses figures in the planning and designing column do not match with the total.

Table 1.4
Status of Completed Swajaldhara Projects

No. of Projects Visited	No. of Projects Completed	Agree-ments Signed	Owner-ship by Gram Panchayat	Owner-ship by Pan-chayat Samities	Owner-ship by Zilla Parisads	Depositing of O & M fund
30	6	6	NIL	NIL	NIL	6

Table 1.5
Mobilization cf local people and planning for Swajaldhara

District	Facilitator					Scheme/ System Planning					Consensus			
	1	*2*	*3*	*4*	*Total*	*i*	*ii*	*iii*	*iv*	*Total*	*a*	*b*	*c*	*Total*
Bolangir	62	40	5	38	100	72	78	76	69	100	27	20	53	100
	(62)	(40)	(5)	(35)		(72)	(78)	(76)	(69)		(27)	(20)	(53)	(100)
Mayurbhanj	75	30	7	35	100	68	66	70	75	100	29	31	50	100
	(75)	(30)	(7)	(35)		(68)	(66)	(70)	(75)		(29)	(31)	(50)	(100)
Khurda	65	42	4	31	100	71	73	75	76	100	28	24	48	100
	(65)	(42)	(4)	(31)		(71)	(73)	(75)	(76)		(28)	(24)	(48)	(100)
Total	202	112	16	104	300	211	217	221	222	300	84	65	151	300
	(67)	(37)	(5)	(35)		(70)	(72)	(74)	(79)		(26)	(22)	(50)	(100)

N.B.:

(A) 1- RWSS, 2- VWSC, 3- NGO/ G.P., 4- A11
(i) Population
(ii) Requirement
(iii) Source
(iv) Others
(a) Motivation
(b) People's interest
(c) Others

(B) Figures in the parenthesis indicate the percentage. Since a single respondent have given multiple responding figures in column 1 & 2 do not match with the respective total.

31

STATE-SOCIETY PARTNERSHIP IN DRINKING WATER MANAGEMENT THE CASE OF *JALANIDHI* PROJECT IN KERALA

T.M. Joseph[*] & Jos Chathukulam*[**]

INTRODUCTION

The advent of the new economic order, manifested in the form of globalization, liberalization, and privatization has brought in a new governance paradigm prescriptively being referred to as good governance. The new system of governance is making the state to loose its pre-eminence as an agent for development. The government has started receding to the background from many of its traditional areas of operation and the non-governmental organizations, community based groups and multinational development agencies have stepped in to fill that void. The thrust of the new approach appears to be one of maximising the growth potential by unleashing outward looking and market driven forces

* Dr. T. M. Joseph, Reader & Head Department of Political Science, Newman College, Thodupuzha.

** Dr. Jos Chathukulam, Director Centre for Rural Management, Kottayam, Kerala.

by making necessary adjustments in the economic structure and the manner of exercise of control over power and resources (Government of India 2002).

BASIC PRINCIPLES OF GOOD GOVERNANCE

The concept of good governance became prominent in the late 1980's simultaneously with the end of cold war. The idea is propelled and propagated by the Bretton Woods institutions-the World Bank, the IMF and the WTO. The notion of good governance is a logical extension of the Structural Adjustment Programme (SAP), proposed by these institutions. Its aim is to complement SAP by political reforms to achieve greater accountability, transparency and efficiency in public service along with the protection of civil and human rights, and to ensure the rule of law through an independent judiciary (Oommen 2003:128).

It appeared when Structural Adjustment Programmes experienced serious implementation difficulties which were then diagnosed as failures of public management owing to unaccountable, patrimonial, and non-transparent governments. Consequently, World Bank (1992:3) defined governance "as the manner in which power is exercised in the management of a country's economic and social resources for development". Good governance, for the World Bank, appeared to be synonymous with sound economic management. In answering to widespread criticism of the international development community that top-down institutional structural reforms lead to institutional vacuum rather than a mobilisation of the societal self-help potentials, the Development Assistance Committee of the OECD formed a Working Group on Participatory Development and Good Governance. It linked good governance to participatory development, human rights and democratisation, and identified the rule of law, public sector management, controlling corruption and reducing excessive military expenditure as important dimensions of good governance (OECD 1995:14).

The new system of governance is making the state to loose its pre-eminence as an agent for development. The government has started receding to the background from many of its traditional areas of operation and the non-governmental organizations, community based groups and multinational development agencies have stepped in to fill that void. The thrust of the new approach appears to be one of maximizing the growth potential by unleashing outward looking and market driven forces by making necessary adjustments in the structure of the economy and the manner of exercise of control over power and resources (Government of India 2002).

UNDP's (1997:4) understanding of the concept of good governance reflects this changing awareness about the relative role of the state:

Governance can be seen as the exercise of economic, political and administrative authority to manage a country's affairs at all levels. It comprises the mechanisms, processes and institutions through which citizens and groups articulate their interests, exercise their legal rights, meet their obligations and mediate their differences. Good governance is, among other things, participatory, transparent and accountable. It is also effective and equitable. And it promotes the rule of law.......Governance has three legs: economic, political and administrative.... Governance encompasses the state, but it transcends the state by including the private sector and civil society organisations.

The good governance agenda underlines the curbing of the role of state and expanding the space for market and competition. It further places emphasis on civil society institutions to strengthen democracy and in constructing an informal sector that can harness people's entrepreneurship through community institutions and inter-personal relationships. That is, governance encapsulates government but it also encompasses non-governmental mechanisms to meet the demands, and aspirations of citizens. In this way, the state and civil society are co-responsible for citizen welfare (Oommen 2003:128-9). In the good governance discourse, democracy emerges as the necessary political framework for successful economic development, and within this discourse democracy and economic liberalism are conceptually linked: bad governance equals state intervention; good governance equals democracy and economic liberalism. (Abrahamsen 2000:51).

In its current usage 'good governance' always includes the following core elements: people-centred, equity, accountability, transparency, participation and consultation in planning and decision-making, effective and efficient public sector management, and civil society involvement. In other words, good governance is pro poor and gender-conscious, is legitimized by participatory processes, undertakes anti-corruption efforts, is bureaucratically accountable, is efficient and effective in the use of resources, and promotes the active involvement of the private sector and the civil society to counteract vested interests. Box 1 provides a full list of properties which have been associated with good governance developed by a UNDP Workshop on Good Governance for Sustainable Development in 1996.

The style of governance has far reaching consequences for people's behavior. Participatory governance gradually leads to more egalitarian institutions which will substitute the rigid hierarchical institutions which captured the thinking and behavior of people.

Box 1
Characteristics of Good Governance

- Participatory
- Sustainable
- Legitimate and acceptable to the people
- Transparent: Promotes equity and equality
- Able to develop the resources and methods of governance
- Promotes gender balance
- Tolerates and accepts diverse perspectives
- Able to mobilize resources for social purposes
- Strengthens indigenous mechanisms
- Operates by rule of law
- Efficient and effective in the use of resources
- Engenders and command respect and trust
- Accountable
- Able to define and take ownership of national solutions
- Enabling and facilitative
- Regulatory rather than controlling
- Able to deal with temporal issues
- Service-oriented

Though the essence of the new paradigm could be seen in the efforts made in the west to dismantle the welfare state in the 1980's, the clear conceptualization and articulation of the paradigm emerged only at the end of 1990's. The emergence of this new paradigm of governance has serious implications for both developed and developing nations (Chandhoke 2003). In the developed countries of the west, the role of welfare state has been declining under the influence of the neo-liberal Reagan Thatcher ideology, which advocates a drastic cut in the public spending on welfare activities. In the developing countries, the state is being forced to give more scope for other actors in development with the introduction of globalization and structural adjustment programmes. As a result, there is a growing pluralisation of state with multiple actors or agencies spread over from local to global level trying to both supplement and supplant the conventional state in its basic role of development.

Good Governance, according to World Bank, is characterized by 'predictable, open and enlightened policy making; a bureaucracy imbued with a professional ethos; an executive arm of government accountable for its actions and a strong civil society participating in public affairs; and all behaving under the rule of law'. The four major areas of concern, identified by the World Bank to be achieved in good governance are: (i) improving the public sector management; (ii) ensuring accountability of public and private sectors, (iii) creating appropriate legal framework for development, and (iv) promoting transparency and information (World Bank 1994). All these four dimensions are to work in a mutually reinforcing way leading to establishment of good governance.

The concept of good or effective governance as advocated by World Bank is broadened and adapted to suit the development demands of developing countries with a view to address the vexed problem of poverty. This has been done by integrating political dimensions into the World Bank's concept of good governance. This assumes that good governance goes along with participatory development and democracy. Democratic system which encourages inclusive and participatory institutions and promotes transparent and accountable societies is a major requirement for good governance (Shylendra 2004). It is in this framework, Government of India has outlined its approach to governance for the implementation of the Tenth Five Year Plan (2002-07) in the country.

The Government of India acknowledges the universally accepted features of good governance like the legitimate exercise of political power along with creating administration which is transparent, equitable, participatory and accountable. But the thrust is made on contextualising them to the emerging needs of the country. 'Governance', to Government of India, 'relates to the management of all such processes that in any society, define the environment which permits and enables individuals to raise their capability levels, on the one hand, and to provide to realise their potential and enlarge the set of available choices, on the other' (Government of India 2002:177). Thus, the proclaimed goal of governance is to bring about human development. The processes of governance are managed by multiple institutions or agencies consisting of the state, market and the civil society; each playing a specific and a critical role in a harmonious manner.

In contextualising governance, the Government of India goes further and has identified specific operational strategies that are to be implemented for realising good governance for development. These strategies include: improved people's participation by way of effective decentralisation through local self governments and involving civil society and voluntary organisations to accelerate socio-economic development; ensuring right to information for transparency; reforming civil service to improve accountability and efficiency; administrative procedural reforms for a hassle free public-government interface; improving the fiscal health of government by improving the resource mobilisation through a transparent and equitable tax-administration and by correcting the subsidy and pricing system of various public services; reforming the judicial system for a speedy delivery of justice; improving the co-ordination between different public agencies and departments involved in development for a better synergy and efficiency; systematic and professional programme/project formulation, implementation and monitoring; empowering the marginal and the excluded to equip them to act as pressure groups to resist bad governance and exploitation by vested interests; and using information technology for a transparent and efficient administration (Government of India 2002).

Taking into consideration various dimensions of governance as well as its contextualisation, the following key elements emerge which are also summarised in Table 1 (Shylendra 2004).

(i) The role and responsibilities of the state for development are to be handled by multiple actors like state, civil society, and market institutions whose reach may extend from global to local level, and who may interact with each other more in a non-hierarchical way. Development is no longer the sole prerogative of the state.

(ii) The ultimate goal of governance paradigm is to bring about human development which is to be achieved through the contribution of multiple actors. Efficiency, equity and accountability are the major ways for attaining the goal. While efficiency is to be attained through market mechanisms and administrative reforms; equity is to be ensured through growth and by enlarging choice and opportunities for all sections. Accountability for delivering development is to be brought about through democratic, transparent and participatory decision-making processes.

(iii) Harmonising the relations between multiple actors is the major essence of the new governance paradigm. Different actors and elements of governance are to adapt to the scenario of plurality in order to evolve a system with harmonious interactions. Making the system more predictable and stable through suitable institutional and legal arrangements is the key for such a harmony.

Development is thus perceived to be delivered by multiple agencies including the state. This realisation has been necessitated as the state alone cannot respond to the diverse needs of development. The different actors who come into picture for development include state agencies like local, state and national governments, and multilateral agencies; market agencies like local traders, national and multinational companies; and civil society institutions like community based organisations (CBOs), and national and international NGOs (Table 2). Thus, governance as a paradigm includes real multiplicity or plurality of institutions which are supposed to be acting in unison to influence development at and from different levels. All these actors are to co-ordinate and network with each other more through horizontal linkages which are both non-hierarchical and non-bureaucratic (Chandhoke 2003). Through such a linkage, the sharp difference between the state and non-state is to be blurred or rendered irrelevant for development.

This new paradigm of governance has got special significance in the state of Kerala which has demonstrated a high degree of social and political development. The institution of State has been, traditionally, held in high esteem by the people of Kerala. This, in turn, had the effect of keeping private initiative and volunteerism at a reasonable distance from the public realm. The people, by and large, looked towards the intervention of the government for the betterment of their living conditions and infrastructural development. This is evident from the fact that the private sector undertakings and large scale business establishments were not given a prominent place in the Kerala society till recently. The delivery of welfare measures was considered to be the exclusive domain of the state.

Table 1
Comparative Features of the Existing and New Governance Paradigm

Features	*Existing Paradigm*	*New Paradigm*
Goal	Economic Development	Human Development
Political System democratic	Democratic or non- Participatory	Democratic and
Actors in Development or without Civil Society	State or Market with Society	State, Market and Civil
Legal framework Market	Favouring State or existence of all three actors	Suitable for the co-
Outcome of various transactions	Non-predictable	Predictable
Reach/Jurisdiction of Actors	National/Local	Global, National and Local
Source of Mandate for Actors	National/Local	Global, National and Local
Relationship between Actors	Bureaucratic & Hierarchical	Horizontal/Non- bureaucratic
Purpose of Relationship Market	Control by State or Networking	Co ordination/
Delivery Mechanism and/or Private firms,	Government agencies Private firms and NGOs and/or NGOs	Government agencies,

But with the advent of the new economic order, it is being increasingly realised that the state, though still dominated by the leftist ideology, cannot withstand the global current towards economic integration. As a result, the government has started receding to the background from many of its traditional areas of operation and the non-governmental organizations, community based groups and multinational development agencies have stepped in to fill that void.

One of the reasons for the increased role of these multiple players in development process, in fact, is the decentralization drive that was unleashed in the country for the last one decade.

The historic 73rd Constitutional Amendment Act 1992 which came into being on April 20, 1993, was intended to improve participation of people in the process of development and to transform Panchayati Raj institutions into institutions of self government for rural areas. The PRIs are provided with new opportunities for

Table 2
Actors in Development under Governance Paradigm

Reach/ Jurisdiction	*State*	*Market*	*Civil Society*
Global	Multilateral Agencies (UN, WB, WTO etc.)	Multi-National Companies (MNCs)	International NGOs
National	National/ Regional Governments	Domestic Companies/ MNC Branches	National/ Regional NGOs
Local	Local self government	Local Traders/ Businessmen	Local NGOs/ CBOs/Citizens

Source: H.S. Shylendra (2004): 'The Emerging Governance Paradigm and its Implications for Poverty Alleviation and Equity', Working Paper 182, Institute of Rural Management, Anand.

planning based on local needs and priorities, empowerment of women, SC/ST development, mobilization of local resources, generation of demand for services, participation of women in decision making etc. The Kerala Legislature in follow -up to the 73rd amendment enacted the Kerala Panchayati Raj Act in 1994 which provides for a three tier Panchayati Raj system in the state. The Panchayati Raj Institutions are Gram (village) panchayat, Block panchayat and District panchayat. The basic unit of Panchayati Raj system in Kerala as stipulated by the Kerala Act is a general body of all persons registered as voters in each of the wards (constituencies) of a Gram panchayat termed as "Gramasabha" (Village Assembly). The Gramasabha is a fundamental participatory unit of PRIs in Kerala, constituted at the ward level and there are a total of 13,259 wards. There are 991 Gram panchayats with an average population of 21,600, 152 Block panchayats with an average population of 1,80,900 and 14 District Panchayats with a population of 15,24,000.

The focus of the study to the emerging pattern of governance in Kerala with special attention on the drinking water sector. It has now been experimented in Kerala to address the drinking water problem with an increased involvement of the people and civil society organizations. The experiment is called *Jalanidhi*, a project implemented with the support of the World Bank, Village Panchayat, Beneficiary Groups and NGOs.

BACKGROUND OF THE PROJECT

Kerala is one of the smallest states in India-covering only 1.3% of the total area of the country. However, it has a total population of over 29 million. The population density, at 747 per sq. km, is much higher than the national average of 267 per sq. km. Geographically, Kerala can be divided into three distinct natural divisions- viz., Sandy Coastal Region (<7.6 m above MSL), Midland region (7.5 - 75 m above MSL) and the Western Ghats (> 75m above MSL). Kerala gets on an average of 307 cm rainfall, the bulk of which (70%) is received during the South-West monsoon which sets in by June and extends upto September. The state also gets rains from the North-East monsoons during October to December. Blessed with an abundance of rainfall of about 3000 mm annually, Kerala has 44 monsoon- fed rivers with a minimum length of 15 km. However, given the terrain conditions, run-off is also quite high. The state experiences severe summer from January to May when the rainfall is minimal. Thus, it is an irony of sorts that a green, verdant state like Kerala has a lesser than average per capita availability of potable water, even lesser than that of Rajasthan. Most of the families depend on private water sources, which are usually open wells. Most of these wells dry up by early March and remain so till the monsoons, which start by May end. Sanitation coverage is only 51.36% in the urban areas and 44% in the rural areas. Only 32% of families below poverty line have latrines.

It is estimated that the Government of Kerala has already invested Rs.50000 crores in the drinking water sector. Despite more than adequate investment made by the Government on the water and sanitation sector over the years, only one third of its population is having access to safe drinking water. Some of the reasons cited for this dismal performance were: an unacceptable top down approach to planning, non-involvement of the users, highly inadequate levels of cost recovery, depletion of sources due to overdrawal, frequent breakdowns due to poor operation & maintenance, etc.

In this context, the policy makers in the state have been influenced by the success of a drinking water scheme that was going on in the *Olavanna Panchayat* in *Kozhikode* district. Popularly known as the *Olavanna* model, this single project stands testimony to the success of a programme that is carried out with people's participation. Taking cue from the *Olavanna* model, The Kerala Rural Water Supply and Sanitation Project (*Jalanidhi*) is designed and implemented by the users themselves as against the "top-down approach". This will be owned, operated and maintained by the users themselves on a total cost- recovery basis. The major objective of the organisation being the sustained provision of adequate quantities of safe, drinking water to the rural poor, the focus on ensuring sustainability of source, sustainability of operations, adequacy of water and quality of water are equally important and are designed into components within the project design.

JALANIDHI

The Kerala Rural Water Supply and Sanitation Project (*Jalanidhi*) is supported by the World Bank and is being initially implemented in four districts: *Kozhikode, Palakkad, Malappuram and Thrissur,* but later extended to all districts. The Government has also created an autonomous institution "Kerala Rural Water Supply and Sanitation Agency" to implement this project. The project is implemented through village Panchayats. The total outlay of the project is Rs.450 crores out of which Rs.350 crores are to be repaid to the World Bank within 35 years. The original project is implemented in six years (2001-2006) and expected to cover three lakh households, benefiting a population of over 15 lakh in 4 districts through 2500 piped water supply schemes.

Philosophy of the *Jalanidhi* Project

Demand Driven Approach–Unlike the supply driven approach hitherto followed, this project is based on the need of the people. The Project will be introduced only in areas where interested groups of people show their willingness to participate in the project and abide by the conditions of cost-sharing. The group then gets a legal entity by registering themselves and only then proceeds with the rest of the planning. The source selection, technology selection, purchase, contracting and implementation is done by this registered body, the *beneficiary group,* with technical help from *Support Organisations.* This inculcates a sense of ownership in the people.

Cost Sharing–To ensure stake holding of the project, 15% of the capital costs is borne by the beneficiary community. Of the remaining, the *Gram Panchayat* bear 10% and 75% is the share of the Government.

Cost Recovery–The Beneficiary Groups themselves meet 100% of the recurring costs of operations and maintenance. This lightens the burden on the state exchequer, thereby helping the Government to utilise this money for other priority needs.

Integrated Approach–The objective of the Project being sustainable supply of safe drinking water, sustainability of source, operations, regularity and adequacy of supply and quality of water. This is being met through other well-integrated components. Sustainability of source is ensured through Point- Source recharge measures. Quality is ascertained through a mix of Sanitation & Hygiene promotion and provision of infrastructure like latrines, compost pits etc. Sustainability of system is ensured through community empowerment, capacity building, women empowerment and social mobilisation.

Pro-Poor Approach– Special efforts were made in the project design to include the poor and vulnerable while selecting the user groups. The project was so designed to incorporate the beneficiary contribution of 15% of capital costs either through cash or in kind, as labour. Intra-group subsidisation and even inter- group subsidisation is permitted at the behest of and under the total responsibility of the beneficiary groups.

Women Development Initiatives–Women are the most affected both directly and indirectly, during water shortages. The project makes conscious efforts to mainstream the women users in the planning and decision-making activities. Apart from this, they are also encouraged to form "Thrift & Credit Groups" to help them make the payments towards the recurring expenditures of the water supply system. Income generation activities are also designed in the project where groups of women are given financial assistance and skill development training to start viable micro-enterprises of their choice.

Community Empowerment– Capacity building and equipping the community to operate the project is a major thrust area of this project as this is planned, designed, implemented, owned, and operated by the users themselves. This will not only ensure the involvement of the people but will also chart a new path to community- based approach for meeting any local needs.

Community Contracting–The users themselves are fully involved in all the activities right from identifying their sources, deciding on the technology to be utilised, community contracting and implementation till the operations and maintenance aspects of the schemes. All contracting of goods, works, and services will be done at the user level itself for which adequate training will be provided and guidelines made available.

Utilization of Available Resources–The Schemes already operational in these project areas, will also be rehabilitated, and handed over to the User groups. This will ensure efficient utilization of investments made.

Dovetail with Decentralised Planning–This project will be operationalised through the Gram Panchayats and the beneficiary groups, thereby acknowledging and strengthening the efforts of *decentralised planning* in Kerala.

IMPLEMENTATION

Government of Kerala has overall responsibility for implementing the state project and the national component with the Government of India's Rajiv Gandhi National Drinking Water Mission (RGNDWM). The main implementing agencies are shown in Table 3.

Table 3
Division of Responsibilities under the *Jalanidhi* Scheme

Activity	*Implementing Agency*
Component A (Institution Building)	KRWSA on behalf of its General Body
Component B (Community Development & Infrastructure Building)	Beneficiary Groups, Gram Panchayats, Support Organisations, and KRWSA
Component C (State-wide Sector Development)	Secretary (Irrigation and Water Supply), Govt. of Kerala and KRWSA
Component D (National Sector Development)	Govt. of India's Rajiv Gandhi National Drinking Water Mission

The Project will be implemented in 80 Gram Panchayats (GPs) in 5 successive but overlapping batches consisting of 5, 15, 20, 25 and 15 GPs respectively. Each Batch will follow a 27 month scheme cycle consisting of four distinct phases (Table 4).

Table 4
Plan of Implementation under the *Jalanidhi* Scheme

1.	Pre- Planning Phase	–	3 months
2.	Panning Phase	–	12 months
3.	Implementation Phase	–	8 months
4.	Post- Implementation phase	–	4 months

The main activities and outputs of each phase are as summarized below:

1. Pre-Planning Phase: The main activities during this phase are GP selection, Pre-qualification of Support Organisations and signing of a Planning Phase Tripartite Agreement between the KRWSA, GP and SO.

2. Planning Phase: The main outputs during this phase will be orientation & capacity building of GPs and SOs, BG registration, formation of Beneficiary committees, resource mapping and pre- feasibility studies. These will be done in the first four months of this phase. Subsequent activities are technology selection, preparation of engineering designs and community empowerment plans and finally a GP level Implementation Phase Proposal. The GP, BG and BC capacity building, sanitation & hygiene promotion and women's development programmes are held as parallel activities during the planning phase. The planning phase concludes with the signing of the Implementation Phase Quadrilateral Agreement (IPQA) between the KRWSA, GP, BG and the SO. There is one IPQA per BG. A separate Implementation Phase Tripartite agreement (IPTA) is signed between the KRWSA, GP and the SO for managing SO contracts.

3. Implementation Phase: The main activities during this phase deal with the actual implementation as procurement of materials, construction of schemes as per agreed plans and procedures and management of project funds.

4. Post-Implementation Phase: During this phase, the GPs will provide advisory support to the communities in efficient operation and maintenance of services, on collection of water tariff and book- keeping. The signing of the Implementation Completion Reports (ICRs) in public gathering by KRWSA, GP and BCs mark the formal exit of KRWSA from a project GP. Sustainability monitoring and Impact assessments will be conducted by the KRWSA to assess the project outcome.

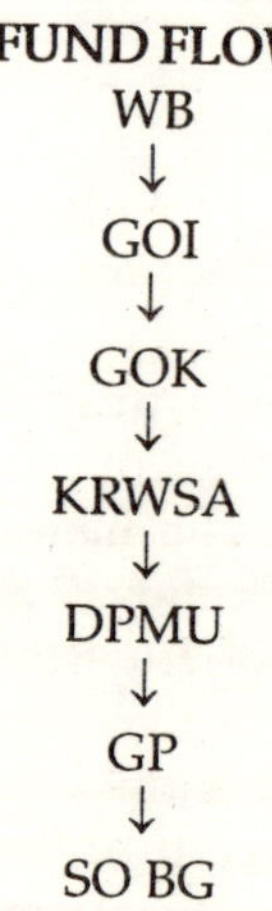

The Multiple Actors involved in the project

Gram Panchayat

The Gram Panchayat (GP) will be the focal point for project activities in the GP area. GPs will be mainly responsible for seeking project assistance following a self-selection process, preparing the implementation phase proposal during the planning phase, and for facilitating project implementation by BGs during implementation phase. GPs will provide counterpart funding to BGs as a percentage of infrastructures building costs. During the post-implementation phase, the GPs will monitor the sustainability of scheme operations and ensure that the BGs satisfactorily discharge their O&M management function including levying and collecting user charges from the beneficiaries to fully recover the recurrent O& M costs.

Beneficiary Groups

The Beneficiary Groups (BGs) will be responsible for planning, technology selection, and constructing their RWSS facilities, providing their part of the capital cost contribution, managing O&M of the improved facilities, and levying and collecting sufficient user charges. Representative beneficiary committees (BCs) will be set up by each of the participating BGs. Their composition, functions, method of selecting members and relationship with the GP and KRWSA have been agreed. The project will make vigorous efforts to maximize women's representation and management role in the BCs. The project rules require that membership of BCs would be at least 50% women, and 20% from socially disadvantaged groups. At least one of the top management position of the BC (President or Treasurer) will be held by a woman.

Support Organisation

The NGOs and private sector consultants will play an important role in the projects. There will be three categories of Support Organisations (SOs). The first will support the GPs, BGs and Beneficiary Committees on a day-to-day basis in planning and implementing the project's activities, and provide brief support to BGs during post-implementation stage to stabilize scheme operations.

The second will be a range of programmes or providing specialized assistance-such as in the preparation of ground water recharge schemes, water quality monitoring, or in developing strategy and materials for sanitation and hygiene promotion programmes.

The third category will be private sector consulting firms which would provide assistance in computerized financial management systems, audits and accounts, design engineering of multi-panchayat water supply schemes, independent construction quality monitoring, and management of information systems.

DPMU

The District Project Management Unit (DPMU) is set up as the district office for the KRWSA project. The main activities of the DPMU are:

- Capacity building of the SO through appropriate training and guidelines.
- Capacity building of the BGs through training organised in support with the respective SO.

- Orientation to the GP, members and staff.
- Confidence building of the BGs through field visit and attending BG meetings by DPMU officials.
- Monitoring of the project activities of SOs and BGs.
- Liaison with the other agencies for effective convergence of various govt. programmes for the benefit of the BGs.

THE STUDY

The study was conducted and focussed on the *Mundathicode Panchayat* in *Trissur* District of Kerala. The importance of Mundathicode is that the first *Jalanidhi* project was launched here. Severely affected by the drinking water shortage, the *Panchayat* received this project with lot of enthusiasm and expectations. As many as 28 beneficiary groups (BGs) were formed in the panchayat and efforts were made to implement the project. All the BGs were studied in detail. Profiles of the BGs were given in Table 5.

The field level data were collected at two levels. Viz. at the leadership level and then at the beneficiaries' level. The leadership level information was collected from the leaders of *Panchayat*, Beneficiary groups, Support Organisation and the KRWSSA. The beneficiary level information is collected through an intensive field work done among 230 beneficiaries of all the 28 BGs.

OBSERVATIONS

Though the water intake has changed, majority of the respondents were not happy with the volume of water they got. As many as 157 people said that they did not get the volume of water that they expected at the launching of the programme, 123 people have opined that there was no sufficient water at the source, while others have attributed certain technical problems like insufficient tank, lesser width of pipe, etc. for the inadequate supply of water.

The technical aspects of the project implementation were supposed to be looked after by the Support Organisation (SO). In the case under study, the S.O. involved is Centre of Science and Technology for Rural Development (COSTFORD), an organization which is widely acclaimed for its commitment to the cause of Rural Development.

Table 5
Beneficiary Groups under the *Jalanidhi* Scheme in *Mundathikode* Panchayat

Sl. No.	*Name of the Beneficiary Group*	*Total No. of Households in the B.G.*	*No. of house-holds surveyed*
1.	Aiswarya	52	05
2.	Anugraha	47	04
3.	Athani Medical College Development	163	15
4.	Bageeratha	98	10
5.	Gandhinagar	38	04
6.	Gramasira	105	10
7.	Gramasree	94	10
8.	Irunnamkallu	41	04
9.	Jawahar	147	15
10.	Kallidathuchira	57	06
11.	Kiliathuchira	64	06
12.	Kottakunnu	49	05
13.	Manakulam	68	07
14.	MKKVS	43	37
15.	Navajyothi	51	05
16.	Peringandoor	87	09
17.	Pheonix	70	07
18.	Pratheeksha	65	07
19.	Puthuruthy Sudhajalam	74	07
20.	Puthuruthy Vadakkummury	48	04
21.	Safron	53	04
22.	Samrudha	148	14
23.	Sangamam	58	06
24.	Sanjeevani	94	09
25.	Souhrudha Nagar	37	03
26.	Subiksha	87	09
27.	Theerthadhara	56	05
28.	Yuvajana Sangam	34	03
	Total	2115	230

One of the measure observation was that water use pattern was considerably changed under the project as is revealed in Table 6.

In order to examine the synergy between various state and non-state actors in pursuit of rural development, the performance of the Support Organisation is brought under close scrutiny. The degree of acceptance of the SO among the beneficiaries is taken as one of the indicators in this regard.

Table 6
Water use Patterns before and after the Launching of the *Jalanidhi* Scheme

Usage of water (in litres)	*Before Jalanidhi* No.	%	*After Jalanidhi* No.	%
100-300	67	29.13	57	24.78
301-500	132	57.39	121	52.60
501-750	31	13.48	52	22.61
Total	230	100	230	100

A question was posed to all the sample beneficiaries that "Are you aware about the existence of a support organization in this project?" It is surprising to note that only 28.70% beneficiaries are aware that a support organization does exist at all. This is supplemented by another answer which revealed that only 30.87% beneficiaries have positive opinion about the job done by the COSTFORD. It has been made clear in the survey that the SO could not effectively integrate with the society in which it is put to work. On a specific question whether the SO has made its presence felt in the Panchayat, only seven percent of the beneficiaries replied in the affirmative. No wonder in 76% of the beneficiaries saying that there is no need for any support organization in this kind of a project.

The apparent problem with the SO is that they did not trust the villagers. On many aspects of the project, the common people have defferent ideas about the implementation process. But the staff of the COSTFORD strictly adhered to the text of their technical knowledge. The local circumstances and wisdom were completely ignored. Shri. Ajith Kumar, a member of Gram Panchayat, for instance, said that the COSTFORD actually worsened the situation. They have constructed the water tank at a much lower terrain, ignoring the views of the Beneficiary Groups. The common men very well knew that the tank should be constructed at a higher plane, to make water available in the remote areas. But since such local wisdom was not heeded to, most of the beneficiaries do not get water at all. Shri Ajith Kumar even suspected financial irregularities in the project. He said that Rs. 35 lakhs were given to the SO in the Mundathikode Panchayat itself. He is of the opinion that the entire project could have been be completed with half of this amount.

It was evident in study that the Village Panchayat, another actor involved in the project was not given the due role in the project implementation other than contributing 10% of the total project expenditure. It did not have any regulatory authority over the SO in designing and implementing the project. But the beneficiaries always consider the Panchayat as their sole authority in terms of rural development. Obviously, the high degree of acceptance that the Panchayat as an institution has got among the beneficiaries.

As many as 208 people (90%) said that they got good support and encouragement from the village panchayat. In fact, none of the beneficiaries had even a minute complaint against the Panchayat. It was an irony that when people put their trust in the Panchayat, it was found itself in a helpless situation. One Panchayat member who does not want to reveal his identity said that the Panchayat has no control over the SO. "In fact, the Panchayat and the SO work at parallel levels. The SO is just a temporary installation of people with no practical experience. They are not competent to carry out this kind of work. They were giving just 1000 rupees to those girls who supervised the work. How much sincerity and commitment can we expect from a diploma holder in civil engineering with such a meagre amount of salary", he asked.

Indirectly the project generates the right kind of trust and cohesiveness among the beneficiaries. Therefore, serious attention was paid in the study to ascertain the level of social capital that has been created as a result of this *jalanidhi* project. The main idea emerged out of the study is that the objective is not achieved. In fact, some of the ideas shared by the leaders of beneficiary groups revealed that the social capital generated was of negative in nature.

Shri Gopi (a traditional Ayurvedic physician by profession), the president of the *Bhagiratha* Beneficiary Group, for instance, said that this project actually cheats the people. It could not solve the water needs of the people. The people were not giving the user fee regularly. He had to spend money from his pocket. He lost his public acceptance even as he took up this responsibility.

The leaders of beneficiary groups were, by and large, not satisfied with the degree of co-operation they got from the beneficiaries. It was difficult for them to collect the users' fee in time. In case someone doesn't get water he/she used to rush to the houses of the president or secretary of the BG with complaints. It becomes a headache for the BG leaders to address these issues. Sheela, an unmarried girl of 22 years who heads the *Puthuruthi*

Sudh Jal, said that she had to suffer a lot of humiliation on account of these complaints. Her parents were not happy with creating such unruly scenes in the house. Even good marriage proposals were called off as this girl is 'a public figure'. She wants to get rid of this responsibility at the earliest. But nobody seems to be willing to take over.

Most of the complaints were due to the forcible opening of locks that regulate the water flow to various outlets. The beneficiaries have a tendency to break open these locks to access more water. The beneficiaries have an understanding among themselves that they should not indulge in such malpractices. But the frequent violations of such understanding indicate the lack of trust that exists among the beneficiaries.

One general trend emerged in the survey is that those who are in the leadership positions do not want to continue any longer. On the other hand, nobody seems to be interested in taking over. This has been made clear from the data collected. In most of the BGs, the leadership positions were not changed since the launching of the scheme. Voting has not been taken place in any of the BGs to choose the office bearers. It indicates the BGs have not yet become vibrant associations.

The general body meetings of the beneficiaries were held rarely. Only seven respondents have replied that the general body is held 2-4 times so far. 23 respondents have replied that it has been held once. The remaining 200 respondents replied that the general body is either not held or they have no idea about it. As many as 226 respondents do not know anything about the executive committee meetings of their BGs. They have no idea of what they are discussing in the executive committee meetings.

Yet again one found emerged in the survey that the political identification of beneficiaries does not affect the smooth functioning of the scheme. As many as 219 respondents were identified with one political party or the other. But only two respondents have said that political identification adversely affected the scheme. Majority of the respondents have replied that the problems emerged in the beneficiary groups are settled within the group amicably. All problems are related to the inadequate availability of water. In any case, the internal disputes and quarrels have not disturbed the implementation of the scheme.

Beneficiaries' identification with the project has been another concern of the study. To ascertain this aspect, questions were asked to test their knowledge about the source of funding for the project. It is noted that 56 respondents had no idea about the sources of funds. Of the total 230 respondents, 174 knew that money comes from the World Bank. But only 82 respondents knew that this money has to be refunded. Even among them only 30 people know that it is the government which should refund this money. But nobody knows the period of repayment and the rate of interest. As many as 217 beneficiaries justified the act of seeking foreign money, since it is a problem of drinking water. The low level of knowledge about the project among the beneficiaries again indicates that the project could achieve only limited success in activating the rural community.

The *jalanidhi* project in Kerala attempts to achieve rural development through an interface between state government, panchayats, community based organizations (CBOs) and non governmental organizations (NGOs). Among these, government and panchayats are state actors and CBOs and NGOs are non-state actors. The study focused on ascertaining the effectiveness of the synergy between the state and non-state actors as it is an emerging pattern of development governance.

Quintessence of the study is that there is no proper coordination between panchayat, CBO and the NGO. While the CBO opines that the panchayat is the main actor, the panchayat takes NGO as the main actor. In fact, as per the script, it is the NGO which has to supervise and implement the project. Therefore, the panchayat is not supposed to interfere in the day-to-day administration of the project. But the people are unaware about it. There is a wide divergence between the perception of the beneficiaries about the role of the panchayat and its assigned role. On many occasions when the beneficiaries do not get water, the beneficiaries rush to the houses of panchayat members with complaints. But the panchayat members were helpless as they were not given any meaningful role in the design of the scheme. The designing of the scheme is left exclusively with the SO. But the SO is neither accountable to the panchayat nor to the people. After the completion of the project, the SO withdraws from the scene and the project is to be then managed by the beneficiary groups themselves. At this stage the people look towards the panchayat as the sole saviour of all their problems.

One of the implications is that there is a gross sidelining of the panchayat in the whole process, though the people lend legitimacy and credibility to the institution. The panchayat's role is limited just to select the SO from the list of SOs approved by the state government and provide 10% financial assistance to the project. It has no control over the SO in the process of designing and implementing the project. At the same time, all complaints against the SO are to be settled by the panchayat. This dichotomy is not visualized in the jalanidhi project.

The SO on the other hand, takes this opportunity to exert their authority over the beneficiaries. Having bestowed with technical knowledge and expertise, they tend to ignore the indigenous knowledge and wisdom of the local people. On many occasions, the local people have convincingly argued for a different design and method of implementation. But they, in turn, were cajoled for their lack of modern knowledge and experience. This had the effect of lessening the credibility of the SO in the eyes of the public.

The beneficiary groups have the good intention of running the project effectively, and thereby solve their water problems. But they were found without proper guidelines and discipline. Driven with their self-interests, they disregarded the interests of their fellow users. This created unnecessary tension and non-cooperation within the beneficiary group, defeating the much desired objective of generating social capital within the society. The study revealed that the already existing amount of social capital is at a loss as they got a new space to quarrel over sharing the facility offered by the project. This makes one to conclude that any good intended programme may result in generating 'negative social capital' if it is poorly designed and implemented.

REFERENCES

Abrahamsen, Rita (2000), Disciplining Democracy: Development Discourse and Good Governance in Africa, Zed Books, London

Chandhoke, Neera (2003), 'Governance and the Pluralisation of the State: Implications for Democratic Citizenship,' Economic and Political Weekly, 38(28): 2957-68.

Government of India (2002), Tenth Five Year Plan (2002-2007): Dimensions and Strategies, Vol. I., Planning Commission, New Delhi.

OECD (1995), Participatory Development and Good Governance, Development Co-operation Guidelines Series, Paris

Oommen T.K. (2003), 'Civil Society and the Goal of Good Governance' in Rajesh Tandon and Ranjita Mohanty ed. Does Civil Society Matter? Sage, New Delhi, pp.125-144.

Rajasekher, D. and R.R. Biradar (2002), People, Government and the NGOs, Working Paper 114, Institute for Social and Economic Change, Bangalore.

Riley, John M. (2002), Stakeholders Participation for Rural Development: Critical Collaboration in State NGO Partnerships, Sage, New Delhi.

Shylendra, H.S. (2004), 'The Emerging Governance Paradigm and its Implications for Poverty Alleviation and Equity', Working Paper 182, Institute of Rural Management, Anand.

UNDP (1997), Governance for Sustainable Development, A UNDP Policy Document, New York.

World Bank (1992), Governance and Development, Washington DC.

World Bank (1994), Governance: The World Banks's Experience, World Bank, Washington D.C.

32

PUBLIC-PUBLIC & PUBLIC PRIVATE PARTNERSHIP IN MAINTENANCE OF DRINKING WATER

G. Palanithurai & R. Ramesh

INTRODUCTION

In India, like in many other countries of the world, drinking water is provided by the public sector, since it is usually considered as a typical public good. The public utility takes charge of the overall cycle of installation of the technical system, water production, distribution, and maintenance. However, the problems associated with water crisis, system operation and maintenance have always remained. Recently, as a movement of reform these functions are partially or fully assigned to private operators through a franchise bidding process. There are various degrees of involvement of the private operator in terms of contract duration, public participation, revenue collection arrangement and so on.

The opposite action of nationalization is privatization. Privatization as solution to the problems associated with water distribution services is proposed and pushed by world bodies such as WTO, IMF and the World Bank through international agreements such as GATTS. The pro-privatization lobby aggressively campaign for water privatization on the grounds that, while water subsidies promote wasteful practices, commodification of water should allow market forces (supply and demand) to set the tariff, which in turn will reduce water consumption and promote water conservation. Furthermore, it is argued that opening this sector to private service providers/contractors will bring in badly needed capital for upgrading and development of infrastructure[1].

This is considered as a corollary to the processes of globalization, liberalization and privatization which are fast gaining ground in several countries of the world. These processes have contributed to a dramatic change in the role of the state[2].

International institutions, governments and multilateral corporations are designing and implementing policies which reduce the role of the state in the provision of basic services and the free access to water resources. This is no longer seen as a responsibility of the state but as the domain of the market forces. In this context, water supply is no longer an obligation of the state to the citizens. It becomes a commodity available only to those who pay for it, even if the price is termed 'user charges'. Corporations and consortiums pressurize various countries to let them supply and sell water, either by themselves or in partnership with public bodies[3].

In India, the revised National water Policy 2002 mentions that "Private sector participation should be encouraged in planning, development and management of water resources projects for diverse uses, wherever feasible. Private sector participation may help in introducing innovating ideas, financial resources and introduce corporate management and improving efficiency and accountability to users. Depending upon the specific situations, various combinations of private sector participation, in building, owning, operating, leasing and transferring of water resources facilities, may be considered".

Encouraging private sector participation in delivering water services and corporate management of water resources naturally provide space to push for corporate control over water resources and services. Although the same water policy 2002 (item no. 12) mentions about making use of local bodies and user groups. In actuality the tilt seems to be towards giving it to the private service providers who are in water business for long. They are from countries such as UK, Germany etc who have come to work in partnership agreement with certain Indian business giants such as Mahindra and Mahindra.

There are Civil Society Organisations that argue control of water by private companies take away this resource from the public and put it in private control for them to make profit. If the government is unable to manage, the best is to delegate it to local bodies or to the CBOs rather than leaving the community members to buy water from private companies. In India, Mehta[4] highlights that most projects with private sector participation purely aim at an increase in production capacity and sale and do not address the main factors of inefficiency of Indian water supply networks.

Rajiv Gandhi Chair for Panchayati Raj Studies of the Gandhigram Rural Institute has attempted to understand how a Public-Private Partnership (PPP) water supply arrangement initiated recently by Government of Tamil Nadu works in Thirupur, Tamil Nadu. It specially focuses on the water service delivery made to the Village Panchayats reportedly covered by the PPP arrangement.

PUBLIC PRIVATE PARTNERSHIP (PPP) IN TIRUPUR, TAMIL NADU–A CASE

As part of the sector reform measures, and private sector participation in service provision, for the first time the Government of Tamil Nadu has gone in for Public Private Partnership (PPP) in providing water for the industries in Tirupur town and for a few villages in Tirupur Block. This is called BOOT (build-Own-Operate-Transfer) model water supply scheme. The estimated cost of the project is Rs. 1,023 crores which includes water supply and sewage systems[5]. This includes both debt and equity capital where USAID and the World Bank have extended financial support to execute this project. Mahindra & Mahindra, Bechtel and the United Utilities of UK are the main contractors besides Hindustan Construction Corporation and Larsen and Toubro Limited. It is called 'Mahindra Consortium for short'.

NTADCL: The consortium has registered a company called New Thirupur Area Development Corporation Limited (NTADCL). The government describes that this is a Special Purpose Vehicle with the primary objective of implementing the Tirupur Area Development Project. This is the first Water Sector project in the country developed under the public-private partnership. It is a 30 year contract arrangement at the end of the 30th year it would be handed over to the government.

Beneficiaries: Under the project, facilities would be constructed to provide water supply access to the dyeing and bleaching industries in Tirupur and the domestic consumers in Tirupur Local Planning Area, comprising the Tirupur Municipality, 15 Village Panchayats and 3 Town Panchayats. In addition, water supply is proposed to be provided to five wayside unions which lie enrooted on the water transmission system corridor[6]. This project is expected to benefit (i) the industries, (ii) the slum dwellers, and (iii) the way-side villages in the same order as mentioned.

Operation & Maintenance: NTADCL would transfer water over a 55 KM long pipeline from Bhavani river and supply 185 million litres of water per day to nearly 1000 textile units and more than 1.6 million residents in Tirupur and its surrounding area. According to the project document, United Utilities and NTADCL will run the joint venture at a fixed 'operation and maintenance fee' that will be recovered entirely from Tirupur municipality. However, in an effort to attract further corporate capital investment in the state, the Government of Tamil Nadu has guaranteed profitability to the investors in the project by creating a hedge fund to pay the interest and operative expenses of the project in the event of a water shortage in Bhavani river[7].

'Users should pay for water on the basis of the quantity used' is one significant move the NTADCL has made. This Project has become operational now. Water supplied to industries is charged Rs. 45/- per 1000 litres. In addition the industries that demand water should pay a deposit of Rs. 25,000 onwards depending on the water demanded, plus they should be able to show a bank guarantee. Water supplied to the Village Panchayats should be charged at Rs. 3.50 per 1000 litres. Village Panchayats (VPs) receive water bills from NTADCL direct.

ANALYSIS AND DISCUSSION

Experience of reforms had clearly shown market failure to exist along with government failure, especially in the provision of public services. Public-Private Partnership (PPP) could be a way to utilize the best of both worlds, while overcoming the specific weakness of each. But for this to be feasible contracts and regulation have to be carefully designed. Well-designed and chosen contracts improve governance. Insights from international experience with PPP in water supply suggest how services from public utilities can be improved with cost saving through the reduction of waste[8]. However, there are varied views about it. Experience records failure stories as much as there are success stories.

THE NEED–RIGHT DICHOTOMY

Access to basic services such as water can either be gained through private contractual arrangement or as entitlements by virtue of citizenship[9]. The concept of water as a basic human right and hence the free access to it as non-negotiable, has been redefined by the World Bank and the UN to mean that water is simply a human 'need'. The difference in interpretation is not merely semantic but very significant. A human need can be fulfilled through the market on the basis of the supply and demand, with no obligation on the part of the state.

On the other hand, in a democratic polity, water as a human right is the responsibility of the state and the fundamental right of the citizen. When the state neglects its duties/obligations, the concept of human right empowers the citizens to have the constitutional and legal remedies to make the state responsible for supplying water for personal consumption and livelihood activities[10]. This is particularly important for women in Indian society, since women are traditionally charged with the responsibility of looking after water needs of the family.

THE CONCEPT OF PAYING FOR WATER

If the right to water is perceived as a contractual rather than a human rights, water services are subject to the laws of demand and supply, which invariably, due to the nature of markets, would be unable to guarantee equality and affordability of access. From human rights lens water is a public entitlement, access to which does not depend on one's ability to pay.

Nevertheless, viewing the water/trade nexus through a human rights lens does not necessarily mean that water services need to be free of charge or state run. In stead, such a lens implies that states, which involve private actors in the provision of basic services, are legally obliged to establish effective and flexible regulatory mechanisms that can secure the progressive realization of the right to water for all people[11]. As far as paying for water in India is concerned the stage we are in is at the bottom most line. It is discussed in detail below under different costs.

DIFFERENT COSTS OF WATER

In various stages of water production and distribution there are several costs involved. They are discussed below with a comparison to the arrangement in NTADCL.

- The prime argument against paying for water is about attaching a price to *water as a resource;*
- Secondly, recovering over a period of time the *capital cost* incurred in constructing the water supply installations and facilities; and
- Thirdly, charging a portion of the *Operation and Maintenance* (O & M) cost of water production and distribution.

One needs to understand the difference among these three types of costs (expenses) to be able to understand and argue what cost does the private service provider or the state is trying to recover from the water user community. Is it charging water as a resource so as to make the citizen realize the scarce nature of the resource?; Is it charging so as to recover the capital cost incurred in constructing the water supply installations and facilities?; Is it charging the Operation and Maintenance expenses?

Charging for water 'viewing water as a resource' raises questions such as rights of water. Charging for water so as to 'recover capital cost incurred in constructing the facilities' is generally viewed wrong because in a welfare state it is a state obligation. Charging the Operation and Maintenance costs is widely accepted and there are some Village Panchayats in Tamil Nadu that has the practice of collecting user charges from house service connection (HSC) holders. It is a responsibility of the local body which is supposed to carry out with the support of the user community. Reports in several village panchayats user charges were collected from all HSC holders, the local bodies would not be in a position to meet the full O & M costs. One component i.e. water pumping charges (electricity bill) would consume all the entire user charge collection in one gulp.

WATER CHARGES IN TAMIL NADU

As far as Tamil Nadu state is concerned, the policy and practice of supplying water to the households rests with the local bodies[12]. The Government of Tamil Nadu from the budgetary allocations made to the TWAD Board-an auxiliary of the Department of Rural

Development, takes care of constructing the installations and incurring the capital cost. Operation and Maintenance (O & M) rests with the local bodies. For the purpose of operation and maintenance, the local bodies through a special Government Order[13] of the Government of Tamil Nadu are permitted to collect user charges (or water tax) to meet the expenditure on O & M.

As per the G.O. and in practice, this water tax is a fixed amount of Rs. 360/- per year collected only from those who hold individual house service connections (HSCs). The common water collection points (otherwise known as Public Fountains) are meant for the poor who cannot afford to pay for water. This positive discrimination is there in favour of the poor. This practice is already in vogue in many of the so-called way-side Village Panchayats in Thirupur. Water is supplied by the TWAD Board from a combined water supply scheme pumped from Mettupalayam (which is locally popular as 2nd Scheme). The Village Panchayats are paying the O & M cost to the TWAD Board. The Panchayats pay at the rate of Rs. 3.50 per 1000 litres. Being subsidized by the Government and it would not cover the entire O & M cost of the scheme. Given the polluted ground water and considering the essentiality of water service the government is subsidizing water. As of now as per the condition, given to the NTADCL, they also charge on par with TWAD Board rates only (i.e. Rs. 3.50 for 1000 litres). The actual rates could be very high as Industries are charged at the rate of Rs. 45 per 1000 litres. But recently, as many of the industries found the rate to be too high, they started going after informal tanker operators. It has resulted in NTADCL to reduce the water price to Rs. 23 per 1000 litres.

Perhaps, with small limitations we can under found kind private sector take up service delivery as: (i) if the community has to pay the entire O & M costs; (ii) fixed payment system may be done away with, and a system of charging for water based on the quantity of water supplied (minus government support) could come into practice; (iii) To be able to pay the amount to the private service provider the local body may have to levy water tax to the poor people, who out of poverty could not afford to have a house service connection (HSC), and use only the public water collection points on the streets.

WHO FOR THE SCHEME IS MEANT FOR?

NTADCL project is expected to benefit (i) the industries (about 4000 units), (ii) the slum dwellers (about 60,000), and (iii) the way-side Village Panchayats (11 VPs) in the same order as mentioned. The analysis of some of the VP functionaries is that the scheme is primarily meant for the industries in Thirupur, secondly for the slum dwellers most of whom are workers in those industries, and the remaining water is distributed to the way-side VPs. The reasons for the variation in the degree of importance are obvious. The industries should survive. The workers in the industries, most of them migrants from poorer rural areas of Tamil Nadu, should be relieved of the water crunch so that they work for the industries. Thirdly, the way-side VPs the water passes through, if by-passed, would become a persistent head ache to the TWAD Board and to Tirupur Municipality by agitations and protests.

The question of some of the VP functionaries in Tirupur is that, going by policy and developmental ethics, water for drinking purposes should get the top most priority. But, industries have got overriding priority in the NTADCL agreement. Clearly, water as a commercial venture (as one of the factors of production with certain money value) is more likely to succeed in industries rather than giving it to the villages for drinking purposes. This proves that the basic tenet of NTADCL like any other private business venture is commercial viability or RoI (Return on Investment) which the Government of Tamil Nadu has paid heed to. They have also made a provision that if any VPs demand more water they can buy it from the company at the rates it is given to the industries (i.e. 1000 @ Rs.45). This raises the issue of rich-poor divide among Panchayats where VPs with sufficient income can afford to buy water for any rates prescribed by the private service provider, but people in relatively poor VPs shall suffer with inadequate water supply.

The VP functionaries show a calculation that indicates that if the NTADCL agreed to supply (on payment of Rs.3.50 per 1000 lr.) all the water a Village Panchayat requires the VP can save a lot of money they spend on EB bill (water pumping charges) and the regular pump maintenance expenses. One of the serious problems of most Panchayats, that is 'fat EB bill', can be brought down spectacularly. But, the quantity supplied, as of now, is not even one fourth of what is required at the Panchayat level. One of the reasons for this is that the NTADCL goes by 1991 census in the year 2006. The VPs are told to take it for granted because the project aims at catering to the water needs of about 4000 garment factories in Tirupur in the first place.

As a strategy of the government to make industries to pollute less, it could be alright if the industries were asked to pay for water, especially those which pollute. The idea could be designed in order to make them work keeping water as a 'ruling constraint' so that they pollute less. It would make the industries to calculate water also as one of the scarce resource like any other resource involved in production of goods. If water was made available free of cost, then its use could be indiscriminate and pollution could also be more. The Village Panchayats would naturally use water only for drinking purposes. Then there is nothing wrong if, either out of corporate social responsibility or through government subsidy, VPs were supplied water at rates much less than what it is given to the industries. As of now, the costs charged by NTADCL are on par with what TWAD Board charges to the Village Panchayats.

ARE THE PANCHAYATS DISCONTENTED ABOUT PPP?

One postulate of this research was that the Village Panchayats might be discontented that water services have been privatized. Response in most of the villages disproved it and portrayed that the Village Panchayats do not mind receiving water from private water service providers. This is especially because of the cost these Village Panchayats are incurring on operation and Maintenance expenses. The Table 1 below shows the current state of O & M expenses.

All the Village Panchayats show only a deficit budget as far as O & M expenses are concerned. Some of the Panchayats do not have any income from water at all. But all of them meet all these four items of expenses stated above. The lion's share is taken away by the Electricity Board by way of EB Bill. There are some Panchayats that are not able to pay the Tank Operators salary regularly. The Panchayats have to await the arrival of government grant to be able to make salary payment for the Tank Operators. In this circumstance, the NTADCL by reducing the pumping charges (EB Bill) is most likely to reduce the water supply related expenses of way-side Village Panchayats. But, they are not fully contended because the NTADCL provides only one third of their

Table 1
Annual Income and Expenditure in the Villages of New Thirupur Scheme

Sl. No.	Villages	Income from Water (Per Year)	Expenditure on Water				Total	Surplus/ Deficit
			Cperator's charge (Per Year)	Electricity Bill (Per Year)	Maintenance (Per Year)	TWAD Payment (Per Year)		
1	Thetti Palayam	8,28,000	1,08,000	18,60,000	9,00,000	3,60,000	32,28,000	24,00,000 (-)
2	Neruperichel	15,12,000	2,16,000	14,40,000	4,20,000	3,60,000	24,36,000	9,24,000 (-)
3	Chetti Palayam	14,40,000	4,32,000	15,00,000	6,00,000	3,60,000	28,92,000	14,52,000 (-)
4	Iduvai	–	43,200	4,80,000	60,000	1,20,000	7,03,200	7,03,200 (-)
5	Mutham Palayam	–	1,36,800	6,00,000	50,000	4,80,000	12,66,800	12,66,800 (-)
6	Aandi Palayam	9,00,000	57,600	6,00,000	30,000	3,00,000	9,87,600	87,600 (-)
7	Mangalam	–	1,08,000	12,00,000	25,000	2,76,000	16,09,000	16,09,000 (-)
8	Puthu Palayam	1,80,000	86,400	18,00,000	84,000	1,80,000	21,50,400	19,70,400 (-)
9	Kaniyan Poondi	90,000	14,400	3,00,000	72,000	2,40,000	6,26,400	5,36,400 (-)
10	Muthali Palayam	–	93,600	12,00,000	50,000	50,000	13,93,600	13,93,600 (-)
11	Murugan Palayam	–	93,600	10,50,000	30,000	1,92,000	13,65,600	13,65,600 (-)

Source: Calculated from the data obtained from respective village panchayats

water requirements. The remaining being TWAD supplied water and the locally pumped water. There are some VPs such as Thottipalayam who are pressuring that the NTADCL should provide for the complete water requirements of Thottipalayam on the grounds that the industries in Tirupur polluted all their ground water.

In fact, this NTADCL arrangement can reduce VP level expenses on water supply, as shown through calculations by some VP functionaries. They state that they are able to save 20 to 30% cost if the NTADCL commit to provide the quantum demanded by the VPs. If this is not done, they would end up paying additionally to NTADCL. Apart from the questions of quantity, affordability and transparency, one apprehension that most VPs raise is of 'accountability'. There is no grievance redressal mechanism to the users. When a grievance is taken to the notice of NTADCL, 'there are only cross-references given and no answers to have the grievance redressed'.

One fact the study has confirmed is that the Civil Society Organizations have a tendency to develop a blanket type understanding of all modes of privatization and raise voice in opposition–no matter whether they are privatization of water resources or water services; no matter whether they are construction related or Operation and Maintenance related; no matter whether they are maintenance contract or part-contracting arrangement.

But the conviction of the opposition by the Civil Society Organisations in Tamil Nadu are clear. They are based on 'commodification of water' and the 'need-right dichotomy'. They have (i) a conceptual understanding and (ii) a practical concern. The conceptual understanding is that water is a 'public good' and it should be delivered by a public sector undertaking and it should not be put in the hands of private companies who would use it for profit-making.

Privatization of water service would most probably distance the poorer Panchayats from being able to get the required quantity of water for its poor residents. In practice at Thiruppur, as of now, the NTADCL is following only the TWAD Board rates in supplying water to the Village Panchayats. Therefore, there is no discontent among the Village Panchayats regarding the water charges, as alleged by users and Civil Social Organisations in other parts of the world (such as Bolivia, for example). In fact, contrary to the anxiety of many, privatization in Tirupur is 'actually reducing the O & M cost' of many of the Village Panchayats, at least for now. We are not sure if this phenomenon would continue in future also.

Is it a temporary arrangement or if it is going to be the permanent arrangement is yet to be as certained. This is because in most countries where water services were privatized it is often the ordinary people who end up paying the increased financial cost resulting from water privatization as the companies try to recoup their investment. This is either through increased prices for water or through state subsidies of the companies[14]. There are industries such as Tamil Nadu Small Industries Promotion Corporation Limited from Erode district that is said to be requesting NTADCL to supply water to its Perunthurai industrial estate[15].

Naturally, for NTADCL providing water to industries would fetch more profits than giving it to the Village Panchayats. Unless the NTADCL has legal binding in the agreement with the Government of TN, they would naturally tend to prefer to supply to industries rather than to the Village Panchayats.

Therefore, the fact is that it is taken as a corporate social responsibility for which the Consortium and the Industries in Thiruppur share the expenses. As of now, although the Panchayats have complaint that the Company does not provide them the required quantity of water, as far as paying a price for water is concerned they do not suffer. They pay what they would otherwise pay the TWAD Board of the Government of Tamil Nadu. Nevertheless, if it would continue in the days to come also or if the Company would revise the rates after some initial months are over is unknown.

The practical concerns are 'equity' and 'accountability'. Water being a basic need and should be accessible to the poor as much as it is available to the rich. But some of the Panchayats (e.g. Kaniyanpoondi) that are unable to pay the TWAD Board (Second Scheme) would not be in a position to pay the NTADCL's Third Scheme. But the fact remains that they need water to supply to the community at least through common collection points, though not through the House Service Connections (HSC).

THE LEAST IMPORTANT CUSTOMER

The question of accountability is another vital aspect that almost all the Panchayats complain about NTADCL. The way NTADCL deals with Village Panchayats is as if the VPs are their least important customer. The transaction is something like "we (NTADCL) would supply a certain quantity of water; you are supposed to pay for it. That's all the dealing is". There is no mechanism to find out if the quantity of water supplied, as claimed by the NTADCL, is true; there is no response when Village Panchayat functionaries want to know on what basis water demand of a Panchayat is calculated (such as Total Population by 55 LPCD etc); There is no grievance redressal mechanism. There are no discussions. The District Collector is said to have signed on behalf of the Village Panchayats. The Panchayats have to be contended with whatever quantity they are told to have been supplied.

BEHIND THE CORPORATE SOCIAL RESPONSIBILITY

That the NTADCL has agreed to provide water to the way-side Village Panchayats is a compulsion warranted by the situation is the opinion of many of the Panchayat functionaries. They have to do it in the name of corporate social responsibility, especially because the main water line from river Bhavani (from where water is pumped) to Tirupur Municipality passes through certain Village Panchayats that are water starving. Bringing water to the industries in Tirupur ignoring the way-side villages would result in the community in the way-side villages get down to tamper with the waterlines to tap water illicitly.

The company needed to avoid such an untoward situation. That is the reason they have, in the guise of corporate social responsibility, agreed to provide a portion of water requirement of the way-side villages on payment on par with the TWAD Board. This is visible from fact they have taken 1991 population as the basis to work out water requirement in these way-side villages and provide only a portion even from that calculation stating that they would get full supply only at the end of the 30 year contract period the company entered into.

As the scheme has become functional only for the past two to three months, the Village Panchayats have not been able to sense it properly how the water allocation is going to be.

Moreover, the recent rains and floods in the state (October–November 2005) have, to a considerable extent, replenished the private water sources and Panchayat water sources in the so-called way-side villages. Water from Mettupalayam (TWAD Scheme II) for these way-side Village Panchayats is supplied regularly. These are some of the reasons that keep the Village Panchayats contended about the supply at the moment. During severe water crisis, if these way-side Panchayats did not get the required water, they would definitely demand and lay hand in the NTADCL water only.

PROVIDING WATER TO THE SLUM DWELLERS IN TIRUPUR

Most of the slum dwellers in Tirupur are from poorer Tamil Nadu villages who migrate to Tirupur especially during agriculturally slake seasons. If, in some year, Tamil Nadu happened to get good rains, nearly a half of the workers would have run back - either informed or uninformed - to take up agricultural pursuits in their native villages and in the neighbourhood and be away from the problems of water scarcity and textile pollution in Tirupur. That would result in labour shortage in the hosiery industries of Tirupur. Although no field studies carried out to prove this point, the number of "Labourers wanted" placards that are hung at the gate of each of the industry in Tirupur during peak agricultural seasons could prove this point. The reasons for keeping employment in the hosiery industry merely as a fall back proposition by many of the migrant workers are: (i) water scarcity in Tirupur industrial area; coupled with the problems of textile chemical pollution in the ground water that is made available for bathing and other domestic use.

NTADCL water is expected to solve this problem, to a considerable extent, so that the workers who have double minds about staying back at Tirupur hosieries, and taking up agriculture in their home place would decide in favour of staying back in Tirupur. This would also serve as a answered to the allegation that people might bring, just in case drinking water provision is completely neglected and water for industrial use is assured.

All the more, this would not amount to blatant disregard for the National Water Policy 2002 which states drinking water should be given top priority, and water for industrial use may be given third on the priority list only after irrigation.

CONCLUSION

Almost all the way-side Village Panchayats covered under the new PPP scheme by the NTADCL are contended about the privatization of water services on the grounds that it has reduced their expenses on O & M of water supply expenses. All the VPs depict only a deficit budget by a big margin from the income sources of water supply, which is their prime botheration as far as water distribution is concerned. But, the new privatized scheme has shown that by providing 1000 litres at the rate of Rs.3.50 it can reduce the EB bill of the Panchayats. Although doubts are expressed that this practice might end up shortly, at the moment it is encouraging to see Village Panchayats recover from the burden of fat EB bills.

The grouse of the Village Panchayats are that the NTADCL does not commit to provide their entire water supply requirements. Similarly, question of accountability has become poor after the PPP arrangement. But for these two important hitches, the Village Panchayats do not find any problem in getting water under the new Public-Private Partnership arrangement in Tirupur.

REFERENCES

Anitha Sampath et al: Water Privatization and Implications in India.

Soumyen Sikdar (2005), Contemporary Issues in Globalisation, Oxford Press, New Delhi.

Keith Gauntlett (2005), 'Water–Life Before Profit' Integral Liberation Vol. 9, No. 2.

A Review of Public-Private Partnership in Water and Environmental Sanitation in India, New Delhi: DFID (1999).

New Tiruppur Area Development Corporation Limited, Minister for Information, Publicity and Local Administration.

Policy Note (2005-2006), Government of Tamil Nadu NTADCL.

Anitha Sampath et al Water Privatization and Implications in India.

Ashim Goyal (2004), 'Getting Water from Public-Private Partnership' India Infrastructure Report.

Lyla Mehta and Birgit Ia Cour Madsen, Poor People's Access to Water, IDS Bulletin 35. 1.

Keith Gauntlett, 'Water–Life Before Profit' Integral Liberation Vol. 9, No.2 June 2005.

Lyla Mehta and Birgit Ia Cour Madsen, Poor People's Access to Water, IDS Bulletin 35.1.

Government of Tamil Nadu, (1994), The Tamil Nadu Panchayats Act.

Government of Tamil Nadu G.O. MS No. 260 RD (WSI) dated 9th Dec. 1998.

Advocacy Internet: Stealing our Water–Implications of GATS for Global Water Resources.

www.psiru.org.

33

COMMUNITY INITIATIVES AND PARTICIPATION OF THE COMMUNITY IN DRINKING WATER SUPPLY: BASIC LESSONS AND EMERGING ISSUES

V. Annamalai & E.V. Prakash Rao

COMMUNITY INITIATIVES: TAMIL NADU

While states responded with sectoral reforms across the country, communities responded on their own. They took advantage of certain pro community initiatives by states like Tamil Nadu in case of "Namakku Naame" (We for Ourselves) Scheme. Government of Tamil Nadu in 1997-98 announced to accelerate the pace of socio-economic development in rural and urban areas with a specific intention of institutionalizing the concept of people's participation in local planning and execution of the developmental process on a continuous and sustainable basis. The basic philosophy of Namakku Naame is that the development at the gross root level has to be self generated, self organized and self sustained. This scheme is to promote an attitude of 'self help among people so that they come forward to fulfill their needs themselves.

The idea is basically to revive the ancient concept of 'Kudimaramathu' (community maintenance) which was in vogue all over Tamil Nadu (still prevalent in some stray pockets) to clear and maintain water courses for irrigation by the community. Each land owning beneficiary under the command area of a irrigation system contributes labour or money for the work in accordance with the size of benefited land holdings. The NNT aimed at extending and enlarging this principle to all community assets creation and maintenance. Instead of Government taking up the activities that could be taken up by the local community, people was encouraged to assess and prioritize such activities, actively contribute and involve themselves for its execution and maintenance to uphold the principle of community participation in development. In the entire process Government is to play a catalytic and supportive role.

Some of the basic features of the scheme are as follows i) Namakku Naame Thittam is applicable to all areas, urban and rural. The basic principle is that the local public identify the activity, contribute and carry out the work themselves. ii) Needs of the local public will vary from place to place. Therefore uniformity or rigidity in selection of activities is not desirable. The projects to be started should be primarily the choice of the people. This scheme can be executed for creation of new assets or for the maintenance of the assets already created. iii) Nature of activities need not be restricted to the works related to the local bodies but any activity irrespective of the department i.e., works related to PWD, Electricity, ADW (SC Welfare), BC Welfare, Highways, Health, Sports etc can also be taken up and executed with their concurrence under Namakku Naame Thittam.

The following are some of the activities which are illustrative but not exhaustive: Construction of buildings, labs, toilets for schools, hospitals and also construction of library rooms, hostels, noon meal centres, TV rooms, PDS shops or any other similar community assets in the villages including creation of drinking water. Creation and improvement of minor irrigation works like desilting of tanks, canals, ponds, ooranies (domestic purpose ponds), construction of sluices and retaining walls, drainage's, check dams etc. Construction of bridges, culverts and formation of new roads, upgradation of roads to metalling or BT or its renewal: Improvement and maintenance of parks, play fields, traffic islands, fountains, street lights etc.

In Andonkoil Panchayat in Karur district people mobilised cash to the tune of Rs 47 lakhs to construct a Over Head Tank and digging a well in the river bed and also laying down pipe line to supply water to the tank as well as to the houses. This was made possible by the NNT scheme where The district collector took active part in the implementation of the project once local people led by the young Village President made efforts to get the scheme approved. Starting from survey of the area up to the distribution of water, committee constituted by the people undertaken the work without involving contractors. Local people did voluntary work for the project. The committee could get enough resources because it is almost part of Karur town, with most of the people from different castes have migrated into the area. Still driven by necessity, many of them contributed Rs. 2000 each to get water in their homes and the others have access to common taps with some marginal amount. The fact that this area is predominated by families owning small-scale industries and thereby reasonably well off, proved to be a strong point in raising such a huge contribution.

Similarly in Rameshwarapatti village in Karur district, consisting of single caste group, people contributed total money involved in the project for laying down of pipeline for drinking water connected to most of the houses. These two cases show that if environment is created it is possible for people respond not only with the funds but also active involvement in implementation and management of the project. The sustainability is also ensured by evolving common resource management principles in terms of capital and O& M costs, regulated use of water and also dealing with free riders and defaulters.

COMMUNITY INITIATIVE: KERALA

Olavanna Gram Panchayat in Kozhikode district had a drinking water scarcity problem. The three rivers including Chaliyar, flowing through the Panchayat are saline. Other non-saline surface water bodies in the Panchayat go dry in February, as summer sets in. The water scarcity forced people (and women in particular) to walk long distances to get water for their daily needs. The water scarcity was especially acute during the drought in 1985. At that time, Kerala Water Authority (KWA) scheme in the village providing uncertain and erratic water supply to just 1,600 of the 7,100 households. The villagers disappointed with the KWA scheme, picketed the Gram Panchayat (GP) office, demanding drinking water schemes. This led to the GP promising to provide drinking water schemes in the village. This commitment led to the commissioning of the first piped water scheme in 1987 in Vettuvedankunnu Ward of the Olavanna Panchayat. This scheme was funded by Government grants to the Block Panchayat. The Vettuvedankunnu scheme consisted of an intake well, an overhead tank, and pipelines to distribute drinking water, through public stand posts, to serve 400 households.

Since late nineteen eighties, the GP and the Block Panchayat have built 18 piped water schemes, which provide drinking water through public taps as well as house taps to 1,362 families. The KWA also commissioned 2 more schemes in 1990 and 1998 which serve 2,400 households. After not happy with the quality of service provided by the KWA and spurred on by the GP President, Mr. Babu Parassari, the Olavanna villagers resolved to meet their drinking water requirements themselves.

The scheme initiated by the GP generated a lot of interest in the 3 villages. The villagers realized that local needs required local solutions. However, given the financial constraints, which were discussed openly in the GP meetings, a retired school teacher had a novel idea in 1989. He collected money from the community and installed a small 1 HP pump with an intake well to service 5 neighboring families in the hamlet of Kambiliparamba. The GP supported this initiative. This project was successful and the other villagers realized that instead of helplessly agitating against the KWA and the GP, they could themselves solve their drinking water problem. This could be done at a low cost and, as they themselves managed the project, they could organize the service according to their requirements. Encouraged by this initiative and supported by the GP President, 54 other households of Kambiliparamba got together in 1989, and, with a contribution of Rs. 4,500 each, formed a registered co-operative Society to provide drinking water for their own needs. This was a piped water supply scheme consisting of a similar intake well, pumpset, overhead tank and distribution system. From 1991 onwards, several such private Societies have been formed and similar small piped water supply schemes commissioned.

The GP has successfully shifted its role from being a provider to a facilitator and it has performed the regulatory function to sustain and encourage this novel project for the last 11 years. The GP does not provide any funds to these Societies, as capital costs or for the operation and maintenance (O&M).

The process of initiating a piped water supply scheme is as follows. After enlisting all households who wish to benefit from a piped water supply scheme, the beneficiaries get together, draft their by-laws and register their co-operative Society (under the Co-operative Societies Act of 1860). Members of the Society are asked to pay their membership fees, which varies from Rs. 4,500 to Rs. 12,500 per household. Land is purchased for the open well and for the overhead storage tank. The location of the well is arrived at by consensus. Through local expertise well and storage tank was constructed and similarly pipelines were also laid. Beneficiary families provide voluntary labor as per the skills required from time-to-time in the scheme. This is in addition to the cash contribution.

The General Body (GB) of each Society, consisting of all users (averaging around 50 households per scheme), elects an Executive Committee (EC) of between 7 and 11 members, including a President, Vice-President, Secretary and Treasurer. The EC is elected annually and runs the day-to-day affairs of the Society. The Treasurer operates the Society's bank account, and is answerable to the EC. The EC has to obtain the permission of the GB to utilize the Society's funds - including the payment for the construction of the water supply scheme. Each year, a General Body Meeting (GBM) is held to scrutinize the accounts, discuss the annual report, and elect the EC for the following year. Transparency in everyday functioning is a critical factor contributing to the sustainability of the Society. Each Society has prepared detailed by-laws for efficient functioning.

Although several residents of Olavanna are fairly well-off by rural standards, there are some relatively poorer families in the GP. The membership to the private Societies is, however, not restricted to the well-off. Even poorer families contributed by paying their initial contribution in installments. However, it needs to be accepted that the really poor have not opted for the private schemes, but are dependent on the GP and the KWA schemes.

The beneficiaries pay all the O&M costs, including the cost of hiring a pump operator and energy costs. The Society makes sure that there is an operating surplus for future repairs and maintenance. Each household is allowed 400 liters per day. Water meters are installed to measure the consumption, and excess consumption attracts a penalty at the rate of Rs. 20 per 1,000 liters.

GRAM PANCHAYAT RUN SCHEMES AT OLAVANNA

In addition to the piped water schemes by private Societies, the GP and the Block Panchayat have built 18 piped water schemes so far, and 12 more are under construction. These schemes provide drinking water through public taps as well as house taps, benefiting 1,362 families. Schemes run by the Olavanna GP also have registered Societies, which function on the same lines as the private schemes. These registered Societies plan, design and construct the drinking water schemes. In these GP schemes, households contribute 25 per cent of the total capital cost of the scheme either in cash, labor or land, while the GP provides the remainder. Poorer and low-caste families benefiting from a Panchayat-run scheme are not required to contribute towards capital costs. Instead, an additional amount ranging from Rs. 300-400 is collected from each of the other contributing beneficiary families. Water supply is limited to 2-3 hours in a day. Once the scheme is constructed, the beneficiaries contribute towards partial O&M. Water meters are installed on house connections, and Rs. 15-20 is collected per month from each family. Excess consumption of water attracts penalty, as in private schemes. Poorer families, who cannot afford to pay the capital cost to have household connections, instead draw water from street taps and pay between Rs.10-15 a month as the O&M costs. The GP pays Rs. 450 per month towards the salary of the pump operators and the cost of fuel. The maintenance of the street taps is done by the local KWA officials, for which the GP pays the KWA an annual fee of Rs. 1,750 per street tap. The GP realizes that this is a high cost and is planning to hand over the entire O&M of the GP schemes to registered Societies of water users, like those which run private schemes.

There are certain key lessons learnt from the experience at Olavanna. First is the willingness to pay by the community. The Olavanna GP is unique in that community groups have paid 100 per cent capital costs, in addition to 100 per cent O&M costs for rural water supply schemes. Secondly "private" schemes are more cost-effective. The difference in the O&M costs of private schemes and GP-run schemes is basically due to the difference in staff salaries at local level. User management has led to user satisfaction. An interesting facet of the sustained service from these schemes is the absence of conflict that might be expected over an issue as important as water supply.

The Panchayat acted as a facilitator and not provider Perhaps the most important lesson of Olavanna is the demonstration that the GP can effectively reduce its role from being a passive provider of public services and can instead become a dynamic facilitator for communities to take on the responsibilities of helping themselves.

COCHIN DECLARATION AND AFTER

The policies and methods regarding water management, which had been part of decentralised planning found place in the Cochin Declaration in 1999. This is based on the realisation that in India many rural water supply services are unsustainable and do not adequately serve the needs of user communities. Hence the following principles should guide the reform process: adopting demand–responsive approaches and the use of participatory processes; changing the role of the government from provider to facilitator: establishing financial viability and sustainability of rural water supply services and promoting integrated water resource management.

The Cochin declaration found its reflections in the creation of Kerala Rural Water Supply and Sanitation Project. According to World Bank " The project will improve water and sanitation services for 1.5 million people, specifically targeting Kerala's poorer and disadvantaged communities in the Kozhikode, Malappuram, Palakkad and Thrissur districts. The project was to assist local and state public agencies in shifting focus from direct service delivery to supporting community groups in planning, constructing and operating piped water supply and environmental sanitation systems".

COMMUNITY PARTICIPATION IN KWSS PROJECT

The World Bank Sponsored KWSS Project at Pookkoottur, Mallapuram District reflected this new approach. The basic characteristics of the project are: demand driven approach, cost sharing, adopting integrated approach, pro-poor approach, women development initiatives, community empowerment, community contracting, utilisation of available resources and dovetailing with decentralised planning. According to Ms. O.T. Maha Devi, Director, Socio-Economic Unit Foundation (Supporting Organisation for the project) Pookkoottur, Mallapuram District, the major positive aspects of the projects are: demand driven and involvement by Gram Panchayat and participation by Beneficiary Groups (BGs); ownership and control over the project by BGs; specific importance to women and gender value; confidence of women BGs increased; each and every member of Panchayat knows what is happening to the project and review of project by Panchayati Raj members; functionality, use and impact (maintenance) is very good coalition: sanitation only for BPL families: waiving of fees of Rs. 2000; deciding BG member; accountability is more in O & M. Originally, 21 groups were formed in the scheme area and now only 19 are functional. At the negative side of things that one has to work with different types of people and that makes community mobilisation difficult. It is difficult to build a self-sustaining groups. There are no blue prints possible. BG appoints Management committee which consists of 10 members from managing the affairs of BG. According to her, there are no possible future problems in O and M. However, if there is any increase in price of power, there is a need for subsidy. The expectation of subsidy also arisen because of there are subsidies given in other sectors.

Experience of Kerala drinking water project shows that by and large the community in the project area is supportive of the reform and are willing to take care of O & M. Specific role for PRIs is envisaged in the project by having a tripartite agreement as the basis between the Gram Panchayat, KRWSSA and a Supporting Organisation which renders technical guidance. The Gram Panchayat facilitates and coordinates the implementation of the project through Beneficiaries' groups on a demand-driven basis. The GP also undertakes to meet a nominal share of the project costs. O & M is the responsibility of the Beneficiaries' Group and the GP is expected to take over all single panchayat schemes of Kerala Water Authority and hand them over to Beneficiaries' Groups within the overall framework of the World Bank project. The CBOs in this project are essentially formed by active involvement of the GP and a harmonious relationship exists between them. The components of the project include aspects of rural sanitation, construction/upgradation of latrines, micro-watershed development in the catchment areas and beneficiaries' say in deciding the source of water. It is to be noted that external funding is a major factor driving this project.

COMMUNITY INITIATIVES AND PARTICIPATION OF COMMUNITIES

In the domain of the drinking water service, the community initiatives seems to be crucial as demonstrated in case of Andonkoil panchayat in Tamil Nadu and Olavanna panchayat in Kerala. In both the cases desperate demand of the people for drinking water led them to contribute money and labour for creation of infrastructure. In both cases, GPs and government played facilitative role by enabling the community efforts to succeed. It did not intervene with the emergence of regulations regarding water distribution and O & M. The capital cost is already by these efforts and part or full costs are met by the beneficiaries. On the other hand, substantial investment has to come from government in case of KWSSA schemes. The participation of the community is limited the stipulations in the scheme and not evolving according to the area. However, not all persons particularly poor are participate in the scheme because not everybody could afford the capital cost and O & M costs. The community participation gets fragmented resource use where water is public resource gets beneficially distributed for people who can afford.

34

HOUSEHOLD PARTICIPATION IN COMMUNITY BASED RURAL WATER SUPPLY SYSTEMS KERALA EXPERIENCE

*Nisha K.R.**

INTRODUCTION

Since water is a basic need, it was assumed for a long time that the responsibility of the provision of water supply should be entrusted to the government. But, the fiscal crisis of the governments combined with structural adjustment programmes compelled most of the developing countries to look for alternatives in water supply management. Community-based development schemes emerged as a solution to the problems which came across in the state management of the resources. Hence, in the wake of decentralized planning process the state governments devolved the responsibility of providing basic necessities to local organizations. Under the

* Doctoral Fellow, Institute for Social and Economic Change, Bangalore-72. *E-mail: krnisha@gmail.com* Currently Working as Faculty, ICFAI Business School, Kochi.

This paper is a part of my Ph.D. thesis under the guidance of Prof: D. Rajasekhar, Center for Decentralization and Development, ISEC, Bangalore.

new initiative of community water supply schemes, the beneficiary groups are responsible for planning, technology selection, and installing rural water supply facilities. Operation and maintenance of the structures created in the projects is also the responsibility of the beneficiaries. In this context, an attempt has been made to answer the question 'what factors affect the beneficiary participation in community-managed rural water supply schemes?'

EVOLUTION OF COMMUNITY-BASED WATER SUPPLY SCHEMES IN KERALA

Households in Kerala traditionally consider own wells, mainly open wells, located in the household premises as the main source of drinking water. These open wells dry up during summer and people have to walk a distance and spent time to collect water. The state government has taken keen interest in tackling the problem of water scarcity since its formation. Government intervention in drinking water supply system in Kerala can be classified under two categories: one, in the form of public/private taps provided either by the Kerala Water Authority (KWA) or Gram Panchayats (GPs) and the other, in the form of public wells, which, in most of the cases, were constructed by the local governments.

Kerala Water Authority is one of the main agencies for the design, construction, operation and maintenance of water supply in the whole state. KWA has been implementing piped water supply schemes based on surface and groundwater sources. It undertakes projects sponsored by multilateral funding agencies through the state and central government. It also executes bilaterally funded projects and accelerated rural water supply schemes on behalf of the government of India.

Decentralized planning process which started in the state in 1996 paved the way for local governments to take up issues related to drinking water. But, even before the implementation of decentralized planning the local bodies in Kerala intervened in the drinking water sector. Gram panchayats constructed public wells and small water supply schemes with technical assistance from Kerala Water Authority. Initiation of people's plan campaign provide more powers to district, block and gram panchayats to take up small water supply schemes with user participation.

Besides, the state government decided to transfer all small rural water supply schemes, within the boundary of one GP, to the local bodies with associated powers to levy and collect user charges for providing water services.

The alternative institutional model introduced a new service delivery mechanism for the rural water supply in the state with the participation of beneficiaries. In Kerala, community-managed water supply schemes were implemented with the help of external funding agencies and central government. External agencies included World Bank and Royal Netherlands Embassy. Even though there was difference in the organisational structure, the process of implementation of water supply schemes was similar in all alternative set-ups. One of the important features of these community-managed water supply schemes was the distribution of responsibility to the users in terms of sharing capital cost and operation and maintenance duties of the schemes after implementation. More often ten per cent of the capital cost was shared by the beneficiaries and divided equally within the group. Contribution to capital cost could be given either in terms of cash or labour. Operation and maintenance (O&M) was fully the responsibility of the beneficiary group. Groups were collecting water tariff from all the users to cover O & M cost of the system. By rule, participation in group meetings and decision-making was mandatory in community-based water supply schemes. But there was not any negative incentive for those who couldn't participate in the group meetings.

THE STUDY

Information for analysing the factors affecting participation was collected through a field survey conducted in Malappuram district in Kerala. Five gram panchayats (GP), one from coastal, three from midland and one from highland were selected for the study. A sample of 310 households was drawn from the government (KWA), panchayat and community-managed water supply systems from the selected GPs. The sample includes 200 households drawn from the community water supply schemes (CWS), distributed among 26 Beneficiary Groups (BGs), 60 households from among the beneficiaries of KWA schemes and 50 from the GP schemes. Data were collected at the household level with the help of formal interview schedules. Additional information regarding the characteristics of the locality[1] was collected through informal discussions with senior citizens in each locality. The following sections of the paper consider the theoretical underpinnings of collective action and participation, factors affecting collective action and the empirical evidence from the field.

COLLECTIVE ACTION *VERSUS* PARTICIPATION

Collective action is defined as 'a function of individuals' incentives to contribute to the maintenance and abide by the rules and regulations, the capacity of the community as a whole to cooperate and to manage the incentives, and the overall policy environment in which the institutions must operate' (McCarthy *et al* 2002: 5). It is argued that collective action is the activity which happens through an organisation (Meinzen-dick *et al* 2002) and people's/user's participation is a necessary condition for the success of collective action (Singh 1992).

Studies on community participated water supply projects have found that demand responsiveness and participation from the part of beneficiaries improve project performance (Isham *et al* 1994; Narayan 1995; Isham and Kahkonen 1999; 2002; Prokopy 2005). A study on financially aided rural water supply projects has found that projects with a high degree of beneficiary participation had a success rate of 68 per cent while projects with low participation from beneficiaries achieved a success rate of only 12 per cent (cited in Skidmore 2000). But critiques argue that participation can have adverse effects in terms of providing overburden on the shoulders of local people and also participation turns out to be mere pretension where local people do not have the ability to influence the outcome of the project (Oakley 1991; Cooke and Kothari 2001, cited in Prokopy 2005).

However, participation has been measured differently in the context of community water supply. Isham *et al* (1994) and Narayan (1995) analysed participation in a hierarchical order with information sharing representing the lower end and decision-making and control representing higher end of the scale. Agarwal (2001) provides a detailed classification of participation in community-based management of natural resources. According to her, participation ranged from mere membership and attendance in the beneficiary group meetings to active involvement in terms of influence in decision-making and interactive participation, which has the potential to empower the beneficiaries.

Prokopy (2005) has argued that in community driven water supply projects contribution of money, labour, materials, etc, could be considered as the low form of participation. For him, households' attendance in the meeting, speaking out at the meeting etc, contributed the middle of hierarchy and participants own initiatives and leadership roles contributed the higher form of participation.

Sara and Katz (1998) have taken demand responsiveness as the factor affecting project outcome instead of participation. They have argued that in demand responsive projects, communities were allowed to make informed choice about whether to participate in project or not. Informed choice measured the degree to which individuals felt involved in decision-making processes and were informed on the consequences of the decisions. Rajasekhar and Veerashekharappa (2003) have also considered participation in group meetings, decision-making and in implementation (in terms of payment of contribution and user fees) as the indicators of participation in water supply projects.

PARTICIPATION IN RURAL WATER SUPPLY SCHEMES IN KERALA

The present study has taken the participation of beneficiary households as the proxy for collective action in community management of water supply. Keeping in mind the existing studies on participation, this study considers the following variables to capture participation.

Attendance in Beneficiary Group (BG) Meetings: Attendance in BG meetings counts the number of members appearing in the group meetings irrespective of the extent of participation. In the case of community based schemes, the attendance of at least one person from the household benefiting from the project in group meetings has been made mandatory. The data reveal that mostly women attended the meetings as the men-folk was away in pursuing livelihood activities. The survey results, which have been presented in Table 1, reveal that a majority of the households (42.6) attended the meetings regularly. Nearly 25 per cent of the beneficiary households never attended the meeting.

Table 1
Distribution of Households (%) by Attendance in BG Meetings

Attendance	*Percentage of households*
Regularly	42.6
Occasionally	32.6
Never attend	24.8
Total	100.0 (310)

Note: Figure in the parenthesis is actual number.

Making Suggestions and Influencing the Decision: If a member makes any suggestions or influences the decisions relating to the development interventions in the group meeting, this can be considered as the higher degree of participation. As far as the sample respondents was concerned, the question asked was "whether they made any suggestion, in general, relating to the scheme or relating to the drinking water' and `what was the extent to which they influenced the decision'. Table 2 reveals that a majority of those attending the meetings were not offering any suggestions in the group meetings. Only 11.3 per cent of the sample respondents reported that they always made suggestions in the group meetings. Percentage of households influencing the decision was even less.

Table 2
Distribution of Households (%) by Extent of Participation

	Making suggestions	*Influencing decisions*
Always	11.3	10.3
Occasionally	26.5	25.2
Never	62.3	64.5
Total	100.0 (310)	100.0 (310)

Note: Figures in the parenthesis are actual numbers.

Table 3
Distribution of Households (%) Influencing Decisions

Influence in decision	*Influence in deciding location of WS*	*Influence in deciding water tariff*
Large influence	5.2	4.2
Medium	8.4	14.2
Little bit	7.7	11.9
Not at all	78.7	69.7
Total	100.0 (310)	100.0 (310)

Note: Figures in the parenthesis are actual numbers.

Influencing the Location of Water Supply and Tariff: It is possible that the beneficiaries (households) influence the decisions regarding specific issues rather than the decisions in general. In the case of decisions relating to water supply, the location of water supply and monthly tariff are important ones. The respondents were, therefore, asked the following questions which specifically related to the above. First, whether the sample households influenced the decisions of location of water supply. Second, whether they influenced the fixation of water tariff rates charged on the users. It was found that, considering all the categories, the percentage of respondents influencing the decisions relating to the fixation of water tariff was high as compared to the decision regarding the location of water supply (Table 3). But, in both the cases, the number of households influencing the decision, to a large extent, was less.

PARTICIPATION INDEX

Two separate participation indices were prepared for the purpose of analysing the variations in participation across the institutional options and factors contributing to these variations. The indices on participation combined two or more qualitative variables. This was done on the assumption that indices gave a better picture on the involvement of beneficiaries in the community based water supply schemes, and they provided better scope for measurement, analysis and reaching good conclusions. For the purpose of constructing indices, weights were assigned to different levels of participation. The weighted sum of different participation level was taken as the participation index[2]. The resultant participation index ranges from 0 to 100 with 100 representing the higher level of participation and 0 representing no participation.

INDEX FOR ATTENDANCE IN THE MEETING

The first index on participation considered mere attendance in the group meeting as well as the extent of making general suggestions in the group meetings. This index was prepared by giving 25 per cent weight to the attendance in meetings and 75 per cent weight to making suggestions[3]. This participation index thus represented the passive form of participation as the variables related to influencing decisions were not included in this index.

The distribution of households by index attendance in the meetings shows that a majority of the households covered by the community scheme were participating in the BG meetings (Table 4). Participation in the case of a majority of the households vary between low and moderate forms, implying that less number of households were involved in making suggestions regularly. Distribution of households by attendance index shows that only 36.5 per cent of the households participated well in community managed water supply schemes.

Table 4
Distribution of Households (%) by Index for Attendance in the Meeting under Different Institutions

Institution	*Index*					*Total*
	No partici-pation (0)	*Low partici-pation (0-25)*	*Moderate partici-pation (25-50)*	*High partici-pation (50-75)*	*Active partici-pation (>75)*	
CWS	6.5	45.5	11.5	20.5	16.0	100.0 (200)
KWA	86.6	6.7	3.3	1.7	1.7	100.0 (60)
GP	24.0	42.0	6.0	24.0	4.0	100.0 (50)
Total	24.8 (77)	37.4 (116)	9.0 (28)	17.4 (54)	11.3 (35)	100.0 (310)

Note: Figures in the parenthesis are actual numbers.

In the case of KWA schemes, there were no formal water users association. But, gram sabhas in the locality provided an opportunity for the users to represent their interests and demands to the authorities concerned. However, only 13 per cent of the households benefiting from KWA schemes either attended or made any suggestion regarding water supply in the gram sabha meetings. A detailed analysis of these households shows that all of them were using public stand posts provided by the KWA. None of the households having private household connection was attending any of the gram sabha meetings. In the case of GP provided water supply schemes, water user's association existed in an informal manner. Seventy-six per cent of the beneficiary households marginally or actively participated in these informal meetings. Among the households participating in group meetings, 42 per cent had low participation indicating that they were merely attending the meetings without making any suggestion.

INDEX FOR DECISION-MAKING

The second index for participation considered only households' participation in terms of their influence in decision-making in the location of water supply and monthly tariff. The index was constructed by giving 25 per cent weight to influence on the location and 75 per cent to the influence on the fixation of monthly tariff of water supply[4]. The decision-making index represented household's participation in decision-making and could be considered as active form of participation. The results show that more than fifty per cent of the households were not involved in the decision-making process (Table 5). Only 6.5 per cent of the households participated actively in decision-making with index value greater than 75.

Low participation in decision-making in community schemes was mainly because of the extended support of NGOs in the implementation process. Field information shows that in a majority of the cases, the NGOs played a major role in the selection of technology for water supply and location of source. In some cases, initially, monthly tariff also was suggested by the NGOs. But, later, the beneficiary groups and committees have been revising the tariff according to the cost operation and maintenance.

Table 5
Distribution of Households (%) by Index for Decision-Making under Different Institution

Institution	*Index*					*Total*
	No partici-pation (0)	*Low partici-pation (0-25)*	*Moderate partici-pation (25-50)*	*High partici-pation (50-75)*	*Active partici-pation (>75)*	
CWS	52.5	13.5	13.0	14.5	6.5	100.0 (200)
KWA	100.0	0.0	0.0	0.0	0.0	100.0 (60)
GP	86.0	8.0	2.0	4.0	0.0	100.0 (50)
Total	67.1 (208)	10.0 (31)	8.7 (27)	10.0 (31)	4.2 (13)	100.0 (310)

Note: Figures in the parenthesis are actual numbers.

In the case of KWA, the beneficiaries did not have any influence on the decision on location of water supply and monthly tariff. Decisions of KWA were taken mainly by the government at the state level. In the gram panchayat schemes also, the participation of households in taking decisions regarding location of water supply and tariff was very meagre. This was mainly because the implementation of panchayat water supply schemes was usually done either through user committee as a convener or through the respective line department (Namboodiri, 2003).

VARIATION IN PARTICIPATION ACROSS GEOGRAPHICAL REGIONS

Participation of households in the community based water supply schemes may vary with geographical area and economic status of the locality because of the variation in water availability and income status. The geographical characteristics of the locality are considered in consistent with the natural divisions found in Kerala. By physical features, the state of Kerala can be divided into three natural divisions: (i) the low lands consisting of coastal areas, (ii) the middle land, and (iii) the high land. Salinity and brackishness were the main two problems faced by coastal households. People in these areas have to spend some time in travelling and queuing in order to collect water. Similarly, people living on the sloping and elevated areas of highland experienced severe water scarcity. The present study, therefore, anticipated that as the necessity of water was more in coastal and highland regions, there might be higher participation (Table 6). In addition to the classification of the locality based on geographical divisions, the study divided the localities based on economic status. Economic status classifies the locality as developed and less developed. Qualitative information on the economic status was collected by considering the locality's access to market, infrastructure facilities like road and electricity, etc.(Table 7).

Cross tabulation results show that households' participation in BG meetings and decision-making did not vary with locality type, implying that households in the coastal, midland and highland areas participated alike in the beneficiary group meetings. But, participation in both BG meetings and decision-making varied with the development status of the locality[5]. It has been shown in Table 6 and 7 that households' participation in group meetings was higher in the less developed locality. The difference in participation between these localities was mainly due to the variation in water availability. It was found that (Table 8) among those households who faced perennial water scarcity (without having own well and private household water connection) a majority belonged to less developed region. Those who faced serious water quality problem also belonged to less developed region.

Table 6
Distribution of Households by (%) Index of Attendance in the BG Meetings and Locality Characteristics

Index of Attendance	*Area*			*Economic status*		
	Coastal	*Midland*	*High-land*	*developed*	*LessDeveloped*	*Total*
No participation	20.0	23.2	36.7	9.9	49.2	24.8
Low participation	34.7	41.9	30.0	44.3	26.3	37.4
Moderate participation	12.6	5.8	11.7	9.4	8.5	9.0
High participation	23.2	16.8	10.0	24.0	6.8	17.4
Active participation	9.5	12.3	11.7	12.5	9.3	11.3
Total	100.0	100.0	100.0	100.0	100.0	100.0
	(95)	(155)	(60)	(192)	(118)	(310)
Pearson Chi-Square (Asymptotic. Sig.)		0.079			0.000	

Note: Figures in the parenthesis are actual numbers.

Table 7
Distribution of Households (%) by Index of Decision-Making and Locality Characteristics

Index of decision-making	*Area*			*Economic status*		
	Coastal	*Midland*	*High-land*	*developed*	*LessDeveloped*	*Total*
No participation	62.1	69.0	70.0	58.3	81.4	67.1
Low participation	15.8	8.4	5.0	14.1	3.4	10.0
Moderate participation	10.5	7.7	8.3	10.9	5.1	8.7
High participation	9.5	10.3	10.0	10.9	8.5	10.0
Active participation	2.1	4.5	6.7	5.7	1.7	4.2
Total	100.0	100.0	100.0	100.0	100.0	100.0
	(95)	(155)	(60)	(192)	(118)	(310)
Pearson Chi-Square (Asymptotic. Sig.)		0.425			0.000	

Note: Figures in the parenthesis are actual numbers.

The study has further analysed the relative importance of other factors which affect collective action in rural water supply schemes. Detailed explanation of the factors has been given in the following sections.

Table 8
Distribution of Households (%) by Water Scarcity and Economic Status of the Locality

Extent of water scarcity	*Economic status*		
	Less developed	*Developed*	*Total*
Perennial scarcity	68.5	31.5	100.0 (130)
Seasonal scarcity	45.3	54.7	100.0 (95)
Bad quality	84.0	16.0	100.0 (50)
No problem	51.4	48.6	100.0 (35)
Total	61.9 (192)	38.1 (118)	100.0 (310)

Note: Figures in the parenthesis are actual numbers.

FACTORS AFFECTING COLLECTIVE ACTION

Factors affecting collective action vary with the nature of the resources and the conditions[6] under which collective action were carried out (Ostrom 1990; Oakerson 1992; Baland and Platteau 1996). Wade (1988) mentioned that organisational success depends on factors like, (i) boundaries of common pool resources, (ii) technology, (iii) relationship between resources and user groups, (iv) characteristics of user groups, (v) noticeability i.e., ease with which rule breaking free riders are detected, and (vi) relationship between users and the state. According to Uphoff (1999), four basic ubiquitous activities of organisation (decision-making, resource mobilization and management, communication, and conflict resolution) were essential for mutually beneficial collective action (MBCA). MBCA reduces transaction costs and increases the probability that individual efforts become effective. Without the above four activities, collective action becomes more difficult and less likely. Oakerson (1992) identifies four sets of factors influencing collective action.

Physical and technical attributes

The physical attributes include, jointness, exclusion and indivisibility. i.e, Jointness means the relative capacity of the resource base to support multiple users at the same time without one interfering with another or diminishing the degree of aggregate level of benefit available to the group, whereas exclusion is the degree to which the commons permit the exclusion of individual users. Indivisibility is the physical boundaries of commons, which determine the minimal scale on which effective coordination can occur. Clearly defined boundaries foster collective action possibilities. In the case of water supply, the above mentioned three attributes differ according to the way in which water is treated, i.e. as public, private or common property good. Drinking water supply service assumes more merit good nature rather than pure public good in most of the countries. The degree to which water can be managed collectively depends on the ability to exclude some, but not others (Narayan 1995). But she has pointed out that the degree of jointness adds complexity to determine the participants in negotiation.

In addition to the resource specific characteristics physical features of the service, like technical defaults and poor quality of construction also affect collective action in community driven projects (Rajasekhar and Veerashekharappa 2003). Narayan (1993: 33) has argued that 'if the users feel that water is not of acceptable quality, delivery through piped water systems is unpredictable or breakdowns are frequent, users are unlikely to pay for or consistently use such water facilities.

Institutional Arrangements

Institutional arrangements are subdivided into three types, i.e,

- operational rules
- conditions of collective choice
- external decision-making arrangements

Even though the institutional arrangements can be differentiated, operational rules are nested in the conditions of collective choice and conditions of collective choice nested in external decision-making arrangements. External arrangements are mainly constitutional in nature. In a similar line, North (1981) has also distinguished different category of institutions as constitutional rules, operational rules and normative behavioural codes. According to him, the constitutional order is, therefore, the first categories of institutions; which specify the fundamental rules for establishing the conditions of collective choice. Olson (1965) has argued that organizational rules ease success in collective action.

Mutual Choice of Strategies and Consequent Patterns of Interaction

Individuals make choices considering the physical features of the commons, characteristics of relevant technology as well as the decision-making arrangements available to govern it. Patterns of interaction emerge from the joint choices made by the members in a group. As per Olson's (1965) 'Logic of Collective Action', large group size and heterogeneity make individual choices more costly and benefits less. But, there are many counter-arguments to Olson's contention that collective action will be difficult under the above conditions (Ostrom 1990; Baland and Platteau 1996).

Another line of argument for collective action failure is that, in heterogeneous group cooperative norms, confidence and trust among members may be low. Ostrom (1990) has observed that size and composition of membership did not have significant effect on cooperative performance (cited in Vedeld, 2000). She has suggested that effective leadership could solve the problem of large groups. From the field setting she pointed out that group size and homogeneity were not a precondition for the success of collective action.

Outcomes and consequences

Equity and effectiveness are two forms of outcomes mentioned by Oakerson (1992). The presence of inequities may lead to the collapse of reciprocity resulting in less efficient use of recourses. The success and sustainability of the community-managed systems depends on the sharing of benefits in an equitable manner (Kerr, 2002). Narayan (1995) has mentioned that, 'over time, as learning takes place, poor outcomes influence the pattern of interaction, which may eventually lead to changes in rules in decision-making arrangements'. Studies on community driven water supply projects have mainly analysed the relation between participation and project outcomes in terms of effectiveness and sustainability. Most of these studies have concluded that participation improves project outcome (Narayan 1995; Sara and Katz 1998; Isham and Kahkonen 1999, 2002; Prokopy 2005). Narayan (1995) has pointed out that the extent of beneficiary participation is determined by the characteristics of both the beneficiaries and the agencies. Two beneficiary characteristics she identified are demand and the degree to which beneficiaries are organised to their role. But, she has not tested empirically the factors affecting participation. Hence, it is worth to identify the factors which give incentive for households to participate in community water supply schemes in order to derive policies to implement the scheme successfully.

Group Characteristics

Consistent with Oakerson (1992) framework, the present study has identified characteristics of the group and the socio-economic background of the beneficiary household as the factors affecting collective action in water supply schemes.

Group size considers the number of members in the community group. Theoretically it is argued that large size of the group adversely affects the collective action in that group (Olson 1965; Ostrom 1990). Hence, the relation between group size and participation is expected to be negative.

Heterogeneity in caste is another important factor which affects collective action. Heterogeneity in the group can be of different nature like heterogeneity in interest, income, political affiliation, caste, class etc (Vedeld 2000). For and against arguments exist in the literature regarding the relationship between heterogeneity and collective action (Ostrom 1990; Baland and Platteau 1996; Vedeld 2000). The present study considers heterogeneity in caste.

Life span of the group measures the duration with which the group has been working. Longer period of working may influence positively to the collective action of the groups. Resilience power to overcome the conflicts and the capacity of learning by doing is the qualities commonly associated with groups of long-term survival. These long-term experiences generally make positive responses from the group members.

Leadership It is argued that problems arise in the group due to large size, and heterogeneity can be solved with the help of effective leadership (Vedeld 2000). The present study has classified a group with active leadership if one or two persons in the beneficiary group took the sole initiative to solve the conflicts and problems arising in the group. It is assumed that active leadership in the group advance the participation of other beneficiaries in the group.

Operational rules and accountability are the arrangements within the institution, which facilitate collective action. Olson (1965) has argued that organizational rules eased the success of collective action. Imposition of punishment for defaulting monthly tariff payment was considered as an indicator of the existence of operational rule in the group. Accountability was measured by combining two variables, namely, the receipt towards monthly tariff payment and maintenance of account book. Existence of rules and norms were expected to have a positive influence on the participation of beneficiaries in the community groups.

Household characteristics

Size of the family: As the size of the family increase the quantity of water needs also increases. Such a need influences the households to participate in the group meetings and decision making. The large family size increase the possibility that at least one member of the household could attend the group meeting regularly. Hence, the relation between household size and participation is assumed to be positive.

Sex of the head of the household and participant: A household being headed by a female has an added disadvantage of earning income in addition to the responsibility of managing the house. So the female headship of the household was supposed to reduce the participation in community groups. Female members of the household usually bear more hardship in fetching water. Therefore, it is assumed that if the participant of the meeting is 'female' participation in the meeting either in terms of making suggestion or influencing in decisions of water supply is higher as compared to men.

Education and Participation in other local organization: it is assumed that better education improves the awareness of public activities and hence the participation. The variable education was measured as the percentage of household members with education higher than 10th standard. Similarly participation in local organization shows peoples' involvement in other organizations like women self help group, gram sabhas and other cultural and political organizations. It represents the strength of social capital in the locality. The study has assumed the percentage of household members involved in at least one of the local organization and assumed to be positively related to the participation.

Water availability of the household: Beneficiaries' participation in group meetings may be influenced by the difficulties that they face in fetching water. The household who did not own well and faced problem to collect water throughout the year is supposed to participate actively in group meetings.

Use of water: The purposes for which a particular household use the water supplied through the different water schemes influence the participation. If the household use the water to a large extent for drinking purpose there is a possibility of higher participation.

Occupation of the household head and per capita income: The occupation of the head of the household indicates household security in earning income. For casual workers like agricultural labourers, employment was not assured throughout the year. As casual workers and daily wage earners, their participation in meetings became more costly compared to those having regular employment. The present study has considered a dummy variable to capture the effects of occupation on participation. Higher per capita income raises the standard of living of the households. Also with higher income, it is assumed that the opportunity cost of participating in the group meeting reduces. Hence, per capita income is expected to have a positive influence on participation.

OUTCOME WATER SUPPLY SCHEMES

Overall satisfaction and frequency of break down in water supply system: Overall satisfaction has been accounted for households' perception about the water supply scheme, whereas, frequency of damage has captured the effectiveness of water supply systems. These variables could be taken as a proxy for outcome of collective action. It has been argued that over a time, poor outcomes influence collective action adversely (Oakerson 1992; Narayan 1995). Hence, it is assumed that frequent problems with the water supply reduce people's incentive to be involved in community schemes.

MODEL

To test the relative importance of the factors which affected community participation in water supply schemes the following model was used.

$$P = f(G, H, O)$$

Where,

P is the participation index

G represents the group characteristics and

H represents the socio-economic characteristic of the household

O represents the outcome of water supply schemes

To asses the effect of each set of factors on participation it has hypothesized that "Collective action in water supply is not associated with the socio economic position of water users and characteristics of the group". Linear regression models were estimated using ordinary least square method with participation index for attendance and decision-making as dependent variable[7]. All the group and household characteristics were taken as independent variables. The description of independent variables used in the model has been provided in Table 9.

In the first model, participation index for attendance was taken as dependent variable. The index shows the extent to which households participated in the group meetings and made suggestions relating to the management of water supply system. The value of the index varies from 0 to 100, with 100 representing 100 per cent participation or active participation in the beneficiary group meeting. For analyzing the factors affecting participation index relating to attendance, only samples drawn from CWS and GP schemes were taken into consideration because participation of the households covered by KWA schemes in group meetings was very negligible. Only 13.4 per cent of the families participated in the group meetings and among the participants half of them had very low participation rate.

In the second model, index of participation in decision-making was taken as dependent variable. Index of decision-making also varies from 0 to 100, indicating higher the value of index more active was the participation in decision-making. This model was applied to the sample drawn only from CWS schemes as the participation index in decision-making was low or nil in the case of households which benefited from GP and KWA schemes.

Initially, both the models were estimated with all the independent variables mentioned above. In the later stage, insignificant variables were omitted[8] from the model by doing a stepwise regression analysis. Estimates of the models have been shown in Table 10 and 11.

Table 9
Description of Independent Variables

	Variable	Description and measurement		Frequency/Mean*	Expected sign
1.	GROUPSIZE	Group size		48	–
2.	HOMOGENEITY	Homogeneity in caste	1 = Homogeneous	22.6	+
3.	LIFESPAN	Number of months since the group has been formed		46	+
4.	LEADERSHIP	Active leadership in the group	1 = Active leadership	38.1	–
5.	IMPOSE	Imposition of punishment	1= Impose punishment		
6.	PUNISHMENT	inthe group		10.6	+
7.	RECEIPT	Getting receipt for tariff payment	1= Yes	51	
8.	FAMSIZE	Size of the family		7	+
9.	SEX- HEAD	Sex of the household head	1= Male	86.5	+
10.	SEX - PARTICIPANT	Sex of the participant in the group meeting	1 = Male	34.5	-
11.	EDUCATION	Percentage of family members with education greater than 10		13.71	+
12.	LO - PARTICIPATION	Percentage of family members participating in local organisation		12.86	+
13.	USE - DRINKING	Water used for drinking purpose	1= Use Water for Drinking to a large extent	42.3	+
14.	WATER	Scarcity of water	1= Perennial scarcity for Water	41.9	+
15.	OCCUPATION	Occupation of head of the household	1 = Casual labour		
16.	PCI	Per capita income		13,264.7	+
17.	SATISFACTION	Overall satisfaction of the scheme	1= Good	73.2	+
18.	OUT – OF - ORDER	Frequency of breakdown	1 = Never	17.4	+

Note: *In the case of qualitative variables the frequency of the variable with value 1 has been given in percentages.

RESULTS

Model 1

$$\text{pindex1} = \alpha_0 + \alpha_1\text{leadership} + \alpha_2\text{receipt} + \alpha_3\text{famsize} + \alpha_4\text{sex\_head} + \alpha_5\text{sex\_participant} + \alpha_6\text{education} + \alpha_7\text{lo\_participation} + \alpha_8\text{use\_drinking} + \alpha_9\text{water} + \alpha_{10}\text{pci} + \alpha_{11}\text{satisfaction} + U_i$$

Table 10

Factors Affecting Participation in Attendance in the Group Meetings

Dependent variable: Participation Index of attendance

Variable	*Coefficient*	*Std. Error*#	*t-Statistic*
C 0.70	9.79	0.07	
1. LEADERSHIP	-5.17	3.89	-1.33
2. RECEIPT	5.12	3.65	1.40
3. FAMSIZE	3.00*	0.64	4.69
4. SEX_HEAD	8.08***	4.84	1.67
5. SEX_PARTICIPANT	11.18**	4.42	2.53
6. EDUCATION	0.30*	0.09	3.33
7. LO_PARTICIPATION	0.51*	0.12	4.41
8. USE_DRINKING	10.40*	3.66	2.84
9. WATER	-19.25*	6.52	-2.95
10. PCI	0.00042**	0.00	2.03
11. SATISFACTION	4.87	3.88	1.26
12. *R-squared*	*0.29*	*F-statistic*	*8.67***
13. Adjusted R-squared	0.25	Log likelihood	-1175.39
Number of observations: 250	*Durbin-Watson stat*		*1.91*

Note: # White Heteroskedasticity-Consistent Standard Errors & Covariance.
*, **, *** Significant at 1 per cent, 5 per cent, 10 per cent level respectively.

Model 2

$$\text{pindex2} = \alpha_0 + \alpha_1\text{lifespan} + \alpha_2\text{impose\_punishment} + \alpha_3\text{famsize} + \alpha_4\text{sex\_participant} + \alpha_5\text{education} + \alpha_6\text{lo\_participation} + \alpha_7 \text{use\_drinking} + \alpha_8\text{water} + \alpha_9\text{pci} + U_i$$

INFLUENCE OF GROUP CHARACTERISTICS ON PARTICIPATION

None of the group characteristics significantly influenced participation of beneficiaries in BG meeting and decision-making except in the case of imposition of punishment. Insignificant relationship of heterogeneity in caste implied that the existence of different castes in a particular group could not influence the beneficiary participation adversely. In other words, there was no caste based discrimination among the beneficiaries in the group.

Table 11
Factors Affecting Participation in Decision-Making

Dependent Variable: Participation Index of Decision-Making

Variable	*Coefficient*	*Std. Error*	*t-Statistic*
C -15.27	13.95	-1.09	
1. LIFESPAN	0.21	0.14	1.46
2. IMPOSE -PUNISHMENT	-13.18***	7.19	-1.83
3. FAMSIZE	2.31*	0.76	3.05
4. SEX-HEAD	7.82	5.91	1.32
5. SEX - PARTICIPANT	10.73**	4.51	2.38
6. EDUCATION	0.34*	0.12	2.99
7. LO -PARTICIPATION	0.51*	0.15	3.29
8. USE -DRINKING	13.76*	4.98	2.76
9. WATER	-17.83**	7.73	-2.31
10. PCI	0.00049***	0.00	1.89
11. R-SQUARED	*0.26*	*F-statistic*	*6.11**
12. ADJUSTED R-SQUARED	*0.22Log likelihood*		*-941.548*
13. DURBIN-WATSON STAT	*1.80*		
14. WHITE HETEROSKEDASTICITY TEST			
15. F-STATISTIC		*1.227524*	*0.158249*
16. OBSERVED R-SQUARED		*78.90032*	*0.194433*

Number of observations: 200

Note: *, **, *** Significant at 1 per cent, 5 per cent, 10 per cent level respectively.

The of the group size was not significantly influencing the rate of participation of beneficiaries. The result of the study differed from that of earlier studies which established that collective action in larger groups became less effective. Active leadership in the group was also not influencing the participation. Among the operational rules, the imposition of punishment and distribution of receipt were not significantly influencing the rate of participation. But, imposition of punishment of discontinuation of water supply or imposition of fine for non-payment of tariff showed a significant and negative relationship with rate of participation in decision-making. This implies that defaulters either did not attend the meetings or did not participate in the group discussions.

The possible reason for insignificance of group characteristics may be the peculiar characteristics of the resource under consideration. In the case of drinking water supply, once the beneficiaries were satisfied with the existing arrangement, their interest in participating in the group meeting reduces, provided that there was no departure from the existing arrangement at any given time.

INFLUENCE OF HOUSEHOLD CHARACTERISTICS

Household characteristics were relatively more influential in the participation indices relating to both attendance and decision-making. Family size was strongly associated with the participation of beneficiaries in group meeting as well as in decision-making. The estimated elasticity of participation (Table 12) with respect to family size shows that as the size of the family increased by one per cent, participation in attendance increased by 0.62 per cent and participation in decision-making by 0.96 per cent. Sex of the household head was another influencing factor affecting participation. The result shows that male headed households participated more in group meetings compared to the female headed households. Rate of participation in attending the meeting of the male headed households was higher by nearly 8 per cent as compared to the female headed households. But, the gender of household head did not influence the rate of participation in decision-making. The sex of the participant was also an influential factor in deciding the rate of participation in attendance and decision-making. The results show that male participants' participation was more in group meeting and decision-making as compared to female participants. This was opposite to the common notion that in water related issues women has more influence as women face more hardship in fetching water. Regression results, however, indicate that although men constitute only 34.5 per cent of the participants, they have more voice as compared to women.

The percentage of family members with education higher than 10th standard and percentage of members involved in local organisation were the other two important variables influencing participation. Both the variables represent the social capital of that household. Social capital is considered as an important social condition that can facilitate coordinated action between individuals

(Blomkvist and Swain 2001). Elasticity of participation with respect to educational level of households show that as the level of education increases by one per cent the rate of participation in attendance increases by 0.12 per cent and influence in decision-making increases by 0.28 per cent. Similarly, increase in the number of family members involved in other local organisation enhances the participation in BG meetings. The elasticity of participation in water user groups with respect to participation in other local organisation shows that one per cent increase in the participation in other local organisations influences the attendance in group meeting by 0.20 per cent and decision-making by 0.40 per cent.

Table 12
Elasticity of Participation in Attendance in the Meeting and Decision-Making

Variable		*Elasticity of participation in attendance in the BG meeting*	*Elasticity of participation in decision-making*
Family size		0.621	0.961
Percentage of family members with education >10		0.123	0.281
Percentage of family members participating in local organisation		0.200	0.396
	Per capita income	0.167	0.398
i	Very Poor	0.025	0.077
ii	Poor	0.089	0.191
iii	Middle income	0.156	0.259
iv	High income	0.263	0.444

The purpose for which water was used had higher influence on participation rate as compared to all other variables. As per the results, if the household was using water for drinking purpose to a large extent, their participation rate in the group meeting increased by 10 per cent. But, the water scarcity of the household decreased the participation rate by 19 per cent. One of the reasons for this may be that a majority of the households facing perennial water scarcity belongs to low income category with a per capita income less than Rs.12,000 (Table 13). Hence, participation of these households in the BG meetings might be costly and involves high opportunity cost for the beneficiaries.

Table 13
Distribution (%) of Household by Water Scarcity and Per Capita Income

Water scarcity	*Per capita income*				*Total*
	Very poor <2500	*Poor 2500-12000*	*Middle 12000-24000*	*High income> 24000*	
Perennial scarcity	1.5	68.5	27.7	2.3	100.0 (130)
Seasonal scarcity	0.0	45.3	43.2	11.6	100.0 (95)
Bad quality	4.0	68.0	20.0	8.0	100.0 (50)
No problem	0.0	34.3	34.3	31.4	100.0 (35)
Total	1.3	57.4	31.9	9.4	100.0 (310)

Note: Figures in the parenthesis are actual numbers.

The above argument can be reinforced by the relationship between per capita income and participation rate. Table 5.15 reveals that as *per capita* income increases participation rate also increase, but by infinitely small per cent. Elasticity of participation with respect to per capita income shows that as income increases by one per cent, rate of participation in attendance increase by 0.17 per cent and influence in decision-making by 0.40 per cent. Computation of elasticity for low, middle and high income categories shows that influence of higher income groups in decision-making was very high as compared to low income categories (Table 12). It shows that those who influenced the decision of location of water supply and tariff, to a large extent, belonged to high income category.

Stepwise regression analysis of both the models show that among different set of variables, household characteristics were the important factors influencing the participation. Among the household characteristics, sex of the participant in group meetings and the educational level of the households were the most important variables influencing the rate of participation and decision-making. None of the outcome variables were significantly influencing the rate of participation in group meeting as well as decision-making.

SUMMARY AND CONCLUSIONS

The study has seriously made efforts to find out the factors affecting community participation in rural water supply schemes using primary data collected from 200 households in Malappuram district, Kerala. Indices of participation were prepared to capture the rate of participation in beneficiary group meetings of community-based water supply schemes. Two forms of indices were constructed-to capture the participation, in general, and the extent of involvement in decision making. In both cases, distribution of households by participation index shows that percentage of households actively participating in group meetings and in decision-making was less.

Separate regression models were estimated in order to identify the factors affecting participation. In the first model, participation index for attendance was taken as dependent variable and variables representing locality, group and household characteristics were taken as explanatory variables, whereas the second model considered participation index for decision-making as dependant variable. Among the three sets of factors, the analysis shows that household characteristics were the most influential in affecting collective action. Within the households' characteristics, sex of the participant as well as the educational level had the most important bearing on participation. The analysis shows that male participants had active participation in group meetings as compared to their female counterpart. Similarly, participation increased with households' level of education and involvement in other local organisations. Per capita income of the household had negligible impact on participation. But, among different income class households, the higher income category of households had a larger influence on the decision-making. Contrary to the expectation, the water scarcity of the household negatively influenced the participation in group meetings. While considering the basic hypothesis of the study, we accept that there is no relationship between community participation and geographical characteristics of the locality. However, beneficiary participation is influenced by household characteristics.

The results of the study shows some deviation from the existing literature on collective action. Studies on collective action and resource management emphasized the importance of group characteristics like, size of the group, homogeneity in the group, presence of active leadership and organizational rules in the success of collective action (Olson 1965; Baland and Platteau 1996; Vedeld 2000). But, none of the group characteristics significantly influenced the participation of beneficiaries in the rural water supply schemes. Physical characteristic of the locality was also not showing any association between participation in the meeting and decision-making. Insignificance of group characteristics in this study may be mainly due to the resource specific features of the product under consideration. In this study, water supply, as a resource, provided less incentive for participation compared to other resources. In water supply context, once the system was installed and functioning without any interruption, it was observed that households' were less interested to participate in the meetings. Increased water demand of the households mainly drives them to participate in the community-managed schemes. As it is mentioned in the beginning majority of the households in the state depend on their own wells for domestic water needs. Piped water supply in most of the cases is only a supplementary to private owned wells of households especially in summer season. This implies that that users' demand is an important factor determining the participation in rural water supply schemes in Kerala. This explains the reason why there was insignificant relationship between participation and group characteristics.

It is considered that if the initiative for collective action in resource management comes from the people themselves there will be better participation. In the context of community based water supply schemes in Kerala, the guidelines for the implementation were provided by the funding agency and the implementation was in the hands of supporting organisations. But, active participation of beneficiaries in community managed water supply programmes in Kerala was less . Under the CWS schemes, local people were given more responsibility in the operation and maintenance without substantial support from the government.

Narayan (1995) has mentioned that the main reason for lack of participation in large scale was the unwillingness to frame the issue in institutional terms. She has argued that boiling down the issue of water supply to merely a technical problem reduce the participation. In such circumstances, participation became an add-on task which could be taken up whenever it was convenient and required for the participants. But community-based water supply institutions can empower the people through developing their capacity in various social interfaces during their participation in these institutions. Especially poor sections of the society can be significantly benefited from the socialising and empowering capacities of these community-based institutions. Hence, considering community-based water supply institutions as an option for social development can lead to better outcomes.

NOTES

1. A beneficiary Group Area Consists of 70 to 150 Households, Including users and Non users of Community Water Supply Schemes. So, one Beneficiary Group area is Considered as a Locality.
2. While the extent to which people influenced the decision-making relating to two aspects important to the water supply was considered in the construction of indices, the influence in decision-making in general was not taken into consideration.
3. Weights were calculated using DEFINITE package by giving first preference ranking to making suggestion in the meeting and second to the mere attendance following the typology of Agarwal (2001). She argues that consultative form of participation (being asked an opinion in specific matters without guarantee of influencing decision) is the higher level of participation compared to passive participation (attending the meeting and listening to decisions taken without speaking up).
4. Decision on the location of water supply was mainly of technical nature whereas decision on monthly tariff reflected the household's willingness and capacity to pay for water. In the case of community managed schemes, engineers of supporting organisations mainly helped the beneficiaries with regard to location of water supply. Decision on the location of water supply was taken once in the entire lifespan of the group. But, the decision on monthly tariff needed frequent revision with the changes in the cost of operation and maintenance, even though the NGOs influenced the decision on monthly tariff in the initial stages. For this reason, the variable influence in the location of water supply was assigned less weight.
5. Pearson Chi-Square value is significant at 1 per cent level. The chi-square test in cross tabulation measures the discrepancy between the observed cell counts and what one would expect if the rows and columns were unrelated. If two sided asymptotic significance of the chi-square statistic is greater than 0.10, it implies that the differences in two variables are due to chance variation.
6. According to Oakerson, operational rules establish conditions of collective choice which allow the group to work appropriately. For him, conditions of collective choice depends on (i) the capacity of individuals to make decisions; (ii) the availability of potential sources of remedy to individuals adversely affected by others; (iii) the

capacity of an affected population to relax the market rule of willing consent iv) the presence of potential veto positions in any process of collective decision-making (1992: 47).

7. In strict statistical terms, this is not advisable for two reasons. First classical linear regression model assumes dependent variable is stochastic in nature. While preparing the index, the present study combined two or more qualitative variables. Combining two or more qualitative variables will lead to loss of randomness associated with the original variables. Second, if the index is used as a dependent variable in regression model, the interpretation cannot be exactly like that in the case of OLS or logit models. The study has continued to use participation as a dependent variable in order to get a comprehensive picture of factors affecting participation and draw policy implications. However, while interpreting the results particular attention has given to address the second problem.

8. Variables were omitted by considering following criteria:1) multicollinearity among the variables; 2) significance of chi-square statistics in cross tabulation of particular variable with dependent variable; 3) contribution of particular variable to the explanatory power of the model (variable is omitted if adjusted R-square increased after omission of that variable from the model).

9. Véron (2000), in the context of Kerala, has pointed out that the failure of community based programmes in the state is due to the insufficient local participation in defining problems, areas of action and project goals. He has pointed out that usually participation implies the implementation of schemes, which are designed at the higher level.

REFERENCES

Agarwal, Bina (2001), Participatory Exclusions, Community Forestry and Gender: An Analysis for South Asia and a Conceptual Framework, World Development, 29 (10): 1623-48.

Baland, J.M. and J.P. Platteau (1996), Halting Degradation of Natural Resources: Is there a Role for Rural Communities? Oxford: FAO/ Clarendon Press.

Bardhan, Pranab (1993), Symposium on Management of Local Commons. *Journal of Economic Perspectives*, 7, (4): 87-92.

Blomkvist, Hans and Ashok Swain (2001), Investigating Democracy and Social Capital in India. Economic and Political Weekly, XXXVI (8): 719-31.

Briscoe, John (1996), Water as an Economic Good: the Idea and What it means in Practice. A Paper Presented at the World Congress of the

International Commission on Irrigation and Drainage, Cairo, September.

Isham, Jonathan, Deepa Narayan and Lant Pritchett (1994), Does Participation Improve Performance? Empirical Evidence from Project Data. World Bank Policy Research Working Paper, No. 1357, Washington D.C: The World Bank.

Isham, Jonathan and Satu Kahkonen (1999), What Determines he Community-based Water Projects. Evidence from Central Java, Indonesia, on Demand Responsiveness, Service Rules and Social Capital. Social Capital Initiative Working Paper, No.4. Washington D C: The World Bank.

Isham, Jonathan and Satu Kahkonen (2002), Institutional Determinants of the Impacts of Community-Based Water Services: Evidence from Sri Lanka and India. Economic Development and Cultural Change, 50(3): 667-91.

Kerr, J. (2002), Watershed Development, Environment Services, and Poverty Alleviation in India. World Development, 38(8): 1387-1400.

KRWSA (2000), Project implementation Plan, Thiruvananthapuram: Kerala Rural Water Supply and Sanitation Agency.

McCarthy, Nancy, Céline Dutilly-Diané, and Boureima Drabo (2002), Cooperation, Collective Action in Natural Resources Management in Burkina Faso: a Methodological Note. CAPRi Working Paper No. 27. Washington D.C: International Food Policy Research Institute.

Meinzen-Dick, Ruth, K.V. Raju and Ashok Gulati (2002), What Affects Organization and Collective Action for Managing Resources? Evidence from Canal Irrigation Systems in India, World Development, 30 (4): 649-66.

Narayan, Deepa (1993), Participatory Evaluation, Tools for Managing Change in Water and Sanitation. World Bank Technical Paper No. 207, Washington D. C: The World Bank.

Narayan, Deepa (1995), The Contribution of People's Participation: Evidences from 121 Rural Water Supply Projects. Environmentally Sustainable Development Occasional Paper Series No.1. Washington D.C: The World Bank.

North, Douglass C. (1981), Structure and Change in Economic History. New York: Norton.

Oarkerson, J.R. (1992), Analyzing Commons: A Framework. In Daniel Bromley (ed.), Making the Commons Work. San Francisco, California: ICS Press.

Olson, Mancur (1965), Logic of Collective Action: Public Goods and the Theory of Groups. Cambridge: Harvard University Press.

Ostrom, Elinor (1990), Governing Commons– The Evolution of Institutions for Collective Action. Cambridge: Cambridge University Press.

Prokopy, Linda, Stalker (2005), The relationship between participation and project outcomes: evidence from rural water supply project in india, World Development, 33 (11) 1801-19.

Rajasekhar, D. and Veerashekharappa (2003), Role of local Organization in Water Supply and Sanitation Sector: a Study in Karnataka and Uttaranchal States, India, World Bank Project on Roles of Rural Local Organization, Bangalore: ISEC.

Saleth, R. Maria and Ariel Dinar (2004), The Institutional Economics of Water: A Cross-Country Analysis of Institutions And Performance. Cheltenham, Edward Elgar publishing for The World Bank.

Sara, Jennifer and Travis Katz (1998), Making Rural Water Supply Sustainable: Report on the Impact of Project Rule. UNDP: World Bank Water and Sanitation Programme.

Singh, Katar (1992), Workshop Report: People's Participation in Natural Resources Management. Working Paper, No.8 Anand: Institute of Rural Management.

State Planning Board (2002), Report on Kerala Water Authority, Thiruvananthapuram: Government of Kerala.

Uphoff, Norman (1999), Understanding Social Capital: Learning from the Analysis and Experience of Participation. In Partha Dasgupta and Ismail Serageldin (eds.), Social Capital: A Multifaceted Perspective. Washington D.C: The World Bank.

Vedeld, Trond (2000), Village Politics: Heterogeneity, Leadership and Collective Action. The Journal of Development Studies, 36(5): 105-34.

Wade R. (1988), Village Republics–Economic Condition for Collective Action in South India. Cambridge: Cambridge University Press.

INDEX

J

K

L